Y0-BCO-429

Menges/Michaeli/Mohren
How to Make Injection Molds

Georg Menges
Walter Michaeli
Paul Mohren

How to Make Injection Molds

Third Edition

HANSER

Hanser Publishers, Munich

Hanser Gardner Publications, Inc., Cincinnati

Translated and revised edition of
"Anleitung zum Bau von Spritzgießwerkzeugen", 5th Edition, 1999
by Prof. Dr.-Ing. Georg Menges, Prof. Dr.-Ing. Walter Michaeli and Ing. Paul Mohren,
Institut für Kunststoffverarbeitung in Industrie und Handwerk
an der Rhein.-Westf. Technischen Hochschule Aachen
(Institute for Plastics Processing, Technical University of Aachen)

Translated by Rolf J. Kahl and Raymond Brown

Distributed in the USA and in Canada by
Hanser Gardner Publications, Inc.
6915 Valley Ave.
Cincinnati, OH 45244, USA
Fax: +1 (513) 527 8950
http://www.hansergardner.com

Distributed in all other countries by
Carl Hanser Verlag
Postfach 86 04 20, 81631 München, Germany
Fax: +49 (89) 99 830-269
http://hanser.de

Library of Congress Cataloging-in-Publication Data

Menges, Georg, 1923–
[Anleitung zum Bau von Spritzgiesswerkzeugen. English]
How to make injection molds. – 3rd ed. / Georg Menges, Walter Michaeli, Paul Mohren;
[translated by Rolf J. Kahl and Raymond Brown].
    p.   cm
Includes bibliographical references and index.
ISBN: 1-56990-282-8
1. Injection molding of plastics. I. Michaeli, Walter. II. Mohren, Paul. III. Title.

TP1150.M3813 2000
668.4'12–dc21   00-046238

Die Deutsche Bibliothek – CIP-Einheitsaufnahme

Menges, Georg:
How to make injection molds / Georg Menges ; Walter Michaeli ; Paul Mohren. [Transl. by Rolf J. Kahl
and Raymond Brown]. – 3. ed.. – Munich : Hanser; Cincinnati : Hanser Gardner, 2000
Einheitssacht.: Anleitung für den Bau von Spritzgießwerkzeugen <engl.>
ISBN: 3-446-21256-6

Typesetting by Cicero Lasersatz, Germany
Printed in Germany by Kösel

Preface to the 3rd Edition

Injection molds are high-precision tools whose smooth-operation on a day-to-day basis is vital to the economic success of many plastics processors. Consequently, errors in design and construction of molds can have grave consequences.

That is where this book comes in. Building on the earlier editions, it draws extensively both on the literature and on the development work carried out at the Institute for Plastics Processing in the Technical University of Aachen with the aid of substantial private and public research funding.

We are especially grateful to those who participated in this edition and to those who laid the foundations for it in earlier editions, namely Dr. H. Bangert, Dr. P. Barth, Dr. W. Hoven-Nievelstein, Dr. O. Kretschmar, Dr. M. Paar, Dr. G. Pötsch, Dr. Th. W. Schmidt., Dr. Ch. Schneider, Prof. E. Schürmann and Prof. S. Stitz.

We are indebted to the co-workers and students in the institute who contributed to the success of this book through their unstinting work and their personal commitment. As they are too numerous to mention, we trust that Ms. G. Nelissen, Ms. I. Zekorn and Mr. W. Okon will accept our thanks on their behalf. Finally, we extend our appreciation to Carl Hanser Verlag, and especially to Dr. W. Glenz and Mr. O. Immel, for taking the manuscript and turning it into such a distinctive and attractive book.

G. Menges
W. Michaeli
P. Mohren

Contents

The following contributors helped to revise and update this new edition:

Chapter 1	Prof. Dr. F. Klocke
	A. Karden
Chapter 2	Prof. Dr. A. Bührig Polaczek
	Prof. Dr. F. Klocke
	A. Karden
	Dr. M. Langen
	Prof. Dr. E. Schmachtenberg
	Dr. M. Polifke
Chapter 3	Prof. Dr. H. Schlüter
	Dr. V. Romberg-Forkert
Chapter 4	P. Niggemeier
Chapter 5	Dr. F. Ehrig
Chapter 6	J. Berthold
	Dr. C. Brockmann
	C. P. Cuttat
	Dr. F. Ehrig
	C. Hopmann
	C. Ronnewinkel
Chapter 7	Dr. A. Rogalla
Chapter 8	Dr. F. Ehrig
	A. Spennemann
	Dr. J. Zachert
Chapter 9	P. Niggemeier
Chapter 11	Dr. F. Ehrig
Chapter 12	Dr. M. Stommel
Chapter 13	Dr. N. Kudlik
Chapter 14	Dr. P. Filz
	H. Genoske
	Dr. A. Biswas
	Dr. K. Schlesinger
Chapter 15	Dr. A. Feldhaus
Chapter 16	Dr. H. Recker
	Dr. O. Schnerr-Häselbarth
Chapter 18	P. Gorbach
Chapter 20	Dr. A. Rogalla
	C. Brockmann
	E. Henze

1 Materials for Injection Molds

The injection molding technique has to meet the ever increasing demand for a high quality product (in terms of both consumption properties and geometry) that is still economically priced. This is feasible only if the molder can adequately control the molding process, if the configuration of the part is adapted to the characteristics of the molding material and the respective conversion technique, and a mold is available which satisfies the requirements for reproducible dimensional accuracy and surface quality. Therefore injection molds have to be made with the highest precision. They are expected to provide reliable and fully repeatable function in spite of being under extreme loads during the molding process, and a long service life to offset the high capital investment. Besides initial design and maintenance while in service, reliability and service life are primarily determined by the mold material used, its heat treatment and the machining operations during mold making [1.1, 1.2].

Injection molding uses almost exclusively high-strength molds made of metals, primarily steel. While the frames are almost always made of steel, the cavities are frequently made of other high-quality materials – metals or nonmetals – and inserted into the cavity retainer plate. Inserts made of materials other than steel are preferably used for cavities that are difficult to shape. They are often made by electrodeposition.

Recently, nonmetallic materials have been growing in importance in mold construction. This is due on the one hand to the use of new technologies, some of which are familiar from prototype production, and especially to the fact that users wish to obtain moldings as quickly and inexpensively as possible that have been produced in realistic series production, so that they can inspect them to rule out weaknesses in the product and problems during later production. The production of such prototype molds, which may also be used for small and medium-sized series, as well as the materials employed, will be discussed later.

An injection mold is generally assembled from a number of single components (see Figure 4.3). Their functions within the mold call for specific properties and therefore appropriate selection of the right material. The forming parts, the cavity in connection with the core, provide configuration and surface texture. It stands to reason that these parts demand particular attention to material selection and handling.

Several factors determine the selection of materials for cavity and core. They result from economic considerations, nature and shape of the molding and its application, and from specific properties of the mold material. Details about the molded part should provide information concerning the plastic material to be employed (e.g. reinforced or unreinforced, tendency to decompose, etc.). They determine minimum cavity dimensions, wear of the mold under production conditions, and the quality demands on the molding with respect to dimensions and surface appearance. The market place determines the quantity of parts to be produced and thus the necessary service life as well as justifiable expenses for making the mold. The demands on the mold material, on its thermal, mechanical, and metallurgical properties are derived from these requirements. Frequently a compromise must be made between conflicting demands.

1.1   Steels

1.1.1   Summary

Normally, steel is the only material that guarantees reliably functioning molds with long service lives, provided that a suitable steel grade has been selected from the assortment offered by steel manufacturers and this grade has been treated so as to develop a structure that produces the properties required in use. This necessitates first of all a suitable chemical composition. The individual alloying elements, according to their amount, have positive as well as negative effects on the desired characteristics. Generally several alloying elements will be present, which can also mutually affect one another (Table 1.1). The requirements result from the demands of the molder and the mold maker. The following properties are expected from steels:

– characteristics permitting economical workability (machining, electric discharge machining, polishing, etching, possibly cold hobbing),
– capacity for heat treatment without problems,
– sufficient toughness and strength,
– resistance to heat and wear,
– high thermal conductivity, and
– corrosion resistance.

The surface contour is still mostly achieved by machining. This is time consuming and calls for expensive machine equipment and results in a surface quality which, in most cases, has to be improved by expensive manual labor. There are limitations to machining because of the mechanical properties of the machined material [1.9]. Steels with a strength of 600 to 800 MPa can be economically machined [1.2] although they are workable up to about 1500 MPa. Because a strength of less than 1200 MPa is generally not sufficient, steels have to be employed that are brought up to the desired strength level by additional treatment after machining, mostly by heat treatment such as hardening and tempering.

Such heat treatment imbues steels with the required properties, especially high surface hardness and sufficient core strength. Each heat treatment involves risks, though (distortion, cracking). Lest molds be rendered unusable by heat treatment, for those with large machined volumes and complex geometries, annealing for stress relief is suggested prior to the last machining step. Eventual dimensional changes from distortion can be remedied in the final step.

To avoid such difficulties steel manufacturers offer prehardened steels in a strength range between 1100 and 1400 MPa. They contain sulfur (between 0.06 and 0.10%) so that they can be machined at all. Uniform distribution of the sulfur is important [1.10]. The higher sulfur content also causes a number of disadvantages, which may outweigh the advantage of better machining. High-sulfur steels cannot be polished as well as steel without sulfur. Electroplating for corrosion resistance (chromium, nickel) cannot be carried out without flaws. In the event of repair work, they cannot be satisfactorily welded and are not suited for chemical treatment such as photochemical etching for producing surface textures.

Table 1.1 Effect of alloying elements on the characteristics of steels [1.3–1.8]

Alloying element → ↓ Property	C	Si	S	P	Cr	Ni	Mn	Co	Mo	V	W	Cu	Ti
Strength	←	←	—	←	←	←	←	←	←	←	←	←	—
Toughness	→	→	→	→	→	←	→	→	→	→	→	—	→
Notched impact strength	→	→	→	→	→	—	—	—	—	—	—	—	—
Elongation	→	→	→	→	→	→	→	—	→	—	→	—	—
Wear resistance	←	—	—	—	←	—	←	—	←	←	←	—	←
Hardenability	—	←	—	—	←	←	←	—	←	—	—	—	—
Hardness	←	←	—	—	←	←	←	←	←	←	←	—	—
Machinability	→	→	←	→	→	→	→	—	—	←	←	—	—
Weldability	→	→	→	→	→	→	→	→	→	→	→	—	—
Ductility	→	→	→	→	→	←	→	→	—	—	→	—	—
Malleability	→	→	→	—	—	—	—	—	→	←	→	—	—
Heat resistance/red hardness	←	←	—	←	←	—	→	←	←	←	←	—	—
Overheating sensitivity	→	→	—	—	←	←	←	→	←	→	←	—	←
Retention of hardness	←	←	—	←	→	—	←	←	←	←	←	—	—
Corrosion resistance	—	←	→	←	←	←	←	←	←	←	←	←	—

In recent years, electric-discharge machining, including spark erosion with traveling wire electrodes, has become a very important method of machine operations for - heattreated steel without sulfur.

If a series of equal cavities of small size has to be made (e.g. typewriter keys), coldhobbing is an economical process. Steels suitable for this process should have good plasticity for cold working after annealing. Therefore soft steel with a carbon content of less than 0.2% is utilized.

After forming, they receive adequate surface hardness through heat treatment. Such hardening is made possible by carburizing. These case-hardening steels are an important group of materials for mold cavities.

Distortion and dimensional changes often occur as a side effect of heat treatment. Dimensional changes are caused by thermal stresses and changes in volume resulting from transformations within the steel. They are unavoidable. Distortion, on the other hand, is caused by incompetent execution of heat treatment before, during, or after the forming process, or by faulty mold design (sharp corners and edges, large differences in cross sections etc.). Deviation from the accurate shape as a consequence of heat treatment is always the sum of distortion and dimensional change. The two cannot be strictly separated. Figure 1.1 points out the various effects which cause configurational changes. Use of steels with a low tendency to dimensional changes reduces such effects to a minimum [1.11, 1.12].

Prehardened, martensitic hardenable, and through-hardening steels should be preferred in order to avoid distortion. Pretreated steels do not need any appreciable heat treatment after forming. The required wear resistance of these steel grades is achieved by a chemical process (e.g. chrome plating) or a diffusion process (e.g. nitriding at temperatures between 450 and 600 °C). Because of the low treatment temperature of 400 to 500 °C for martensitic steels, only small transformation and thermal stresses occur in these steels, and the risks generally associated with heat treatment are slight [1.13]. In through-hardening steels, the heat treatment causes a uniform structure throughout the cross section and thus no appreciable stresses occur.

The range of applications for through-hardening steels is limited, however, since the danger of cracking under bending forces is high, especially in the case of molds with large cavities. Demands for molds with a tough core and a wear-resistant, hardened surface are best met with case-hardening steels (e.g. for long cores or similar items).

High wear in use is most effectively countered with high surface hardness. The best hardening results and a uniform surface quality can be achieved with steels that are free from surface imperfections, are of the highest purity and have a uniform structure. Completely pure steels are the precondition for the impeccably polishable, i.e. flawless, mold surface required for processing clear plastics for optical articles. A high degree of purity is obtained only with steels that are refined once or more. Remelting improves the mechanical properties, too. These steels have gained special importance in mold making for producing cavity inserts.

The maximum abrasive strength is obtained with steels produced by powder metallurgy (hard-material alloys).

Mold temperature and heat exchange in the mold are determined by the plastic material and the respective molding technique. The thermal effect of the mold temperature during the processing of the majority of thermoplastics (usually below 120 °C) is practically insignificant for the selection of the mold material. There are, however, an increasing number of thermoplastics with melt temperatures up to 400 °C that require a constant mold temperature of more than 200 °C during processing. Mold

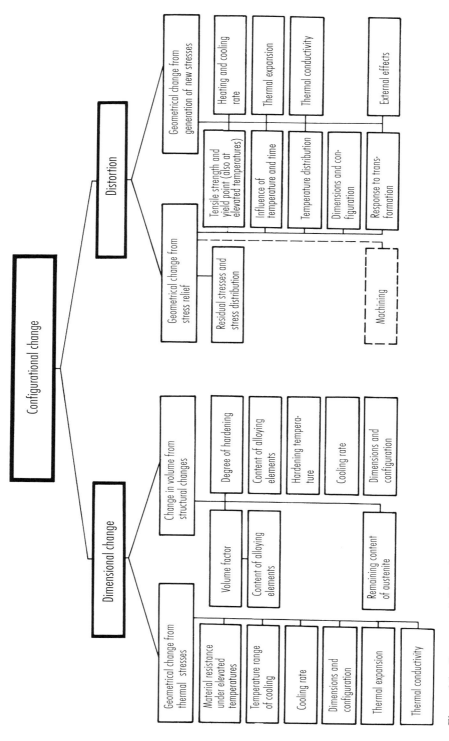

Figure 1.1 Summary of effects that cause configurational changes of steel parts during heat treatment [1.12]

temperatures for thermosets are also between 150 and 250 °C. At this point, the mechanical properties of the mold material are affected. Wear and tendency to distortion increase, creep to rupture data and fatigue resistance decrease [1.14]. This has to be taken into account through selecting the most suitable material. A heat-treatment diagram (hardness versus temperature) can indicate the permissible temperature of use if the latter is taken as 30 to 50 °C below the tempering temperature.

Heat exchange between the solidifying molding and the mold has a substantial effect on the cost of a part. This heat exchange is considerably influenced by the thermal conductivity of the mold material which again is affected by its alloying components. Their different structures give rise to varying thermal conductivity. Notch sensitivity can be countered to a certain degree with case hardening or nitriding because such treatment causes compressive stresses in the surface [1.15]. Nevertheless, attention should be paid to avoiding notches during the design and manufacturing stages.

Some plastics release chemically aggressive substances during processing, such as hydrochloric acid, acetic acid or formaldehyde. They attack the mold surface if it is not protected with a deposit of hard chromium or nickel.

Since such deposits have a tendency to peel off if the molds are improperly designed (e.g., shape, sharp corners) or handled, corrosion resistant steels should be used for making molds which are employed under such conditions. Then no further precautions against possible corrosion from humidity or coolant are necessary.

The demands listed so far are partially contradictory. Therefore the mold designer and mold maker have to select that steel grade which is best suited for a particular job.

The following steel grades are presently offered for producing cavity inserts:

– case hardening steels,
– nitriding steels,
– through-hardening steels,
– tempered steels for use as supplied,
– martensitic-hardening steels,
– hard-material alloys,
– corrosion-resistant steels,
– refined steels.

The most important steels for mold making are shown in Table 1.2, along with their composition, heat treatment and application areas. A more detailed characterization of these groups of steels is provided below.

1.1.2   Case-Hardening Steels

Case-hardening steels best meet the qualifications for mold making. They are not expensive, and it should not come as a surprise that their share is about 80% of the total steel consumed in mold making [1.16] (This figure includes consumption for base and clamp plates). Unalloyed or low-alloyed steels offer a special advantage. Through case hardening, carburization or cementization (so-called because cementite is formed during subsequent hardening), a mold surface as hard as glass is generated and, at the same time, a tough, ductile core. The hard surface renders the mold wear resistant and the tough core confers resistance to shock and alternate loading [1.17].

There are other criteria favoring the application of case-hardening steels over highcarbon and through-hardening steels. Easy machining and, if properly produced and treated, very good polishing are especially worth mentioning. In carburization, there is the possibility of performing local hardening by covering certain areas. However, the extensive heat-treatment procedure required for carburization cannot prevent dimensional changes from occurring, with the result that additional outlay is required before the molds can be used [1.18]. Another advantage over the other steel groups is the low strength after low-temperature annealing. Case-hardening steels are therefore suitable for hobbing (see Section 2.3). This process is particularly suitable for small cavities and multi-cavity molds with a large number of equal cavities.

1.1.3 Nitriding Steels

Basically all steels which contain nitride-forming alloying elements can be nitrided. Such alloying elements are chromium, molybdenum, vanadium and preferably aluminum, which especially favors formation of nitrides. These steels absorb nitrogen from the surrounding medium by diffusion into their outer layer. This process can take place in a salt bath, in gas, powder or in the plasma of a strong corona discharge (ionitriding) at temperatures between 350 and 580 °C. Temperature and time are determined by the individual medium. This procedure causes the above mentioned alloying elements to form nitrides. They provide the steel with an extraordinarily hard and wear-resistant surface with a Brinell hardness between 600 and 800, the value depending on steel grade and process. The greatest hardness is not at the immediate surface but a few hundredths of a millimeter deeper. Therefore the mold should be appropriately larger and the dimensions corrected by grinding after nitriding [1.19]. Ionitrided molds do not need this posttreatment, and it should not be done, which is a special advantage of this process.

Nitriding has the following advantages:

– After nitriding, there is no need for heating, quenching or annealing since nitriding bestows the hardness direct.
– Nitrided parts are free from distortion because they are heated only to about 550 °C.
– The hardness of the nitrided layer is not affected by heating up to 500 °C (retention of hardness).
– Nitrided molds are thus suitable for processing thermosets and such thermoplastics that are shaped at high mold temperatures.
– Nitriding steels yield an extremely hard and at the same time wear-resistant outer layer having good surface slip.

Disadvantageous is the low extent to which the nitrided layer is anchored to the base material as the hard layer can peel off under high surface pressure [1.20].

1.1.4 Through-Hardening Steels

Through-hardening steels increase their hardness by the formation of martensite, which results from rapid quenching. The mechanical properties depend on the quenching medium and the cooling rate. Quenching media are water, oil or air. Water works fastest and has the most drastic effect, while oil and air are milder. Thermal conduction, among other factors, determines the cooling rate, too. The heat conduction depends on the

Table 1.2   Steels for injection molds [1.18, 1.26, 1.28]

Steel type	AISI No.	Composition (%)			Thermal conductivity (W/(m · K))	Thermal expansion (10⁻⁶ K⁻¹)
Carbon steel	1020	0.18–0.23 C		0.30–0.60 Mn	46.7	11–15
		0.04 P	0.05 S			
	1030	0.28–0.34 C		0.60–0.90 Mn	46.7	14.9
		0.04 P	0.05 S			
	1040	0.37–0.44 C		0.60–0.90 Mn	46.7	
		0.04 P	0.05 S			
	1095	0.90–1.03 C		0.30–0.50 Mn	43.3	11–14
		0.04 P	0.05 S			
Alloy steel	4130	0.28–0.33 C		0.30–0.60 Mn	46.7	
		0.20–0.35 Si		0.80–1.10 Cr		
		0.15–0.25 Mo		0.035 P		
		0.04 S				
	4140	0.38–0.43 C		0.75–1.00 Mn		
		0.20–0.35 Si		0.80–1.00 Cr		
		0.15–0.25 Mo		0.035 P		
		0.04 S				
	6150	0.48–0.53 C		0.70–0.90 Mn	60.6	10–12
		0.20–0.35 Si		0.80–1.10 Cr		
		0.15 V	0.035 P	0.04 S		
	8620	0.18–0.23 C		0.70–0.90 Mn		
		0.20–0.35 Si		0.40–0.70 Ni		
		0.40–0.60 Cr		0.15–0.25 Mo		
		0.035 P	0.04 S	46.7		
Tool steels Shock-resisting steels:	S1	0.50 C	0.75 Si	1.25 Cr	62.3	11–13
		2.50 W	0.20 V			
	S7	0.50 C	0.70 Mn	0.25 Si		14.9
		3.25 Cr	1.40 Mo			
Cold-work steels: – oil hardening	O1	0.90 C	1.20 Mn	0.50 Cr		
		0.50 W	0.20 V			
– medium alloy,	A2	1.00 C	1.00 Mo	5.00 Cr		
– air hardening	A4	0.95 C	2.00 Mn	0.35 Si		
		2.20 G	1.15 Mo			
– medium alloy,	A6	0.70 C	2.00 Mn	1.00 Cr		
– air hardening		1.00 Mo				
– high carbon,	D2	1.50 C	1.00 Mo	12.00 Cr		
– high chromium		1.00 V				
Hot-Work steels: – chromium base	H13	0.35 C	0.40 Mn	1.00 Si	24.6	12–13
		1.40 Mo	5.00 Cr	1.00 V		
– tungsten base	H23	0.30 C	12.00 Cr	12.00 W		
Special purpose steels: – low-alloy	L6	0.75 C	0.75 Mn	0.90 Cr		
		1.75 Ni	0.35 Mo			
Mold steels: – low carbon	P2	0.07 C	0.20 Mo	2.00 Cr		
– medium alloy	P20	0.35 C	0.80 Mn	0.50 Si	29.0	12.7
		0.45 Mo	1.70 Cr			
Stainless steel: (martensitic)	420	0.15 C (min.)	1.00 Mn	1.00 Si	23.0	11–12
		12.00–14.00 Cr				

Such tables are subject to change from time to time with new steels added or others eliminated, and composition of steels is sometimes altered. Current publications (AISI/SAE) should be consulted if latest information is desired.

surface-to-volume ratio of the mold and the alloying elements added to the steel. Ni, Mn, Cr, Si and other elements lower the critical cooling rate and, therefore, permit throughhardening of larger cross sections [1.17].

The hardening process consists in preheating, heating to prescribed temperature, quenching with formation of a hard martensitic structure and then normalizing to improve toughness. Because of the low toughness of through-hardening steels, molds with deep cavities have a higher risk of cracking.

Unlike tempering, normalizing reduces the hardness only slightly. The temperatures for normalizing are between 160 and 250 °C. Besides improving toughness, normalizing has the effect of reducing stresses. (Occasionally this treatment is, therefore, called stress relieving but it should not be confused with stress-relieving annealing.)

Through-hardening steels exhibit very good dimensional stability when heat-treated. Because of their natural hardening capacity they have high compressive strength and are especially suitable for molds with shallow cavities where high pressure peaks may be expected. They can also be recommended in molds for insert molding (with possible high edge pressure) and, due to good wear resistance and high normalizing temperatures, for processing thermosets [1.20, 1.21].

As far as the mechanical properties are concerned, through-hardening steels have a homogeneous structure. Major postmachining does not remove outer layers having special strength as is the case with case-hardening steels. Since the introduction of electrical discharge machining (EDM), the use of through-hardening steels has been steadily gaining in importance.

1.1.5 Heat-Treated Steels

These steels are tempered by the manufacturer and can therefore be used as supplied without the need for further heat treatment.

After hardening, these steels are submitted to a tempering process. At temperatures above 500 °C, the martensite decomposes into carbide and alpha iron. This causes a reduction in hardness and strength of these steels and, at the same time, an increase in toughness. Ductility and toughness increase with rising tempering temperatures; hardness and strength, however, decrease.

Through judicious choice of temperature (usually constant) and duration of tempering (1 to 2 hours), it is possible to obtain a certain degree of toughness, the exact value depending on the strength. 1200 to 1400 MPa yield strength can be assumed as the upper limit. Steels of higher strength can no longer be economically machined [1.22].

To improve the machining properties, sulfur (0.06%) is added to the heat-treated steels. However, this reduces the scope for electroplating the mold surfaces, such as by hard chroming. Similarly, photo-etchability is severely restricted by the manganese sulfides emitted at the surface.

These disadvantages are compensated by treating the steels in some cases with calcium as they are being made. The sulfur content can thereby be reduced (< 0.005%; these steels are said to be highly desulfurized) and the machinability and etchability improved simultaneously [1.18, 1.23].

Heat-treated steels are preferably employed for medium-sized and large molds. They have the additional advantage that corrections are more easily accomplished if deemed necessary after a first trial run [1.2, 1.25].

1.1.6   Martensitic Steels

Martensitic steels combine extreme strength and hardness with the advantage of simple heat treatment. They are supplied in an annealed state. Their structure consists of tough nickel martensite with a strength of 1000–1150 MPa. To an extent depending on their strength, their machinability is comparable to that of tempered steel. It takes about 10 to 20% more machine time than with mild steels.

After machining, molds are subjected to a simple heat treatment that harbors no risks. They are heated up to temperatures between 480 and 500 °C, kept at this temperature for 3 to 5 hours and slowly cooled in still air. No normalizing is done.

Due to the low hardening temperature, no distortion is to be expected. There is only a slight dimensional change from a shrinkage of 0.05 to 0.1% on all sides. The wear resistance of the mold surface can be improved even further by a diffusion process such as nitriding. Care should be taken not to exceed 480 to 500 °C when doing this. The preceding hardening step can be omitted if diffusion treatment occurs. The extraordinary toughness of martensitic steel at the high hardness of 530 to 600 Brinell is remarkable [1.2, 1.13, 1.20, 1.25].

The use of martensitic steels is recommended for smaller cavity inserts with complex contours that exhibit large differences in cross sections and detached thin flanges.

The use of other steel grades for such molds would definitely result in distortion. It should be noted that hardened martensitic steels can be easily welded using materials of the same kind without the need for preheating.

1.1.7   Hard Mold Alloys

Hard mold alloys are steels produced by a powder-metallurgical process that have a high proportion of small carbides which are uniformly embedded in a steel matrix of different composition, primarily a chromium-molybdenum-carbon matrix with added cobalt and nickel [1.18, 1.26–1.28]. These practically isotropic steels are produced in a diffusion process under pressure and temperature from homogeneous alloying powders of maximum purity. To an extent depending on the steel grade and manufacturer, these steels have a titanium carbide content of 50 vol.%. This high carbide content renders the steels extremely wear resistant. They are thus particularly suitable for processing wearpromoting compounds (thermosets and reinforced thermoplastics) and for various tool parts subjected to increased wear, such as nozzles, gate systems, etc.

Hard mold alloys are supplied in a soft-annealed state and may be machined. After heat treatment (6–8 hours storage at 480 °C), the hardness increases to 60–62 Rc. To increase the hardness further, storage may be combined with a nitriding treatment. The surface hardness increases as a result to 72–74 Rc [1.27].

Due to their isotropic structure and low coefficient of thermal expansion, hard mold alloys undergo extremely little distortion during heat treatment.

1.1.8   Corrosion-Resistant Steels

Some polymeric materials release chemically aggressive substances during processing that attack steel and harm the surface, e.g. by causing rusting. One way of protecting molds against corrosion is afforded by applying a protective electroplating coat (chromnickel plating). Complicated geometries make it difficult, however to apply uniform coatings, particularly on the corners and edges. It is therefore possible that the shape of the moldings will change as a result. In addition, protective electroplating coats tend to peel off very easily. If the molds are likely to be used in a corrosive setting, it is advisable to use corrosion-resistant steel.

The corrosion resistance of such steels is due to their having been alloyed with chromium. Given a chromium content of at least 12%, very dense, strongly adhering, invisible layers of chromium oxide are formed in contact with atmospheric oxygen or other oxygen-providing media. They protect the steel against corrosion. In this case the crystalline structure is a solid solution of iron and chromium. Adding other alloying elements to steel, especially carbon, may partially reduce the corrosion resistance. Carbon has the tendency to react with chromium and form ineffective chromium carbide. Only in the hardened state does carbon remain bound to the iron atoms or in solution, thereby allowing the chromium to fully play its anti-corrosive function. Through heat treatment above ca. 400 °C, part of the carbon escapes from the solution and can form chromium carbide. Thus, carbon binds part of the chromium, preventing it from performing its protective role. Because of the required mechanical properties, however, carbon cannot be dispensed with mold in stainless steels for mold making [1.28]. Therefore hardened chrome steel is always the best choice.

The corrosion resistance also depends on the quality of the mold surface. A rough surface presents a larger area for attack than a smooth, highly polished one. Chrome steel with only 13% chromium, for instance, is corrosion resistant only if the surface is highly polished. In addition, the plastics processor himself should make every effort to protect his molds from corrosion by cleaning them before storage.

Steels with 17% chromium and martensitic structure have better corrosion resistance but tend to coarseness and formation of soft ferritic parts in their crystalline structure. Only the hard material alloys (tool steels with the maximum carbide content produced by powder metallurgical processes) are not susceptible to these problems (see also Section 1.1.7).

1.1.9 Refined Steels

The cosmetic appearance of a plastic molding depends largely on the mold's surface quality. Particular attention to this fact should be paid when selecting a steel for mold inserts that will be employed for molding transparent articles such as lenses etc.

The better the steel can be polished, the higher is the surface quality of the mold. How well steels can be polished is determined by their degree of purity. This degree depends on the amount of nonmetallic inclusions in the steel, such as oxides, sulfides and silicates [1.19]. These inclusions, unavoidable in an open-hearth steel, can be eliminated by remelting. In modern practice three methods with self-consuming electrodes are applied: melting in an electric-arc vacuum oven, in an electron beam oven or by electro-slag refining.

For refining in an electric-arc vacuum oven, a normally produced steel rod as self-consuming electrode is burnt off in a cooled copper die under high vacuum (10^{-1} or 10^{-3} Pa).

In an electron beam oven, the rod is melted in an ingot mold by an electron beam.

Electro-slag refining has gained greatest importance in recent years. In this process, the electrode, which is in the form of a completely alloyed block, dips into electrically conductive liquid slag in a water-cooled mold and is melted [1.29]. The melt dripping from the electrode passes through the liquid slag and is purified by respective metallurgical reactions [1.30, 1.31].

Refined steels have the following advantages over those conventionally produced:

– more homogeneous primary structure and extensive freedom from ingot liquation and internal defects from solidification,
– less crystal liquation and therefore more homogeneous microstructure,
– reduced quantity and size and a more favorable distribution of nonmetallic inclusions, such as oxides, sulfides and silicates [1.32].

Remelted and refined steels are the purest steels presently on the market. Therefore they can be very well polished.

1.2   Cast Steel

The production of molds from forged or rolled steel is relatively expensive because of the high labor costs for machining and the expenses for the necessary equipment which is partly special machinery. In addition, there are high material losses from cutting operations, which can amount to 30 to 50% for large cavities [1.19]. Besides this, stylists and consumers demand a molding whose surface quality or texture cannot be produced in a cavity by conventional cutting operations. Solutions can be found in physical or chemical procedures or precision casting.

The most common casting steel grades and casting processes are discussed in detail in Sections 2.1.1–2.1.3. A summary of cast steel alloys is shown in Table 1.3.

1.3   Nonferrous Metallics

The best known nonferrous metals employed in mold making are:

– copper alloys,
– zinc alloys,
– aluminum alloys,
– bismuth-tin alloys.

1.3.1   Copper Alloys

The importance of copper alloys as materials for mold making is based on their high thermal conductivity and ductility, which equalizes stresses from nonuniform heating quickly and safely. The mechanical properties of pure copper are moderate. Although they can be improved by cold rolling or forming, they generally do not meet the demands on a mold material.

Therefore, copper serves in mold making at best as an auxiliary material, for instance, as components of heat exchangers in molds cast from low melting alloys. Beryllium-copper-cobalt alloys, however, are important materials for mold cavities.

Table 1.3   Cast steels for injection molds; composition – treatment – properties [1.38]

Carbon steels	AISI 1020	0.18–0.23% C 0.30–0.60% Mn 0.04% P 0.05% S annealed BH 122, normalized BH 134 Welding is readily done by most arc and gas processes. Preheating unnecessary unless parts are very heavy
	AISI 1040	0.37–0.44% C 0.60–0.90% Mn 0.04% P 0.05% S normalized and annealed BH 175 normalized and oil quenched BH 225 Welding can be done by arc, resistance and gas processes. To decrease cooling rate and subsequently hardness preheating to above 150 preferably 260 °C. Final heating to 600–650 °C restores ductility and relieves stress
Alloy steels	AISI 1330	0.28–0.33% C 1.60–1.90% Mn 0.20–0.35% Si 0.035% P 0.04% S normalized BH 187, normalized and tempered BH 160 Can be relatively easily welded with mild-steel filler metal
	AISI 4130	0.28–0.33% C 0.40–0.60% Mn 0.20–0.35% Si 0.80–1.10% Cr 0.15–0.25% Mo 0.035% P 0.04% S annealed BH 175 Welding same as 1330
	AISI 4340	0.38–0.43% C 0.60–0.80% Mn 0.20–0.35% Si 1.65–2.00% Ni 0.70–0.90% Cr 0.20–0.30% Mo 0.035% P 0.04% S normalized and annealed BH 200 quenched and tempered BH 300 Welding calls for pre- and postheating and filler metals of the same mechanical properties as the base metal. Stress relieving is desirable especially in repair work
	AISI 8630	0.28–0.30% C 0.70–0.90% Mn 0.20–0.35% Si 0.40–0.70% Ni 0.40–0.60% Cr 0.15–0.25% Mo normalized BH 240, annealed BH 175 Welding preferably with filler metal of the same chemical analysis. No preheating needed for sections up to 0.5 in. but stress relieving after welding
Heat-resistant steel	AISI 501	0.10% C 1.00% Mn 1.00% Si 4.00–4.00% Cr 0.40–0.65% Mo Welding similar to 8630

1.3.1.1 Beryllium-Copper Alloys

As with other alloys, the mechanical and thermal properties depend on the chemical composition of the alloy. With increasing beryllium content, the mechanical properties improve while the thermal properties deteriorate. Beryllium-copper alloys with more than 1.7% beryllium have prevailed in mold making. They have a tensile strength of up to 1200 MPa and can be hardened up to 440 Brinell. Hardening up to 330 to 360 Brinell is generally sufficient for practical purposes. In this range the material is very ductile, has no tendency to fragility at the edges, and is readily polishable [1.39].

Tempering, which provides a homogeneous hardness, can be done in a simple oven [1.40].

Alloys with a beryllium content below 1.7% are only used for functional components, such as heat conductors in heat exchangers, because of their diminished strength. Table 1.4 summarizes the technical data of some beryllium-copper alloys.

Molds made of beryllium copper are sufficiently corrosion resistant and can be chrome or nickel plated if needed. Protective coating is mainly done by electroless nickel plating today. It has the advantage of a more even deposit and a hardenability of up to 780 Brinell with an appropriate heat treatment at about 400 °C. Besides this, a nickel coating is less prone to cracking than a hard chrome coating [1.40]. In contrast to cast steel, beryllium-copper alloys are practically insensitive to thermal shock.

Mechanical damage can be repaired by brazing or welding. Beryllium-copper rods with about 2% beryllium are preferably used for welding with 250 Amps [1.39, 1.40]. The molds are appropriately preheated to 300 °C [1.40].

Molds of beryllium copper can be made by machining or casting. Cold-hobbing is only conditionally suitable for shallow cavities. The alloy can be better formed by hothobbing in a temperature range of 600 to 800 °C followed by solution annealing.

Beryllium-copper alloys are employed as materials for molds or functional components wherever high heat conductivity is important either for the whole mold or part of it. The temperature differential between cavity wall and cooling channel is reduced. This results in higher output with the same, or often better, quality with the same cooling capacity. These alloys also offer an advantage if high demands on surface accuracy have to be met. They are well suited for casting with patterns having a structured surface such as wood, leather or fabric grain.

1.3.2   Zinc and its Alloys

High-grade zinc alloys for casting are used in mold making only for prototype molds or molds for small production runs because of their inferior mechanical properties. They are more frequently utilized for blow molding or vacuum forming where molds are not under high mechanical loads. Zinc alloys, like copper alloys, are characterized by high thermal conductivity of about 100 W/(m · K).

Molds of zinc alloys are mostly cast. The low casting temperature (melting point ca. 390 °C, casting temperature 410 to 450 °C) is a special advantage. It allows patterns to be made of wood, plaster, or even plastics, in addition to steel. Patterns of plaster and plastics can be made especially easily and fast. High-grade zinc alloys can also be cast in sand and ceramic molds. Die casting, although possible, is used only rarely.

The excellent behavior of zinc alloys during casting produces smooth and nonporous surfaces of intricate and structured surfaces [1.39]. Molds, especially plaster molds, must be adequately dried before use (plaster for several days at 220 °C), since otherwise a porous and rough cavity surface results from the generation of steam [1.21].

Molds of high-grade zinc alloys can also be made by cold-hobbing. The blank is preheated to 200 to 250 °C before hobbing. Even deep cavities are produced in one step without intermediate annealing. Hobs with a Brinell hardness of 430 are employed. Vertical motion is sometimes combined with rotation. With such hobs cavity inserts for molding helical gears can be made [1.21].

Another process using high-grade zinc alloys is metal spraying (Section 2.2).

Because of low mechanical strength only cavity inserts are made of high-grade zinc alloys. These inserts are fitted into steel bases, which have to resist the forces from

Table 1.4 Technical data of beryllium-copper [1.41]

Designation	C 17 200	C 17 300	C17 000	C 17 510	C 17 500
Composition	Be 1.80–2.00% Co/Ni 0.20% min. Co/Ni/Fe 0.6% max., Cu balance	Be 1.80–2.00%, Co/Ni 0.20% min., Co/Ni/Fe 0.6% max., Pb 0.20–0.6%, Cu balance	Be 1.60–1.79% Co/Ni 0.20% min., Co/Ni/Fe 0.6% max., Cu balance	Be 0.20–0.60% Ni 1.40–2.20% Cu balance	Be 0.40–0.75% Co 2.40–2.70% Cu balance
Density kg/m^3	8250		8415	8610	8775
Therm. conductivity W/(m · K)	130		130	260	260
Therm. expansion 10^{-6} K^{-1}	17.5		17.5	17.6	18
Specifie heat capacity J/(kg · K)	420		420	420	420
Electric resistivity 10^{-8}Ω · m	7.7		7.7	3.8	3.8
Modulus of elasticity GPa	131		128	138	138
Tensile strength [MPa]	I 415–585 II 620–895 III 1140–1310 IV 1275–1480		I 415–585 II 620–895 III 1035–1310 IV 1170–1450	I 240–380 II 450–550 III 690–830 IV 760–900	
Yield strength [MPa]	I 140–415 II 515–725 III 1000–1200 IV 1140–1380		I 140–205 II 515–725 III 860–1070 IV 930–1140	I 140–310 II 340–515 III 550–690 IV 690–830	
Elongation [%]	I 35–60 II 10–20 III 3–10 IV 2–5		I 35–60 II 10–20 III 4–10 IV 2–5	I 20–35 II 10–15 III 10–25 IV 10–20	
Rockwell hardness	I B45–80 II B88–103 III C36–41 IV C39–44		I B45–85 II B91–103 III C32–39 IV C35–41	I B25–50 II B60–75 III B92–100 IV B95–102	

Notes: I: Solution heat treated
II: Cold drawn
III: Solution heat treated and age hardened
IV: Cold drawn and age hardened

clamping and injection pressure during the molding process. The most common highgrade zinc alloys, known under their trade names Zamak or Kirksite, are summarized in Table 1.5.

1.3.3   Aluminum Alloys

For a long time, aluminum was not used to any major extent for making injection molds. However, extensive improvements in properties, particularly mechanical properties, have led to this material being used more and more often in recent times. The main properties in favor of using aluminum for mold making are:

– low density,
– good machinability,
– high thermal conductivity, and
– corrosion resistance.

Aluminum alloys are available as casting and forging alloys. The casting alloys are not of much significance. Casting is difficult. Perfect molds presuppose correct choice of alloy, the appropriate melting and casting technique and a mold design amenable to casting. It is therefore advisable to collaborate with the manufacturer at the design stage.

More important for mold making are the forging alloys which, in the form of heathardening, high-strength aluminum-zinc-magnesium-copper alloys, have already shown their mettle in aircraft construction. These are now available commercially in the form of complete mold structures as well as plates that have been machined on all sides and as blanks [1.44]. The chemical composition and physical properties of a typical aluminum alloy used in mold making are shown in Table 1.6.

Aluminum molds weigh less than steel molds due to their specific weights (between 2.7 and 2.85 g/cm^3, depending on alloying additives). Unfortunately this "positive characteristic" is not fully recognized because the individual plates, which are components of a mold, have to be about 40% thicker than steel plates because of their lower mechanical strength (modulus of elasticity is only roughly 30% that of steel; Table 1.6). In spite of this, aluminum molds weigh about 50% less than steel molds. This is a considerable advantage during mold making and assembly as well as later on during setup in the molding shop. With smaller molds, it is frequently possible to do without expensive lifting devices such as cranes and lift trucks.

Additional advantages result from good machinability, which can allow cutting speeds five times as fast as those for steel [1.44]. This holds particularly true if the recommendations of the producer are taken into account and those tools are used which are especially suited for aluminum. Distortion from machining is not to be expected because aluminum has very little residual stresses due to special heat treatment during production (the wrought products are hot-rolled, then heat-treated and stretched until stress-free). Aluminum can be worked with cutting tools as well as shaped by means of EDM.

Electrode material made of electrolytic copper or copper containing alloys are primarily used for this procedure. Here, too, the high eroding speed (6 to 8 times that of steel) is economically significant. Cut or eroded surfaces can be polished with standard machines and wheels and made wear resistant by chrome plating or anodizing. Service lives of up to 200,000 shots per mold have been achieved with aluminum molds, the

Table 1.5 High-grade zinc alloys for injection molds [1.42, 1.43]

Designation	Density (kg/m³)	Melting point (°C)	Shrinkage %	Thermal expansion (10⁻⁶/K)	Tensile strength (MPa)	Elongation (% in 50 mm)	Brinell hardness BH	Compressive strength (MPa)	Shear strength (MPa)
Zamak	6700	390	1.1	27	220–240	1–2	1000	600–700	300
Kirksite A	6700	380	0.7–1.2	27	226	3	1000	420–527	246
Kayem	6700	380	1.1	28	236	1.25	1090	793	–
Kayem 2	6600	358	1.1	–	149	Very low	1450–1500	685	–

Table 1.6 Chemical composition and physical properties of a typical heat-hardening aluminum forging alloy used for mold making [1.44, 1.45, 1.47]

Chemical composition

A.A. No.	Composition (wt. -%)								
	Cu	Cs	Fe	Mg	Mn	Si	Ti	Zn	Others
3.4365	1.2–2.0	0.18–0.28	0.5	2.1–2.9	0.3	0.4	0.2	5.1–6.1	Ti + Zr = 0.25

Physical properties

A.A. No.	Density	Yield strength	Ultimate tensile strength	Modulus of elasticity	Coefficient of thermal expansion	Thermal conductivity
	kg/dm³	N/mm²	N/mm²	MPa	W/(m °C)	W/(K · m)
3.4365	2.82	410–530	480–610	72000	23.7	153 (0–100 °C)

number depending on the application [1.48, 1.49]. Surface treatment will be discussed later.

The thermal conductivity of the materials used for the molds is critical to injection molding. When aluminum alloys are used, it ensures good, rapid heat distribution and dissipation. This shortens cycle times and may lead to enhanced quality of moldings.

In the presence of atmospheric oxygen, a dense, strongly adhering oxide layer about 0.001 µm thick forms on the surface of the aluminum [1.44, 1.46]. This layer protects the underlying materials, particularly in dry environments, against atmospheric attack. If it is destroyed or removed during machining, it re-forms anew spontaneously. It is damaged by acids formed from salts and gases in contact with moisture (water of condensation). If this kind of stress is expected during operation, surface treatment is advisable. Good results in this connection have been produced by

– chrome plating,
– nickel plating,
– anodizing, and
– PVD coating.

Surface treatment enhances not only corrosion resistance but also wear and abrasion resistance and sometimes facilitates demolding. Layer thicknesses and hardnesses obtained with the aforementioned coating methods are summarized in Table 1.7.

Table 1.7 Coating processes for aluminum and properties obtained [1.45, 1.47]

Process	Layer thickness	Hardness
Chrome plating	20–200 µm	900–1100 HV
Nickel plating	35–50 µm	500–600 HV
Anodizing	30–60 µm	350–450 HV
PVD coating	5–10 µm	

Due to the positive aspects for mold making described above, mold plates already milled and surface-ground to finished dimensions, with the corresponding bores for guide systems right through to complete mold structures, are now commercially available.

Designs combining aluminum with steel have yielded good results. This construction has the advantage of allowing the more resistant steels to be used for areas subject to high wear and abrasion. It therefore combines the advantages of both materials.

The service life of aluminum molds depends on the plastic to be processed (reinforced or unreinforced), its processing conditions (injection pressure, temperature) and the geometry of the molding. The literature contains reports of molds that have produced 15,000–200,000 moldings. Aluminum is of little or no suitability for processing thermosets on account of the high processing temperatures and the associated thermal stress [1.45, 1.49].

1.3.4 Bismuth-Tin Alloys

Bismuth-tin alloys are marketed under the trade name Cerro alloys. They are relatively soft, heavy metals, which generally react to shock in a brittle manner but exhibit plasticity under constant loading. The strength of these alloys increases with aging [1.51].

Table 1.8   Bismuth-tin alloys (Cerro® alloys) for injection molds [1.52]

Designation (registered trademark)		Cerrotru	Cerrocast
Density	kg/dm³	8.64	8.16
Melting point or range	°C	138	138–170
Specific heat	kJ/kg	1.88	1.97
Thermal expansion	10⁻⁶	15	15
Thermal conductivity	W/(K · m)	21	38
Brinell hardness	BHN	22	22
Ultimate tensile strength	MPa	56	56
Elongation (slow loading)	%	200	200
Max. constant load	MPa	3.5	3.5
Composition % Bi % Sn		58 42	40 60

Bismuth-tin alloys are compositions of metals with low melting points (between 40 and 170 °C depending on composition). They are suitable for normal casting as well as for die or vacuum casting. They can also be used with a special spray gun. Those alloys which do not change their volume during solidification are particularly suited for making molds.

Because of their moderate mechanical properties, bismuth-tin alloys are primarily used for prototype molds in injection molding. More common is their use for blow molds, molds for hot forming and for high-precision matrices in electrolytic depositing. Besides this, they are used as a material for fusible cores, which will be discussed in more detail later on (Section 2.10). Table 1.8 shows the physical and chemical Properties of some Cerro alloys.

1.4   Materials for Electrolytic Deposition

Electrolytic deposition of metals has two applications in mold making. A distinction is made between "decorative" plating and electrolytic forming. The techniques are similar but their end products and their intended use are very much different. In plating, a thin layer, usually about 25 μm thick, is deposited. This layer is intended to protect the metal underneath from corrosion, to facilitate demolding, and to reduce forming of deposits and facilitate the cleaning of molds.

To do this, the electrolytic deposit must adhere well to the substrate.

In electrolytic forming, a considerably thicker layer is deposited on a pattern that has the contours and dimensions of the later cavity insert. The thickness of the deposit is arbitrary and limited only by production time. After the desired thickness is obtained and further work in the form of adding a backing is completed, the structure should detach easily from the pattern.

There are a number of materials that can be electrolytically deposited. The most important ones are nickel and cobalt-nickel alloys. Nickel is the most frequently - employed metal for electrolytic forming because of its strength, toughness, and corrosion resistance. Besides this, the process is simple and easily controlled. Hardness, strength,

Table 1.9   Characteristics of electro-deposited nickel [1.53, 1.55]

Hardness	4500–5400 MPa
Tensile strength	360–1510 MPa
Typ. strength for molds	1400 MPa
Yield strength	230–640 MPa
Typ. strength for molds	460 MPa
Elongation	2–37%
Typ. elongation for molds	10%
Temperature resistance	max. 300 °C
Corrosion and wear resistant	
Fine, ductile grain	

ductility, and residual stresses can be varied within a broad range by selecting certain electrolytic solutions and process conditions [1.53, 1.54].

Even the hardest forms of electrolytically deposited copper and iron are too soft for cavity walls. Copper is primarily used for backing an already formed nickel shell. The rate at which copper can be deposited is a very desirable feature of this procedure.

By contrast, electrolytically deposited chromium is so hard that postoperations such as drilling holes for ejector pins are impractical. In addition chromium exhibits high residual stresses, which can easily lead to cracks in the chrome layer. Therefore chromium is used only for protective coatings in mold making.

1.5   Surface Treatment of Steels for Injection Molds

1.5.1   General Information

As already explained in Section 1.1, materials for mold making have to exhibit distinctive properties. Frequently, a compromise becomes necessary because the properties of steels depend largely on their chemical composition and the alloying components affect each other.

Therefore it is reasonable that steel producers and users, together with plastics processors, keep looking for suitable processes to improve the quality and particularly the service life of injection molds. This can be done by numerous surface treatments. The goal of all these processes is to improve

– surface quality,
– fatigue and wear resistance,
– corrosion resistance,
– sliding ability, and
– to reduce the tendency to form material residues and deposits in the mold.

The surface properties of components of injection molds can be considerably modified by suitable machining, controlled heat treatment or a change in the alloying elements in the surface layers, by diffusion or build-up. Special demands can thus be met.

Before the details of some selected procedures are discussed below, it must be pointed out that the application of the various processes calls for specialist knowledge and, in some cases, considerable technical resources. Therefore such work is ordinarily done by specialists. Steel producers also provide advice in this respect.

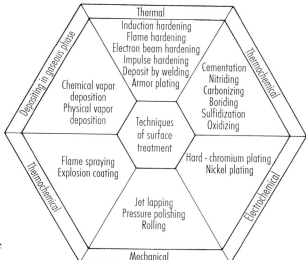

Figure 1.2 Techniques of surface treatment [1.56, 1.57]

Currently, the most common surface treatments for mold steels are shown in Figure 1.2. Some are presented in more detail below.

1.5.2 Heat Treatment of Steels

In traditional heat-treatment processes changes in the crystalline structure are caused by simple heating and cooling. This confers distinct properties on the steel. To this end, the steel producers provide special heat-treatment information showing the feasibility of treatments and the resulting properties. Such "simple" treatments do not require – as long as small dimensions are considered – a great amount of technique. They only assume a theoretical understanding of the reactions occurring during heat treatment. Annealing, hardening and tempering are considered "simple" treatments.

1.5.3 Thermochemical Treatment Methods

Unlike their thermal counterparts, thermochemical methods utilize chemical elements that diffuse from a gaseous, liquid or solid state into the material surface to produce a hard, wear-resistant layer [1.56]. A distinction is drawn between

– carburizing,
– nitriding, and
– boriding.

1.5.3.1 Carburizing

In carburizing, the carbon content of layers close to the surface of steel, having a low carbon content of ordinarily less than 0.25%, and therefore not soluble is increased to as much as 0.9% at temperatures between 850 and 980 °C. If quenching is done immediately from the carburizing temperature, this process is called direct hardening [1.58]. Carburizing produces a hard surface, which is supported by a soft, tough interior core [1.58].

After hardening the steel is tempered. The tempering temperature is determined by later application or temperature at which the mold is used [1.56] and determines the hardness of the surface. Tempering at temperatures between 100 and 300 °C yields a hardness of 600 to 700 Brinell.

1.5.3.2 Nitriding

Nitriding (see also Section 1.1.3) can be done in a salt bath, in gas, or powder. The medium determines the temperature and duration of the process. The aim of all processes is to saturate the outer layer of the workpiece by diffusion of nitrogen so as to increase hardness, wear resistance or corrosion resistance.

In nitriding in a bath, the molds are preheated to 400 °C. The nitriding itself is done at 580 °C. The nitriding time depends on the desired depth of nitriding, but two hours are generally sufficient. A special form of nitriding in a bath is the Tenifer treatment, which is characterized by an aerated nitriding bath of special composition. The required surface hardness for injection molds is likewise achieved after two hours at a temperature of 570 °C [1.59].

Much longer nitriding times are necessary for gas nitriding. The desired surface hardness takes at least 15 to 30 hours treatment, the exact time depending on the steel grade. The treatment is performed at temperatures between 500 and 550 °C. It is possible to selectively nitride certain areas by partially covering the surface with a coating of copper or nickel or with a special paste [1.59].

When nitro-carburizing in gas, a reaction component that provides carbon is used in addition to ammonia. This enriches the connecting zone with carbon. Applicable components are endogas, exogas or combinations thereof as well as natural gas or liquids that contain carbon dioxide. The treatment temperatures are about 570 °C, and the time is 2 to 6 hours.

Activator content, nitriding time and temperature determine the quality of the results of powder nitriding for a particular steel grade. It is carried out at temperatures between 450 and 570 °C [1.59].

Ionitriding is nitriding in the plasma of a high-amperage corona discharge. It causes nitrogen to be deposited on the mold surface. The hardening depth can range from a few micrometers to 1 mm. The greatest hardness is achieved here immediately at the surface, reaching 800 Brinell for some steel grades. Posttreatment of the surface is, therefore, unnecessary. The treatment is done in the temperature range between 350 and 580 °C. Treatment times start at a few minutes and are practically unlimited (20 min to 36 h) [1.60].

Distortion of molds through nitriding or ionitriding does not normally occur. After heat treatment, molds are obtained which are tough and free of stresses, have high surface hardness and improved corrosion resistance.

Nitriding steels are annealed before delivery and can be machined without difficulties.

1.5.3.3 Boriding

In boriding, layers close to the surface are enriched with boron. The result is a very thin but extremely hard (1800 to 2100 HV 0.025 [1.56]) and wear-resistant layer of iron boride, which is interlocked with the base metal. Boriding can be effected in a solid compound, through treatment in gas or in borax-based melts, both with and without electrolyte.

Boriding is carried out in a temperature range of 800 to 1050 °C. The usual duration is between 15 minutes and 30 hours. Treatment time, temperature and base material determine the thickness of the layer. Thicknesses of up to 600 µm can be achieved. Partial boriding is feasible.

After boriding, workpieces can be heat-treated to confer a higher "load capacity" on the base material. The use of a heat-treated steel is a precondition for this. The temperature depends on the base material. This treatment should be limited to workpieces with a medium layer of boride (100 to 120 µm). Thicker layers are liable to crack [1.61].

Borided surfaces usually have a dull-gray appearance, and build-up of the boride layer on the surface is possible, the extent depending on the processing conditions. Therefore, a finishing operation in the form of grinding, polishing, lapping or honing is often necessary.

1.5.4 Electrochemical Treatments

During processing, some polymeric materials release chemically aggressive substances, mostly hydrochloric or acetic acid. In such cases, the molds are frequently protected by electrolytic plating with chromium or nickel.

Not only is corrosion resistance enhanced, but also the antifriction properties and, in the case of chromium, the wear resistance as well.

This plating is permanently effective only if the thickness of the deposit is uniform and sharp edges in the mold are avoided. Nonuniform thickness and sharp edges cause stresses in the protective layer, which can lead to peeling off under loads. The risk of a nonuniform deposit is particularly great in molds with intricate contours (undercuts etc.). Plated thin flanges subject to bending stresses are very susceptible to cracking.

Before protective plating by electrolytic depositing can be performed, the molds have to be finished accordingly. Build-up and surface quality of the deposit depend on the quality of the base material. Ground, or even better, polished and compacted surfaces give the best results.

1.5.4.1 Chrome Plating

The thickness of electrolytically deposited layers of chromium depends on the current density and the temperature of the electrolyte. Their hardness also depends on the temperature of the heat treatment after plating. Common thicknesses lie between 5 and 200 µm, in special cases between 0.5 and 1 mm [1.56]. A surface hardness of 900 HV 0.2 is obtained. Because the chrome layer is deposited electrolytically, it builds up in lumps and needs to be reground to the requirements imposed on the quality of the coated workpiece.

1.5.4.2 Nickel Plating

A distinction is drawn between electrolytic and chemical nickel plating. The properties of the deposits also depend on the processing parameters. Nickel deposits are relatively soft and therefore not wear resistant.

1.5.4.3 NYE-CARD Process

In order to remedy this disadvantage, procedures have been developed, such as the NYE-CARD process, with which 20 to 70% by volume silicon particles of 25 to 75 μm are incorporated into a nickel-phosphorus layer. The nickel phosphorides contain 7 to 10% phosphorus.

The temperatures to which the material to be coated is exposed are less than 100 °C during the process. If subsequent heat treatment to improve adhesion is intended, the temperature should be about 315 °C. Besides steel, materials such as aluminum or copper alloys can be coated. It makes sense, however, to employ hardened base materials due to of the low thickness of the deposit, and to bring the surface quality of the base up to the required standard.

1.5.4.4 Hard Alloy Coating

Hard alloy coatings are electrolytically deposited protective layers of either tungsten-chromium (Hardalloy W) or vanadium-cobalt (Hardalloy TD). They are applied to a surface that has previously been smoothed and compacted by an ion beam [1.62].

1.5.5   Coating at Reduced Pressure

Common coating techniques based on the principle of deposition from the gas phase are chemical vapor deposition (CVD) and physical vapor deposition (PVD) and their respective variants. Both techniques have specific advantages and disadvantages. Because CVD requires high temperatures above the retention of hardness for tool steels, steel materials require subsequent heat treatment. However, this may lead to imprecise dimensions and shapes due to distortion. PVD techniques do not require subsequent heat treatment. However, CVD layers generally yield higher adhesive strength and depth of penetration into deep, narrow openings.

1.5.5.1 CVD Process

The CVD process (chemical vapor deposition) is based on the deposition of solids from a gaseous phase by a chemical reaction at temperatures above 800 °C [1.56]. It is shown schematically in Figure 1.3.

In a CVD process, carbides, metals, nitrides, borides, silicides or oxides can be deposited on the heated surface of molds at temperatures between 800 and 1100 °C. Depending on location and mold type, layers having a thickness of 6 to 30 μm and a strength of up to 4000 MPa may be deposited, e.g. a coating of titanium carbide with a thickness of 10 μm. The deposits faithfully copy the surface of the mold, that is, traces from machining of the surface such as scratches and grooves are not hidden by a CVD process. Therefore the mold surface must already have the same quality before coating that is expected from the finished tool [1.64].

The high temperatures needed for this process cause the base material to loose hardness and strength. This disadvantage must be compensated by another heat treatment and associated hardening of the base material. Appropriate steel grades should be used.

Because every heat treatment involves risks such as distortion, efforts are being made to lower the process temperatures with a view to improving the technique. Temperatures of 700 °C and lower are being discussed [1.65].

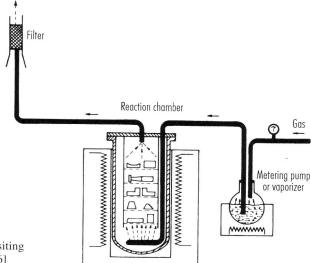

Figure 1.3   Equipment for depositing titanium carbide (schematic) [1.56]

1.5.5.2   PVD Process

The term PVD (physical vapor deposition) summarizes all those coating techniques by which metals, their alloys and their chemical compounds such as oxides, nitrides, and carbides can be deposited in a vacuum under the simultaneous effects of thermal and kinetic energy through particle bombardment [1.56]. The process is shown schematically in Figure 1.4.

PVD techniques include

– deposition through evaporation in high vacuum,
– ion plating,
– sputtering [1.56].

In contrast to the CVD process, the coating of molds by this physical procedure occurs at temperatures between 500 and 550 °C. This temperature is in many cases below the tempering temperature of the base material and so further heat treatment (with the associated risk of distortion) after coating does not become necessary. This process is suitable for all tool steels [1.64].

For the PVD technique, the quality and cleanliness (freedom from rust and grease) of the mold surface before coating are also crucial to the bond and surface quality after coating.

A variant of PVD, namely PVD arc coating, makes it possible to deposit virtually any coating material in monolayer and multilayer systems. The coating materials are generally chosen on the basis of attainable hardness, coefficient of friction, corrosion resistance and cost. Candidate coating materials or systems include TiN, TiC, TiCN, TiAlN, CrN and CrAlN [1.66, 1.67]. The thinnest hard layers are obtained by ion implantation, e.g. of nitrogen or carbon [1.68].

PVD coatings have the advantage of being independent of the surface contour for all practical purposes. They do not affect either the fineness of shape or the dimensional

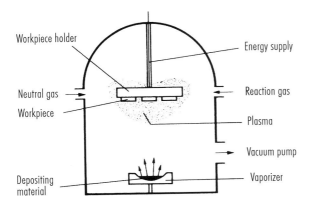

Workpiece holder

Neutral gas

Workpiece

Depositing
material

Energy supply

Reaction gas

Plasma

Vacuum pump

Vaporizer

Figure 1.4   Basics of the PVD
technique [1.56]

accuracy. Changes in dimensions are of the order of less than 5 µm and roughness values of less than 0.5 µm are attainable [1.37].

So far, titanium nitride (TiN) layers have frequently been used in practice. Titanium is vaporized in vacuum at temperatures of 550 °C, and together with the nitrogen present, forms a golden wear-resistant deposit up to 5 µm thick on the metal surface [1.36].

The use of PVD layers greatly reduces deposit formation and makes it possible to perform less risky cleaning of the surface.

The use of PVD layers can in some cases considerably extend the service life of the molds (by as much as 20 times). Greatly extended service lives have been demonstrated in molds exposed to corrosive attack, e.g. from the processing of flame-retardent polyamide or acetal [1.57]. A further major advantage is a lowering of release forces.

1.6   Laser Surface Treatment

Laser surface treatment methods rank among the special finishing procedures. Aside from its widespread use for cutting and welding, laser surface treatment is finding more and more widespread application in industrial practice. Laser surface treatment methods can be divided into two groups, namely thermal and thermochemical. Laser treatment therefore presents an alternative to conventional hardening techniques for increasing the service life of molds with high component costs.

Thermal variants include laser hardening and laser remelting while thermochemical methods include laser alloying, laser dispersing and laser coating. With the thermochemical methods, the material properties are affected not only by the heat treatment but also by the addition of extra materials. The outer layer properties are improved in respect of the mechanical, tribological, thermochemical and chemical wear resistance of the surface [1.35].

Laser sources nowadays are industrial carbon dioxide lasers rated between 1 and 25 kW and Nd:YAG lasers rated at 0.5 to 3 kW. The beam is focused on the surface of the workpiece, which becomes heated by absorption. The extent of absorption or reflection depends on the wavelength, the material's properties and the texture of the workpiece. Heat incorporation also depends on the intensity of the beam and the exposure time. The exposure time is derived from the rate at which the beam is moved across the surface of the workpiece. Processing usually occurs in adjacent tracks that may or may not overlap. It is thus possible to perform large-scale and partial surface coating [1.35].

1.6.1   Laser Hardening and Re-Melting

All steel and cast-iron materials with a carbon content exceeding 3% can be laser hardened since they harden martensitically. In laser hardening, the workpiece is heated to temperatures above the austenitizing temperature; at the surface, heating is due to absorption of the infrared laser beam and down to a certain depth in the lower zones, due to heat conduction. This heating process occurs so quickly that the cooling which occurs immediately afterwards creates an extremely steep temperature gradient between the workpiece surface and the rest of the workpiece. As the laser beam is moved on, the absorbed quantity of heat is dissipated so quickly inside the workpiece that the critical cooling rate for martensite formation is exceeded (self-quenching) [1.34].

Laser hardening is a partial hardening process in which the focused beam is guided back and forth across the area to be hardened. The advantage of this is that selective hardening can be performed at points where increased wear is expected. As a result, hardening can be performed with very little distortion, zones of partial hardening can be reproducibly created and the workpiece suffers only minor thermal stress. However, the disadvantage is that large areas have to be hardened in several steps.

Laser re-melting is primarily used for cast materials. The heating and subsequent - selfquenching generate supersaturated solutions, metastable phases and amorphous structures. Re-melting generates homogeneous, extremely fine-grained structures in the outer layer of the component. These structures are notable for their high strength and, at the same time, high toughness [1.35].

1.6.2   Laser Alloying, Dispersing, and Coating

Laser alloying is used for completely dissolving additional materials into the base material. Convection and diffusion processes in the melting bath lead to homogeneous mixing of the base and additional materials. The high cooling rates ensure that the additives stay in solution even after cooling. The additive elements and compounds primarily increase the wear and corrosion resistance.

By contrast, laser dispersing does not dissolve the additive but rather keeps it finely dispersed in the base structure. Standard high-melting or dissolving-resistant additives for laser alloying are SiC, TaC, TiC, VC and WC.

Finally, the aim of laser coating is not to cause any mingling of the base and additive materials. A homogeneous, strongly adhering layer is generated on the substrate material. The additive material is completely melted on top and the base structure is only melted in the narrow edge zone to create a metallurgical bond. The most common additive materials are usually low-melting Ni or Co alloys.

1.7   Electron Beam Hardening

In principle, electron beams may be used instead of laser beams. Consequently, electron beam hardening works on the same principle as laser hardening. Bombardment by the electron beam causes rapid heating up and subsequent self-quenching of the outer layer of the workpiece. Since an electron beam is extremely diffuse under ambient conditions and the necessary intensity cannot be attained, processing can only be carried out in high-vacuum chambers. This restriction increases the processing time because a high vacuum has to be produced after the vacuum chamber has been set up. In addition, the cramped conditions of the electron beam equipment restricts the size of the components that can be processed. For these reasons, this process has so far found little industrial use.

1.8   Lamcoat Coating

The Lamcoat process was developed in the USA for smooth, sliding surfaces. In it, a soft layer based on tungsten disulfide is applied mechanically to the mold surface at room temperature. To this end, the mold is first thoroughly cleaned in an alcoholic ultrasonic degreasing bath by means of special high-pressure micro-sprays (no material ablation occurs). Then the tungsten disulfide is applied manually in a spray booth with the aid of dry, purified compressed air at high pressure. The coating material does not accumulate on the surface, however, but penetrates extensively into it to form molecular bonds with the base material. Excess material does not material. As a result, a very thin layer measuring 0.5–1.5 µm is formed that cannot be made any thicker.

Studies in the USA have shown that a Lamcoat coating reduces friction by up to 70%. This leads to lower injection pressures and increases flow paths by up to 10%. Further positive results are shorter cycle times and a 30–70% increase in mold service life [1.33]. This coating technology is currently offered by two companies in Germany.

References

[1.1]   Becker, H. J.: Herstellung und Wärmebehandlung von Werkzeugen für die Kunst-stoffverarbeitung. VDI-Z, 114 (1972), 7, pp. 527–532.
[1.2]   Krumpholz, R.; Meilgen, R.: Zweckmäßige Stahlauswahl beim Verarbeiten von Kunststoffen. Kunststoffe, 63 (1973), 5, pp. 286–291.
[1.3]   Kuhlmann, E.: Die Werkstoffe der metallverarbeitenden Berufe. Girardet, Essen, 1954.
[1.4]   Fachkunde Metall. 50th Ed., Verlag Europe-Lehrmittel Nourney, Vollmer GmbH & Co., Haan Gruiten, 1990.
[1.5]   Werkstoffkunde für Praktiker. 3rd Ed., Verlag Europe-Lehrmittel Nourney, Vollmer GmbH & Co., Haan Gruiten, 1989.
[1.6]   Werkstofftechnik für Metallberufe. Verlag Europe-Lehrmittel Nourney, Vollmer GmbH & Co., Haan Gruiten.
[1.7]   Das ist Edelstahl. Thyssen Edelstahl Werke AG. Publication, 1980, 2, Krefeld.
[1.8]   Werkstoff + Werkzeug. Thyssen Edelstahl Werke AG. Publication, 1984, 3, Krefeld.
[1.9]   König, W.: Neuartige Bearbeitungsverfahren. Lecture, Production techniques, 11, Tech. University Aachen, Laboratory for Machine Tools and Plant Management.
[1.10] Auswahl und Wärmebehandlung von Stahl für Kunststoff-Formen. Arburg heute, 5 (1974), 7, pp. 16–18.
[1.11] Dember, G.: Stähle zum Herstellen von Werkzeugen für die Kunststoffverarbeitung, Kunststoff-Formenbau, Werkstoffe und Verarbeitungsverfahren. VDI-Verlag, Düssel-dorf, 1976.

[1.12]   Maßänderungsarme Stähle. Company brochure, Boehler.

[1.13]   Becker, H. J.: Werkzeugstähle für die Kunststoffverarbeitung im Preß- und Spritzgieß-verfahren. VDI-Z, 113 (1971), 5, pp. 385–390.

[1.14]   Catić, I.: Kriterien zur Auswahl der Formnestwerkstoffe. Plastverarbeiter, 26 (1975), 11, pp. 633–637.

[1.15]   Werkstoffe für den Formenbau. Technical information, 4.6, BASF, Ludwigshafen/Rh., 1969.

[1.16]   Treml, F.: Stähle für die Kunststoffverarbeitung. Paper given at the IKV Aachen, November 15, 1969.

[1.17]   Malmberg, W.: Glühen, Härten und Vergüten des Stahls. Springer Verlag, Berlin, Göttingen, Heidelberg, 1961.

[1.18]   Mennig, G.: Mold Making Handbook, 2nd ed., Carl Hanser Verlag, Munich, 1998.

[1.19]   Stoeckhert, K.: Werkzeugbau für die Kunststoffverarbeitung. 3rd Ed., Carl Hanser Verlag, Munich, Vienna, 1979.

[1.20]   Weckener, H. D.; Häpken, H.; Dörlam, H.: Werkzeugstähle für die Kunststoff-verarbeitung. Stahlwerke Südwestfalen AG. Information, 15/75, Hüttental-Geisweid, 1975, pp. 31–40.

[1.21]   Spritzgießen von Thermoplasten. Company brochure, Farbwerke Hoechst AG., Frankfurt, 1971.

[1.22]   Illgner, K. H.: Gesichtspunkte zur Auswahl von Vergütungs- und Einsatzstählen. Metalloberfläche, 22 (1968), 11, pp. 321–330.

[1.23]   Bauer, M.: Kunstofformenstählen – Moderne Werkzeugstähle. Paper presented at the 8th Technical Conference on Plastic, Würzburg, September 24–25, 1997.

[1.24]   Auswahl und Wärmebehandlung von Stahl für Kunststoff-Formen, Arburg heute, 5 (1974), 8, pp. 23–25.

[1.25]   Auswahl und Wärmebehandlung von Stahl für Kunststoff-Formen, Arburg heute, 5 (1974), 7, pp. 16–18.

[1.26]   Dittrich, A.; Kortmann, W.: Werkstoffauswahl und Obernächenbearbeitung von Kunststoffstählen. Thyssen Edelst. Technical Report, 7 (1981), 2, pp. 190–192.

[1.27]   Frehn, F.: Höchst-Carbidhaltige. Werkzeugstähle für die Kunststoff-Verarbeitung. Kunststoffe, 66 (1976), 4, pp. 220–226.

[1.28]   Boehler Company. Publication, Düsseldorf, 1997.

[1.29]   Becker, H. J.; Haberling, E.; Rascher, K.: Herstellung von Werkzeugstählen durch das Electro-Schlacke-Umschmelz-(ESU)-Verfahren. Thyssen Edelst. Technical Report, 15 (1989), 2, pp. 138–146.

[1.30]   Weckener, H. D.; Dörlatu, H.: Neuere Entwicklungen auf dem Gebiet der Werkzeug-stähle für die Kunststoffverarbeitung. Publication, 57/67, Stahlwerke Südwestfalen AG, Hüttental-Geisweid, 1967.

[1.31]   Stähle und Superlegierungen. Elektroschlacke umgeschmolzen. WF Informations MBB, 5 (1972), pp. 140–143.

[1.32]   Verderber, W.; Leidel, B.: Besondere Maßnahmen bei der Herstellung von Werkzeug-stählen für die Kunststoffverarbeitung. Information, 15/75, Stahlwerke Südwestfalen AG, Hüttental-Geisweid (1975), pp. 22–27.

[1.33]   Mikroschicht stoppt teuren Verschleiß. VDI-Nachrichten, 7. 8. 1998.

[1.34]   Meis, F. U.; Schmitz-Justen, C.: Gezieltes Härten durch Laser-Licht. Industrie-Anzeiger, 105 (1983), 56/57, pp. 28–31.

[1.35]   König, W.; Klocke, F.: Fertigungsverfahren. Vol. 3: Abtragen und Generieren. Springer Verlag, Berlin, 1997.

[1.36]   Wild, R.: PVD-Hartstoffbeschichtung – Was Werkzeugbauer und Verarbeiter beachten sollten. Plastverarbeiter, 39 (1988), 4, pp. 14–28.

[1.37]   Bennighoff, H.: Beschichten von Werkzeugen. K. Plast. Kautsch. Z., 373 (1988), p. 13.

[1.38]   Spezialgegossene Werkzeuge. Information Stahlwerke Carp & Hones, Ennepetal, 1156/4, 1973.

[1.39]   Merten, H.: Gegossene NE-Metallformen. Angewandte NE-Metalle – Angewandte Gießverfahren. VDI-Bildungswerk, BW 2197. VDI-Verlag, Düsseldorf.

[1.40]   Beck, G.: Kupfer Beryllium, Kunststoff-Formenbau, Werkstoffe und Verarbeitungs-verfahren. VDI-Verlag, Düsseldorf, 1976.

[1.41]   Gegossene Formen. Company brochure, BECU, Hemer-Westig.

[1.42]   Richter, F. H.: Gießen von Spritzguß- und Tiefziehformen aus Feinzinklegierungen. Kunststoffe, 50 (1960), 12, pp. 723–727.

[1.43]   Wolf, W.: Tiehzieh-, Präge- und Stanzwerkzeuge aus Zinklegierungen. Z. Metall für Technik, Industrie und Handel, 6 (1952), 9/10, pp. 240–243.

[1.44]   Erstling, A.: Erfolge mit Aluminium im Werkzeug- und Formenbau. Werkstoff und Innovation, 11/12/1990, pp. 31–34.

[1.45]   Menning, G.: Mold Making Handbook, 2nd ed., Carl Hanser Verlag, Munich, 1998.

[1.46]   Erstling, A.: Aluminium für Blasform Werkzeuge. Reprint of Plastverarbeiter. Vol. 1, 1995, pp. 76–80.

[1.47]   Prospectus, Almet, Fellbach.

[1.48]   Erstling, A.: Aluminium im Spritzgießwerkzeugbau. Kunststoffe, 78 (1988), 7, pp. 596–598.

[1.49]   Zürb, G.: Vorteilhafter Aluminium-Einsatz im Werkzeug und Formenbau. Stahlbauer. Vol. 4, 1986.

[1.50]   Erstling, A.: Sparen mit Alu. Form und Werkzeug. November 1997, pp. 40–51.

[1.51]   Cerro-Legierungen für schnellen Werkzeugbau. Plastverarbeiter, 25 (1974), 3, pp. 175–176.

[1.52]   Eigenschaften und Anwendungen der Cerro-Legierungen. Prospectus, Hek GmbH, Lübeck.

[1.53]   Winkler, L.: Galvanoformung – ein modernes Fertigungsverfahren. Vol. 1 and 11. Metalloberfläche, 21 (1967), 8, 9, 11.

[1.54]   Watson, S. A.: Taschenbuch der Galvanoformung mit Nickel. Lenze, Saulgau/Württ., 1976.

[1.55]   Galvanoeinsätze für Klein- und Großwerkzeuge. Prospectus, Gesellschaft für Galvanoplastik mbH, Lahr.

[1.56]   Kortmann, W.: Vergleichende Betrachtungen der gebräuchlichsten Oberflächenbehandlungsverfahren. Thyssen Edelstahl. Technical report, 11 (1985), 2, pp. 163–199.

[1.57]   Walkenhorst, U.: Überblick über verschiedene Schutzmaßnahmen gegen Verschleiß und Korrosion in Spritzgießmaschinen und -werkzeugen. Reprint of 2nd Tooling Conference at Würzburg, October 4–5, 1988.

[1.58]   Rationalisierung im Formenbau. Symposium report, 1981, VKI. Kunstst. Plast., 29 (1982), 172, pp. 25–28.

[1.59]   Werkzeugstähle für die Kunststoffverarbeitung. Publication, 0122/4. Edelstahlwerke Witten AG, Witten, 1973.

[1.60]   Ionitrieren ist mehr als Härten. Publication, Kloeckner Ionen GmbH, Köln.

[1.61]   Publication. Elektroschmelzwerk Kempten GmbH, Munich, 1974.

[1.62]   Publication. Hardalloy W. and T. D., IEPCO, Zürich, 1979.

[1.63]   Piwowarski, E.: Herstellen von Spritzgießwerkzeugen. Kunststoffe, 78 (1988), 12, pp. 1137–1146.

[1.64]   Ludwig, J. H.: Werkzeugwerkstoffe, ihre Oberflächenbehandlung, Verschmutzung und Reinigung. Gummi Asbest Kunstst., 35 (1982), 2, pp. 72–78.

[1.65]   Ruminski, L.: Harte Haut aus Titancarbid. Bild der Wissenschaft, 2 (1984), p. 28.

[1.66]   Müller, D.; Härlen, U.: Verschleißschutz an Werkzeugen für die Blechbearbeitung. In: Leistungssteigerung in der Stanztechnik, VDI Bildungswerk, Düsseldorf, 1996.

[1.67]   Leistungsfähige Fertigungsprozesse – Lösungen für den Werkzeugbau. In: Produktionstechnik: Aachener Perspektiven, VDI-Verlag, Düsseldorf, 1996.

[1.68]   Frey, H.: Maßgeschneiderte Oberflächen. VDI-Nachrichten-Magazin, 1988, 2, pp. 7–11.

2   Mold Making Techniques

Injection molds are made by a highly varied number of processes and combinations thereof.

Figure 2.1 demonstrates the relative costs for cavities made from various materials. Accordingly, steel cavities appear to be many times more expensive than those made of other materials. In spite of this, a cavity made of steel is normally the preferred choice. This apparent contradiction is explained with the consideration that the service life of a steel mold is the longest, and the additional costs for a cavity represent only a fraction of those for the whole mold.

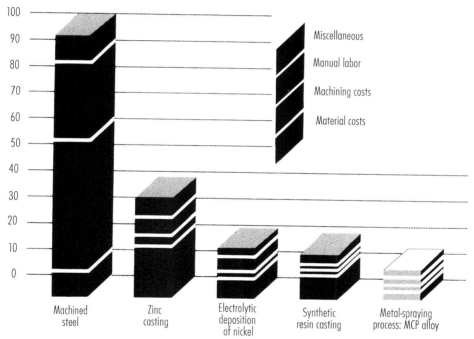

Figure 2.1   Comparison of costs: production methods in mold making [2.1]

Cavities made by electrolytic deposition as well as other procedures, which cannot be done in-house, call for additional working hours until the mold is finally available. This may be rather inconvenient. The making of an electrolytically deposited insert takes weeks or even months. A cavity made of heat-treated steel can be used for sampling without problems and still be finished afterwards. The high production costs also justify the application of a superior material because its costs are generally only 10 to 20% of the total mold costs.

In spite of all the modern procedures in planning, design, and production, mold making calls for highly qualified and trained craftsmen and such personnel are in short supply nowadays. Thus, the production of molds always poses a bottleneck.

It is clear, therefore, that only up-to-date equipment is found in modern mold-making facilities such as numerically-controlled machine tools. With their help one tries to reduce the chances of rejects or to automate the working process without human operator (e.g. EDM).

2.1 Production of Metallic Injection Molds and Mold Inserts by Casting

The production of mold inserts, or whole mold halves by casting, attained a certain preeminence in some application areas for a time. The reason was that the casting process offers suitable alloys for nearly every type of application and that there are hardly any limits concerning geometry. Molds requiring extensive machining could therefore be made economically by casting. Another application area is the simple, more cost effective production of injection molds for low production runs and samples, particularly of non-ferrous metals. Only a brief account of the casting methods for producing mold inserts is provided below. Readers requiring more detailed information are referred to the literature at the end of the chapter.

2.1.1 Casting Methods and Cast Alloys

Of the numerous casting methods available [2.2, 2.3], variants of sand casting and precision casting are used to make mold inserts. The choice of casting method depends on the dimensions of the mold, the specified dimensional tolerances, the desired faithfulness of reproduction and the requisite surface quality.

After casting, the mold essentially has the contours necessary for producing the molding. For large molds cast in one piece, the heat-exchange system can be integrated directly by casting a tubing system or by means of a special arrangement of recesses at the rear through which the temperature-control medium can flow freely.

Generally, the inner contours of the mold (the mold recesses) are cast slightly larger and so require only a minimal amount of additional machining. Another critical factor is the requirements imposed on the surface quality of the molding. Any posttreatment of the surfaces (e.g. polishing) that may be necessary is performed by the same methods as in conventional mold making. Grained and textured surfaces such as can be produced by precision casting mostly do not require posttreatment. With cast molds, just as in conventional mold making, incorporation of holes for ejector pins, sprue bushings and inserts as well as the fitting of slide bars and the application of wear-resistant protective layers are all performed on the cast blank.

The metallic casting materials suitable for mold making fall into two groups:

– ferrous materials (cast steel alloys, cast iron materials), and
– non-ferrous materials (aluminum, copper, zinc and tin-bismuth alloys).

Only cast steel will generally satisfy the mechanical demands of mold inserts that are required for more than just experimental and low-production runs. Furthermore, only steel has an adequate degree of polishability. Many of the steel grades successfully employed in mold making are amenable to casting. However, it must be borne in mind that castings always have a coarse structure that is not comparable to the transformation structure of forged or rolled steels. At the macroscopic level, castings have different primary grain sizes between the edge and core zones. There is limited scope for using subsequent heat treatment to eliminate the primary phases that settle out on the grain surfaces during solidification. For these reasons, when making cast molds, it is best to use steel grades that have little tendency to form coarse crystals or to separate by liquation [2.4]. Some common cast steel grades are shown in Table 1.3.

Not only does thermal posttreatment bring about the improvement in structure mentioned above but it also enhances the mechanical properties, and the necessary notch resistance and stress relief are obtained. The strength, which depends on the carbon content, is lower than that of rolled or forged steel, and so too are the toughness and ductility [2.5]. However, they meet the major demands imposed on them. The service life of cast steel molds depends on the wear resistance and, under thermal load, on the thermal shock resistance. Given comparable steel grades, the thermal shock resistance of cast steels is generally lower than that of worked steels.

Mold inserts of copper and aluminum alloys are made both by casting and machining. Refined-zinc cast alloys for injection molding are used only for making mold inserts for experimental injection, for the production of low runs and for blow molding molds. Refined-zinc cast alloys, like copper alloys, have excellent thermal conductivity of 100 W/(m · K). The mold-filling characteristics of zinc alloys so outstanding that smooth, pore-free surfaces are even obtained in the case of pronounced contours with structured surfaces [2.6]. The most common refined-zinc alloys, sold under the names Zamak, Kirksite and Kayem, are summarized in Table 1.5.

Tin-bismuth alloys, also called Cerro alloys, are comparatively soft, heavy, low-melting metals (melting point varying according to composition between 47 and 170 °C) [2.7]. Particularly suitable for mold making are the Cerro alloys that neither shrink nor grow during solidification. Due to their moderate mechanical properties, Cerro alloys in injection molding are only used for molds for trial runs or for blow molds. Moreover, they serve as material for fusible cores. The physical and mechanical properties of some Cerro alloys are shown in Table 1.8.

2.1.2 Sand Casting

This process is used to produce medium-to-large molds weighing several tons per mold half. It consists of three major production steps:

– production of a negative pattern (wood, plastic, metal),
– production of the sand mold with the aid of the negative pattern, and
– casting the sand mold and removing the cooled casting.

The negative pattern is made either direct or from an original or positive master pattern. To an extent depending on the shape, dimensions, alloy and sand-casting method, allowance must be made for machining and necessary drafts of 1° to 5°. When making the pattern, allowance for shrinkage has to be made. To determine the shrinkage, the dimensional change of the cast metal from solidification and cooling and the shrinkage of the plastics to be processed in the mold have to be taken into account (does not apply to wooden patterns) [2.8]. Typical allowances for a number of cast metals in sand casting are listed in Figure 2.2. The exact measurements in each case depend on the casting method, part size and part shape. These should be set down in the design phase after consultation with the foundry.

The casting mold is made by applying the mold material to the pattern and solidifying it either by compaction (physically) or by hardening (chemically).

A wide range of synthetic mold materials of varied composition is available [2.2]. Washed, classified quartz sand is the predominant refractory base substance. For special needs, e.g. to prevent high-alloy casting materials (cast-steel alloys) from reacting with the melt, chromite, zirconium or olivine sand may be used. The binders used for mold sands are organic and inorganic. The inorganic binders may be divided into natural and synthetic types. Natural inorganic binders are clays such as montmorillonite, glauconite, kaolinite and illite. Synthetic, inorganic binders include waterglass, cement and gypsum. Organic binders are synthetic resins such as phenol, urea, furan and epoxy resins. In practice, the molds are made predominantly of bentonite-bound (a natural inorganic binder) mold materials (classified quartz sand) that have to be mechanically compacted in order that adequate sag resistance may be obtained.

After the mold has been produced, the pattern is removed. To an extent depending on the requirements imposed on surface texture and alloy, the finished sand mold may or may not be smoothed with a facing. After casting, the finished mold is more or less ready. The sand mold is destroyed when the mold is removed.

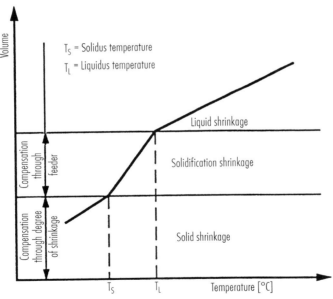

Material	Solidification shrinkage in %	Shrinkage in %
Cast iron:		
With lamellar graphite	−1 to 4	0.9 to 1.1
With spheroidal graphite	1 to 6	0.8 to 0.9
Malleable cast iron	5.5 to 6	0.5 to 1.9
Cast steel	5 to 6	1.5 to 2.8
Aluminum base	5 to 6	0.9 to 1.4
Copper base	4 to 8	0.8 to 2.4

Figure 2.2   Shrinkage on solidification and shrinkage for different casting alloys

A somewhat different procedure is employed in the lost foam method. In this, a polystyrene foam pattern is embedded in sand, remains in the casting mold, and is gasified by the casting heat only when the liquid metal is poured into the casting mold. Polystyrene patterns may be milled from slab material (once-off production) or foamed in mold devices (mass production). Since the pattern is generally only used a few times to make inserts for injection molds, the use of CAD interfaces can allow a polystyrene pattern to be milled quickly and cost effectively.

The major advantage of cast molds is the fact that the mold is ready for use almost immediately after casting. Posttreatment is limited, especially if a heat-exchange system has already been integrated by embedding a prefabricated tubing system before casting.

2.1.3 Precision Casting Techniques

Precision casting is used for mold inserts that must satisfy particularly high demands on reproducibility. The techniques are eminently suitable for fine contours and, owing to the very high reproducibility, for the faithful reproduction of surface structures, such as that of wood, leather, fabrics, etc.

A number of different types of precision casting process exist [2.9] that vary in the sequence of processes, the ceramic molding material and the binders employed. Mold inserts are usually made by the Shaw process (Figure 2.3) or variants thereof. For molding, a pattern is required that already contains the shrinkage allowance (see also Section 2.1.2). The patterns are reusable and so further castings can be made for replacement parts. This pattern forms the basis for producing the ceramic casting mold (entailing one, two or more intermediate steps, depending on method chosen). The liquid ceramic molding compound usually consists of very finely ground zirconium sand mixed with a liquid binder. After the mold has been produced, it is baked for several hours at elevated temperatures. It is then ready to be used for casting. After casting, the ceramic mold is broken and the part removed.

Precision-cast parts may be made from the same molding steels employed for making injection molds, but all other casting alloys may also be used. Posttreatment of precision-cast parts is generally restricted to the mounting and mating surfaces, as well as all regions that comprise the mold parting surface.

2.2 Rapid Tooling for Injection Molds

Time and costs are becoming more and more important factors in the development of new products. It is therefore extremely important for the injection molding industry to produce prototypes that can go into production as quickly as possible. To be sure, rapid prototyping is being employed more and more often but such prototypes frequently cannot fully match the imposed requirements. Where there is a need for molds that are as close to going into production as possible, rapid tooling (RT) is the only process by which the molds can be made that will enable the production of injection molded prototypes from the same material that will eventually be used for the mass-produced part. Rapid tooling allows properties such as orientation, distortion, strength and long-term characteristics to be determined at an early stage in product development. Such proximity to series production, however, also entails greater outlay on time and costs. For this reason, RT will only be used where the specifications require it.

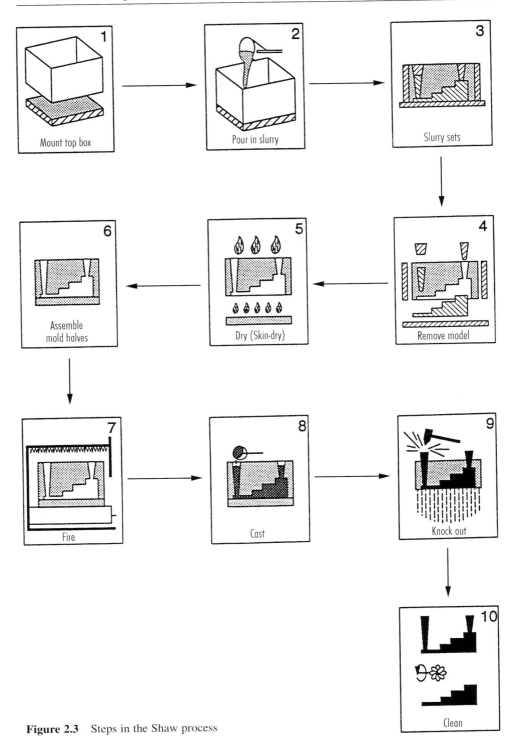

Figure 2.3   Steps in the Shaw process

2.2.1   State of the Art

A breakdown of all RT techniques is shown in Figure 2.4. The material additive processes lead fastest to moldings and are therefore the most promising. These also include new and further developments in RP. Examples of such techniques are laser sintering, laser-generated RP and stereolithography, which enable mold inserts to be made directly from a three-dimensional CAD model of the desired mold.

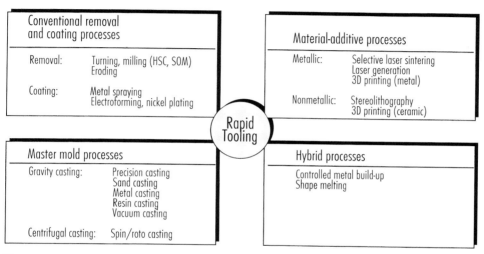

Figure 2.4   Classification of RT processes

RT also covers conventional processes for removal or coating. These include high-speed cutting (HSC) [2.10] with direct control through the processing program generated by the computer from the CAD model; erosion with rapidly machining graphite electrodes; and metal spraying, which has been used for decades in mold making.

Master molding techniques like precision and resin casting may be considered as belonging to RT. These process chains become rapid tooling techniques when an RP technique is used to produce the necessary master mold.

Whereas, in the techniques mentioned so far, the prototype mold is produced either directly by means of a material additive method or in several processing stages, the so-called hybrid techniques integrate several such stages in one item of equipment. These processing stages are a combination of processes from the other three groups (conventional, master-molding, and material additive techniques). Because hybrid techniques combine sequential processes in unit, they can be just as fast as the material additive techniques. All these techniques are still in one development, however.

For a better understanding of the diverse processes and combinations involved, the different RT techniques will be presented and discussed in this chapter.

Figure 2.5 illustrates the general procedure for RT. All methods rely on the existence of consistent 3D CAD data that can be converted into closed volume elements. These data are processed and sliced into layers. 2D horizontally stacked, parallel layers are thus generated inside the computer that, with the aid of a technique such as laser sintering,

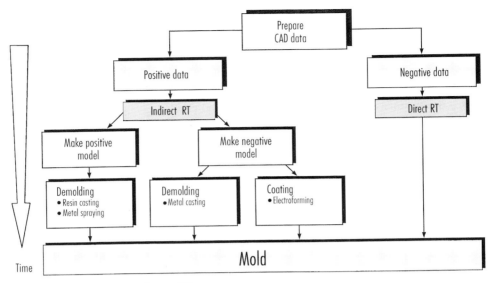

Figure 2.5   Basic approaches to RT

can be successively created within a few hours without the use of tools or a mold [2.11]. Furthermore, there is generally no need for supervision by an operative.

To an extent depending on the principle underlying the chosen RT method, either a positive pattern, i.e. the molding to be fabricated later, or a negative pattern (the requisite mold geometry) is produced. Once a physical positive pattern has been produced, usually any number of moldings may be made by master-molding and coating techniques, which will ultimately lead to a prototype mold after one or more stages. Examples of such process chains are resin casting and metal spraying.

The systematic use of 3D CAD systems during design affords a simple means of generating mold cavities. Most 3D CAD systems already contain modules that can largely perform a conversion from positive to negative automatically.

Once the data have been prepared thus, there are two possibilities to choose from. One is to create a physical model of the moving and fixed mold halves (negative patterns) so as to make a certain number of moldings that will lead to a mold. Metal casting is an example of this. Alternatively, the negatives, perhaps made by stereolithography, may be electrostatically coated. This process chain is shorter than the molding chain just mentioned.

Because the possibilities presented so far involve a sequence of different processes, they are known as indirect tooling. By contrast, direct tooling involves using the generated negative data without intervening steps to produce on an RP/RT system, as is the case with selective laser sintering. Although this is undoubtedly a particularly fast option, the boundary conditions need closer examination. Some of the resultant molds entail laborious postmachining, which is more time consuming.

A major criterion other than subdivision into direct and indirect RT is the choice of tooling material. These are either metallic or so-called substitute materials, the latter usually being filled epoxy resins, two-component polyurethane systems, silicone rubber [2.12] or ceramics.

The more important and promising RT methods are presented below.

2.2.2   Direct Rapid Tooling

The goal of all developments in the field of RT is automated, direct fabrication of prototype molds, whose properties approach those of production parts, from 3D CAD data describing the mold geometry. This data set must already allow for technical aspects of molds, such as drafts, allowance for dimensional shrinkage and shrinkage parameters for the RT process.

Processes for the direct fabrication of metallic and nonmetallic molds are presented below. In either case, the mold may be made from the CAD data direct.

2.2.2.1   Direct Fabrication of Metallic Molds

Direct fabrication of prototype molds encompasses conventional methods that allow rapid processing (machining) of, e.g. aluminum.

2.2.2.1.1 Generative Methods

A common feature of generative methods for making metallic molds is that the workpiece is formed by addition of material or the transition of a material from the liquid or powder state into the solid state, and not by removal of material as is the case with conventional production methods.

All the processes involved here have been developed out of RP methods (e.g. selective laser sintering, 3D printing, metal LOM (Laminated Object Manufacturing), shape melting, and multiphase jet solidification) or utilize conventional techniques augmented by layered structuring (laser-generated RP, controlled metal build-up).

In selective laser sintering (SLS) of metals, a laser beam melts powder starting materials layer by layer, with the layer thickness varying from 0.1 to 0.4 mm in line with the particle size of the metal powder [2.13, 2.14]. The mold is thus generated layer by layer.

Sintering may be performed indirectly and directly. In the indirect method (DTM process), metal powder coated with binder is sintered in an inert work chamber (e.g. flooded with nitrogen). Heated to a temperature just below the melting point of the binder, the powder is applied thinly by a roller and melted at selected sites. The geometry of the desired mold inserts is thus obtained by melting the polymer coating. The resultant green part, which has low mechanical strength, is then heat-treated. The polymer binder is burned out at elevated temperatures to produce the brown part, which is then sintered at a higher temperature. At an even higher temperature again, the brown part is infiltrated with copper (at approx. 1120 °C), solder alloy or epoxy resin [2.15], this serving to seal the open pores that were formed when the polymer binder was removed (Figure 2.6).

In direct laser sintering manufacture, metals are sintered in the absence of binder (EOS process). The advantage of not using coated powders is that the laborious removal of binder, and the possibility of introducing inaccuracies into the processing stage, can be dispensed with. Nevertheless, the part must be infiltrated since it has only proved possible so far to sinter parts to 70% of the theoretical density [2.16]. After infiltration, posttreatment is necessary and generally takes the form of polishing.

Aside from pure metal powders and powders treated with binder, multicomponent metallic powders are used. These consist of a powder mixture containing at least two metals that can also be used in the direct sintering process. The lower-melting component provides the cohesion in the SLS preform and the higher-melting component melts in the furnace to imbue the mold with its ultimate strength. Candidate metals and

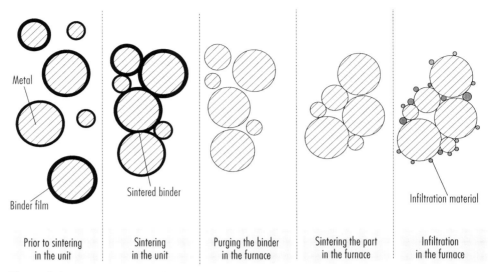

Metal

Sintered binder

Binder film

Infiltration material

| Prior to sintering in the unit | Sintering in the unit | Purging the binder in the furnace | Sintering the part in the furnace | Infiltration in the furnace |

Figure 2.6   Indirect sintering followed by infiltration

metal alloys for direct and indirect sintering are, according to [2.17]: aluminum, aluminum bronze, copper, nickel, steel, nickel-bronze powder and stainless steel. The maximum size capable of being made by laser sintering is currently $250 \cdot 250 \cdot 150$ mm^3.

Another way to apply metal is by laser-generated RP. Powder is continually added to the melt in a movable process head [2.18]. The added material combines with the melted material on the preceding layer. The layers can be added in thicknesses of 0.5 to 3 mm. Metal powder is blown in and melted in a focused laser beam. As the process head moves relative to the work surface, fine beads of metal are formed. The materials used are chrome and nickel alloys, copper and steel. Laser-generated RP is not as accurate as laser sintering and can only generate less complex geometries due to the process setup.

A further development of laser-generated RP is that of controlled metal build-up [2.19]. This is a combination of laser-generated RP and HSC milling (Figure 2.7). Once a layer 0.1 to 0.15 mm thick has been generated by laser, it is then milled. This results in high contour accuracy of a level not previously possible with laser-generated RP. The maximum part size is currently 200 mm^3 for medium complexity. No undercuts are possible.

Other processes still undergoing development are shape melting and multiphase jet solidification [2.20]. Both processes are similar to fused deposition modeling, which is an RP process [2.21]. In shape melting, a metal filament is melted in an arc and deposited while, in multiphase jet solidification, melt-like material is applied layer by layer via a nozzle system. Low-melting alloys and binders filled with stainless steel, ceramic or titanium powder are employed. As in SLS, the binder is burned out, and the workpiece is infiltrated and polished. However, the two processes are still not as accurate as SLS.

3D printing of metals is now being used to fabricate prototype molds for injection molding, but it is not yet commercially available. Figure 2.8 illustrates how the 3D printing process works. After a layer of metal powder has been applied, binder is applied selectively by means of a traversing jet that is similar to an ink jet. This occurs at low

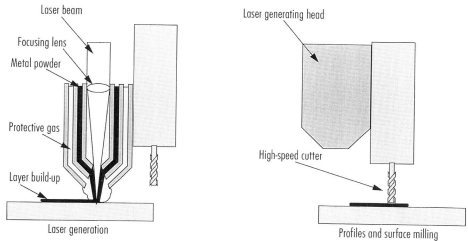

Bild 2.7   Controlled metal build-up after [2.19]

temperatures because only the binder has to be melted. Whereas local heating in direct laser sintering can cause severe distortion, this effect does not occur in 3D printing. Once a layer has been printed, an elevator lowers the platform so that more powder can be applied and the next layer generated. The coating is 0.1 mm thick. When steel powder is used in 3D printing, bronze is used for infiltration. Shrinkage is predictable to ± 0.2%. Part size is still severely restricted by the equipment and currently cannot exceed an edge length of 150 mm [2.21].

Another process currently being developed for the fabrication of metallic molds is that of metal LOM in which metal sheets of the same thickness are drawn from a roll, cut out by laser and then joined together. The joining method is simply that of bolting, according

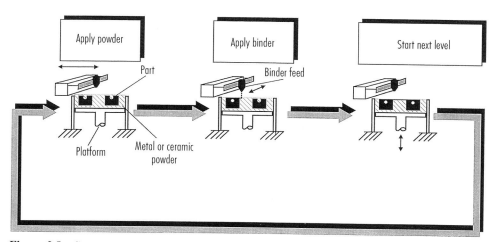

Figure 2.8   Steps in 3D printing

to [2.22]. So far, molds made in this way have only been used for metal shaping and for injection molding wax patterns for precision casting. The advantage of molds joined by bolts is that the geometry can be modified simply by swapping individual metal sheets.

The variant developed by [2.23] is a combination of laser cutting and diffusion welding. Unlike metal LOM and most other RT processes, which grow the layers at constant thickness, this process variant allows sheets of any thickness to be used. As a result, simple geometric sections of a mold may be used as a compact segment, a fact which allows RT only to be used where it is necessary and expedient. Possible dimensional accuracy is in the order of 0.1%. Due to the process itself, it is never lower than   ± 0.1 mm in the build direction. The tolerances of laser cutting are from 0.001 to 0.1 mm. Unlike most of the processes mentioned so far, this process imposes virtually no restrictions on part size [2.24].

Direct RT processes are still in their infancy. Apart from selective laser sintering and 3D printing of metals, all the processes discussed in this section are still being developed and so are not yet available on the market. This explains why stereolithography, despite the fact that it is a direct fabrication process involving nonmetallic materials, is virtually the only one used for these purposes. Because it has constantly evolved over the last 10 years and is offered by many service providers, it is readily available. Moreover, many large companies are in possession of stereolithographic equipment and still elect to use it for making prototype molds.

2.2.2.1.2 Direct Fabrication of Nonmetallic Molds

While most direct methods for making metallic molds require posttreatment (infiltration and mechanical finishing), the production of molds from auxiliary materials largely dispenses with this need.

Stereolithography (STL) is based on the curing of liquid, UV-curing polymers through the action of a computer-controlled laser. The laser beam traverses predetermined contours on the surface of a UV-curable photopolymer bath point by point, thereby curing the polymer. An elevator lowers the part so that the next layer can be cured. Once the whole part has been generated, it is postcured by UV radiation in a postcuring furnace [2.24].

STL's potential lies in its accuracy, which is as yet unsurpassed. Because it was the first RP process to come onto the market, at the end of the 1980s, it has a head-start over other technologies. Ongoing improvements to the resins and the process have brought about the current accuracies of 0.04 mm in the x- and y-axes and 0.05 mm in the z-axis.

The process was originally developed for RP purposes but is also used for rapid tooling of injection molds because of its accuracy and the resultant good surfaces which it produces. When STL is used to make a mold cavity, the mold halves are generated on the machine and then mounted in a frame. Usually, however, the shell technique is employed. In this, a shell of the mold contour is built by STL and then back-filled with filled epoxy resin [2.25]. The use of the shell technique to produce such an RT mold is illustrated in Figure 2.9.

Parts made by stereolithography feature high precision and outstanding surface properties. Unlike all other direct methods for making metallic molds, no further treatment is necessary other than posttreatment of the typical step-like structure stemming from the layered build-up by the RP/RT processes. This translates to a considerable advantage time-wise, particularly when the mold surfaces must be glossy and planar. The downside is the poor thermal and mechanical properties of the available resins (acrylate, vinyl ether, epoxy), which cause the molds to have very short service

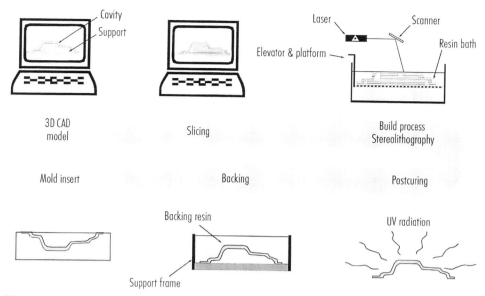

Figure 2.9 The shell technique for generating an STL mold

lives. The best dimensional and surface properties are obtained with epoxy resins; the use of particularly powerful lasers makes for faster, more extensive curing of the resin even during the stereolithography process, and this in turn minimizes distortion [2.26].

Although STL has primarily been used for RP, the number of RT applications is on the increase. It is used to make molds for casting wax patterns as well as for injection molding thermoplastics. Such molds serve in the production of parts for a pilot series, which can yield important information about the filling characteristics of the cavities. Moreover, it is even possible to identify fabrication problems at this very early stage.

Ceramics are other materials used for direct rapid tooling. Bettany [2.27] has reported on the use of ceramic molds for injection molding. They are employed in the 3D printing process described in the previous section as well as in ballistic particle manufacturing (droplets of the melted material are deposited by means of piezoelectric ink-jet nozzle). The advantage over metallic molds is the high strength of ceramic molds. This comes particularly to the fore when abrasive, filled polymers are processed.

2.2.3 Indirect Rapid Tooling (Multistage Process Chains)

An RT chain is defined here as a succession of individual molding stages. The use of such a molding chain leads from a master pattern to a cavity that may be used for injection molding. In the sense of this definition, intermediate stages such as machining or simple assembly of already finished cavity modules do not count as individual links in this chain. A good RT chain is notable on the one hand for having a minimum number of molding stages (chain links). The lower the number of molding stages, the more accurately the part matches the master pattern and the faster a prototype mold can be made. Every intermediate pattern can only be as good as the pattern from which it proceeds. Consequently, the attainable tolerances become greater and the surface quality diminishes from casting to casting. On the other hand, each RT chain must finish with a cavity that withstands the high mechanical and thermal loads that occur in the injection molding process. The bottom line for all RT molding chains is therefore to have as few links as possible so as to end up with an injection mold whose strength and quality somewhat exceed requirements.

Indirect rapid tooling may be effected with a positive or a negative pattern. While these patterns serve as the master patterns for casting processes, using the virtual negative pattern and new RT process chains can dispense with the master pattern and enable a cavity to be made directly in sand or ceramic slip for casting metals.

2.2.3.1 Process Chains Involving a Positive Pattern

Rapid manufacture of prototype molds using the shortest possible process chain frequently involves using RP to make a positive pattern. These patterns can be made by any means, i.e. also conventionally.

Casting is frequently employed in the production of prototype molds. This master-pattern process entails observing the ground rules for designing cast parts. These include:

– avoidance of accumulation of material,
– avoidance of major changes in cross-section, of thin flanges (1.5 mm minimum) and of sharp edges (minimum radius of 0.5),
– avoidance of vertical walls (1% min. conicity) [2.28].

The simplest, and at the same time a very common process, is that of making a silicone rubber mold, starting from an RP pattern. The pattern is equipped with gate and risers and fixed in a frame. Once the parting line has been prepared, liquid silicone resin is poured over the pattern in a vacuum chamber. It is not possible to use this type of mold to injection mold prototypes in production material. This process is known as soft tooling since prototypes with certain heat or mechanical resistance can be cast in two-component polyurethane resins [2.24]. Not only can the Shore A hardness be adjusted to the range 47–90, heat resistance of up to 140 °C and resin strengths of up to 85 Shore D are possible. This is the only casting method in which the ground rules mentioned above can be largely ignored, due to the use of yielding silicone rubber. Nor is any shrinkage allowance required (1:1 reproduction).

Resin casting is illustrated in Figure 2.10. The RP pattern is embedded as far as the parting line and fixed in a frame. After delineating, a thixotropic surface resin (gelcoat) is usually applied and cooling coils are incorporated. An aluminum-filled epoxy mold-casting resin is then used for back-filling. When the molds are being designed, allowance must be made not only for a design suitable for plastics but particularly for shrinkage by the resin (range: 0.5–1.5%).

The service life of the mold depends greatly on the injected material and the processing parameters during injection molding.

The process allows metal inserts and slide bars to be embedded (Figure 2.11).

The coating processes employed are familiar from conventional mold making. They include flame spraying, arc spraying, laser coating, and plasma and metal spraying. Because both flame spraying and plasma spraying entail temperatures of 3,000 °C and above, it is necessary to create a heat-resistant positive pattern. For this reason, we shall in the context of rapid production only discuss metal spraying with a metal-spraying pistol. In this process, two spray wires are melted in an arc and atomized into small

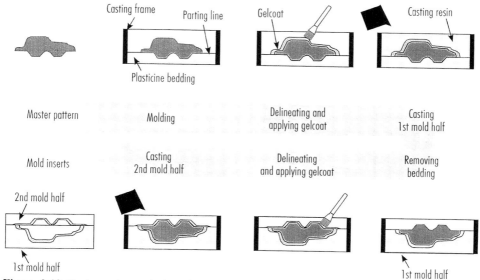

Figure 2.10 Resin casting technique for making a mold

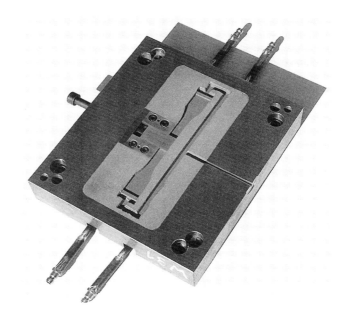

Figure 2.11 A sliding split
mold cast in resin

particles in the presence of compressed air (Figure 2.12). When the particles impinge on
the surface of an RP positive pattern, a liquid film forms that solidifies instantaneously.

The homogeneity of the 1.5–5 mm thick layer depends on the temperature and the
distance of the nozzle from the pattern. Since the particles are cooled immediately on

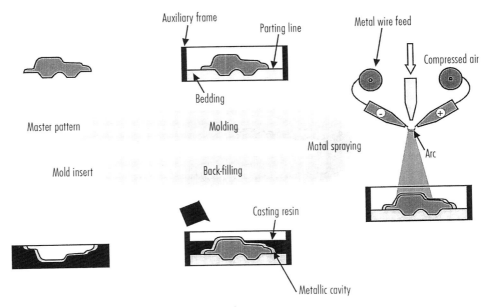

Figure 2.12 Principle underlying metal spraying

contacting the pattern from approx. 2000 °C to 60 °C, wooden patterns, for instance, may be used in addition to RP materials [2.29]. A shell made in this way only needs to be back-filled with, e.g. casting resin.

Another coating method is the long-established electroforming, which has frequently been used in the past [2.30] for high-quality injection molds (Figure 2.13). Electroforming is the most accurate method of reproducing surface texture in metal [2.31].

The RP pattern is first coated with silver or graphite to render it electrically conducting [2.32]. In an electroplating bath, individual metals are successively or simultaneously electrolytically deposited, the pattern being coated with the corresponding material. The result is shells 4–5 mm thick that may be built up of different alloys or metal layers. Electroforming with nickel yields the best results due to such good material properties as high strength, rigidity and hardness (e.g. NiCo alloy, up 50 RC hardness), its compatibility with the base material and its good corrosion resistance.

For large parts, RP master patterns are made and then coated. It is essential beforehand to make a heat-resistant mold of this pattern as the thermal expansion of the stereolithographic resin could cause excessive distortion.

Electroforming reproduces the finest of details, but the part frequently has to remain in the electroplating bath for several days. The layer thickness is much more homogeneous than that produced by manual metal spraying. The maximum part size is restricted by the size of the electroplating bath.

2.2.3.2　Process Chains Involving a Negative Pattern

All of the process chains below begin with the creation of an RP pattern of the mold (negative pattern). To produce a purely metallic prototype mold, several casting

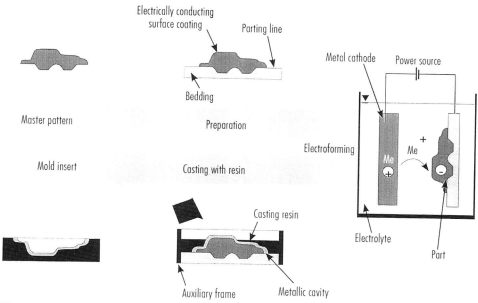

Figure 2.13   Principle underlying electroforming

techniques may be used in addition to RT. This will often considerably shorten the process chains, as will be demonstrated below.

Of greatest economic importance and hence the most widespread casting technique is that of investment casting, which normally employs patterns of investment wax [2.33]. The range of possible processes for creating these patterns has been extended by the advent of RT.

Thus, it is possible with the aid of selective laser sintering, fused deposition modeling and ballistic particle manufacturing to fabricate patterns from investment wax direct [2.12]. The stereolithography technique, given suitable software, makes it possible to produce hollow-structure patterns, e.g. by means of Quick-Cast (a 3D system) or the shell-core technique (EOS) [2.34]. In conjunction with a special illumination technique, these epoxy resin STL parts can be employed as expendable patterns for investment casting. To this end, only the molding shell is built up from the resin; the inner construction consists of a large number of honeycomb-shaped chambers all joined to each other. The density of the mold part is now only 20% that of the solid part, but has excellent strength values and a very good surface finish [2.35]. Very low internal stresses occur so that the mold is extraordinarily accurate and dimensionally stable. For investment casting, the vent holes of the STL part are sealed with investment wax. During burning out of the STL part, the part gasifies almost residue-free (residual ash content approx. 2 mg/g). A further possibility is direct production of a gasifiable pattern to produce sintered patterns of polycarbonate. These are sturdier and less heat-sensitive than wax patterns.

All the patterns made like this are surrounded with a ceramic coating. This is achieved by immersing the pattern into a ceramic slip bath and subsequently covering it with sand. This process is repeated until the desired coating thickness of the refractory ceramic shell is achieved. After this, the mold part must dry before it is burned in excess oxygen at

1,100 °C. During firing, the master pattern gasifies and so the corresponding materials can then be cast in the resultant ceramic mold. After drying, the ceramic body is smashed to yield the desired part. It is important for the quality of the cast part that the wax pattern be totally and uniformly wetted when first immersed in the ceramic bath.

Since the cast material shrinks on cooling down in the ceramic mold, the master pattern must be correspondingly larger than the original. Additionally, shrinkage in each RP process employed, as well as of the ceramic shell, must be considered. Distortion of the mold shells must also be expected, and must be rectified. Prototypes made by investment casting can accommodate high loads, have a high workpiece accuracy and good surface quality. A serious disadvantage of the process is the long drying time of the ceramic shell of up to one week. The lowest wall thickness that can be produced is 1.5 mm. Attainable surface roughness quoted in the literature ranges from mean values of 5.9 to 23 μm [2.36].

Investment casting is used for making metallic prototype molds, inserts and metallic pilot parts. The process is particularly suitable for cylindrical cores. Figure 2.14 shows an example of an investment-cast mold of complex geometry that was successfully injection molded.

The process chain of investment casting can be shortened considerably by the application of 3D printing. In this case, a virtual negative pattern is required instead of a physical negative one. The 3D printing process described above thus allows direct production of the ceramic shell. This is referred to as direct shell production casting (DSPC) [2.37]. Pouring in the metal and deforming are the only steps necessary.

As in investment casting, evaporative pattern casting employs expendable patterns that remain in the mold and evaporate without residue when the hot metal is poured in [2.38]. Very high accuracy can be achieved with this technique.

A one-part, positive master pattern of a readily evaporative material (EPS foam) is modeled in sand. After compaction of the sand, high-melting metal can be poured

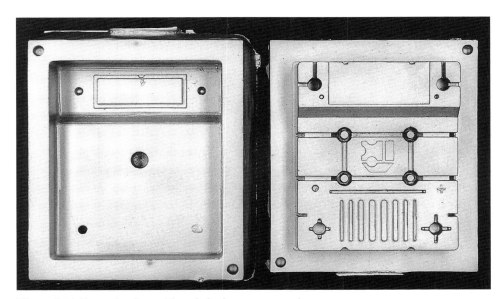

Figure 2.14 Example of a mold made by investment casting

directly into the mold. The gas produced by decomposition of the pattern can escape readily because the sand is porous. A material frequently employed in the field of RT is the light metal Zamak, a zinc-aluminum-copper alloy that is easy to posttreat [2.28].

An RT operation based on evaporative pattern casting is that of expandable pattern casting (EPC). To enable fast production of the desired component, an auxiliary mold is instead constructed which may be generated by any RP method. In the EPC process, first polystyrene beads are prefoamed to a specified density by slow heating to 110 °C. The beads are then foamed in a mold. The individual polystyrene beads are thermally bonded and welded together. Finally, the part is covered with a ceramic coating. Hot metal is then poured into the mold, causing the polystyrene to evaporate at the same time. A good dimensional stability and reusability of the ceramic sand are notable features of this process.

An even simpler and therefore shorter process chain is the direct production of the polystyrene master pattern via RP by means of the Sparx process. Foamed polystyrene film in a so-called hot-plot machine is cut with the aid of a plotter and bonded onto the preceding layer. The material is gasifiable and therefore suitable for evaporative pattern casting [2.25].

A sinter process which still requires upstream molding steps is the Keltool technique. In this process, unlike the process chains described above, a higher strength copy of a mold pattern is made. The goal here is to convert patterns made of a low-strength RP material into metal parts.

To prepare an injection mold, a pattern of the mold half is generated first by any RP process.

As in the silicone casting process, a highly heat-resistant RTV silicone that can be demolded after curing is poured around the master pattern (Figure 2.15). An epoxy binder that is very highly filled with a metal alloy in powder form is then poured into this mold.

The result is a stellite green part of the cavity to be made. As with selective laser sintering, the binder is thermally desorbed, and infiltration performed, with the polymeric binder being replaced by a copper/zinc (Cu/Zn) alloy (Figure 2.16). The surfaces can then be machined [2.39].

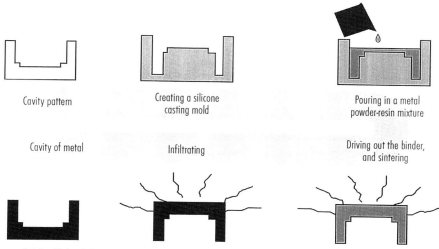

Cavity pattern

Creating a silicone
casting mold

Pouring in a metal
powder-resin mixture

Cavity of metal

Infiltrating

Driving out the binder,
and sintering

Figure 2.15   Principle underlying the Keltool process

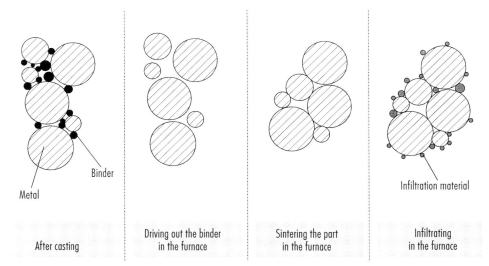

Metal

Binder

Infiltration material

After casting | Driving out the binder in the furnace | Sintering the part in the furnace | Infiltrating in the furnace

Figure 2.16   Processes occurring in the Keltool process

2.2.4   Outlook

Many of the direct RT operations proposed here are still in development and are subject to size limitations, for example. Nevertheless, all process chains presented here are generally available on the market and may be used for rapid prototyping of molds.

Finally, an overview of all the processes discussed here is presented for comparison purposes (Figure 2.17). Many of the metallic processes are not yet commercially available on the market. Only selective laser sintering, metal spraying and electroforming are available. The Keltool process is also available but it is only widespread in the American market at this time. Aside from complexity and stability, however, major criteria are attainable surface quality, the availability and the price, which can vary extensively according to geometry.

Therefore, despite their speed, when using all these operations, it is always best to estimate whether it is more economical to avail of RT or whether it might not be better to employ a conventional process instead.

2.3   Hobbing

Hobbing is used for producing accurate hollow molds. It comes in two variants, cold-hobbing, the more common, and hot-hobbing.

Cold-hobbing is a technique for producing molds or cavities without removing material. A hardened and polished hob, which has the external contour of the molding, is forced into a blank of soft-annealed steel at a low speed (between 0.1 and 10 mm/min). The hob is reproduced as a negative pattern in the blank. Figure 2.18 demonstrates the process schematically. The technique is limited by the maximum permissible pressure on the hob of ca. 3,000 MPa and the yield strength of the blank material after annealing. The

Figure 2.17 Survey of rapid tooling processes

Process	Metallic	Nonmetallic	Commercially available	In Development	Complexity	Durability
Selective laser sintering	●		●		High	High
Controlled metal build-up	●			●	Low	High
Shape melting/multiphase jet solidification	●			●	Medium	High
3D-metal printing	●			●	High	High
Metal laminated object manufacturing	●			●	Medium	High
Laser cutting/ diffuse welding	●			●	Low	High
Direct stereolithography		●	●		Low	Low
Resin casting		●	●		Medium	Medium
Metal spraying	●		●		Medium	Medium
Electroforming	●		●		Medium	High
Investment casting	●			●	High	High
Keltool	●		●		High	High

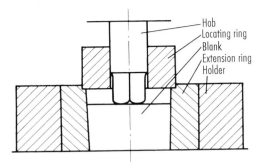

Figure 2.18   Schematic presentation of hobbing [2.43]

best conditions for cold-hobbing are provided by steels annealed to a low strength of 600 MPa.

The yield strength after annealing depends primarily on the content of alloys dissolved in ferrite and on the quantity and distribution of embedded carbides [2.42, 2.43]. In accordance with their hardness after annealing (Brinell hardness) and their chemical composition, materials commonly used for cold-hobbing fall into three categories (Section 1.1.2). Figure 2.19 shows the attainable relative hobbing depth as a function of hobbing pressure and hardness after annealing. The nondimensional hobbing depth t/d is the ratio between the depth t and the hob diameter d of a cylindrical hob. If the hob has a different cross section, e.g. is square or rectangular, then $t/1,13\sqrt{A}$, where A is the cross-sectional area [2.43]. The hobbing depth can be increased beyond the dimensions shown in Figure 2.19 by certain steps. Strain-hardening of the blank material occurs with increasing depth. This strain-hardening is neutralized by intermediate annealing (recrystallization). Thereafter, the hobbing process can continue until the

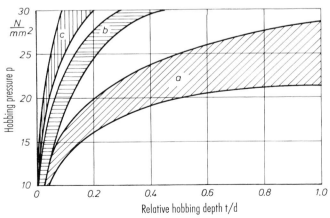

Figure 2.19   Pressure for hobbing common tool steels with a cylindrical hob (d = 30 mm ø) and cylindrical blank (D = 67 mm ø, h = 60 mm), hob velocity v = 0.03 mm/s, hob is copper plated, lubrication with cylinder oil
a Steel with 300 to 400 BHN,
b Steel with 500 to 600 BHN,
c Steel with 600 to 700 BHN [2.43]

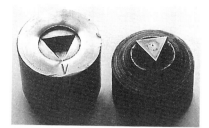

Figure 2.21   Mold insert made by hobbing (left) and matching hob (right) [2.41]

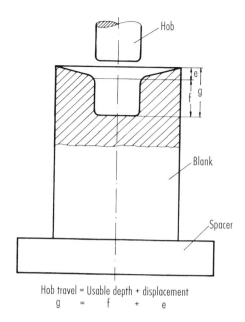

Figure 2.20   Relationship between hob travel, usable depth and depth of displaced material [2.42]

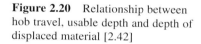

maximum load on the hob is reached again. Care must be taken to avoid the formation of scale during annealing, because only clean surfaces permit optimum hobbing results. The hobbing depth may also be increased by preheating the blank. Depending on material and preheating temperature, 20 to 50% more hobbing depth can be attained. Finally, recesses can ease the flow of the material and result in increased hobbing depth [2.43].

During hobbing, the rim of the blank is pulled in. This indentation has to be machined off afterwards and must be considered when the hobbing depth is determined. Figure 2.20 shows the correlation between hob travel, usable depth and indentation. A hobbed mold insert, not yet machined, is shown in Figure 2.21.

The surface quality of hob and blank is of special significance for hobbing. Only impeccably polished surfaces do not impede the flow of the material and they prevent sticking and welding. For the same reason, attention has to be paid to sufficient lubrication. Molybdenum disulfide has proved to be an effective lubricant, while oil usually does not have adequate pressure resistance. To reduce friction, the hob is frequently copper plated in a solution of copper sulfate after having been polished [2.42–2.44].

Besides surface quality of hob and blank, the dimensions of the blank are also important for flawless flow of the material. For cold-hobbing into solid material, the original height of the blank should not be less than 1.5 to 2.5 times the diameter of the hob [2.42, 2.44]. The diameter of the blank, which has to correspond to the size of the opening in the cavity retainer plate, should be double the diameter of the hob.

Cold-hobbing is generally used for low cavities with little height. It offers several advantages over other techniques. The hob, which constitutes the positive pattern of the final molding, can often be made more economically than a negative pattern. With a hob, several equal mold inserts can be made in a short time. Because the fibers of the material are not cut, unlike the case for machining operations, the mold has a better surface quality and a long service life.

Limitations on cold-hobbing result from the mechanical properties of hob and blank and therefore the size of a cavity.

2.4   Machining and Other Material Removing Operations

2.4.1   Machining Production Methods

Machining production methods may be divided into processes with geometrically defined cutter (turning, milling, drilling, sawing) and geometrically undefined cutter (grinding, honing, lapping). The machinery, frequently special equipment, has to finish the object to the extent that only little postoperation, mostly manual in nature (polishing, lapping, and finishing), is left.

Modern tooling machines for mold making generally feature multiaxial CNC controls and highly accurate positioning systems. The result is higher accuracy and greater efficiency against rejects. The result of a survey [2.45] shows NC machining as having just a 25% share compared to 75% for the copying technique, but this does not hold true for modern tool shops and the fabrication of large molds.

Nowadays, heat-treated workpieces may be finished to final strength by milling (e.g. Rm up to 2000 MPa). Various operations, e.g. cavity sinking by EDM, can be replaced by complete milling operations and the process chain thus shortened. Furthermore, the thermal damage to the outer zone that would otherwise result from erosion does not occur. Hard milling can be used both with conventional cutting-tool materials, such as hard metals, and with cubic boron nitride (CBN). For plastic injection molds, hard metals or coated hard metals should prove to be optimum cutting-tool materials.

Machining frees existing residual stresses. This can cause distortion either immediately or during later heat treatment. It is advisable, therefore, to relieve stresses by annealing after roughing. Any occurring distortion can be compensated by ensuing finishing, which usually does not generate any further stresses.

After heat treatment, the machined inserts are smoothed, ground and polished to obtain a good surface quality, because the surface conditions of a cavity are, in the end, responsible for the surface quality of a molding and its ease of release.

Defects in the surface of the cavity are reproduced to different extents depending on the molding material and processing conditions. Deviations from the ideal geometrical contour of the cavity surface, such as ripples and roughness, diminish the appearance in particular and form "undercuts", which increase the necessary release forces.

There are three milling variants:

– three-axis milling,
– three-plus-two-axis milling and
– five-axis milling (simultaneous).

Competition has recently developed between high-speed cutting (HSC) and simultaneous five-axis milling. HSC is characterized by high cutting speeds and high spindle rotation speeds. Steel materials with hardness values of up to 62 HRC can also be machined with contemporary standard HSC millers [2.46]. HSC machining can be carried out as a complete machining so that the process steps of electrode manufacturing and eroding can be dispensed with completely. In addition, better surface quality is often achieved, and this allows drastic reduction in manual postmachining [2.47].

For the production of injection and die-casting molds, a combination of milling and eroding may also be performed. The amount of milling should be maximized since the machining times are shorter on account of higher removal capability. However, very complex contours, filigree geometries and deep cavities can be produced by subsequent spark-erosive machining. Often, field electrodes are used [2.48]. The electrode can, in turn, be made from graphite or copper by HSC (for details of the production method for micro cavities, see Sections 20.1.2–20.1.2.6).

2.4.2   Surface Treatment (Finishing)

In many cases, and by no means exclusively for the production of optical articles, the condition of the cavity surface (porosity, ripples, roughness) is crucial to the quality of the final product. This has a decisive effect on the time needed for mold making and thus on the costs of the mold. Moreover, the ease with which the molding can be released and deposits from thermosets and rubber are affected.

Mirror-finish surfaces require the greatest amount of polishing and facilitate demolding. As opposed to these are untreated cavity surfaces for the production of moldings which do not have to meet optical requirements. Here release properties are the criterion governing the condition of the cavity surface. This also applies to textured surfaces.

The texture determines the ease of demolding and calls for more draft than for polished molds if the texture forms "undercuts", as when grooves run across the direction of demolding. Some polishing procedures will now be presented below.

2.4.2.1   Grinding and Polishing (Manual or Assisted)

After the cavity has been completed by turning, milling, EDM, etc., the surfaces generally have to be smoothened by grinding and polishing until the desired surface quality of the moldings is obtained and release is easy. Even nowadays, this is still mainly done manually, supported by electrically or pneumatically powered equipment or with ultrasonics [2.49–2.51].

The sequence of operations, coarse and precision grinding and polishing, are presented in detail in Figure 2.22.

Coarse grinding produces a blank-metal, geometrically correct surface with a roughness of $R_a < 1$ µm, which can be finished in precision-grinding step or immediate polishing [2.52].

Careful work and observance of some basic rules can yield a surface quality with roughness heights of 0.001 to 0.01 µm (see Table 2.1) after polishing. A precondition for this, of course, is steels that are free from inclusions and have a uniform fine-grained structure, such as remelted steels (Section 1.1.9).

A disadvantage of manual finishing processes is that they are personnel-intensive and that they do not guarantee reproducible removal. Machine-assisted removal with geometric undefined cutter (grinding, honing, lapping) has nonetheless been unable to make a breakthrough. These techniques have major kinematic and technologicial restrictions in the case of complex, 3D contours.

Some of the fully-automatic polishing processes presented here have also exhibited considerable shortcomings. For this reason, they are almost exclusively used in

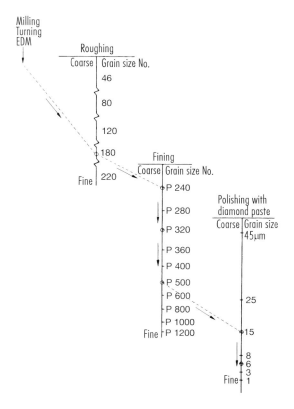

Figure 2.22   Steps of the mechanical surface treatment [2.52]

combination with manual mechanical polishing methods. They are presented here briefly, for the sake of completeness.

2.4.2.2 Vibratory Grinding

Vibratory or slide grinding is an alternative to the conventional rotary barrel process. The workpieces are placed in a container which is subsequently filled with a mixture of granulated zinc, water, alumina as polishing medium, and a wetting agent or anti-rust compound until the pieces are completely covered. Then the container is set into vibrating motion. This presses and thoroughly mixes the mixture against the walls of the molds. Thus, a kind of wiping action occurs that smooths the walls. A distinct disadvantage of this technique is pronounced abrasion of protruding edges. These have to be covered for protection [2.53]. Limitations on this process are imposed by the size and weight of the molds.

2.4.2.3 Sand Blasting (Jet Lapping)

Sand blasting is of the best known and most common procedures. For mold making, it is modified such that the blasting medium is a water-air mixture containing fine glass beads. Mold surfaces are treated with this mixture under a pressure of 500 to 1000 kPa.

This levels out any unevenness, such as grooves. The attainable surface quality is not comparable to that of surfaces treated mechanically. The roughness height is about 5 μm [2.53]. The application of this technique appears to make sense only for flat parts. Disadvantages are non-reproducible removal and relatively low dimensional stability.

2.4.2.4 Pressure Lapping

This process is a variant of jet lapping and also known as "extrude-honing". It is limited to the treatment of openings. As the name indicates, it has found special significance in the fabrication of profile-extrusion tools where arbitrarily shaped openings with the lowest of cross sections have to be polished.

The procedure uses applications a pasty polishing compound of variable viscosity that contains silicon carbide, boron carbide or diamond grits of various sizes depending on the dimension of the opening. The compound is moved back and forth and average roughness heights of $R_a = 0.05$ μm are achieved in no time [2.54 to 2.56]. The process is done automatically and requires only a short set-up time.

2.4.2.5 Electrochemical Polishing

With electrochemical polishing, or electro-polishing in short, the top layers of a workpiece are removed [2.57]. The process is based on anodic metal machining and therefore qualifies as a "cold" process. Thus, the workpiece does not become thermally stressed; see also Section 2.6. The process works without contact between workpiece and mold, so no mechanical loading occurs. Since removal only occurs at the workpiece, the workpiece is subjected to virtually no abrasion [2.58].

Through the removal of material, leveling of the surface of the workpiece occurs. High dimensional and molding accuracies, as well as good surface properties, can be achieved by electrochemical polishing. The aim is often to remove impurities introduced into the outer surface layer during preceding machining processes. Further advantages of the operation are reproducible removal and the resultant high degree of automatability [2.58].

Defects in the steel, such as inclusions and pores, are exposed. Therefore, the materials to be electrochemically polished must be of high purity. Various steels, especially the usual carbon steels, cannot be optimally electrochemically polished [2.53].

2.4.2.6 Electric-Discharge Polishing

Electric-discharge polishing is not essentially a new or independent procedure. It is an extension of electric-discharge machining (Section 2.5.1) and immediately follows erosive fine finishing. Thus, erosion and polishing are done on the same equipment using the set-up. Consequently, to an extent depending on the level of surface finish required, it can replace time-consuming and costly manual postmachining.

In electric-discharge polishing, the discharge energies are very much reduced, e.g. through lower discharge currents, relative to electric-discharge fine finishing. As a result, removal rates are low and so electric-discharge polishing is also a time-consuming finishing process. Because electric discharge polishing works on the principle of removal by heat, thermal damage is done to the outer zone. The outer zone can be minimized but it can never be removed completely.

The structure of surfaces after electric-discharge polishing characterized is by rows of adjoining and superimposed discharge craters similar to that of electric-discharge

Table 2.1 Steps for grinding and polishing operations [2.52]

Roughing Grain size no. 180	Fining Grain size 200–600	Finishing Diamond flour or paste, 0.1–180 µm
– Grinding operations must not develop so much heat that structure and hardness of the material are affected. Therefore it is important to select the correct grinding wheel and appropriate cooling. – Only clean wheels and stones which are not clogged should be used. – The workpiece has to be carefully cleaned after each application of a compound, before the next compound is applied. – If the operation is done by hand, a change of direction is essential to avoid unevenness or scratches. – One should work with one grain size in one direction, then with the next size in an angle of 30 to 45 ° until the surface does not exhibit anymore traces of the previous direction. The same procedure has to be repeated with the following grain size. – After traces have disappeared, continue each operation for the same time to make sure that the cold-deformed layer is removed. R_a 1 µm	– Only clean and unclogged tools should be used. – Add ample coolant to prevent heating of the surface and to flush chips. – Grain size of tools depends on previous roughing and intended polishing. – With every change of grain size, workpiece and hands have to be cleaned to prevent larger grains interfering with finer size. – This procedure becomes even more important with decreasing grain size. – Pressure should be distributed uniformly when working manually. Scratches and cold-deformed layers from the preceding grain size have to be removed before switching to the next size. Large, plane faces should not be worked on with abrasive paper. Abrasive strones reduce the danger of creating waviness. R_a 0.1 to 1 µm	Steps for manual polishing of fixed workpiece: – Workpiece has to be carefully cleaned. A pin-head-size amount of diamond paste is applied with a polishing stick of desired hardness and moved back and forth until cutting starts. Then thinner is added and polishing continued until all marks from previous operation have disappeared. – Careful cleaning of workpiece and hands. Then one uses either a polishing tool of the same hardness with a finer paste or a softer tool with the same paste and works in an angle of 30 to 45° to the preceding direction. Thus the end of each step can be easily recognized. – One continues with these operations until the desired result is obtained. Steps for manual polishing of rotating workpieces: – When working the inside of an object the speed has to be reduced with increasing hole size. – The polishing stick is moved back and forth to remove chips from the hole. Special adjustable tools for polishing bores are available. For polishing the outside of cylindrical workpieces special lap rings can be employed. R_a 0.001 to 0.1 µm

machining. Here, however, they are shallow, largely circular and all of about equal size. The surface roughness of so polished molds is about $R_a = 0.1$ to 0.3 µm with a diameter of the discharge craters of about 10 µm. These patterns are in the range of finely ground surfaces and meet the requirements of mold making in many cases. Thus, it is possible to forgo manual polishing, which is difficult with complex geometries [2.57, 2.60]. The necessary time is 15 to 30 min/cm², the exact pattern depending on shape and size.

Hence, electric-discharge machining allows molds to be machined completely in one set-up by means of roughing, prefinishing, fine finishing and polishing. However, the workable area is limited in this process. Furthermore, electric-discharge polishing is very time-consuming. On account of the thermal removal principle of electric-discharge machining, a thermally damaged outer zone always remains on the workpiece. This can be minimized by electric-discharge polishing, but can never be removed completely.

2.5 Electric-Discharge Forming Processes

Modern mold making would be inconceivable without electric-discharge equipment. With its help, complicated geometric shapes, the smallest of internal radii and deep grooves can be achieved in one working step in annealed, tempered and hardened steel with virtually no distortion [2.58, 2.61]. The process is contactless, i.e. there is a gap between the tool and the workpiece. Material removal is heat-based, requiring electric discharges to occur between tool and workpiece electrode [2.58]. (For method of producing microcavities, see Section 20.1.2–20.1.2.6).

2.5.1 Electric-Discharge Machining (EDM)

Electric-discharge machining is a reproducing forming process, which uses the material removing effect of short, successive electric discharges in a dielectric fluid. Hydrocarbons are the standard dielectric, although water-based media containing dissolved organic compounds may be used. The tool electrode is generally produced as the shaping electrode and is hobbed into the workpiece, to reproduce the contour [2.58].

With each consecutive impulse, a low volume of material of the workpiece and the electrode is heated up to the melting or evaporation temperature and blasted from the working area by electrical and mechanical forces. Through judicious selection of the process parameters, far greater removal can be made to occur at the workpiece than at the tool, allowing the process to be economically viable. The relative abrasion, i.e., removal at the tool in relation to removal at the workpiece, can be reduced to values below 0.1% [2.48, 2.58].

This creates craters in both electrodes, the size of which are related to the energy of the spark. Thus, a distinction is drawn between roughing (high impulse energy) and planing. The multitude of discharge craters gives the surface a distinctive structure, a certain roughness and a characteristic mat appearance without directed marks from machining. The debris is flushed out of the spark gap and deposited in the container. Flushing can be designed as a purely movement-related operation. This type of flushing is very easy to realize since only the tool electrode, together with the sleeve, has to lift up a short distance. This lifting movement causes the dielectric in the gap to be changed. Admittedly, this variant is only really adequate for flat cavities. For complex contours, pressure or suction flushing by the workpiece or tool electrodes would need to be

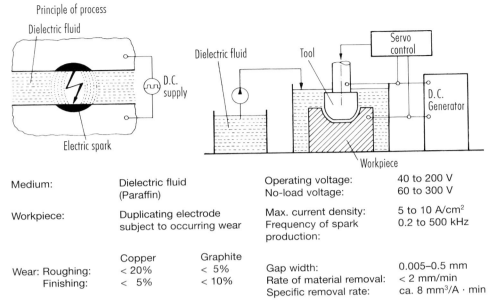

Medium:	Dielectric fluid (Paraffin)		Operating voltage: No-load voltage:	40 to 200 V 60 to 300 V
Workpiece:	Duplicating electrode subject to occurring wear		Max. current density: Frequency of spark production:	5 to 10 A/cm² 0.2 to 500 kHz
	Copper	Graphite		
Wear: Roughing: Finishing:	< 20% < 5%	< 5% < 10%	Gap width: Rate of material removal: Specific removal rate:	0.005–0.5 mm < 2 mm/min ca. 8 mm³/A · min

Figure 2.23   Principle of electrical discharge machining [2.62, 2.63]

superimposed [2.58]. Polarizing of workpiece and tool depends on the combination of materials employed, and is done such that the largest volume is removed from the workpiece [2.62]. The underlying principle of EDM is demonstrated in Figure 2.23.

In plain vertical eroding, the eroded configuration is already dimensionally determined by the shape and dimensions of the electrode. Machining of undercuts is not feasible. The introduction of planetary electric-discharge machining has now extended the possibilities of the erosion technique. It is a machining technique featuring a relative motion between workpiece and electrode that is achieved by a combination of three movements, vertical, eccentric and orbital [2.63]. The planetary electric-discharge machining is also known as the three-dimensional or multi-space technique [2.64]. Figure 2.24 shows the process schematically.

The technological advantages of planetary electric-discharge machining are presented in Figure 2.25. This technique now allows undercuts to be formed in a cavity [2.63, 2.64]. A further, major advantage is that, through compensation of the undersized electrode, it is possible to completely machine a mold with just one electrode.

Basically, all good electrical conductors can be employed as electrodes if they also exhibit good thermal conductivity. In most cases, the melting point of these materials is high enough to prevent rapid wear of the tool electrode [2.66]. Nowadays, graphite and copper electrodes are used for steel, and tungsten-copper electrodes for hard metals.

The electrodes are made by turning, planing or grinding, the mode of fabrication depending on the configuration, required accuracy, and material. High-speed cutting can be used to optimize fabrication of graphite or copper

Because of the high demands on the surface quality of injection molds and the wear on the electrodes, several electrodes are used for roughing and finishing cavity walls, especially for vertical eroding. Thus, microerosion permits a reproducing accuracy of

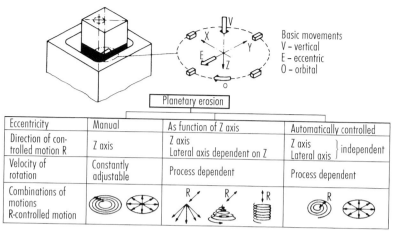

Basic movements
V – vertical
E – eccentric
O – orbital

Eccentricity	Manual	As function of Z axis	Automatically controlled
Direction of controlled motion R	Z axis	Z axis Lateral axis dependent on Z	Z axis Lateral axis } independent
Velocity of rotation	Constantly adjustable	Process dependent	Process dependent
Combinations of motions R-controlled motion			

Figure 2.24 Basic movements during planetary erosion [2.63]

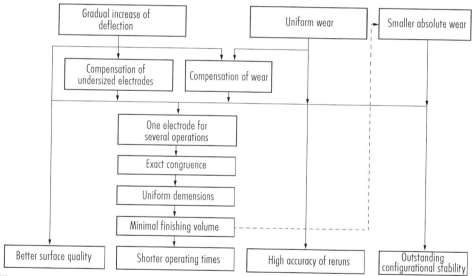

Figure 2.25 Technological advantages of planetary erosion [2.65]

1 μm and less, with roughness heights of 0.1 μm. A mold made by this technique usually only needs a final polishing [2.67]. In some cases, this is not sufficient, however, e.g. for the production of optical parts or for cavities whose surface must be textured by etching.

In spark erosion, the structure of the surface is inevitably changed by heat. The high spark temperature melts the steel surface and, at the same time, decomposes the high-molecular hydrocarbons of the dielectric fluid into their components. The released carbon diffuses into the steel surface and produces very hard layers with carbide-forming elements. Their thickness depends on the energy of the spark [2.61]. Moreover, a concentration of the electrode material can be detected in the melted region [2.63]. Between the hardened top layer and the basic structure there is a transition layer [2.66]. The consequences of this change in structure are high residual tensile stresses [2.68] in the outer layers that can result in cracking and may sometimes impede necessary posttreatment, e.g. photochemical etching.

Nevertheless, the EDM process has found a permanent place in mold making nowadays. Some molds could not be made without it. Crucial advantages of it are that materials of any hardness can be processed and that it lends itself to the fabrication of complex, filigree contours.

A further advantage is that it works automatically and without supervision and is very precise and troublefree. Therefore modern electric-discharge machines are numerically controlled with four-axial screen control by dialogue. To better automate the process, the machinery is sometimes equipped with automatic tool and/or workpiece changing devices. Thus, pallet loading and pallet displacement can be arranged such that it is possible to handle pallet in several coordinates in the fluid. Startup and exact machining can be done without supervision and the work can continue on several workpieces without operator.

2.5.2 Cutting by Spark Erosion with Traveling-Wire Electrodes

This is a very economical process for cutting through-holes of arbitrary geometry in workpieces. The walls of the openings may be inclined to the plate surface. Thanks to the considerable efficiency of this process, low cavities are increasingly being cut directly into mold plates.

Cutting by spark erosion is based on the same principle of thermal erosion that has been used in EDM for some time (see Section 2.5.1). The metal is removed by an electrical discharge without contact or mechanical action between the workpiece and a thin wire electrode [2.69]. The electrode is numerically controlled and moved through the metal like a jig or band saw. Deionized water is the dielectric fluid, and is fed to the cutting area through coaxial nozzles. It is subsequently cleaned and regenerated in separate equipment. Modern equipment has 5-axis CNC controls with high-precision positioning systems [2.48].

Deionized water has several advantages over hydrocarbons. It creates a wider spark gap, which improves flushing and the whole process; the debris is lower, there are no solid decomposition products and no arc is generated that would inevitably result in a wire break [2.70]. In addition, there is a lower risk of emissions.

Figure 2.26 depicts the principle of cutting by spark erosion.

Standard equipment can handle complicated openings and difficult contours with cutting heights up to 600 mm. The width of the gap depends on the diameter of the wire electrode and is determined by the task at hand. It is common practice to use wire with a diameter of 0.03 to 0.3 mm [2.69]. The wire is constantly replaced by winding from a reel. Abrasion and tension would otherwise cause the wire to break. Furthermore, the cuts would not be accurate as the wire diameter would become progressively shorter.

The maximum cutting speed of modern machines is roughly 350 mm^2/min. With the aid of so-called multi-cut technology (principal cut and several follow-up cuts), surfaces with a roughness height of $R_a = 0.15$ μm can be achieved [2.48].

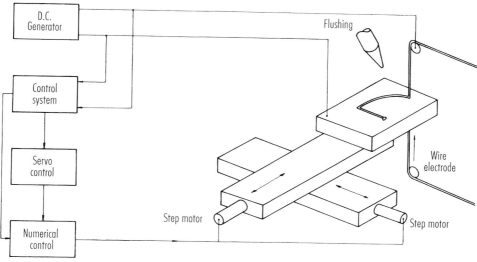

Figure 2.26   Principle of machine control for electric-discharge band sawing with wire electrodes [2.70]

As with conventional EDM, the workpiece is subjected to thermal load that can lead to structural changes in the layers near the surface. Mechanical finishing of the eroded surfaces may be advisable in such cases [2.62].

2.6   Electrochemical Machining (ECM)

This material-removal process employs electrolysis to dissolve a metal workpiece. The dissolution is caused by an exchange of charges and materials between the workpiece, produced as anode, and the tool, produced as cathode, under the force of an electric current in an electrolyte that serves as the effective medium [2.71].

The process is a non-contact one, i.e. a machining gap remains between the workpiece and the tool. Since only the metal anode is removed, the ECM process is virtually abrasive-free. Moreover, ECM is a "cold" process in which the workpieces are not subjected to heat [2.48, 2.58].

This process has some advantages over the EDM process, such as no hardening of the surface, no wear of electrodes, and high removal rates, but it also has serious drawbacks [2.72]. The equipment is very expensive and is only suitable for larger series of the same configuration because of the cost- and time-consuming fabrication of anodes. Such series are rare in the case of making cavities for injection molds.

2.7   Electrochemical Material Removal-Etching

For decorative or functional purposes, a surface is very often textured. This is either done for cosmetic reasons, for obtaining a more scratch and wear resistant surface (e.g. leather or wood grain) or a better hand. Flow marks (weld lines, streaking) can be hidden, too [2.73, 2.74].

The previous, mostly mechanical and predominantly manual, procedures often did not allow imaginative designs. Only the chemical process has opened up new possibilities for the designer.

The basis of this process is the solubility of metals in acids, bases and salt solutions. Metallic materials dissolve as a result of potential differences between microregions of the material or between material and etching agent (Figure 2.27). The metal atoms emit electrons and are discharged as ions from the metal lattice. The free ions are used up by reducing processes with cations and anions present in the etching agent. The removed metal combines with anions to form an insoluble metal salt, which has to be removed from the etching agent by filtering or centrifuging [2.62].

The exact composition of the etching agent is generally a trade secret of the developer. Almost all steels, without restriction on the amount of alloying elements such as nickel or chromium (including stainless steel), can be chemically machined or textured. Besides steel molds, those made of nonferrous metals can also be chemically treated [2.75]. Particularly recommended are the tool steels listed in Table 2.2.

Table 2.2   Steels for chemical etching [2.74, 2.75]

AISI-SAE steel designation	General characteristics
AISI S7	Shock-resisting tool steel
AISI A2	Medium-alloy tool steel
AISI H13	Hot-work tool steel
AISI P20	Medium-alloy mold steel
AISI 420	Stainless steel

The surface finish that can be achieved by chemical material removal or etching depends mostly on the material and its surface conditions and, of course, on the etching agent. Uniform removal is only achieved with materials that have a homogeneous composition and structure. The finer the grain of the structure, the smoother and better the etched surface will turn out. Therefore molds are frequently heat-treated before etching. The depth of heat treatment should always be greater than the depth of etching. If this is not the case, the heat-treated layer may be penetrated. This would result in very irregular etching. Adequate layers are obtained by a preceding case hardening [2.74, 2.75].

As already mentioned, the initial roughness of the mold plays an important role as regards the surface finish after etching. Non-permissible traces from machining are not covered up but remain hazily visible. Before etching, the surface should be well planed with an abrasive of grain size 240. The permissible depth of etching depends on the injection molding processing conditions. The speed of material removal is determined by the etching agent, the temperature and the type of material. It is generally 0.01 and 0.08 mm/min, and increases with rising temperature [2.62].

Basically there are two procedures employed for etching, namely dip etching and spray etching (Figure 2.27). Both have advantages and disadvantages. With dip etching, molds of almost any size can be treated in simple, cost-effective equipment. Difficulties arise from the need for disposing of the reaction products and constantly exchanging the etching agent near the part surface. It is easier to remove the reaction products in spray etching and maintain a steady exchange of the agent on the part surface. The process

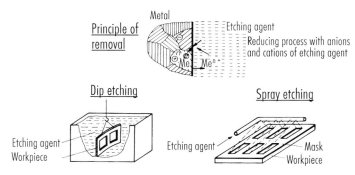

Medium: Aqueous solutions of e.g. HCl, HNO$_3$, H$_2$SO$_4$, NaOH
Rate of material removal: 0.01 to 0.08 mm/min
Surface quality: R$_0$ = 1 to 15 µm
Generation of shape: By masking, time controlled immersion, or removal of piece
                                 from etching agent

Figure 2.27   Material removal by chemical dissolution [2.62]

itself, however, takes considerably more effort and the equipment is more expensive. The etching agent is pressurized and sprayed through nozzles against the surface to be etched. Any masks for areas not to be etched must not be destroyed when hit by the spray, or lifted, permitting the agent to act underneath.

A number of techniques have been developed for masking areas where no material should be removed. They depend on the kind of texture to be applied and range from manual masking to silk-screening, and photochemical means. The last of these allows high accuracy of reproduction to be achieved [2.74]. The metal surface is provided with a light-sensitive coating, on which the pattern of a film is copied. Figure 2.28 shows this procedure schematically. A texture made in this way is correct in details and equally well reproducible. Therefore the process is particularly interesting for multicavity molds. A broad range of existing patterns is offered on the market nowadays.

2.8   Surfaces Processed by Spark Erosion or Chemical Dissolution (Etching)

With the help of modern process techniques – spark erosion and especially photochemical etching – almost any desired surface design can be obtained.

Both procedures give mold surfaces a characteristic appearance. Spark-eroded molds exhibit a mostly flat structure with the rim of the discharge crater rounded. Etched surfaces are different. Their structure is sharp-edged and deeper. In both cases the structure can be corrected by subsequent blasting with hard (silicon carbide) or soft (glass spheres) particles and thus adjusted to the wishes of the consumer. With hard particles, the contour is roughened, and with soft ones, it is smoothed.

Each plastics material reproduces the surface differently depending on viscosity, speed of solidification and processing parameters such as injection pressure and mold temperature.

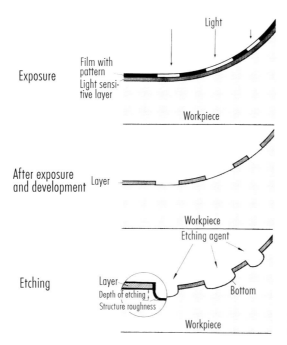

Exposure

Film with pattern
Light sensi-
tive layer

Light

Workpiece

After exposure
and development   Layer

Workpiece
Etching agent

Etching   Layer
Depth of etching
Structure roughness   Bottom

Workpiece

Figure 2.28   Photochemical etching (schematic) [2.74]

As a rule, the lower the melt viscosity, the greater the accuracy of reproduction. Consequently, materials with a low melt viscosity reproduce a mold surface precisely and with sharp edges. Very mate surfaces that are also mar-resistant are the result. Materials with a high melt viscosity form a more "rounded" mold surface that is shiny but sensitive to marring. Higher processing parameters, such as mold temperature, injection speed and cavity pressure, reproduce delicate structure of the mold surface more precisely and give this surface an overall matter appearance. This also means that complex and complicated parts with a large surface and those with large differences in wall thickness show a uniform surface only if the melt is under the same conditions at all places of the cavity.

With this, dimensions and positions of gate and runner gain special significance. Given unfavorable gate position, poor reproduction and increasing shine can be observed in areas far from the gate. The reason for this is that the melt further away from the gate has already cooled and therefore the pressure is too low to reproduce the structure in detail.

Textured surfaces act like undercuts during demolding; they obstruct the release process. Therefore, certain depths dependent on the draft of the wall must not be exceeded during etching or spark erosion. It is important whether the texture runs perpendicular, parallel or irregularly to the direction of ejection. As a rule of thumb, the depth of etching may be 0.02 mm maximum per 1° draft [2.74, 2.75].

For spark-eroded molds, the draft x° for some materials dependent on the roughness can be taken from Table 2.3. These values are valid only for cavities and not for the core of a mold since the molding shrinks onto it during cooling. If it has to be etched at all, the depth must be lower or the draft greater. If the recommended values cannot be adhered to, different mold-wall temperatures should be applied to try and shrink out the molding from the undercut. This can also be accomplished by removing the core first, and allowing the molding to shrink towards the center and out of the texture (e.g. ball pen covers). A precondition for this is a greater draft at the core than at the outer contour [2.76, 2.77].

Table 2.3   Minimum draft x° depending on roughness average values (R_a) of etched surface structure [2.76] (For glass-reinforced materials, one step higher)

R_a	Draft x°		
µm	PA	PC	ABS
0.40	0.5	1.0	0.5
0.56	0.5	1.0	0.5
0.80	0.5	1.0	0.5
1.12	0.5	1.0	0.5
1.60	0.5	1.5	1.0
2.24	1.0	2.0	1.5
3.15	1.5	2.0	2.0
4.50	2.0	3.0	2.5
6.30	2.5	4.0	3.0
9.00	3.0	5.0	4.0
12.50	4.0	6.0	5.0
18.00	5.0	7.0	6.0

2.9   Laser Carving

Now about 10 years old, laser carving has advanced to the stage of already being used in preliminary injection molding trials. It is marketed under the name LASERCAV [2.78]. The beam of a laser is bundled by means of appropriate lenses and focused precisely on the object for machining. A power density of more than 2000 W/mm² is generated at the focal point. This leads to peak temperatures of approx. 2500 °C in steel. At the same time, the instantaneous focal point is exposed to a gas atmosphere that has such a high oxygen content that the steel burns spontaneously at this spot. If the beam is now moved along the steel surface, a bead of iron oxide is formed that detaches from the underlying steel surface on account of the heat stress generated. Increasing the power of the laser beam in the focal spot causes the surface beneath it to melt as well. This melt can also be blown away in the form of glowing droplets by the gas jet.

The diameter of the beam in the focal spot and thus the width of the processed tracks is 0.3 mm. A distance of 0.05 to 0.2 mm between tracks is standard. This offset of 0.05 mm yields a surface roughness of r_A of 1.5 µm. This is roughly the same surface quality as yielded by erosion finishing. The cavity is machined layer by layer, the layer thickness usually ranging from 0.05 to 0.2 mm. A special control device ensures that the penetration depth of the beam remains at the predetermined value (e.g. as pre-set by the NC program). Attainable tolerances are 0.025 mm. The particular advantage of this technique is that the NC program for guiding the laser beam is obtained directly from the virtual image of a molding or cavity that has been generated by a CAD program and transferred via the stereolithography interface of the CAD system (for processes for producing microcavities, see Sections 20.1.2–20.1.2.6).

2.9.1   Rapid Tooling with LASERCAV

This direct way of programming straight from the computer offers for the first time the possibility of taking tool materials, any kind of alloyed steel of any hardness, other metals or ceramics and working up the desired shape directly, without intervening material steps. Consequently, this process can be expected to supersede most of the rapid tooling processes developed in recent years. Although the surface quality and the size of the possible die and cavities do not yet satisfy all demands, it may be expected that this process, when combined with other machining processes such as grinding, eroding, or milling, will satisfy all requirements. The advantages that accrue thereby extend far beyond merely speeding up the process, because it is possible for the first time to use the same material that will be used to mass produce the tool later. Moreover, in many instances, it will likely also be used in mass production if design changes are not needed. The next few years will show just how much the relatively expensive investment will pay off and how competitive the process will be.

2.10   Molds for the Fusible-Core Technique

In the injection molding of technical plastics parts, the mold parts designer is continually faced with the problem of incorporating undercuts into the part such that they will demold properly. Growing technical and design requirements make the problem of demolding a major part of the design phase. Often, the desire for optimum design has to give way to demoldability. Moldings that feature complex undercuts, or that represent a 3D hollow body, can be made by 2 different fabrication techniques: the shell technique or the fusible-core technique. In the shell technique, the molding is built up from two or more parts, known as shells. The shells are made by means of conventional tooling and machine technology, either in the same mold or in two different molds. The shells are then joined by means of screws, snap-on connections, bonding or welding in a further step to form a mold part. Another method of joining is to mold material around a flange.

The housing for a water pump is shown in Figure 2.29. This is notable for the fact that it is designed as a single part and thus has a conventionally non-demoldable internal geometry. The inside surface of the part is indistinguishable from the outside one.

The production of a part like this requires a method that allows the demolding of internal geometries that cannot be conventionally demolded. The fusible-core technique is one such method. The various stages are shown in Figure 2.30.

A metal core consisting of a low-melting metal alloy is inserted into an injection mold and plastic material is molded around it. The surface of the core forms the internal contour of the mold part. The mold part is demolded with the core inside it and transferred to a heated melting medium. The core melts completely and runs out of the mold part, without causing damage. The liquid core material can then be used to make another core. The cores are made with the aid of a core-casting machine in what is known as the lowpressure casting process.

Figure 2.29 One-piece housing for a water pump, made by the fusible-core technique [2.79]

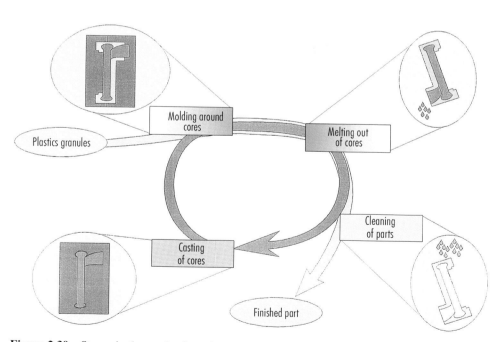

Figure 2.30 Stages in the production of moldings made by the fusible-core technique [2.80]

Because the production process involves several stages, the plastic and the core material must satisfy a large number of requirements. Apart from withstanding the high pressures acting on it when the plastic is injected, the core must resist the temperature stresses involved when material is molded around it. Premature melting of the core during this process causes flaws in the molding. If the core material is too soft, the core might be shifted and the walls of the molding may vary in thickness. If the core material is too brittle, it might fail when the mold is closed or when material is molded around it, thereby making reliable production difficult.

The need to melt the core imposes further requirements. For one thing, the core material must not be damaged during melting in order that continuous reuse in the cycle may be ensured. Also, the mold and the plastic must not be damaged by the melting-out step.

Throughout the production cycle, the various materials, plastic, core material, melting medium and also the mold materials are in constant interaction. This interaction is influenced by such process parameters as temperature, pressure and speed. A further influence that must be mentioned is the geometry of the part and of the core. It is therefore evident that the fusible-core technique is an elaborate, complex production method.

The great advantage of making moldings by the fusible-core technique as opposed to that of multiple shells is that production occurs "in cast". As a result, the part is more homogeneous, has a more accurate contour and does not have any weak zones caused by joints. Further advantages are:

– greater design scope,
– complex geometries can be realized as a single part,
– dimensionally stable internal and external contours, with high surface finish
– more simple mold design, and
– increased parts functionality through integration of insert parts (part-in-part technique).

These production methods are thus particularly suitable where high demands are imposed on the strength, level of seal, and dimensional accuracy of the part. Disadvantages are the apparent high costs of manufacture due to the necessity of making and removing the core.

However, it is precisely the use of a core that remains in the molded part during demolding that gives the designer much more freedom when designing parts. Further, the fusible-core technique affords a means of simplifying mold designs for complex contours. Whereas to demold these complex contours by the traditional mold-making method would necessitate a large number of highly elaborate, perhaps interpenetrating, ejectors and cores, the use of fusible cores can simplify mold design.

Different variants of this production process have been developed since the start of the 1980s that permit mold parts featuring complex, smooth internal geometries to be produced with high dimensional stability. All these processes have the same basic idea: the manufacture of injection-molded parts with lost cores, comparable with sand casting of metallic materials. These processes are:

– the fusible-core technique,
– the dissolved core technique, and
– the salt core technique.

From today's point of view, the fusible-core technique has become the established mass-production method. The reasons for this are the superior mechanical properties of the core material, advantages in separating the core material and plastic part and the simpler process technology involved in reusing the core material.

The fusible-core technique was developed into a large-scale production process in the 1980s for the manufacture of intake systems for combustion engines [2.81–2.84]. However, it has been in existence since the early 1960s. An example of such an intake system is shown in Figure 2.31.

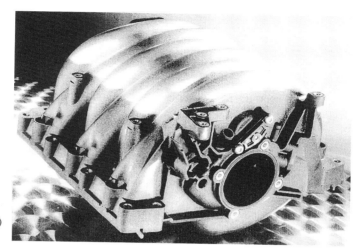

Figure 2.31 Intake system for a 6-cylinder Audi engine (Photo: Mann + Hummel)

2.10.1 Molds for Sheathing the Fusible Cores

Aside from insertion of the core, the production sequence for sheathing the core or cores is identical with that of conventional injection molding. After the core has been inserted, the mold is closed and locked. The plastic is injected into the cavity and around the core. When the cooling period has elapsed, the plastic part is demolded with the core inside it. The cycle begins again with the insertion of a new core into the open mold.

In the injection-molding cycle, injection of the polymer melt is the crucial phase because this is when the core is subjected to the maximum stress, both thermal and mechanical. Hot polymer melt impinges on a comparatively cold core and cools instantaneously at the phase boundary.

The three key requirements imposed on the process by sheathing are summarized in Figure 2.32.

Even though the fusible-core technique is a special technique, the necessary molds can be completely built up from the usual standard parts for mold making. The moldmaking materials and the machining methods are no different in the case of molds for the fusible-core technique than they are for conventional molds. The use of standard parts can render the building up of a mold cost effective and efficient. The mold for sheathing the fusible cores essentially has the same construction as a conventional injection mold. The pump housing shown in Figure 2.29 will be used to illustrate this. The injection mold used for the housing is shown in Figure 2.33. It consists of clamp plates, a hot runner, the two mold plates with inserts and cooling channels as well as the

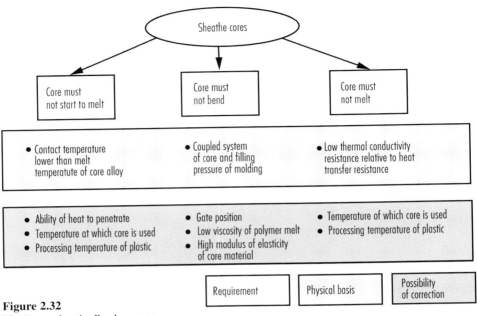

Figure 2.32
The "core sheating" subprocess

ejector unit. Centering and locating of the plates beneath each other are effected by means of suitable guide elements.

The three fusible cores required for the part are inserted, in this example, into the opened mold on the ejector side. Positioning of the fusible cores is effected by means of a conical pivot. An essential function of the injection mold is to securely locate the core. The core must be inserted in precisely the right position without play and fixed in position by the core mounting.

The closing movement of the mold must not cause the core to move or fall out of the mold.

It is absolutely imperative when designing the core to include areas that are not sheathed by plastic because it is only in these areas that the core can be located and fixed into position in the mold. "Free floating" of the core in the mold, i.e. without locating positions, is not feasible from a fabrication point of view. The molding must therefore have gaps through which the core protrudes or through which the mold can protrude into the core. These areas are where the core is located and fixed in position in the mold.

The most favorable design in this area features a conical locator for the core in the mold as this also allows simultaneous centering in the mold (Figure 2.34, left). When providing for a conical locator for the core, it is beneficial if there is also a conical core mounting on the opposite side of the core, so that the core is firmly clamped in the cavity.

A bolt entering from the side can prevent the core from inadvertently falling out. This is not necessary if the injection molding machine has a vertical clamping unit since gravity prevents the core from falling out or slipping. Experience has shown that this measure is not necessary for small cores, even if they are incorporated in the clamping side of the injection mold. However, this presumes that the mold closes smoothly.

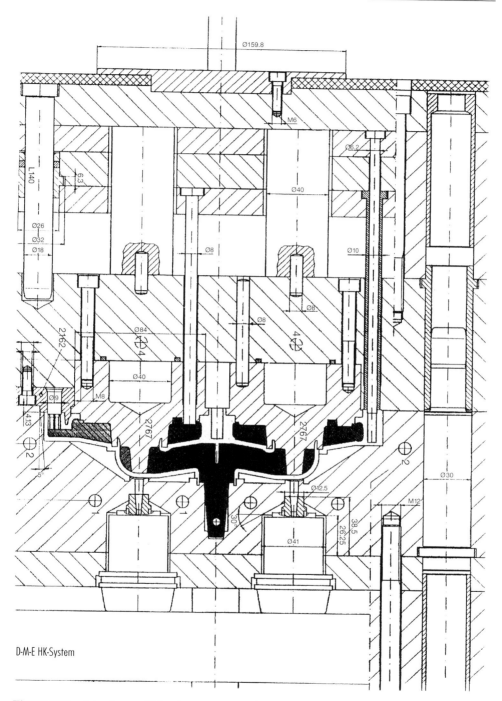

Figure 2.33   Injection mold for the water-pump housing shown in Figure 2.29
(■ = fusible core)

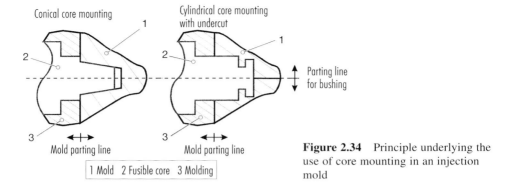

Figure 2.34   Principle underlying the use of core mounting in an injection mold

A different design for the mold bearing is shown on the right side of Figure 2.34. Here, the core mounting is formed as a cylinder with a peripheral groove into which a divided bushing latches. The advantage of this is that the core cannot fall out of the mold as the bushing closes. Furthermore, this is particularly recommended when a core mounting is to be incorporated only at one position in the mold. There are a large number of other variants by which the core can be fixed securely in the injection mold.

When wall thickness is a critical condition, the core mounting is best generously dimensioned since the heat can flow out of the core via the surface area of the core mounting. In this design, it is also advantageous to provide intensive cooling in the region of the core mounting, e.g. by means of a surrounding cooling channel.

2.10.1.1 Gating the Molding

The design of the gate system and the type of gating for the molding are chosen by the same systematic method employed for conventional injection-molding machines. When the position of the gate is chosen, it should be borne in mind that all of the heat is introduced into the mold through the gate cross-section. Consequently, the frozen wall thickness at this position is very low, so there is a high temperature gradient on the mold side. If the gating for the molding is positioned so that material is injected directly onto the core, there is the danger that the core will start to melt in this region. There are two reasons for this. First, the temperature in the gate is always the processing temperature, and may in fact even be higher due to a local temperature increase caused by shear heating. Second, the gate area is always the region of maximum filling pressure in the molding. The simultaneous action of pressure and temperature can cause the core in this region to be "washed out". These flaws in the core later manifest themselves as indentations in the melted-out part. Direct gating onto the core is therefore to be avoided.

2.10.1.2 Thermal Considerations Concerning Mold Design

Not long after the melt makes contact with the core, a so-called contact temperature is established that must be lower than the melting temperature of the core material. The contact temperature is influenced by such material values as ability of heat to penetrate, and the temperatures of plastic and core immediately prior to first contact.

In the holding-pressure phase, the core is completely sheathed in plastic. The heat introduced by the plastic is dissipated into the mold and the core. However, the heat flow from the plastic into the core and from the plastic into the mold are not the same because the core, unlike the mold, must store the heat. The consequence of this is that the temperature of the core rises during the holding-pressure and cooling period. In order that the core may be prevented from melting during sheathing, under no circumstances must the temperature of the core be allowed to exceed the melting temperature of the core material.

Due to the rise in temperature of the fusible core, the temperature gradient on the core side is flatter than on the mold side. Compared with a conventional mold, this means that a mold for use in fusible-core technology must dissipate more heat across the surface of the mold. For this reason, the temperature-control system of the injection mold here must have greater dimensions than in a conventional injection mold.

2.10.1.3 Core Shifting

Core shifting plays a critical role in fusible-core technology. As a result of the filling behavior of the mold, asymmetric injection or eccentric positioning of the core, the filling process necessarily generates a quite substantial lateral load on the core. This, in turn, causes deformation of the core from its fixed position in the mold. Low deformations of the core cause the molding to have irregular wall thicknesses and large deformations may cause the core to penetrate the wall of the molding. The problem of core shifting is extremely pronounced with curved cores.

Compared with classical construction materials for injection molds, the core material tends to be softer and its modulus of elasticity is 13 times lower than that of steel. It therefore has low rigidity, a property crucial to core shifting. Material data are not the only important factors – the geometry also plays a major role. Factors here include clamping of the core and the pressure profile on the core surface. No generalizations can be made about the extent of core shifting. The rheological properties of the melt and mechanical effects on the core need to be taken into account. Bangert [2.85] has proposed the scheme shown in Figure 2.35 as an aid.

In this scheme, the filling phase is broken down into small discrete time steps in which the rise in pressure per time step is calculated. Two characteristic flow path lengths, l_1 and l_2, are determined and the effective pressure profile on the core is calculated. From this, the core shift is determined and the geometry of the flow channel around the resultant bending line of the core is modified. For the next time step, the pressure rise is determined and the program sequence loops until the core is completely surrounded. The difference in the characteristic flow path lengths, l_1 and l_2, is the maximum distance between the preceding and the following melt front for asymmetric flow around the core. Knaup [2.86] has used this scheme to calculate the influence of different gate positions on the core shift of an intake system. The results are presented in [2.86]. By changing the position of the gate, it proved possible to virtually eliminate core shift.

2.10.1.4 Venting

In simple molds, the air displaced by the incoming melt has adequate scope for escaping from the cavity, e.g. via the ejector pins, the mold parting line or via joints in mold inserts. Generally, no extra measures are needed to ensure that the air escapes from the cavity.

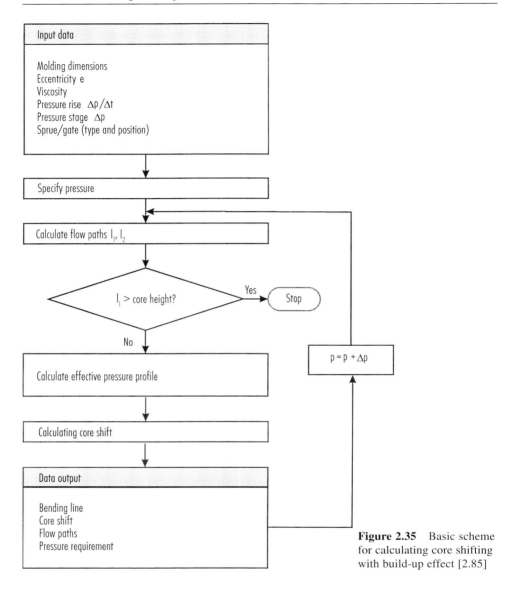

Figure 2.35 Basic scheme for calculating core shifting with build-up effect [2.85]

When metal or nonmetal cores are used and design is poor, there may be areas in which the air is trapped during the filling phase and unable to escape. As a result, molded parts may be incomplete at these areas. Furthermore, the pressure here may cause the air to heat up so much that the plastic burns. This is called the diesel effect.

With conventional mold designs, the consequence of the diesel effect is often undesirable side-effects, ranging from the need for regular, careful removal of the combustion residue right through to irreparable damage to the mold due to corrosion by the combustion residue. When metal cores are used, this also leads to undesirable side-effects. Local overheating causes the constituents of the plastic melt to burn and these deposit themselves on or oxidize the core material. Damage is caused to the core material, which should not be reused.

Trapped air in conventional molds, as well as in molds for fusible-core technology, should generally be avoided through optimum positioning of the gate or appropriate mold construction. Design features for venting are presented in [2.87, 2.88] (see also Chapter 7).

2.10.2 Molds for Making the Fusible Cores

The fusible-core technique is a fabrication method that requires a lost core. Therefore the first thing to do is make this core. Figure 2.36 shows the process cycle involved in making metal cores. This process cycle is similar to the process for classical injection molding of plastics. However, because different materials are used, the various process steps differ from those of injection molding.

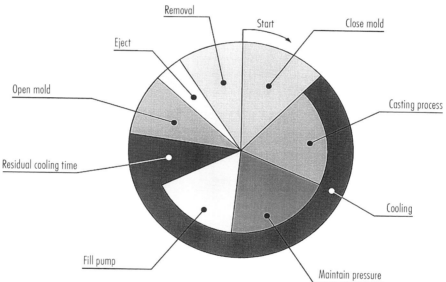

Figure 2.36 Sequence of steps in casting cores

At the start of the cycle, the casting mold is closed and locked. Then the casting process is commenced. Unlike the injection molding of polymeric materials, the very much lower viscosity of the metal alloy requires only low filling pressures (< 1 MPa) and the resultant lower closing force of the mold. The melt is introduced through heated tubes or pipes into the mold by means of a cylinder piston pump. When the casting mold has been volumetrically filled, and provided a needle valve nozzle is used, the casting pressure can be switched off. Because the volume of the alloy does not change during solidification, no holding pressure of the kind required for injection molding is needed. If, however, an open nozzle is used, the casting pressure must be maintained until the alloy in the gate area is cold. If the pressure is released too soon, unfrozen core material flows back into the conveying system and the core might possibly be partially hollow. After the gate has been closed or has frozen, the pressure on the casting system is released and the conveying pump is filled again. After the residual cooling time, the casting mold is opened and the core is demolded and removed from the casting mold by robot or manually.

2.10.2.1 Core Material

Nowadays, the core material is usually an eutectic alloy of tin and bismuth. The special attraction of the tin/bismuth eutectic alloy is that solidification is virtually shrinkage-free. Tin contracts, whereas bismuth expands in volume by 3.3% [2.89]. The internal dimensions of the casting mold or so-called metal mold are thus almost identical with the outer dimensions of the casting after demolding. Table 2.4 shows selected properties of an eutectic alloy of tin and bismuth.

Table 2.4   Selected properties of a eutectic alloy of tin and bismuth

Property		Unit	Value
Density at 20 °C		g/cm^3	8.58
Specific heat capacity	Solid	kJ/(kg K)	0.167
	Liquid		0.201
Melt enthalpy		kJ/kg	44.8
Thermal conductivity at 20 °C		W/(m K)	18.5
Viscosity		m Pa s	2.1

2.10.2.2 Construction of a Casting Mold

The construction of a casting mold for fusible cores is essentially the same as that of an injection mold. By way of example, Figure 2.37 shows the casting mold used for the fusible cores for the water-pump housing. The mold also comprises clamp plates, the two mold plates with mold inserts and cooling channels, as well as the ejector unit. Alignment and location of the plates beneath each other are effected by means of suitable guide elements, which are not shown in this cross-section.

Casting molds for the fusible-mold technique impose high requirements on the gap dimensions of the cavity. The reason for this is the low viscosity of the metallic melt, which is comparable to that of water. To use an analogy, the casting mold is to be filled with water without the water running out of the gaps in the parting line or the inserts and ejector pins.

2.10.2.3 Gating Systems

Since the casting molds are maintained at a temperature between 30 and 80 °C, the contact temperature for first contact between the hot metal melt and the cold mold wall is always lower than the melting temperature of the core material. Through the low contact temperature, and the allowance made for the forces of inertia and gravity, the first requirement concerning the filling phase of the casting mold presents itself. The gate

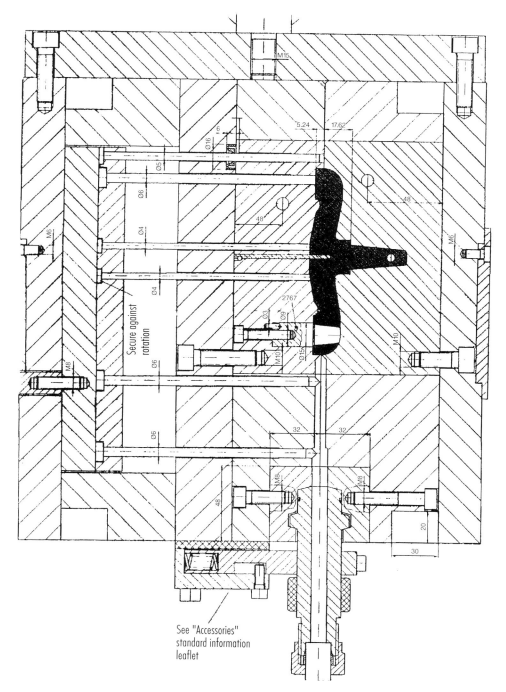

Figure 2.37   Diagram of the casting mold for the fusible cores used to make the water-pump housing shown in Figure 2.29 (■ = fusible core)

must be positioned at the lowest point of the cavity. Further, it is crucial that the geometry inside the core is arranged so that casting is always effected against gravity. A cast "dropping" onto the mold wall will freeze immediately and manifests itself as an undesirable mark on the surface of the core. A high-quality surface therefore requires slow, steady casting.

As in the injection molding of polymeric materials, hot-runner systems are used in core-casting molds. A distinction is made between open nozzles and needle valve systems. The casting mold shown in Figure 2.37 has an open hot runner, and a manifold in the form of a cold runner. The disadvantage of this design is that the fusible cores must be machined mechanically in the gate region of the cold runner after demolding. This is both time-consuming and the dust from grinding represents a loss of core material.

Casting molds in a modern production system for exhaust equipment feature needle valve nozzles whose geometry matches that of the fused core and are designed as a hot runner. The gate does not have to be posttreated, a fact that considerably facilitates fully automated production.

2.10.2.4 Thermal Considerations Concerning the Core-Casting Mold

Particularly with voluminous fusible cores of the kind used for intake systems, the cycle times for core casting are much longer than those for injection molding. Therefore, in this application, two core-casting machines feed one injection-molding machine. The reason for the long cooling times is the thermal design of the core-casting molds and the material employed.

The casting mold is usually made of steel. The tool steels usually employed have a thermal conductivity in the range 20 to 30 W/(m · K). Case-hardened steels and several annealed steels are notable for having a higher thermal conductivity of ranging from 40 to 46 W/(m · K). Even for these steels, the thermal diffusivity is roughly in the range of the tin/bismuth alloy. The thermal diffusivity, as a measure of the speed at which a temperature jump is introduced into a material, is shown in Figure 2.38 for a range of materials.

It can be seen that the thermal diffusivity of a case-hardened steel is roughly in the range of an alloy of tin and bismuth. To shorten the cooling time during core manufacture, it would therefore be sensible to use a material for the metal mold that has a much greater thermal diffusivity. Copper is renowned for its excellent thermal properties and would in theory be an ideal candidate. However, the use of copper as a mold material is ruled out because the liquid tin penetrates into the copper at the grain boundaries, forming an alloy through the creation of an intercrystalline phase. This causes the copper to physically dissolve [2.90] and visible damage in the form of washout is caused to the mold.

A suitable coating process, such as electrolytic nickel plating or chrome plating, may be used to apply a thin passive layer that suppresses alloy formation and hence greatly reduces the rate of corrosion. The reason for this is a very thin but dense semiconducting film of oxide on the metal surface that displaces the electron potential extensively in the positive direction. This property, which is known as chemical passivity, is possessed mainly by the transition metals, whose principle characteristic is that their electron configuration has vacant d orbitals. They include the metals in the iron and platinum group, chromium, molybdenum, tungsten, titanium and zirconium. Some other metals, such as aluminum, also exhibit chemical passivity [2.91]. For this reason, and because of its thermal conductivity, aluminum is also a very interesting mold material. So far, no experience has been gained of the compatibility of these materials in various combinations.

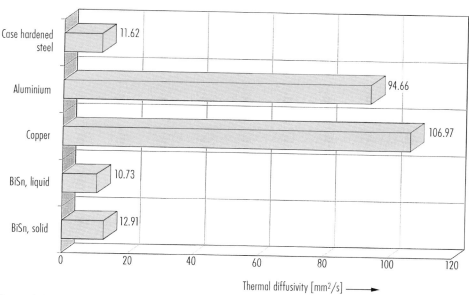

Figure 2.38   Thermal diffusivity of various materials

2.10.2.5 Demolding Cast Fusible Cores

Once the cores have cooled to the demolding temperature in the casting mold and the mold has opened, they can be demolded from the metal mold by means of an ejector. Unlike polymeric materials, the fusible cores do not shrink to the contour of the mold.

Drafts of 2–3° on the "nozzle side" and smaller drafts on the ejector side, prove to be helpful demolding aids.

The introduction of small recesses that forcibly demold the fusible cores has turned out to be inappropriate because they rapidly clog during production and thus become ineffective. Furthermore, the approach adopted in conventional injection molding of using different temperatures on the "nozzle side" and ejector side has also not proved useful. The reason is that the alloy does not shrink during cooling.

References

[2.1]   Werkzeugbau nach dem MCP/TAFA. Prospectus, HEK, Lübeck.
[2.2]   Flemming, E.; Tilch, W.: Formstoffe und Formverfahren. Deutscher Verlag für Grunstoffindustrie, Leipzig, Stuttgart, 1993.
[2.3]   Spur, G.; Stöferle, Th.: Handbuch der Fertigungstechnik. Vol. 1: "Urformen". Carl Hanser Verlag, Munich, 1981.
[2.4]   Zeuner, H.; Menzel, A.: Herstellung und Anwendung gegossener Werkzeuge. Rheinstahl-Technik, 3/71, pp. 91–96.
[2.5]   Kloos, K. H.; Diehl, H.; Nieth, F.; Tomala, W.; Düssler, W.: Werkstofftechnik, Dubbels Taschenbuch für den Maschinenbau. 15th Ed., Springer, Berlin, Heidelberg, New York, Tokyo, 1983, p. 294 ff.
[2.6]   Merten, M.: Gegossene NE-Metallformen – Angewandte NE-Metalle – Angewandte Gießverfahren. VDI-Bildungswerk, BW 2197, VDI-Verlag, Düsseldorf.

[2.7]   Cerro-Legierungen für schnellen Werkzeugbau. Plastverarbeiter, 25 (1974), 3, pp. 175–176.

[2.8]   Menden, A.: Gießerei Molellbau. Giesserei-Verlag GmbH, Düsseldorf, 1991.

[2.9]   Clegg, A. J.: Precision Casting Process. Pergamon Press, Oxford, New York, Frankfurt, 1991.

[2.10]   Chock, J.: Zauberformel HSC. Maschine + Werkzeug, 3/97, pp. 10–12.

[2.11]   Duretek, I.; Santek, D.: Der schnelle Weg. Plastverarbeiter, 47 (1996), 11, pp. 44–56.

[2.12]   Dusel, K. H.; Eyerer, P.: Materialien für Rapid Tooling Technologien. User Meeting, Report 4, Dresden, 1996, pp. 246–256.

[2.13]   Santek, D.; Duretek, I.: Materialien für Rapid Prototyping – schneller Modellbau aus CAD-Daten. Österreichische Kunstoffzeitschrift, 27 (1996), pp. 82–85.

[2.14]   Haferkamp, H.; Bach, F. W.; Gerken, J.: Rapid Manufacturing of Metal Parts by Laser Sintering. 28th International Symposium on Automotive Technology and Automation, ISATA; Stuttgart, 1995, pp. 459–474.

[2.15]   Eckstein, M.: Rapid Metal Prototyping. 41st International Colloquium, TU, Ilmenau, September 23–26, 1996, pp. 383–386.

[2.16]   Lorenzen, J.; Breitinger, E.: Rapid Tooling – Verfahren zur schnellen Herstellung von Prototypenwerkzeugen (Chapter 8). Paper presented at the Conference on: Verkürzung von Entwicklungszeiten durch Rapid Prototyping, EUROFORUM, Köln, October 24–25, 1996.

[2.17]   Shellabear, M.: Bindemittel ade – Neue Wege beim Lasersintern. Laser Praxis, June 1995, p. 48.

[2.18]   Klocke, E.; Nöken, S.: Verfahren und Prozeßketten zur Fertigung metallischer Bauteile und Werkzeuge. Spektrum der Wissenschaft, April 1995, pp. 97–99.

[2.19]   Klocke, E.; Clemens, U.: An RP network for product development Prototyping. Technology International Annual Review, UK & International Press, Surrey, UK, 1997.

[2.20]   Gasser, A.: Herstellung metallischer Bauteile durch Selektives. Lasersintern und Lasergenerieren. NCG-Conference. Arbeitskreis Rapid Prototyping, LBBZ, Aachen, December 1995.

[2.21]   Sachs, E.; Cima, M.; Allen, S.: Tooling by Three Dimensional Printing – A Desktop Manufacturing Process. 28th International Symposium on Automotive Technology and Automation, ISATA, 1995, Stuttgart, pp. 405–420.

[2.22]   Thomson, G. A.; Pridham, M. S.: RP and Tooling with a high powered laser. 1st National Conference on Rapid Prototyping and Tooling, Buckinghamshire College, UK, November 6–7, 1995.

[2.23]   Wiesner, R.; Eckstein, M. et al.: Rapid Prototyping durch Laserschneiden und Diffusionsschweißen. Proceedings of the 6th European Conference on Laser Treatment of Materials, ECLAT. Vol. 2, September 16–18, 1996.

[2.24]   Gebhardt, A.: Rapid Prototyping – Werkzeuge für die schnelle Produktentwicklung. Carl Hanser Verlag, Munich, 1996, pp. 23–26.

[2.25]   Langen, M.: Einsatz der Stereolithographie für das Rapid Tooling in der Spritzgießverarbeitung. 3rd User Meeting: Intelligente Produktionssysteme – Solid Freeform Manufacturing, Dresden, September 1995, pp. 255–265.

[2.26]   Schulthess, A.: Future Developments in Cibatool SL-Resins. 6th European Stereolithography User Meeting, Strasbourg, France, November 7–8, 1994.

[2.27]   Bettany, S.; Cobb, R. C.: A Rapid Ceramic Tooling System for Prototype Plastic Injection Moldings. 1st National Conference on Rapid Prototyping and Tooling, Buckinghamshire College, UK, November 6–7, 1995.

[2.28]   Auf dem Weg zum Serienteil. Werkzeug & Formenbau, November 1995, pp. 44–48.

[2.29]   Gernot, T. H. C.: Beiträge zum Simultaneous Engineering bei der Produkt- und Prozeßplanung für die Spritzgießfertigung. Dissertation, RWTH, Aachen, 1994.

[2.30]   Bocking, C. E.; Dover, S. J.: Rapid Tooling Using Electroforming. European Action on Rapid Prototyping, EARP, No. 6, August 1996, pp. 8–9.

[2.31]   Prospectus. Galvanoform Company. Gesellschaft für Galvanoplastik mbH, Lahr, 1996.

[2.32]   Greul, M.: Prototype Moulds by using FDM and Electroplating. European Action on Rapid Prototyping, EARP, No. 6, July 1995, pp. 1–2.

[2.33]   Flimm, J.: Spanlose Formgebung. 6th Ed., Carl Hanser Verlag, Munich, 1996, pp. 20–35.

[2.34] Andre, L.: Investment Casting and Ceramic Moulds. Using the Quick Cast Process. 6th European Stereolithography User Meeting, Strasbourg, France, November 7–8, 1994.

[2.35] Tsang, H.; Bennet, G.: Rapid Tooling-direct use of SLA moulds for investment casting. First National Conference on Rapid Prototyping and Tooling, Buckinghamshire College, UK, November 6–7, 1995.

[2.36] Sprunk, J.: Feinguß für alle Industriebetriebe. 2nd Ed., Industrie-Werkstätten GmbH, Bochum, 1987.

[2.37] Evans, B.: Production Outsourcing in Metal Casting. Rapid Prototyping & Manufacturing. Conference, Dearborn, MI, USA, April 22–24, 1997, pp. 177–205.

[2.38] Clerico, M.; Lanzoni, C.; Minks, H.: Ort an Expandable Pattern Casting. 3rd User Meeting: Intelligente Produktionssysteme – Solid Freeform Manufacturing, Dresden, September 1995, pp. 307–316.

[2.39] Connelly, R.: Verbal information. Becton Dickinson, Raleigh, USA, May 1997.

[2.40] Langen, M.: Einsatz des Rapid Prototyping und Rapid Tooling im Rahmen eines Simultaneous Engineering in der Spritzgießverarbeitung. Dissertation, RWTH, Aachen, 1998.

[2.41] Galvanisiergerechtes Gestalten von Werkstücken. Report. International Nickel, Düsseldorf, 1968.

[2.42] Kalteinsenken von Werkzeugen. VDI-Richtlinie, 3170, 1961.

[2.43] Hoischen, H.: Werkzeugformgebung durch Kalteinsenken. Werkstatt und Betrieb, 104 (1971), 4, pp. 275–282.

[2.44] Stoeckhert, K.: Formenbau für die Kunststoffverarbeitung. 3rd Ed., Carl Hanser Verlag, Munich, 1979.

[2.45] Piwowarski, E.: Erstellen von Spritzgießwerkzeugen. Kunststoffe, 78 (1988), 12, pp. 1137–1146.

[2.46] HSC-Fräszentrum bearbeitet Stahl bis 62 HRC. In: Maschinenmarkt, Würzburg, 102 (1996), 36.

[2.47] Malle, K.: Von der HSC-Produktidee zum Bearbeitungszentrum. VDI-Z, 138 (1996), 6.

[2.48] Eversheim, W.; Klocke, F.: Werkzeugbau mit Zukunft – Strategie und Technologie. Springer Verlag, Berlin, 1997.

[2.49] Prospectus. Joisten und Kettenbaurn, Berg.-Gladbach.

[2.50] Prospectus. Novapax, Berlin.

[2.51] Ultraschall-Läpp- und Poliergerät. Kunststoffe, 68 (1978), 12, p. 818.

[2.52] Polieren von Werkzeugstählen. Prospectus. Uddeholm, Hilden.

[2.53] Schmidt, P.: Verfahren zum Polieren von Spritzgießformen. Paper presented at the 3rd Plastics Conference at Leoben, 1975.

[2.54] Preßläppen – ein Verfahren zur Erzielung glatter Oberflächen in Werkzeugen für die Kunststoffverarbeitung. Kunststoffe, 73 (1983), 10, p. 567.

[2.55] Prospectus. Deploeg, Technik BV, Helmond, Holland.

[2.56] Extrude-Hone – ein Preßläppverfahren zum wirtschaftlichen Polieren, Entgraten und Abrunden. Betrieb + Meister, 4 (1985), p. 16.

[2.57] Hornisch, R.: Alternative zur mechanischen Oberflächenbearbeitung. Oberfläche + Jot. 1982, 2, pp. 24–29.

[2.58] König, W.; Klocke, E.: Fertigungsverfahren. Vol. 3: Abtragen und Generieren, Springer Verlag, Berlin, 1997.

[2.59] Jutzler, W. I.: Das funkenerosive Polieren. Ind.-Anz., 105 (1983), 56/57, pp. 17–19.

[2.60] Verbal information. Laboratorium für Werkzeugmaschinen und Betriebslehre. RWTH, Aachen.

[2.61] Steiner, J.: Senkerodieren von Stahl im Werkzeubau. Kunststoffe, 76 (1986), 12, pp. 1193–1194.

[2.62] König, W.: Neuartige Bearbeitungsverfahren. Laboratoriurn für Werkzeugmaschinen und Betriebstehre. Lecture, RWTH, Aachen.

[2.63] Kortmann, W.; Walkenhorst, U.: Funkenerosion von Werkzeugstählen und Hartstoffen. Thyssen Edelstahl. Technical Report, 7 (1981), 2, pp. 200–201.

[2.64] Schekulin, K.: Rationalisierung des Kunststoff-Forinenbaus durch neue Erodier-Technologien. Kunststoffberater, 27 (1982), 11, pp. 22–29.

[2.65] Schaede, J.: Planetär-Erodieren. VDI-Bildungswerk, BW 4409, VDI-Verlag, Düsseldorf.

[2.66]   Hermes, J.: Funkenerosion – ein Verfahren zum Herstellen von Raumformen. VDI-Z, 112 (1970), 17, pp. 1188–1192.

[2.67]   Genath, B.: Die Mikroerosion erzielt hohe Genauigkeiten. VDI-Nachr., No. 6, 7.2.1973.

[2.68]   Höpken, H.: Extreme Oberflächentemperaturen bei der Funkenerosion und ihre Folgen. Information, Stahlwerke Südwestfalen, 13/1974, p. 16.

[2.69]   CNC Feinschneiderodieren für den Miniaturbereich, 116 (1983), 6, pp. 64–66.

[2.70]   Weck, M.; König, W.: Abtragende Bearbeitungsverfahren. Laboratory for Machine Tools and Plant Management, RWTH, Aachen.

[2.71]   Elektrochemische Bearbeitung – Anodisches Abtragen mit äußerer Stromquelle. VDI-Richtlinie, 3401, 1, 1970 and 2, 1972.

[2.72]   Pielorz, E.: Elektrochemisches Senken (ECM). Paper given in Haus der Technik in Essen, 1984 (available from Köppern Company, 4320 Hattrigen/Ruhr).

[2.73]   Wagner, U.: Strukturierte Formoberflächen im Spritzguß, Plastverarbeiter, 24 (1973), 6, pp. 1–3.

[2.74]   Fotostrukturen als Oberflächedessin. Prospectus. Wagner Graviertechnik, Öhringer, 1975.

[2.75]   Lüdemann, D.: Oberflächen-Strukturieren, Kunststoff-Forinenbau, Werkstoffe und Verarbeitungsverfahren. VDI-Verlag, Düsseldorf, 1976.

[2.76]   Schauf, D.: Die Abbildung strukturierter Formnestoberflächen durch Thermoplaste. Paper and reprint, 2nd Tooling Conference at Würzburg, October 4 and 5, 1988.

[2.77]   Schauf, D.: Die Formnestoberfläche. Technical Information, Bayer AG, Leverkusen, 1983.

[2.78]   Berührungsloses Fräsen mit Hilfe eines Lasers. LCTec GmbH, Laser und Computertechnik, Tirolerstr. 85, D-87459, Pfronten.

[2.79]   Schmachtenberg, E.; Polifke, M.: Vom Saugrohr zum Pumpengehäuse. Kunststoffe, 86 (1996), 3, pp. 323–326.

[2.80]   Schmachtenberg, E.; Polifke, M.: Schmelzkerntechnik – Ein Sonderverfahren des Spritzgießens. Contribution on occasion of the honory colloquium for Dr. F. Johannaber, Essen, 26. 1. 1996.

[2.81]   Altmann, O.; Jeschonnek, P.: Entwicklung neuer Technologien für hochbeanspruchte Hohlkörper aus Kunststoff im Motorenbereich. Kunststoffe im Fahrzeugbau – Technik und Wirtschaftlichkeit, VDI-Verlag, Düsseldorf, 1988, pp. 245–281.

[2.82]   Haldenwanger, H.-G.; Mineif, R.; Arnegger, K.; Schuler, S.: Kunststoff-Motorbauteile in Ausschmelzkerntechnik am Beispiel eines Saugrohres. Kunststoffberater, 9/1987.

[2.83]   Jeschonnek, P.: Von der Bauteilentwicklung bis zur Großserienfertigung (Vol. 1). Plastverarbeiter, 42 (1991), 9, pp. 166–173.

[2.84]   Jeschonnek, P.: Von der Bauteilentwicklung bis zur Großserienfertigung (Vol. 2). Plastverarbeiter, 42 (1991), 10, pp. 64–67.

[2.85]   Bangert, H.: Systematische Konstruktion von Spritzgießwerkzeugen und Rechnereinsatz. Dissertation, RWTH, Aachen, 1981.

[2.86]   Knaup, J.: Werkzeugverformung beim Füllvorgang vorausberechnen. Carl Hanser Verlag, CAD CAM CIM, May 1994, pp. CA 64–CA 68.

[2.87]   Hartmann, W.: Entlüften des Formenhohlraumes. Spritzgießwerkzeuge. Verein deutscher Ingenieure VDI-Ges. Kunststofftechnik (ed.), VDI-Verlag, 1990.

[2.88]   Menges, G.; Mohren, P.: Spritzgießwerkzeuge, 3rd Ed., Carl Hanser Verlag, Munich, Vienna, 1991.

[2.89]   Eigenschaften und Anwendungen von niedrigschmelzenden MCP-Legierungen. Information, HEK, Lübeck, May 1995.

[2.90]   Bergmann, W.: Werkstofftechnik. Vol. 1: Grundlagen. Carl Hanser Verlag, Munich, 1984.

[2.91]   Wranglén, G.: Korrosion und Korrosionsschutz – Grundlagen, Vorgänge, Schutzmaßnahmen, Prüfung. Springer-Verlag, Berlin, 1985.

3 Procedure for Estimating Mold Costs

3.1 General Outline

Injection molds are made with the highest precision because they have to meet a variety of requirements and are generally unique or made as very few pieces.

They are to some extent produced by very time- and cost-consuming procedures and are therefore a decisive factor in calculating the costs of a molded product. The mold costs for small series often affect the introduction of a new product as a deciding criterion [3.1]. In spite of this, in many shops, calculation not have the place to which it should be entitled.

The respective mold costs are often not computed at all but estimated based on experience or in comparison with molds made in the past. This is also a consequence of the fact that the number of orders is only 5% of the number of quotations. The necessarily resulting uncertainties in such a situation are compensated by an extra charge for safety, which is determined by subjective criteria [3.2]. This leads to differences in quotations which render the customer uncertain.

Therefore the goal of a procedure to estimate mold costs must

– raise the certainty and accuracy of a cost calculation,
– reduce the time consumption for the calculation,
– make it possible to calculate costs of molds for which there is not yet any experience,
– ensure a reliable cost calculation even without many years of experience [3.3].

Extreme caution is in order if molds are quoted considerably less expensive than the result of such a calculation would call for. Crucial working steps may have been omitted, which would result in irreparable shortcomings in use.

3.2 Procedures for Estimating Mold Costs

Mold cost can be computed in two different ways, either on the basis of the data of production planning or based on a forecast procedure.

The first procedure assigns costs to each working step and to the used material. The high accuracy of this procedure is opposed by many disadvantages and difficulties. The method is time consuming and requires from the accountant detailed knowledge of working hours and costs in mold making. Besides this, it can be applied only after the mold design has been finalized.

A basis for estimating the costs of injection molds was worked out by the Fachverband Technische Teile im Gesamtverband Kunststoffverarbeitende Industrie (GKV) (Professional Association Technical Parts in General Association of the Plastics Processing Industry) [3.4]. This should facilitate estimating the costs of molds. It is based on practical experience, e.g. working time for runners (Figure 3.1). If such costs

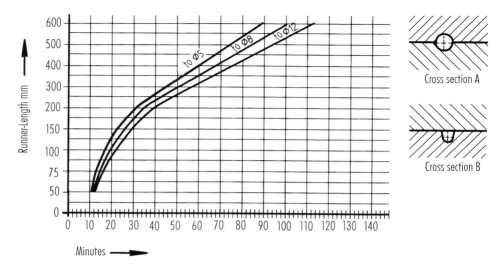

Figure 3.1 Time for machining runners [3.4]. The stated times relate to milling in one platen only (cross section B), excluding set-up time. The set-up time is 30 min for platen up to 100 mm diagonal or in diameter, 35 min for up to 250 mm and 40 min for up to 500 mm

are combined with those of standard mold bases and other standards taken from the catalogues of producers of standards and the costs of outside work and design, then the costs of an injection mold is the result. A form as it is also proposed by GKV is best used for the compilation (Figure 3.2).

In mold-making shops the quotations are generally determined with the help of prognosis procedures. From the literature two general basic methods for predicting costs are known (Figure 3.3) [3.5]: cost function and costs similarity. The first method, the cost function starts with the assumption that there is a dependency between the costs of a mold and its characteristics. This dependency is expressed in a mathematical function. The characteristics are the independent variables or affecting quantities, which determine the costs.

The second method is the costs similarity. Starting with an injection mold to be calculated and its characteristics, another existing mold with similar characteristics is looked for in the shop. The costs for this mold are generally known and can now be used for the new object. In doing so one can fall back on existing data such as the system of classification, which is described in [3.6].

Both methods have their advantages and disadvantages. The cost function provides accurate results only if the affecting quantities have nearly the same effect on costs. This is rarely the case with the variety of injection molds today [3.3].

With the similarity method one can only fall back on molds which are designed in the same way and, thus, have similar cost-effective quantities. To make use of the specific benefits of both methods a combination of them presents itself (Figure 3.4). This can be achieved by grouping similar injection molds or structural components of the same kind together and determining a cost function within each group [3.5].

There is a proposal [3.7], therefore, to divide the total calculation into four cost groups related to their corresponding functions (Figure 3.5).

Machine group	Drilling					Milling					Turning, copying, engraving					Grinding					EDM				Finishing		
	1	2	3	4	5	1	2	3	4	5	1	2	3	4	5	1	2	3	4	5	3	4	5	2	3		

Mold costs (Estimated/Final)

Product _____

Comments _____

Material _____

Mold class semiautomatic ☐ automatic ☐

inj md* ☐ com md*** ☐ inj com md**** ☐   Machine type ☐

Delivery time ____ weeks ____ months

Gating: Hot runner ☐ Sprue ☐ Pinpoint ☐ Tunnel ☐ Edge gate ☐
others

Mold base designation

Material a ____ kg. each ____ $
 b ____ kg. each ____ $
 c ____ kg. each ____ $

Accessories a ____ $
 b ____ $
 c ____ $
 d ____ $
 e ____ $

Outside labor (Hardening, Electrolyt. proc. Surface treatm.) a ____ $
 b ____ $
 c ____ $

Mold making ____ hrs each

Subtotal ____ $

Machine group 1 total ____ hrs each ____ $
Machine group 2 total ____ hrs each ____ $
Machine group 3 total ____ hrs each ____ $
Machine group 4 total ____ hrs each ____ $
Machine group 5 total ____ hrs each ____ $
Additional set-up time ____ hrs each

Total hours ____ hrs

Sum

Additional charges Overhead ____ % Charge for risks ____ % Return ____ % Total ____ %

Costs of sampling

Total costs

Net sales price

Customer

Customer No _____

Drawing No _____

Index _____

Inquiry No ____ of Offer No. of

Mold No _____ Cav

Order No _____

Date _____ Name

Machine group
Pattern
Template
Auxiliary equipment
Hob
Electrodes
Inserts, stationary side
Mold inserts a / b / c
Cavity
Inserts, movable side a / b / c
Cavity
Slides
Interlocks
Ejectors
Latches
Cooling, heating
Installing hydraulics
Installing electrics
Safety mechanism
Standard runner systems
Assembly
Total working hours

Smoothing, planing, polishing, fitting, blasting, general bench work

* = injection mold   ** = compression mold   *** = injection-compression mold   **** = injection-compression mold

Figure 3.2 Blank form for computing mold costs as suggested by IKV (Institute for Plastics Processing) [3.4]

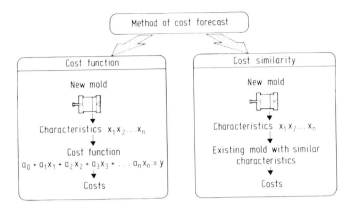

Figure 3.3 Method of cost forecast [3.3]

Costs are determined for each cost group and added to the total costs. The systematic work on the individual groups and the additive structure reduces the risk of a miscalculation and its effect on the total costs [3.7].

In the following the individual groups for estimating costs are presented in detail.

3.3   Cost Group I: Cavity

With cost group I the costs for making the cavity are computed.

They are essentially dependent on the contour of the part, the required precision and the desired surface finish. The costs are determined by the time consumption for making the cavity and the respective hourly wages.

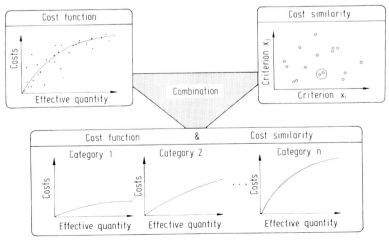

Figure 3.4 Combination of cost function and cost similarity [3.3]

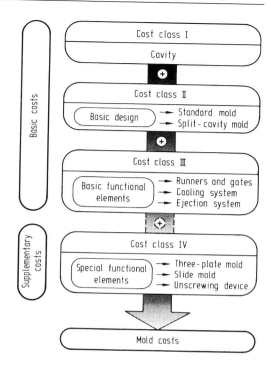

Figure 3.5 Cost classes for estimating mold costs [3.3]

They generally result from

$$C_C = (t_C + t_E) \cdot S_{MW} + C_M \tag{3.1}$$

C_C ($)   Costs for cavity,
t_C (h)   Time spent on cavity,
t_E (h)   Time spent on EDM,
S_{MW} ($/h) Average machine and labor costs,
C_M ($)   Additional material costs (inserts, electrodes, etc.) can often be neglected in face of total costs.

3.3.1   Computation of Working Hours for Cavities

The time t_C required to produce a cavity can be calculated on the basis of statistically or analytically determined parameters.

$$t_C = \{(C_M \cdot (C_D + C_A))\,C_P \cdot C_S + C_C\} \cdot C_T \cdot C_{DD} \cdot C_N \text{ [h]} \tag{3.2} [3.8]$$

C_M Machining procedure,
C_D Depth of cavity,
C_A Surface area of cavity,
C_P Shape of parting line,
C_S Surface quality,
C_C Number of cores,
C_T Tolerances,

C_{DD} Degree of difficulty,
C_N Number of cavities.
C_D, C_A, C_C are real working hours, the remainders are time factors.

The following correlations are used to determine the individual times or time factors.

3.3.2   Time Factor for Machining Procedure

The shares of individual kinds of machining for making the male and the female half of a cavity are identified as a percentage and multiplied by the machining factor f_M (Table 3.1). This factor has been found in practice and is a quantity for the speed differences of various techniques in machining the contours of cavities.

$$C_M = \sum_{i=1}^{n_M} f_{Mi} \cdot a_i \qquad (3.3)$$

Herein is

$$\sum_{i=1}^{n_M} a_i = 1 \qquad (3.4)$$

C_M Time factor machining procedure,
f_{Mi} Machining factor (Table 3.1),
a_i Percentage of the respective machining procedure,
n_M Number of machining procedures.

Table 3.1   Machining factor f_{Mi}

Milling	EDM	Duplicate milling	Turning	Grinding	Manual labor
0.85	1.35	1.0 to 1.35	0.4	0.8 to 1.2	0.8

3.3.3   Machine Time for Cavity Depth

If one looks at a molding above and below a suitably selected parting line, one has to distinguish between elevations (E) and depressions (D).
    The time consumption resulting from the depth of the cavity is determined by the mean of elevations and depressions above (1) and below (2) the parting line. In doing so it seems practical to establish the elevations as their projected area on the plane through the parting line. If the core is not machined from the solid material but made as an insert, the result for one cavity half is

$$C_{D(1)} = \frac{\sum_{i=1}^{n_E} (m_{Ei} + m_{Di}) \cdot f_{EPi}}{m_R \cdot n_E} \qquad \left.\begin{array}{l} \text{Elevation with depression;} \\ \text{core machined} \\ \text{from solid} \end{array}\right\}$$

$$= \frac{\sum_{i=1}^{n_E} m_{Ei} \cdot f_{EPi}}{m_R \cdot n_E} \quad \text{(elevation)}$$

$$= \frac{\sum_{i=1}^{n_D} m_{Di} \cdot f_{DPi}}{m_R \cdot 2n_D} \quad \text{(depression)}$$

Elevation with depression; core made as insert

(3.5)

$C_{D(1)}$ Time consumption for one half of cavity [h]
m_E Height of elevation   [mm]
m_D Depth of depression   [mm]   } of molding
n_E Number of elevations   [–]
n_D Number of depressions   [–]
m_R Averageremoval = [1 mm h^{-1}],
f_{EP} Ratio between area of elevation   } and
f_{DP} Ratio between area of depression } projected area

$C_{D(2)}$ is computed analogously

$$C_D = C_{D(1)} + C_{D(2)} \tag{3.6}$$

C_D Time consumption for depth of cavity [h].

3.3.4 Time Consumption for Cavity Surface

The surface of the cavity or the molding respectively is the second basic quantity after the depth which affects the machine time directly. It is

$$C_S = f_S \cdot A_M^{0.77} \quad [h] \tag{3.7}$$

with score factor share of turning f_S

$$f_S = (1 - 0.5a_T) \cdot 0.79 \quad [h] \tag{3.8}$$

C_S Time consumption for cavity surface [h]
A_M Surface area of molding [mm^2 · E-02]
a_T Turning as share of machining [–]

3.3.5   Time Factor for Parting Line

Steps in the parting line are considered by the time factor C_P.

Table 3.2   Time factor for parting lines C_P

Number of steps	C_P for plane faces	C_P for curved faces
0	1.00	1.10
1	1.05	1.15
2	1.10	1.20
3	1.15	1.25

3.3.6   Time Factor for Surface Quality

The quality of the surface is as important for the appearance of the molding as for its troublefree release. The surface quality factor c_s is affected by the roughness height, which can be achieved with certain machining procedures. It can be taken from Table 3.3.

Table 3.3 Factor for surface quality C_S

Surface quality	Roughness µm	Quality factor C_S	Note
Coarse	$R_a \geq 100$	0.8–1.0	Faces transverse to demolding direction
Standard	$10 \leq R_a < 100$	1.0–1.2	EDM roughness
Fine	$1 \leq R_a < 10$	1.2–1.4	Technically smooth
High grade	$R_a < 1$	1.4–1.6	Superfinish

3.3.7   Machining Time for Fixed Cores

The machining and fitting of cores in both cavity halves is covered by the time factor C_C. This work becomes more difficult with increasing deviation of the fitting area from a circular shape. The contour factor is multiplied by the number of cores with equal fitting area.

$$C_C = \sum_{i=1}^{j} t_B \cdot f_{CF} \cdot n_i \quad [h] \tag{3.9}$$

C_C Machine time for fixed cores [h]
t_B Time base = 1 [h]

f_{CF} Contour factor (Table 3.4) [–]
n   Number of cores with equal fitting area [–]
j   Number of different fitting areas [–]

Table 3.4 Contour factor for cores f_{CF}

Contour factor f_{CF}	Fitting area	
1	Circular	○
2	Angular	□
4	Circular, large	○
8	Angular, large	□
10	Curved contour	⧢

3.3.8 Time Factor for Tolerances

Close tolerances raise costs. To produce moldings economically no closer tolerances should be considered than necessary for the technical function.

A standard for making precision molds implies that mold tolerances should not exceed about 10% of those of the finished molding. The factor for dimensional tolerances C_T comprises the expected expenditure for required accuracy and posttreatment (Figure 3.6).

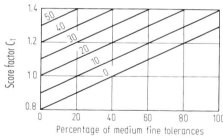

Figure 3.6 Time valuation for tolerance requirements [3.7]

Close tolerances as well as critical tolerances for bearings (centricity, accuracy of angle, parallelism, flatness, freedom from displacement) considerably increase the time needed for producing the cavity.

3.3.9 Time Factor for Degree of Difficulty and Multifariousness

A departure from an average degree of difficulty (C_{DD} = 1) is considered the special effort for an extreme length/diameter ratio of cores, their large number in a small area and complex surfaces. For large plane parts without openings the time factor is reduced (C_{DD} < 1). Table 3.5 shows relevant criteria with their corresponding factor.

Table 3.5 Time factor for degree of difficulty and multifariousness C_{DD}

C_{DD}	Difficulty	Criteria	
0.7	Very simple	**Standard molding**	Large, plane areas, circular parts
0.8	Simple		Rectangular parts, areas with some openings; mount depth/diameter: L/D $\leqq$ 1
1.0	Medium	**Technical molding**	Circular and angular openings, L/D = 1
1.2			Shift possible, L/D $\approx$ 1–5 small parts
1.4	Difficult	**Precision molding**	High density of cores L/D $\approx$ 5, complex surface
1.6	Extremely difficult		Very high density of cores 5 $\leqq$ L/D $\leqq$ 15, complex spherical faces

3.3.10 Time Factor for Number of Cavities

For a larger number of equal cavity inserts or several equal cavities an allowance per cavity has to be considered based on the fabrication in series. The correlation between the time factor C_N and the number of cavities n_c is presented with Figure 3.7.

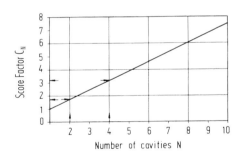

Figure 3.7 Score factor for number of cavities [3.7]

3.3.11 Computation of Working Hours for EDM Electrodes

Because the geometry of the electrode surface corresponds to the contour of the part, the working time can be calculated in the same manner as was done for the cavities (Equation 3.2).

$$t_C = \{(C_M \cdot (C_D + C_A \cdot a_E)) \cdot C_S + C_C\} \cdot C_T \cdot C_{DD} \cdot C_N \quad [h] \tag{3.10}$$

C_M Like 3.3.2,
C_D Like Section 3.3.3 (elevations and depressions have to be exchanged),
C_A Like Section 3.3.4,
a_E Share of EDM for producing cavity,
C_S = 1.3,
C_C Like Section 3.3.7,

C_T Like Section 3.3.8,
C_{DD} Like Section 3.3.9,
C_N Like Section 3.3.10.

3.4   Cost Group II: Basic Molds

The basic mold retains the cavities, the basic functional components (runner, heat-exchange and ejector system) and any necessary special functional elements (three-plate mold, slides, unscrewing unit). As far as self-made basic molds are concerned, it is practical to distinguish different quality grades.

A basic mold of grade I is for a small number of moldings with little precision, for test series etc. It is not hardened.

A basic mold grade II has case-hardened plates, additional alignment, heat insulation on the stationary half and, if of round design, is equipped with three leader pins. It is assumed to produce technical parts and medium-sized quantities.

A basic mold grade III is largely hardened and made with large quantities, high precision and reliability in mind [3.7].

However, injection molds are largely built with mold standards. Thus the total costs of the basic mold are primarily the costs for the readily usable standards, expenses for specific machining operations not included (Figure 3.8). It is suggested, though, to consult the catalogs of suppliers for up-to-date prices and the availability of mold bases of a different design like such with floating plates, etc.

Standard basic molds are all of the highest quality and differ only in the steel grade being employed, which affects the service life of the mold, its polishability or its corrosion resistance. The costs presented in Figure 3.9 are based on the use of AISI 4130 type steel. The bases are supplied preheat-treated and precision ground.

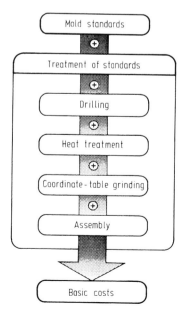

Figure 3.8 Total basic costs [3.3]

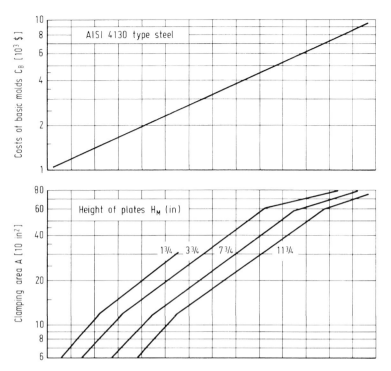

Figure 3.9 Costs of basic molds [3.7]

3.5   Cost Group III: Basic Functional Components

Runner, heat-exchange and ejection systems are basic functional components and therefore by necessity part of every injection mold. If individual elements are related to these basic components, their costs, including additional expenses, can be determined in a more general way. Thus, listing them is all that is needed for the calculation, since their dimensions have only a modest effect. An extensive use of standards is assumed like in cost group II. Figure 3.10 presents the factors of influence for cost group III.

3.5.1   Sprue and Runner System

The type of runner system is determined by economic requirements, part geometry and quality demands.

The costs for sprue gates, disk gates, tunnel gates, and edge gates can be calculated with

$$C_G = t_G \cdot S_{MW}   [\$] \tag{3.11}$$

t_G Machine hours for machining gate, Table 3.6 [h],
S_{MW} Average machine and labor costs [$/h].

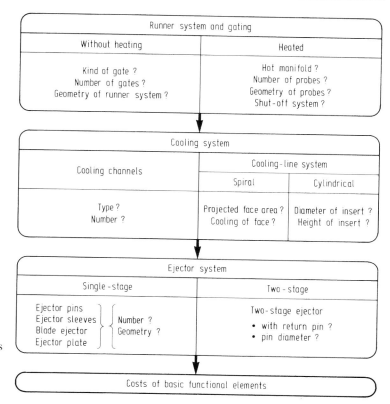

Figure 3.10 Effects on cost class III: Basic functional elements [3.3]

It is assumed that all steps using the same machining operation are made in one pass. Setup time has already been considered in cost group II.

Table 3.6 Working time for the machining of gates [3.4, 3.9]

Type of gate	Time				
Sprue gate	Contained in mold base				
Sprue with n pinpoint gates	n	1	2	3	4
	t (min)	35	50	65	70
Disk gate	30 min				
Tunnel gate	15 min				
Edge gate	$t = (0.35 \cdot b + 50) \cdot i$ b= width of gate (mm) t = (min)				

Edge gate (continued)						
	n	1	2	3	4	5
	i	1	1.4	1.8	2.2	2.5
	n number of gates					

3.5.2 Runner System

Cost for runners are largely determined by their necessary length [3.4]:

$$C_R = g_R \cdot l_R \cdot S_{MW} \ [\$] \tag{3.12}$$

C_R Costs of runners [\$],
S_{MW} Average machine and labor costs [\$/h],
l_R Length of runner [mm],
g_R Correction factor for diameter of runner,
          0.14 min/mm for d_R = 5 mm
          0.16 min/mm for d_R = 8 mm
          0.18 min/mm for d_R = 12 mm,
d_R Diameter of runner [mm].

If runners are machined into both mold plates, the costs can be doubled. The setup time can also be neglected (Section 3.5.1).

3.5.3 Hot-Runner Systems

The total costs of a hot-runner manifold can be determined with Equation (3.13):

$$C_{HR} = \{(C_{BHR} + g_A \cdot A) + n_N \cdot (C_N + 225 \ \$ + C_{NS})\} \cdot g_G \ [\$] \tag{3.13}$$

C_{HR} Total costs of hot runner and assembly [\$],
C_{BHR} basic costs for hot runner [\$],
g_A Area coefficient [8 · E-03 \$/mm²],
A Clamping area,
n_N Number of nozzles [–],
C_N Cost of one nozzle [\$],
C_{NS} Cost of a shut-off nozzle [\$],
g_G 1.1 to 1.2 for molding glass-reinforced plastics,
           1.0 for molding unreinforced plastics.

Hot-runner systems are mainly made with standards today. Therefore basic hot-runner costs and the costs for nozzles can be taken from catalogs of the producers of standards. Added to the basic costs are the costs for floating plates, the manifold (I-, X-or H-shaped), the material for assembly and sealing as well as assembly and machining.

Because of the variety of systems and their related demands, it is not possible to break down all the costs here in detail.

3.5.4 Heat-Exchange System

For a given number of cooling lines the costs of the system C_H can be calculated with

$$C_H = k_D \cdot n \cdot S_{MW} \ [\$] \tag{3.14}$$

k_D Factor for degree of difficulty allows for mold size, shape of channel (Table 3.7),
n Number of cooling lines (without connection).

Table 3.7 Coefficient of difficulty for machining cooling lines

Coefficient k_D	Clamping area A (10^2 cm²)				
	4.00	6.25	9.00	12.25	16.00
Straight bore	0.41	0.45	0.50	0.56	0.60
Oblique bore	0.68	0.75	0.83	0.93	1.00
Inserting helical core or heat pipe	0.81	0.90	0.99	1.11	1.20

3.5.5 Ejector System

The costs for standard parts (ejector pins, ejector sleeves, blade ejectors, return pins, etc.) are easily and best taken from an up-to-date catalog of a suitable supplier of standard mold components.

The costs for holes, for attachment of ejector elements in ejector and ejector retainer plate and adjustment to the geometry of the molding are determined with Equation (3.15):

$$C_{EM} = S_{MW} \sum_{i=1}^{5} \frac{d_i - l_{Gi}}{1850 \frac{mm^2}{h}} + 0.8h \cdot n_i \cdot r_H \quad [\$] \tag{3.15}$$

d   Diameter of ejector element,
l_G Guided length of ejector element,
n   Number,
r_H Difficulties with machining guide holes (partly according to [3.11]),
$r_H = 1$ For ejector pins and shoulder pins,
$r_H = 2$ For sleeve and blade ejectors,
$r_H = 0.2$ for return pins.

3.6   Cost Group IV: Special Functions

Undercuts produced by the runner system or the part itself obstruct demolding. They usually call for a special mold design.

The costs for the occurring special functions such as three-plate molds, slides, unscrewing devices have to be determined and added to the previously computed basic costs. The factors affecting costs in cost group IV are presented in Figure 3.11. Since the units for special functions can again be made with standards, it is possible to compile diagrams as in cost group II, which also consider costs for machining and assembly. They permit the additional cost to be established in relationship with the clamping area.

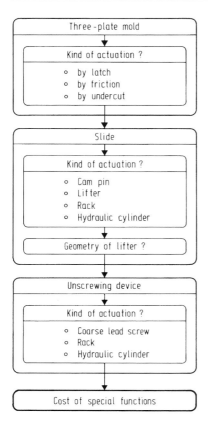

Figure 3.11   Effects on cost class IV: Special functional elements [3.3]

3.7   Other Cost Calculation Methods

Apart from the methods presented here, there are a number of other ways of calculating the costs of injection molds. Important here are primarily those methods which are based on the similarity of molded parts and molds [3.12, 3.14].

3.7.1   Costs Based on Similarity Considerations

Similarity costing is based on the principle that similar molds or similar molded parts cause similar costs. This is a systematic approach to what is actually standard practice. It requires a database containing molds that have already been costed and which are used for performing the new costing.

It must be borne in mind that molds of similar construction are not necessarily used for making similar molded parts. Equally, it is possible for similar molded parts to be made by means of different and thus less similar molds, as is the case for different positions of the parting surfaces. The following basic possibilities exist:

– search for similarly constructed molds,
– search for molds with similar cavity,
– search for similarly constructed molds with similar cavity, and
– search for certain molds, if a similar mold is known.

Easy-to-use search algorithms allow users to search the mold data stored in the computer for a mold whose structure and/or cavity match that of the mold being costed. The existing calculation can then be adapted for the new mold [3.12].

The search for similarly constructed molds may be performed with the aid of the search criteria shown in Figure 3.12. Not all features need to be used during the search, which can be conducted at various levels of detail to suit the case in hand.

To determine molds with similar cavity, resort may be made to the dimensionless Pacyna characteristics, named after Pacyna, who developed them originally for classifying castings. They are used to make a description of the workpiece which focuses on the shape of the workpiece. It involves "absolute criteria in which a cube with the same material volume as the workpiece to be classified serves as a reference body" [3.13]. These three Pacyna characteristics are shown in Figure 3.13. The definition is logarithmic. The various characteristics are [3.12]:

Type of mold
– Normal mold, split mold...

Type of gating
– Hot runner, cold runner...
– Number of gate systems, gates
– Type of gate (pin-point, diaphragm...)

Type of temperature control
– Cooling medium
– Core cooling, cavity cooling, surface cooling

Type of demolding system
– Type of demolding (jaws, splits...)
– Actuating devices (e.g. pneumatics on mold)
– Number of parting lines

Size of mold
– Number of cavities
– Length, breadth, height

Miscellaneous
– Centering
– Assignment to one injection molding machine
– Existing measuring devices
– Handling systems employed

Figure 3.12 Searching for similar mold design

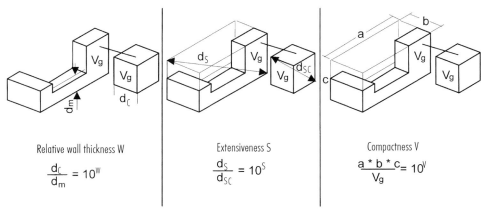

Relative wall thickness W

$$\frac{d_C}{d_m} = 10^W$$

Extensiveness S

$$\frac{d_S}{d_{SC}} = 10^S$$

Compactness V

$$\frac{a * b * c}{V_g} = 10^V$$

Figure 3.13 Pacyna characteristics

Extensiveness, S

This is the ratio of the maximum dimension of the workpiece to the maximum dimensions of the reference cube. The maximum dimension of the workpiece is taken to be the spatial diagonal of the cuboid whose edges match the dimensions of the workpiece in length, breadth and height, or, expressed in other words, the spatial diagonal, d_S, of the smallest possible rectangular packing of the part.

Relative Wall Thickness, W

This parameter is obtained by dividing the edge length, d_C, of the reference cube by the mean wall thickness, d_m, of the workpiece.

Compactness, V

This characterizes the ratio of the packaging volume, V_P, to the workpiece volume, V_g, which by definition matches that of workpiece and reference cube.

Since the Pacyna characteristics are dimensionless, the molded part volume is included as an additional characteristic for characterizing the absolute size of a molded part.

Similarity costing requires the existence of an extensive database built up from similar molds before costing can begin. Furthermore, it is important to carry out selective post-costing of molds that have already been made. For a pure similarity costing, however, a relatively homogeneous product range needs to exist in order that adequate information about similar molds may be made available. If a mold is bigger than or has a different construction than the usual product range, it cannot be costed by this approach alone. For this reason, it is advisable to combine similarity costing with other costing methods.

3.7.2   The Principle behind Hierarchical Similarity Searching

Brunkhorst's [3.14] hierarchical similarity search may be used to apply the similarity search to not only the entire mold but also the various functional elements. The search is first performed for the entire mold. When the mold with the greatest similarity has been found, another similarity search is performed for the various functional groups. This approach allows elements from different molds to be combined and the mold can be made as detailed as desired. Figure 3.14 shows the structure of a mold for this hierarchical similarity search.

Again, hierarchical similarity searching requires the existence of an extensive database.

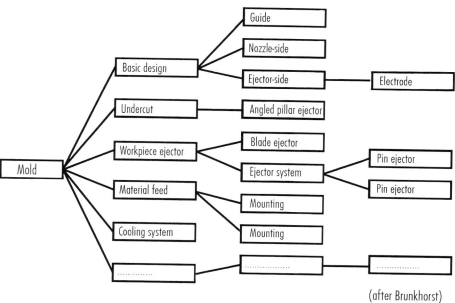

(after Brunkhorst)

Figure 3.14   Search structure for hierarchical similarity searching

References

[3.1]   Schneider, W.: Substitutionssystematik. Unpublished report, IKV, Aachen, 1979.
[3.2]   Proos, G.: Confessions of a Mold Maker. Plast. Eng., 36 (1980), 1, pp. 29–33.
[3.3]   Menges, G.; v. Eysmondt, B.; Bodewig, W.: Sicherheit und Genauigkeit erhöhen. Plastverarbeiter, 38 (1987), 3, pp. 76–80.
[3.4]   Kalkulationsgrundsätze für die Berechnung von Spritzgießwerkzeugen. Published by Fachverband Technische Teile in the Gesamtverband Kunststoffverarbeitende Industrie e.V. (GKV) in cooperation with the study group Werkzeugkalkulation of the Fachverband Technische Teile in the GKV, Frankfurt, 1988.
[3.5]   Evershein, W.; Rothenbücher, J.: Kurzkalkulation von Spannvorrichtungen für die mechanische Fertigung. ZWF, 80 (1985), 6, pp. 266–274.
[3.6]   Walsche, K.; Lowe, P.: Computer-aided Cost Estimation for Injection Moulds. Plast. Rubber Intern., 9 (1984), 4, pp. 30.

[3.7]   Schlüter, H.: Verfahren zur Abschätzung der Werkzeugkosten bei der Konstruktion von Spritzgußteilen. Dissertation, Tech. University, Aachen, 1982.

[3.8]   Formberechnungsbogen, Anlage 1 zu Kalkulations-Grundsätze für die Berechnung von Spritzgießwerkzeugen. Fachverband Technische Teile im Gesamtverband Kunststoffverarbeitende Industrie e.V. (GKV) in cooperation with the study group Werkzeugkalkulation of the Fachverband Technische Teile in the KGV, Frankfurt/M., February 1980.

[3.9]   Informations from misc. mold makers.

[3.10] Catalogs and informations from misc. producers of standards.

[3.11] Ufrecht, M.: Die Werkzeugbelastung beim Überspritzen. Unpublished report, IKV, Aachen, 1978.

[3.12] Grundmann, M.: Entwicklung eines Kalkulationsinstrumentariums für Spritzgießbetriebe auf der Basis von Ähnlichkeitsbetrachtungen, Dissertation, RWTH, Aachen, 1993.

[3.13] Pacyna, H.: Die Klassifikation von Gußstücken, Gießerei-Verlag GmbH, Düsseldorf, 1969.

[3.14] Brunkhorst, U.: Angebots- und Auftragsplanung im Werkzeug- und Formenbau, Arbeitskreis Technische Informationssysteme im Werkzeug und Formenbau, IPH, Hannover, September 21, 1995.

4   The Injection Molding Process

The injection molding process is one of the key production methods for processing plastics. It is used to produce molded parts of almost any complexity that are to be made in medium to large numbers in the same design. There are major restrictions on wall thickness, which generally should not exceed a few millimeters, and on shape – it must be possible to demold the part. This will be discussed later.

The advantages of this process are:

– direct route from raw material to finished part,
– very little finishing, or none at all, of molded parts,
– full automatability,
– high reproducibility,
– low piece costs for large volumes.

4.1   Cycle Sequence in Injection Molding

The molded parts are produced discontinuously in cycles. The sequence of one cycle is shown schematically in Figure 4.1.

The raw material, usually in the form of pellets, is fed into the plasticating unit where it will be melted. The plasticating unit is generally a single-screw extruder in which the screw reciprocates coaxially against a hydraulically actuated cylinder. The continually rotating screw plasticates the pellets to form a melt that is transported forward by the rotation. Because the injection nozzle is still closed during plastication, the melt is pushed to the front of the screw. As a result, the screw is pushed to the right against the resistance of the barrel, which is called the back pressure.

At the start of the cycle, the mold is closed by actuating the press, which on an injection molding machine is called the clamping unit. Before the melt, which is generated in and supplied by the plasticating unit (in a precise, metered quantity), is injected into the closed mold, the plasticating unit traverses against the mold, causing the injection nozzle of the plasticating unit to press against the sprue bushing of the mold. The pressure with which the nozzle is pressed against the sprue bushing must be adjusted in such a way that the joint remains sealed when the melt is injected afterwards. At the same time, the nozzle is opened and the melt can be pushed from the front of the barrel into the cavity of the mold.

As the cavity is filled, pressure builds up inside. This is counteracted by pressing the clamping unit against the mold under as much clamping force as possible to prevent melt from escaping out of the cavity through the mold parting lines.

The connection between the mold and plasticating unit is maintained until the filling process is complete. Generally, however, filling of the cavity does not mean the end of the process because the melt changes its volume on solidifying (freezing). In order that

Stage 1: Injection

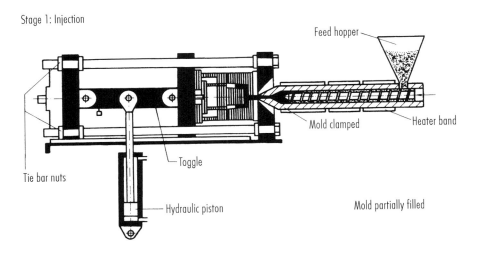

Stage 2: Holding pressure and plastication

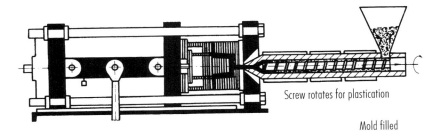

Stage 3: Ejection

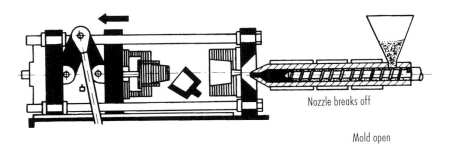

Figure 4.1   The injection molding cycle [4.1–4.3]

either more melt may be forced in to make up the difference in volume or to prevent the melt from running out of the mold, the connection must be maintained until the melt has frozen in the gate. The connection is broken by screwing back the plasticating unit, and closing the injection nozzle. Detaching of the nozzle causes thermal isolation between mold and plasticating unit because these are at totally different temperatures.

Since the plasticating process requires a certain amount of time, as soon as the nozzle is detached and closed, the plasticating unit usually starts rotating, drawing in – metering – more material, melting it and moving it to the front.

When the molding (molded part) has solidified to the extent that it can retain its shape without external support, the clamping unit opens the mold and the molding is pushed out of the cavity by ejectors.

The cycle then repeats. Figure 4.1 shows the order in which the processes occur. The basic cycle described here may vary for other materials and processes.

4.1.1   Injection Molding of Thermoplastics

When thermoplastics are heated, they experience a change of state; they turn soft and melt, becoming flowable. When cooled down, they solidify again. This is the reason that plasticating units are operated hot and molds are operated cold when working with thermoplastics. Generally, the temperature difference is more than 100 °C. The thermoplastic materials developed for injection molding generally constitute relatively low-viscosity melts with the result that injection times are short and low clamping forces are needed.

The injection mold should remove the heat from the material fast and steadily. Therefore, the cooling system has to be carefully designed. The coolant – usually water, provided the mold temperature is below 100 °C – flows through channels around the cavity. For reasons of economics, such as the quality of the molded parts, which depends heavily on uniform heat flow in the mold, the cooling circuit is monitored very precisely and cooling equipment is used to ensure that the coolant is always at the same temperature.

Molded parts requiring no machining can only be produced if all joints and mold parting lines are so well sealed that melt is unable to penetrate and harden there. Otherwise, flash would be formed and machining become necessary. To this end, all joint gaps must remain smaller than 0.03 mm even during full injection pressure, until the melt has solidified. These requirements are particularly demanding where large molded parts and large injection molding machines are involved as the molds must be extremely rigid and the clamping units must function very precisely; the rigidity of the clamp plates is particularly critical.

4.1.2   Injection Molding of Crosslinkable Plastics

These plastics only attain their final molecular structure by crosslinking under heat. For this reason, they must be kept at as low a warm temperature as possible in the plasticating unit, i.e. the viscosity must be just low enough that the cavity is filled. This prevents premature crosslinking from interfering with complete formation of the molded part. The plasticating unit is therefore usually kept below 100 °C and frictional heat is minimized through the use of compressionless screws.

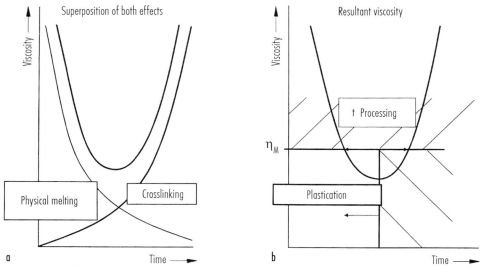

Figure 4.2   Viscosity function of cross-linking molding compounds – processing limits

The mold, on the other hand, is at such a high temperature that a reaction, and therefore crosslinking, occurs rapidly. There is a limit on the upper temperature because no thermal damage may be done to the surface of the molded parts.

Figure 4.2 shows the change in viscosity of such plastic materials and which cycles they occur in.

4.1.2.1   Injection Molding of Elastomers

Elastomeric materials, such as rubber, have virtually the same molecular structure when supplied as when they are in their final state. The only effect of heat is to generate a wide-meshed lattice in which adjacent molecules are chemically bound to each other. Consequently, the change in volume that accompanies crosslinking is slight.

To prevent elastomers from crosslinking before entering the mold, the plasticating units are generally kept at below 100 °C.

A number of modern, synthetic elastomers are an exception here, however, e.g. liquid rubbers.

Because the elastomer is heated up by more than 60 °C in the cavity, its volume increases despite the fact that crosslinking is occurring at the same time; the result is that high cavity pressures are generated. Since, on contact with the hot wall of the cavity, the materials undergo a decrease in viscosity before they crosslink, the gaps of the parting lines must be smaller than 0.02 mm if flash is to be avoided. This is generally beyond the realms of possibility, especially with large molds, and so flash is often unavoidable.

4.1.2.2   Injection Molding of Thermosets

These materials are supplied in a low-molecular state for injection molding. They are, in addition, mostly filled with mineral or wood powder, or fibers and have a relatively high viscosity in the injection unit at the low temperatures permitted there (< 120 °C).

Here, too, the temperature of the molds is about 100 °C higher than that of the plasticating unit. Narrow-meshed crosslinking results in rapid solidification. The crosslinking process releases heat that has to be dissipated. These materials become particularly fluid when they come into contact with the hot cavity wall. Therefore, gaps along the parting lines have to be less than 0.15 mm wide if flash is to be avoided.

4.2 Terms Used in Connection with Injection Molds

The terminology used in this book corresponds largely to that shown in Figure 4.3. These terms are established in practice. There also exists an ISTA booklet (International Special Tooling Association) which deals with the terminology of components of injection molds.

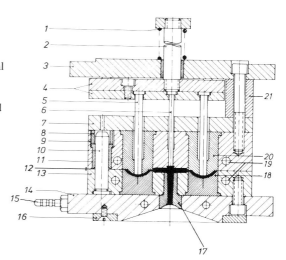

Figure 4.3 Designations for components of an injection mold (typical European design) [4.4]
1 Compression spring, 2 Ejector bolt, 3 Movable clamping plate, 4 Ejector and ejector retainer plates, 5 Ejector pin, 6 Central sprue ejector, 7 Support plate, 8 Straight bushing, 9 Cavity retainer plate, 10 Leader pin, 11 Shoulder bushing, 12 Parting line, 13 Cavity retainer plate, 14 Stationary clamping plate, 15 Plug for cooling line connection, 16 Locating ring, 17 Sprue bushing, 18 Cavity insert, 19 Cooling line, 20 Cavity insert, 21 Support pillar

4.3 Classification of Molds

Depending on the material to be processed one frequently talks about

– injection molds (for thermoplastics),
– thermoset molds,
– elastomer (rubber) molds,
– structural-foam molds.

Because all these molds are not basically dissimilar, different criteria will be used here for classifying molds. They are based on different functions.

4.4   Functions of the Injection Mold

For the production of more or less complicated parts (moldings) in one cycle, a mold containing one or several cavities is needed. The mold has to be made individually in each case. The basic tasks of a mold are accommodation and distribution of the melt, shaping and cooling of the material (or adding activating heat for thermosets and elastomers), solidification of the melt, and ejection of the molding. All these tasks of a mold can be accomplished with the following functional systems:

– sprue and runner system,
– cavity (venting),
– heat exchange system,
– ejection system,
– guiding and locating system,
– machine platen mounts,
– accommodation of forces,
– transmission of motions.

Figure 4.4 demonstrates these functions with a simple mold for a tumbler.

Besides forming the part, the mold has another important function; demolding the part. From an economic viewpoint the cycle should be as short as possible, but from the aspect of quality, ejection, especially of complex moldings, has to be reliable without damage to either part or cavity.

The design of an ejection system depends on the configuration of the molding [4.7]; one distinguishes parts

– without undercuts,
– with external undercuts,
– with internal undercuts.

A number of design possibilities arise from this distinction as well as another important classification. From the fact that moldings can be pushed out, stripped off, unscrewed, torn off, cut off, one can recognize the demand for a classification with respect to the

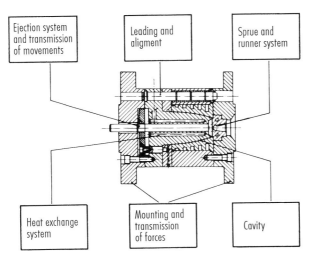

Ejection system and transmission of movements

Leading and aligment

Sprue and runner system

Heat exchange system

Mounting and transmission of forces

Cavity

Figure 4.4   Breakdown of the functions of an injection mold [4.5]

demolding system. This classification is justified because it immediately allows the necessary amount of work to be recognized, which affects costs. It also indicates the feasible size and number of cavities as a result of space requirements.

4.4.1   Criteria for Classification of Molds

The previously itemized groups of functions can be classified according to mold design and the characteristics of the molded parts (Table 4.1). The characteristics of the parts

Table 4.1   Design characteristics (Characteristics depending upon design and characteristics determining design) [4.5]

Characteristics dependent upon design	Characteristics dependent upon molding
Transmission of motion Ejection system (partly) Number of parting lines Number of floating plates Alignment Transmission of forces Mounting to machine platen	Cavity Cavity layout Sprue and runner system (partly) Heat-exchange system Slides and lifters Ejection system (partly)

Table 4.2   Distinction of molds according to primary design features [4.5]

Distinction according to	Influencing factors	Design version	Mold designation
Number of parting lines	Geometry of molding Number of cavities Type of gating Ejection principle	Two-plate mold Three-plate mold Stripper plate (two parting lines)	Standard mold Mold designed for tearing off molding Stripper mold Stack mold
Ejection system	Shape of molding Plastic material Processing parameters Lot size Position of molding relative to parting line	Slides Split cavity Unscrewing device Stripper plate	Slide mold Split-cavity mold Unscrewing mold Stripper mold
Heat-exchange system	Injection molding machine Cycle time Plastic material Economics	Hot manifold Insulating runner	Hot-runner mold Insulated-runner mold
Transmission of forces	Rigidity of mold Geometry of molding Injection pressure (spec.) Plastic material	Split cavity Interlock machined out of the solid material Leader pins	Split-cavity mold Standard mold

can vary within one group of mold types; design features are invariable within one group and therefore of general validity for one and the same type.

Another distinction according to primary design features is represented in Table 4.2. This demonstrates how mold types may result from different design criteria and their associated effects.

Designations of molds are not always uniform in literature and common use. They are mostly based on specific components or demolding functions, or indicate the potential for a particular application. Table 4.3 lists criteria leading to mold designations.

Table 4.3 Criteria leading to a characteristic mold designation [4.5]

Designation	Criteria
1. Standard mold	Simplest design ("standard"): one parting line; one-directional opening motion, demolding primarily by gravity, with ejector pins or sleeves
2. Slide mold	One parting line; opening motion in main direction and transverse with slide actuated by cam pin
3. Stripper mold	Similar to 1., but demolding with stripper plate
4. Mold designed for cutting off molding	Similar to 1., but separation of runner and molding by cutting with additional plate moving transverse (like 3.)
5. Split-cavity mold	One parting line; opening motion in main direction and transverse; cavity halves slide on inclined planes and can withstand lateral forces
6. Unscrewing mold	Rotational motion for automatically demolding a thread is mechanically actuated
7. Mold designed for tearing off molding	Two parting lines for demolding runner and molding separately after they have been torn apart; one-directional opening motion in two stages
8. Stack mold	Cavity plates stacked with several parting lines
9. Insulated-runner mold	Two parting lines; no conventional runner system but channels with enlarged cross section permitting formation of a hot core insulated by a surrounding frozen skin
10. Hot-runner mold	Runner is located in an electrically heated manifold
11. Special molds	Combinations of 2. to 10. for moldings with special requirements which do not permit a simple solution

A classification of molds with regard to the demolding system results in the basic mold types shown in Figures 4.5 and 4.6. Molds with a relatively complex design such as cut-off, stack, hot-runner, insulating-runner and other special molds can be integrated into this system. Besides this, a statistical analysis [4.8] has demonstrated that predominantly "simpler" molds are presently in use.

Figures 4.5 and 4.6 clearly summarize what has been described so far. The basic categories are presented in the following sequence:

– schematic diagram,

- major components,
- characteristics,
- moldings,
- opening path,
- example.

Standard mold	Mold with stripper plate	Slide molds	
MS PL SS	MS PL SS	MS PL SS	Schematic diagram
a Clamping plate MS b Ejector system c Cavity d Sprue e Clamping plate SS	a Clamping plate MS b Stripper plate c Cavity c Sprue e Clamping plate SS	a Ejection system b Cam pin c Cavity d Slide e Sprue	Major components
Most simple sesign; Two mold halves; One parting line; Opening in one direction; Demolding by gravity, ejector pins or sleeve	Design similar to standard mold but with stripper plate for ejection	Design similar to standard mold but with slides and cam pins for additional lateral movement	Characteristic
For all kinds of moldings without undercut	For cup like shaped moldings without undercut	For parts with under-cuts or external threads	Moldings
①	①	② ③ ①	Opening path
			Example

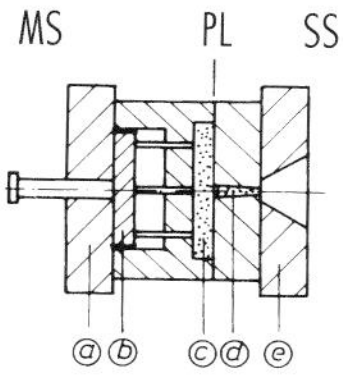

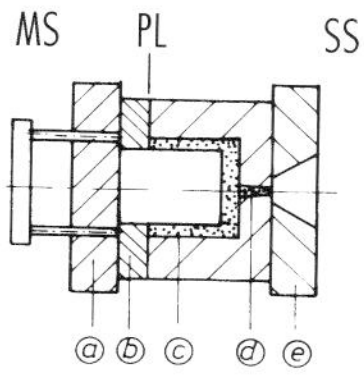

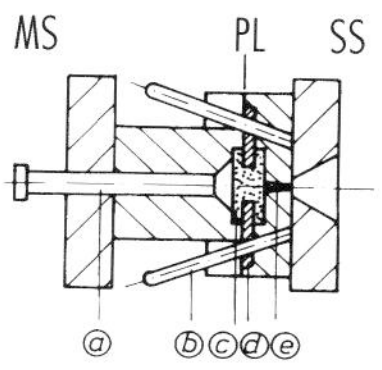

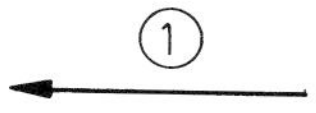

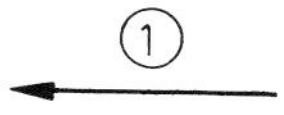

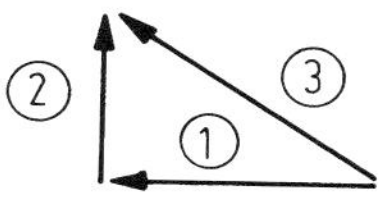

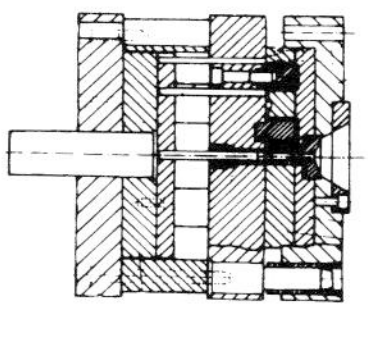

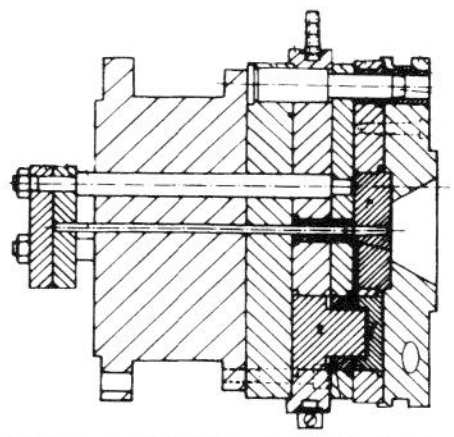

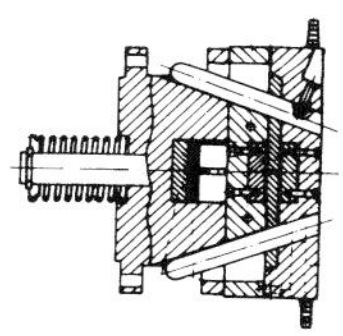

Figure 4.5 Basic categories of injection molds [4.5]
MS = movable side, SS = stationary side, PL = parting line

Split cavity mold	Mold with unscrewing device	Three-plate mold	
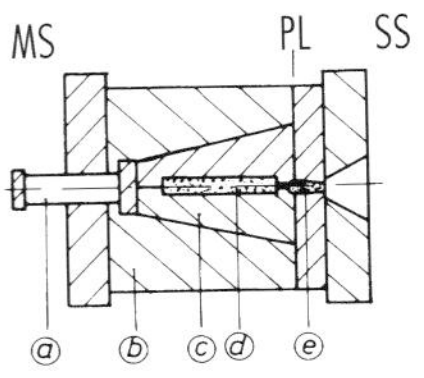	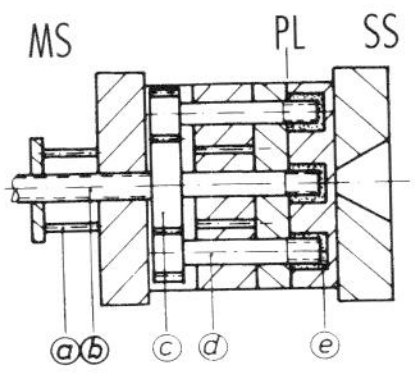	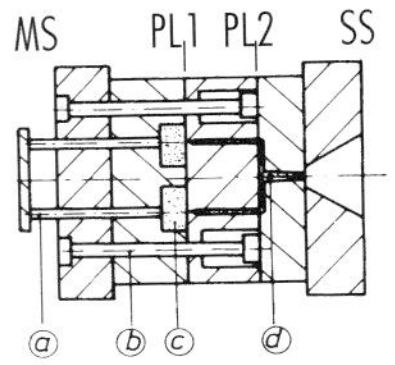	Schematic diagram
a Ejector system b Retainer block c Split cavity block d Cavity e Sprue	a Ejector system b Lead screw c Gear d Core e Cavity	a Ejection system b Stripper bolt c Cavity d Slide e Sprue and runner	Major components
Design similar to standard mold but with split cavity block for moldings with undercuts or external threads	Thread-forming core is rotated by built-in and mechanically actuated drive	Two parting lines; Movement of floating plate actuated by latch or stripper bolt; Two-step opening movement	Characteristic
For oblong or wide moldings with undercuts or threads	For moldings with internal or external threads	Automatic separation of molding and runner	Moldings
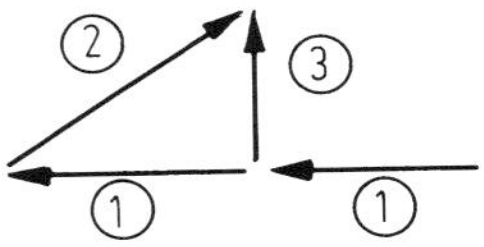	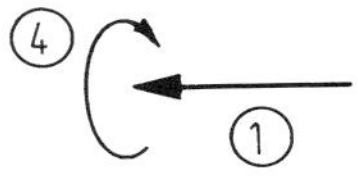		Opening path
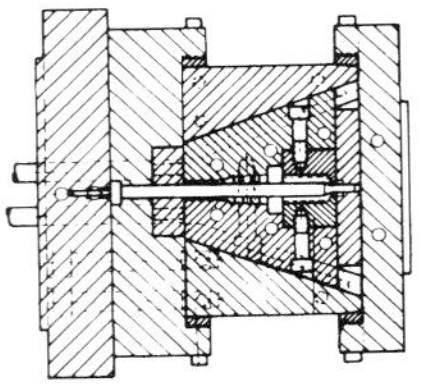	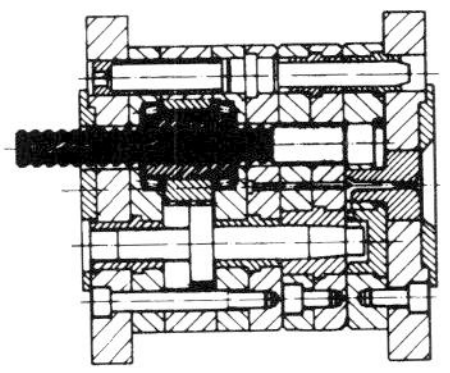	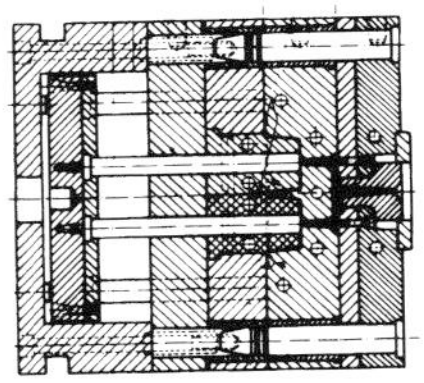	Example

Figure 4.6   Basic categories of injection molds [4.5]
MS = movable side, SS = stationary side, PL = parting line

The schematic presentation should illustrate the principle of each group.

The row "moldings" (Figures 4.5 and 4.6) provides only an indication of the possibilities provided with such molds. The design examples are taken from the references [4.9, 4.10].

The numbers in row "opening path" refer to the sequence and directions of motions and stand for:

1. Main opening movement: guiding motion.
2. Movement between guide and slide: relative motion.
3. Movement of slide during demolding: absolute motion.
4. Movement of unscrewing core: relative rotation.

4.4.2 Basic Procedure for Mold Design

It is advisable to proceed with any mold design systematically because a mold and its operation have to meet a variety of conditions. Figure 4.7 demonstrates how interrelated the conditions are and which boundary and secondary conditions have to be met by the main function. This statement becomes even more evident with an example. The path of decisions to be made by the designer is exemplified by a flow chart for the design of a standard mold for producing several covers simultaneously (Figure 4.8a–h). It is suggested that this path be traced step by step to get a feel for the logic of the procedure.

4.4.3 Determination of Mold Size

The size of a mold depends primarily on the size of the machine. Frequently an existing machine or a certain machine size poses an important limitation, to which the design engineer has to submit.

Such limitations are

- shot size, the amount of melt that can be conveyed into the mold with one stroke of the screw or the plunger,
- plasticating rate, the amount of plasticated material the machine can provide per unit time,
- clamping force, which has to compensate the reactive force from maximum internal cavity pressure,
- maximum area of machine platen given by the distances between tie bars (Figure 4.18) maximum injection pressure.

4.4.3.1 Maximum Number of Cavities

At first the maximum theoretical number of cavities is calculated [4.4]

$$N_1 = \frac{\text{max. shot size } S_v \text{ in cm}^3}{\text{volume of part and runner } M_v \text{ in cm}^3} \tag{4.1}$$

This computation assumes utilization of the whole maximum shot size of the machine computed from screw diameter and displacement. It is not a wise practice, however, for reasons of quality (uniform melt, adequate cushion for holding pressure) to select the maximum quantity.

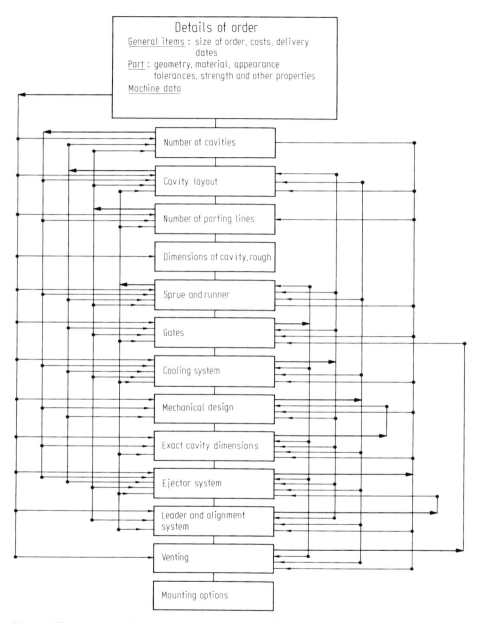

Figure 4.7   Algorithm for a mold-design procedure [4.11]

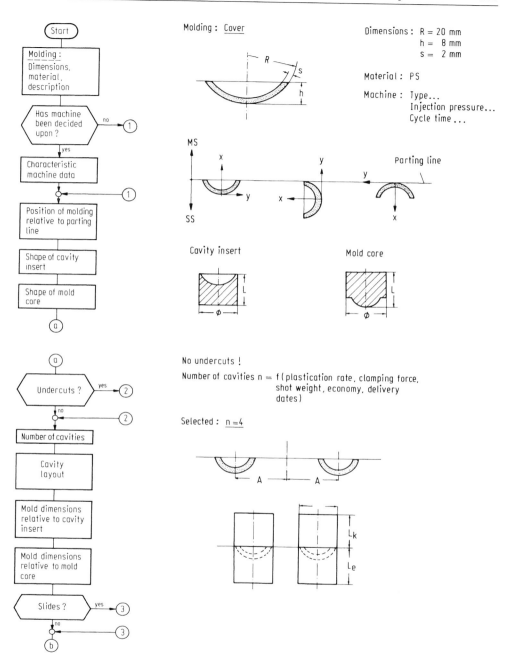

Figure 4.8a–b   Design example: Standard mold [4.5]
MS = movable side, SS = stationary side
(Continued on next page)

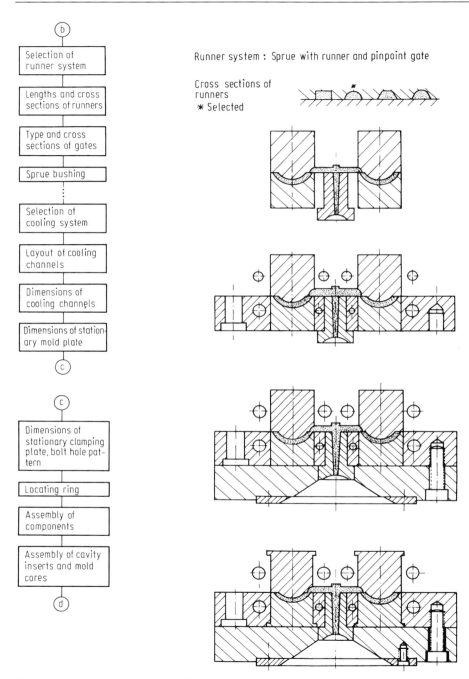

Runner system : Sprue with runner and pinpoint gate

Cross sections of runners
* Selected

Figure 4.8b–d   Design example: Standard mold (continued)
(Continued on next page)

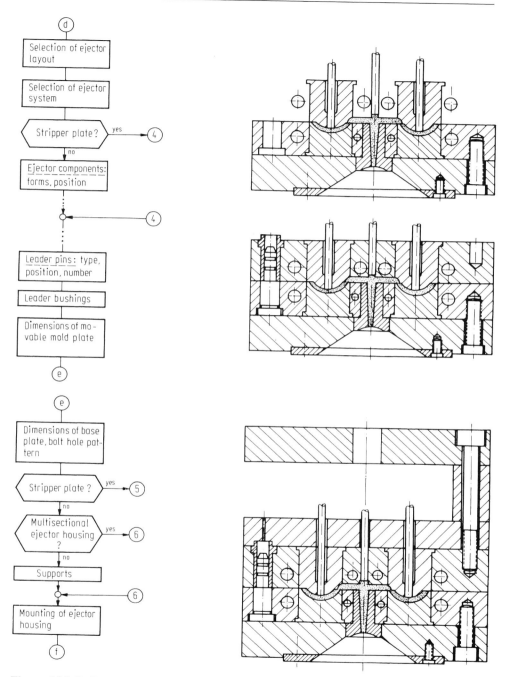

Figure 4.8d–f   Design example: Standard mold (continued)
(Continued on next page)

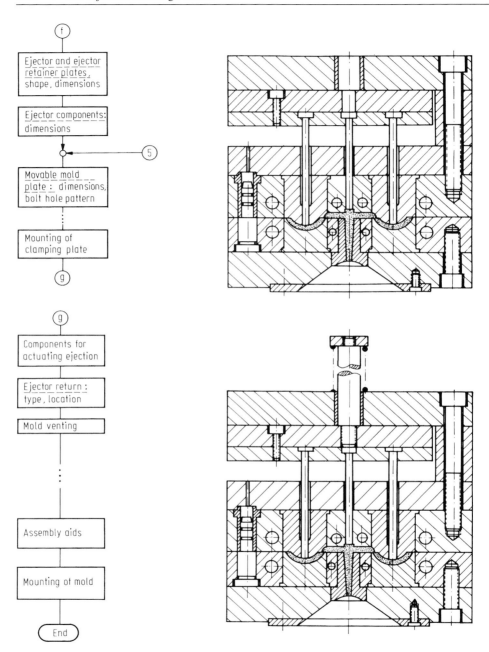

Figure 4.8f–g   Design example: Standard mold (continued)

The number of cavities for thin-walled parts can furthermore be determined by the plasticating rate of the machine

$$N_2 = \frac{\text{plasticating rate R in cm}^3/\text{min}}{\text{number of shots Z/min} \cdot (\text{part} + \text{runner volume in cm}^3)} \qquad (4.1a)$$

Modern reciprocating-screw injection molding machines have such a high plasticating rate that the number of cavities N_2 should only be checked for thin-walled parts with large shot size. An empirical rule states

$$0.4 \, N_1 \leqq N_2 \leqq 0.8 \, N_1 \qquad (4.2)$$

4.4.3.2 Clamping Force

The minimum clamping force is derived from the reactive force of the cavity, which results from the projected area of all cavities and runners and the maximum cavity pressure:

$$F = A \cdot p \qquad (4.3)$$

Herein F is the reactive force, A the projected areas of cavities and runner system and p the cavity pressure. Depending on material and part the cavity pressure is between 20 and 100 MPa, proper processing assumed. Faulty operation can rise this pressure up to the full injection pressure. It is advisable, therefore, to calculate with the maximum injection pressure of the machine and the total projected area that can be covered with melt.

$$F_{max} = A_{max} \cdot p_{inj} < F_{clamp} \qquad (4.4)$$

4.4.3.3 Maximum Clamping Area

This area is determined by the distances between the tie bars (Figure 4.18). Generally one avoids the additional trouble of pulling tie bars. Therefore between the largest mold dimension should be about 10 mm smaller than the distance the corresponding tie bars. Clamping units are built to withstand the maximum cavity pressure that can be expected. Machines for processing foam with low pressure can have light-duty clamping units or larger clamping platens and wider distances between tie bars. Care should be taken that the plates do not bend more under loads than several micrometers. Otherwise the admissible gap width of the parting line cannot be maintained even if the molds themselves are sufficiently rigid. In this respect today's machinery is frequently undersized.

4.4.3.4 Required Opening Stroke

The opening stroke has to be adequately long to permit troublefree ejection from molds with very long cores (example: mold for bucket). Minimum requirement is a stroke of more than twice the length of the core.

On the other hand, a stroke that is longer than needed uses up cycle time, which has to be kept short for reasons of costs.

The opening stroke can certainly be adjusted but the investment for a more than normal stroke is high. Therefore, one has proposed [4.12] to tilt the mold during the opening stroke with an auxiliary equipment and to demold then (Figure 4.9).

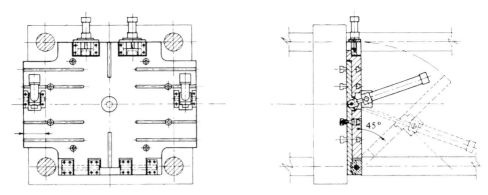

Figure 4.9   Device for tilting the mold during demolding [4.12]

4.4.4   The Flow Length/Wall Thickness Ratio

Another criterion pertaining to the machine is the ratio between flow length and wall thickness. According to Hagen-Poiseuille's law the ratio between flow length L and the square of the wall thickness of the molding H^2 is determined by the injection pressure p_{inj}, a quantity of the machine, and the viscosity of the melt if the velocity of the melt flow is given.

For thermoplastics there are certain optimum values for the velocity [4.13], which are determined by the orientation, to which the molecules are subjected. They are around $v_{inj} \sim 30$ cm/s.

Mostly, however, empirical data are used, which are provided by the raw-material suppliers for their products in the form of flow-length/wall-thickness diagrams (Figure 4.10). The data for this presentation have been established for each material by experiment.

They present common data for the fabrication of a molding characterized by the maximum flow length of melt in the cavity and the related (thinnest) section thickness and are purely empirical. The flow length/wall thickness ratio derived from Hagen-Poiseuille's law is in accordance with the similarity principle:

$$\frac{L}{H^2} = \frac{\Delta p}{32 \cdot \varphi \cdot \bar{v}_F \cdot \eta_{a\ eff}} \tag{4.5}$$

wherein
L   Flow length,
$H = 2\ W \cdot T/(W + T)$ the hydraulic radius with width (W) and thickness (T),
φ 1.5 for width much larger than thickness (almost always the case),
$\bar{v}_F$ Velocity of the flow front. Qualitatively preferable value: ca. 30 cm/sec,
Δp Maximum injection pressure; for common machines: ca. 120 MPa.

For estimates, the "apparent effective viscosity" is

for amorphous materials:
$\eta_{a\ eff} = 250$ to 270 Pa $\cdot$ s (max. error + 10%, with a melt temperature $T_M = T_E$ (freeze temperature) + 150 °C).

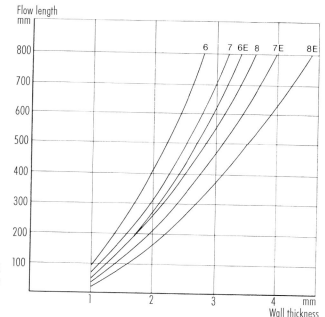

Figure 4.10 Relationship between wall thickness and flow length for a number of PMMA molding resins (Degussa Corporation) [4.14] They meet the requirements of DIN 7745.

There are two series of grades: Standard grades 6, 7 and 8, and E-grades with higher molecular weight.

for crystalline materials:

$\eta_{a\ eff}$ = 170 Pa · s (max. error ±5%, with a melt temperature $T_M = T_E$ (freeze temperature) + 250 to 375 °C).

With the flow length for common molds and machines (120 MPa injection pressure) can be estimated

for amorphous materials

$$L = \frac{10^6 \cdot H^2}{12 \cdot 260} = 325 \cdot H^2 \ (cm) \tag{4.6}$$

for crystalline materials

$$L = \frac{10^6 \cdot H^2}{12 \cdot 170} = 500 \cdot H^2 \ (cm) \tag{4.7}$$

(L and H are in cm)

If the material is filled with short glass fibers or powder, 30% by volume, one can assume a reduction factor of 2, that is the flow length is reduced to one half.

4.4.5 Computation of Number of Cavities

The first design step is to determine the number of cavities. Technical (in form of the available machine equipment and the required quality and costs) as well as economic (in form of the delivery date) criteria are considered. To facilitate this multi-level decision a flow chart (Figure 4.11) is proposed, which exemplifies a suitable procedure [4.6, 4.16].

Beginning with the assumption that the cost price of an article is closely related to the method by which it is manufactured, one can conversely conclude that cost calculation

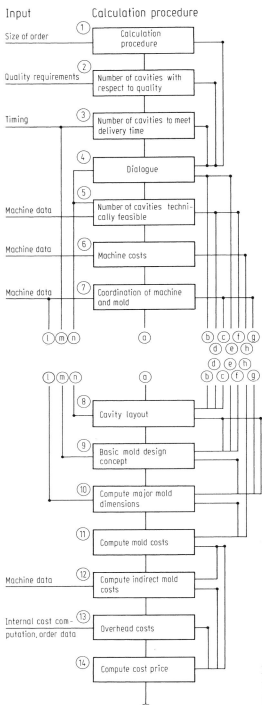

Input   Calculation procedure

Figure 4.11   Algorithm for establishing
the optimum number of cavities [4.16]
(Applied expressions are explained in the
text)

should already start at the design stage, especially in mold design, in order to find an optimal solution. It is practical to break down the whole procedure into several known partial or elementary functions [4.17], which can replace the overall function with respect to cause and effect. Then these partial functions have to be evaluated with the help of a value analysis. An important point is the source of costs. Figure 4.12 shows the costs to be considered here allocated to cost groups.

Summation of all individual costs or cost groups as demonstrated in Figure 4.12 results in the cost price. The trend of these costs is schematically shown in Figure 4.13 [4.3, 4.18–4.21].

Since it is not known in advance which combination of numbers of cavities, mold system and injection molding machine results in the lowest costs, these three quantities have to be varied within certain limits, which have to be determined.

First the number of cavities has to be narrowed down. This is done with the first 5 steps of the algorithm in Figure 4.11, which because of their significance will be discussed in more detail later. If there is no information concerning quality demands (step 2) or delivery time (step 3), one can commence with a practically sound number of cavities in step 1 [4.20], which only depends on the order size (Figure 4.14). One can see that the curve does not begin below n = 10,000 parts. In fact, smaller numbers are not profitable because the expenses for the mold would result in a high amortization charge. This charge increases with decreasing lot size and can render the article uneconomic.

According to this experience [4.20], a single cavity mold is the most economic solution for lot sizes up to 100,000 parts as long as the required delivery date does not command otherwise. If best quality and positive availability is kept in mind, then one can generally only agree with this rule. This diagram [4.20] counts even more, the more problematic processing of a material becomes.

Group of costs	Costs
Machine costs	Depreciation and amortization Interest Maintenance Building Wages Energy Cooling water
Cost related to product	Material Postoperation
Mold costs	Mold material Machining Design Outside labor
Indirect mold costs	Setup Sampling Maintenance
Overhead	Production Mold making Design Material Administration and sales

Figure 4.12 Summary of costs – optimum number of cavities [4.16]

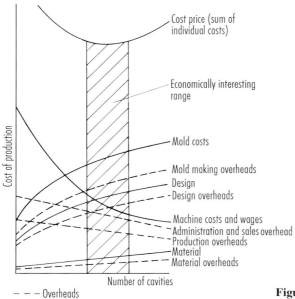

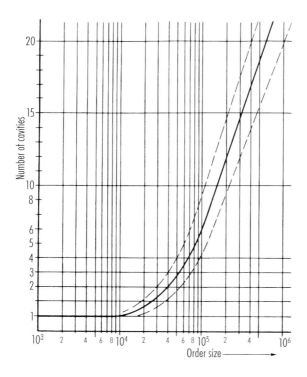

Figure 4.13   Costs related to number of cavities [4.16]

Figure 4.14   Correlation between number of cavities and order size [4.20]

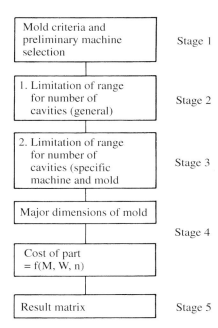

Mold criteria and preliminary machine selection	Stage 1
1. Limitation of range for number of cavities (general)	Stage 2
2. Limitation of range for number of cavities (specific machine and mold	Stage 3
Major dimensions of mold	Stage 4
Cost of part = f(M, W, n)	
Result matrix	Stage 5

Figure 4.15 Algorithm for determining the optimum number of cavities (Combination: Mold-Machine) – abridged [4.6]

If there are additional requirements, a further examination is worthwhile. A proposal from [4.6] is presented with Figure 4.15, which can be followed.

4.4.5.1 Algorithm for the Determination of the Technically and Economically Optimum Number of Cavities

If the optimum number of cavities is determined from a technical and economic point of view, the most favorable combination of mold and machine is found at the same time, because a realistic definition of this number can only be made with regard to the characteristics of the molding machine and a suitable mold principle for the part in mind [4.16].

Starting with an inquiry for the production of injection molded parts, which contains in essence

– part geometry,
– molding material,
– demands on the part,
– lot size,
– delivery date.

At least the following parameters have to be laid down by the molder and mold maker in their quotations:

– number of cavities n,
– mold principle W,
– major mold dimensions,
– number m and types M of molding machines,
– mold costs,
– costs of parts S (n, W, m, M).

The listed parameters can only be established together because they are mutually interdependent. For instance, there is a connection between the number of cavities n and the number m as well as the type of molding machine M. This dependency is a result from the lot size and the delivery date on one side and the technical necessities of processing (plasticating rate, shot size, etc.) and the machine data on the other side.

Depending on the kind of gating and its location the mold principle also depends on the number of cavities, for instance with a change from a single-cavity to a two-cavity mold. The major mold dimensions depend on the number of cavities, the mold concept and the machine type. Conversely there may be a dependency of the mold concept on the major mold dimensions, if e.g., because of high lateral opening forces, a thick leader pin for a slide mold should become necessary, which cannot be realized any more after dimensioning. Thus, the use of a split-cavity mold would make more sense.

Mold and part costs are directly or indirectly dependent on the remaining parameters laid down in the quotation.

In order to find the most favorable number of cavities (mold-machine combination) with justifiable effort, the following course of action is proposed. Figure 4.15 shows an abridged version of the flow chart of an algorithm, with which one can determine the technically and economically best combination of mold and machine for the part to be produced. In step 1, the part is analyzed and all feasible mold conceptions are procured. Furthermore a preliminary selection of machines is made, that is those molding machines are considered among which the finally selected one will be found. Subsequently, in step 2, the first limitation of the scope of cavity numbers is accomplished based on criteria, which depend on part data in the first place. After this, in step 3, further narrowing down of the number of cavities follows after a review of the essential technical criteria. In step 4, the part costs can be calculated after the major mold dimensions have been computed. In doing this the number of cavities is modified for a certain machine and certain mold and after this, machine and mold are likewise modified. By this variation, in step 5, matrix of results is obtained for the part costs. It should make visible the most economical, although not necessarily the technically best combination of machine, mold principle, and number of cavities.

Figure 4.16 presents the flow chart of the algorithm in more detail. In step 1, all feasible mold design modes are determined according to an analysis of the part (Section 4.4.1 to 4.4.3). The most important angles for this are the creation of the system and location of gating as well as the way of demolding. It is the goal of this stage to specify all feasible alternative mold principles. Restrictions to the number of cavities for design reasons have to be noted for the respective principle.

Subsequently, in step 2, those machines are singled out from the whole machine equipment, from which one is later selected to do the job. Narrowing down the whole spectrum of machinery to those machines which may be considered, is done by schedule or from experience. This considerably reduces the total effort in finding the most favorable combination of machine and mold in the following steps.

In step 3, practical number of cavities is established from experience with similar parts.

To increase the significance of such a statement it makes sense to carry out an analysis concerning the dependence of the number of cavities on the lot size for a certain family of moldings subdivided in to groups of similar size as the molding. Such data can be entered in diagrams and, thus, expressed mathematically. With this, a preliminary estimate of the number of cavities is available. Step 4 deals with the qualitative number of cavities. With more than one cavity in a mold there are no suitable processing

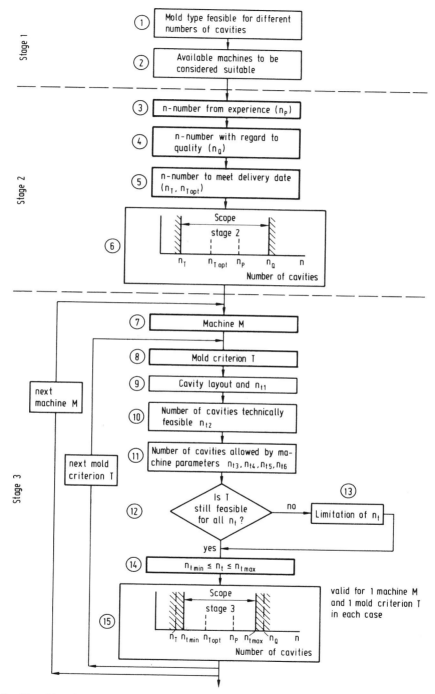

Figure 4.16   Algorithm for determining the optimum number of cavities (Combination: Mold-Machine) [4.6] (continued on next page)

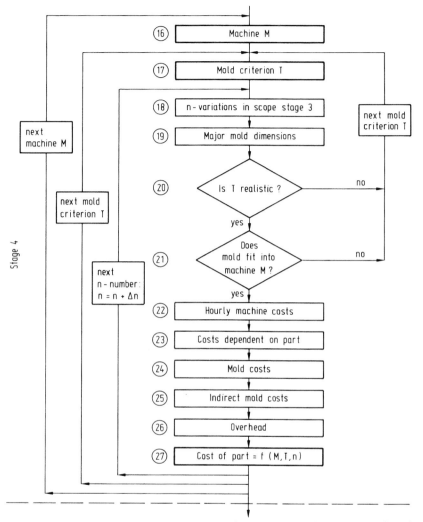

Figure 4.16   Algorithm for determining the optimum number of cavities (continued)
(continued on next page)

conditions, strictly speaking, that would allow all parts to be molded to perfection at the
same time. This reveals a connection between the required quality of a molding and the
feasible maximum number of cavities. Even for larger lot sizes a decision has to be
eventually made in favor of a smaller number of cavities thans technically feasible to
obtain high-quality parts. In [4.22] a rating of quality (accuracy of shape and
dimensions) into four groups is proposed. In addition to this a classification into families
of parts should be made. A generally accepted quantitative statement about the
dependency of the degree of quality on the number of cavities can hardly be made. Such
dependencies, however, can be easily found by measuring weights and dimensions of
moldings from multi-cavity molds practically employed in the shop.

Machine M	Mold criterion T	Number of cavities n	Cost of part S (M,T,n)
M_1	T_1	n_1	S (1,1,1)
		n_2	S (1,1,2)
		n_3	S (1,1,3)
		$\vdots$	
	T_2	n_1	S (1,2,1)
		n_2	S (1,2,2)
		n_3	S (1,2,3)
		$\vdots$	
	$\vdots$		
M_2	T_1	n_1	S (2,1,1)
		n_2	S (2,1,2)
		n_3	S (2,1,3)
		$\vdots$	

Figure 4.16 Algorithm for determining the optimum number of cavities (continued)

Subsequent to the number of cavities, with respect to quality, step 5 decides on the number of cavities needed to comply with the delivery date. It must not fall short, so that the order can be produced within the available time span. The time for handling the whole order t, is composed of

$$t_U = t_{Des} + t_{MM} + t_M \tag{4.8}$$

t_{Des} Time for mold design,
t_{MM}, Time for mold making,
t_M Time for molding order.

The time for mold design t_{Des} is regarded independent of the number of cavities, while the time for making the mold t_M, can be taken as recedingly increasing with the number of cavities [4.22]. It can be characterized by this approximate equation:

$$t_{MM} = t_{c1} \cdot n^{0.7} \tag{4.9}$$

t_{c1} Time for making a single cavity mold,
n Number of cavities (exponent 0.7 from empirical data).

Based on the time t_M for molding the order (working hours), the minimum number of cavities for meeting the delivery date can be determined with

$$n_D = \frac{K_R \cdot t_{cycl} \cdot L}{t_M} \tag{4.10}$$

K_R Factor for rejects,
t_{cycl} Cycle time,
L   Lot size.

Besides this number of cavities there is another optimal number with respect to timing, n_{topt} for which the minimum of the time t_U, is entered. This time depends with $t_{MM}(n)$ and $t_M(n)$ on the number of cavities, too. By equating the first derivative of the function $t_U = f(n)$ with zero the number of cavities which is an optimum with regard to time is obtained [4.22]

$$n_{topt} = \frac{K_R \cdot L \cdot t_{cycl}}{0.7\, t_{cl}} \qquad (4.11)$$

A first operating range for the number of cavities to be expected can be established with np, n_Q, n_D, and n_{topt} in step 6. The lower limit is given by n_D, the upper one by n_Q. The numbers n_{topt} and n_p provide additional nonobligatory information.

In stage 3 the first machine of the preselected machinery (step 2) and the first mold concept from all feasible ones (step 1) is brought up for further study by steps 7 and 8. With step 9 the cavity layout and the technical number of cavities n_{t1}, (space on the platen) is established. On principle, only a symmetrical layout should be allowed, otherwise the forces on mold and tie bars cannot act uniformly. Figure 4.17 presents a layout in series and a circular layout.

In [4.16, 4.23] cavity layout is dealt with in more detail. Most important criteria for the layout are sufficient rigidity between the cavities and space for the heat-exchange system. The available clamping area with the dimensions W_v and W_h is determined by the distances between the tie bars. The possibility of removing a tie bar should be reserved for special cases (Figure 4.18).

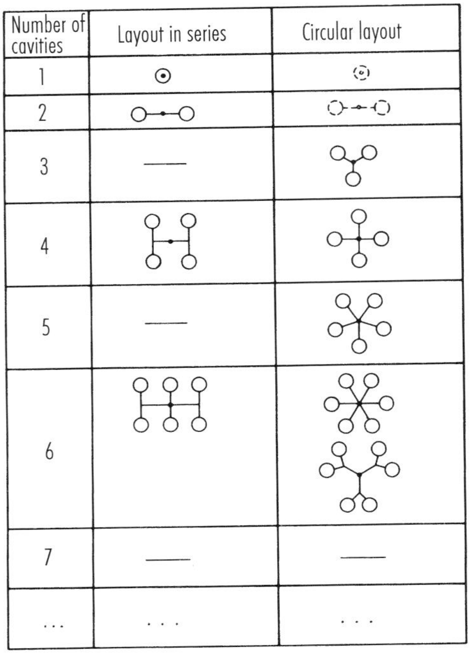

Figure 4.17   Cavity layouts with one parting line [4.18]

a)
W_h
W_h
W_h
Tie bars
Available area
Removable tie bar
b)
W_h
W_h

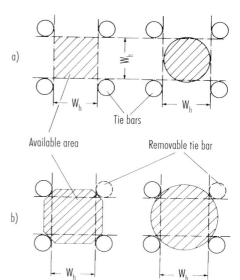

Figure 4.18 Number of cavities technically feasible – distance between tie bars [4.16]
a Mold located between tie bars,
b One tie bar has to be removed for installation

The cavity layout is established for the number of cavities $n_{t1} = n_Q$. Figure 4.19 demonstrates this operation to find n_{t1}, by superimposing the layout on the mold platen with regard to the dimensions of the molding.

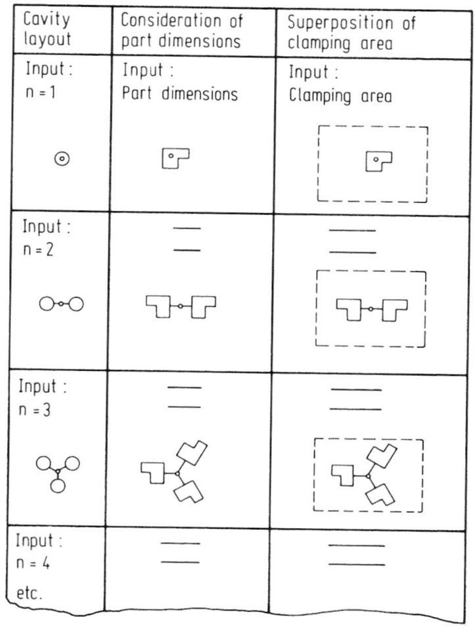

Cavity layout
Consideration of part dimensions
Superposition of clamping area
Input: n = 1
Input: Part dimensions
Input: Clamping area
Input: n = 2
Input: n = 3
Input: n = 4 etc.

Figure 4.19 Number of cavities n_{t1} technically feasible (Clamping area) [4.6, 4.16]

Cavity layout
Consideration of part dimensions
Superposition of flow limit
Input: n = 1
Input: Part dimensions
Input: Flow limit L_{max}
L_{max}
Input: n = 2
L_{max}
Input: n = 3
L_{max}
Input: n = 4 etc.

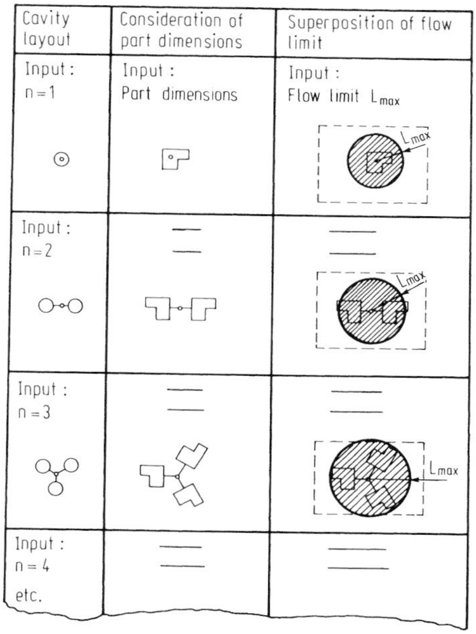

Figure 4.20 Number of cavities n_{t6} technically feasible (Rheology) [4.6, 4.16]

If there is sufficient space on the mold platen for the cavity number n_Q then $n_{t1} = n_Q$; if not then the next lower number is selected and the layout determined again. The wanted number is found when the available space is just enough for all cavities.

With step 10 the number of cavities n_{t2} is established with respect to the available injection pressure. The previously determined cavity number n_{t1} is taken and examined whether for $n_{t2} = n_{t1}$, the maximum injection pressure of the machine is sufficient to fill the cavities. If this is not so, then n_{t2}, is reduced so long until it meets the demand.

One possible way to examine the requirement is an estimation of the pressure demand of cavity and gate, and deducting this from the maximum injection pressure of the machine. The remaining pressure is available for nozzle and runner system. This pressure allows to compute the maximum length L_{max} of the runners, which can be entered into the plane of the parting line (Figure 4.20).

Besides this optical check there is a second more general possibility of establishing the pressure demand of the whole system with separate programs [4.22, 4.24].

With step 11, the technical cavity numbers n_{t3} to n_{t6} are determined.

n_{t3} is based on clamping force,
n_{t4} is based on minimum shot capacity,
n_{t5} is based on maximum shot capacity,
n_{t6} is based on plasticating rate.

The numbers n_{t3}, n_{t5} and n_{t6} are determined in a similar way as n_{t2} in step 10, always beginning with the largest number so far, while the cavity number n_{t4} is raised step by step, starting with n_D (step 6), until the actual shot capacity is larger than the minimal one.

With step 12 and 13, the numbers of cavities within the range established so far are tested as to whether or not they can be attained with the mold concept M of step 8. Based on the completed steps 1 to 13, a range of cavity numbers $n_{t\,min}$ to $n_{t\,max}$ can be specified with steps 14 and 15, which meet the demands on quality and timely delivery on one side and can, on the other side, be technically realized with the mold concept W and the machine M. With this in stage 4 of the flow chart, a field of operation has become available, in which the number of cavities can be modified and a calculation of the economics is carried out, that is, the costs for producing the part can be computed.

For the same machine M and the same mold concept W (steps 16 and 17), the major mold dimensions are determined before the economics are calculated (steps 22 to 27). It should be understood that not only those data and dimensions are meant which can affect the mold concept, but also all important external geometry data such as dimensions of plates and height of the mold.

With the cavity number $n_{t\,min}$ the major mold dimensions are established in step 19. If these dimensions are defined, one has to examine the concept with regard to its feasibility of realization (e.g. can part dimensions be achieved? – can demolding forces with ejection system be accomplished?).

Subsequently, one checks with step 21 whether the mold still fits the machine. Mold height, opening and ejection stroke are compared for the first time and the dimensions of the mold platen with the machine data for the second time.

Figure 4.21 presents the system of calculating costs on which steps 22 to 27 are based. The hourly machine costs (step 22) include:

– depreciation,
– interest,

– cost of maintenance,
– cost of locality,
– cost of energy,
– cost of cooling water,
– share of hourly wages.

The part-dependent costs (step 23) are material and finishing costs.

Machine	M
Mold type	T
Number of cavities	n
Time of machine run	T_R
Order size	S
Hourly machine costs	C_{MH}
Costs related to part	C_P
Mold costs	C_T
Indirect mold costs	C_{TI}
Overhead	C_O
Production costs per part $$C(M, T, n) = \frac{C_{MH} \cdot T_F + C_P + C_T + C_{TI} + C_O}{S}$$	

Figure 4.21   Calculation of production costs per part [4.6]

With the given mold principle (step 17) and the knowledge of the major mold dimensions (step 19), the mold costs can be estimated according to different procedures [4.16, 4.25, 4.26]. With reference to [4.16] the mold costs are composed of

– cost of design,
– cost of material,
– cost of manufacturing,
– cost of outside manufacturer.

The indirect mold costs comprise

– cost of sampling,
– cost of setup,
– cost of maintenance.

The costs determined so far are directly assigned costs.

By apportioning the overhead of secondary accounts to the main account the total overhead costs per order are identified.

Then, the cost of production for the part can be computed with step 27 according to Figure 4.21.

Depending on the interests of the particular company, other calculation procedures can or must be applied.

After all computation runs for the field of operation in stage 4 (n-variations) have been completed, the next mold concept is treated on the same machine. The calculations are carried out for other machines from step 2 in a similar way.

As the total result, all costs for a molding are finally presented in step 27 and can be entered in step 28. The most economical number of cavities, that is the combination of machine-mold-cavity number with the lowest part cost can be found there.

The handling of such an extensive number of data and the execution of the numerous individual calculations is best done by computer.

4.4.5.2   Costs for Sampling, Setup, and Maintenance

Some experience can be taken from the literature.

Sampling comprises the first startup of a mold for a trial run after completion. It is obvious that time consumption for this depends to a high degree on the soundness of the design and the precision of mold making. Experience affects the time consumption to not a small degree. If modification of the mold should become necessary, and gates, heat exchanger or ejectors have to be relocated, then a considerable amount of working hours has to be added to another sampling run. Even more critical is the need for a new cavity because of a faulty assumption of shrinkage. In such a case, more than one sample run may be needed. Therefore, the information in Figure 4.22 should be looked upon as optimal data. For this reason an experienced mold maker will estimate the time from a trial run of a mold until its availability for production a multiple of the time shown in Figure 4.22. It can easily reach several weeks for larger and complicated molds.

It was demonstrated, however, that considerably shorter sampling times were needed until availability of the mold if the mold design was supported by a simulation program such as CADMOULD or MOLDFLOW [4.27, 4.28].

In contrast to this, the setup time can be estimated far more precisely although, here too, great differences can be observed depending on how the setup is organized and carried out. Burgholf [4.30] examined setup procedures in his thesis and found up to

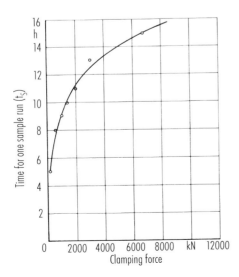

Figure 4.22   Time for sampling [4.29]

400% difference in comparable equipment and molds depending on the organization of this job.

Modern molding shops, working fully automatically, such as Netstal, a Swiss company, accomplish a change of mold and material in a maximum of 30 minutes [4.31].

Of course, molds for such an installation call for a particularly high precision, which exceeds by far what is common today [4.32].

Rapid-clamp systems which are frequently offered these days, also permit a shortening of these times. The special advantage of this equipment for medium-sized to larger molds is that due to the setup of preheated molds, the time till operation approaches zero. In the case of manual setup, molds are not preheated for reasons of accident prevention.

Finally, maintenance costs have to be considered. Figure 4.23 can serve as a reference. Setup, sampling, and maintenance costs can be determined with the following equations.

Setup Costs

Setup costs are calculated from [4.34]

$$C_{SU} = t_{SU} \cdot (C_{MH} + n \cdot C_{SH} + m \cdot C_H) + a_{CH} \cdot t_{SU} \cdot (q \cdot p) \tag{4.12}$$

with

t_{SU} Setup time (h),
C_{MH} Machine cost per hour (without labor costs),
n Number of setup personnel,
C_{SH} Hourly wage of setup personnel,

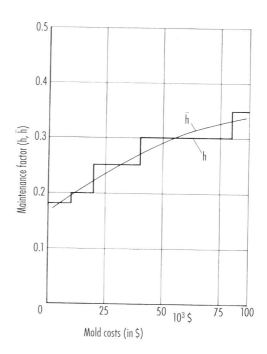

Figure 4.23 Factor for maintenance expenses [4.33]

Mold costs (in $)

m   Number of auxiliary personnel,
C_H Hourly wages for auxiliary personnel,
q   Plastificating rate,
p   Cost of the follow-up material,
a_{CH} Additional time for material change.

Costs of Sampling

$$C_S = t_P \cdot (C_{MH} + C_H) + c \frac{3600 \cdot t_S}{t_{cycl}} \cdot (V_P + V_R) \cdot \rho_M \cdot Pr \qquad (4.13)$$

with
t_P Time for sampling (h),
C_{MH} Machine costs per hour (without labor costs),
C_H Hourly wages for sampling,
c   Time effectiveness,
t_{cycl} Cycle time (s),
V_P Volume of molding (dm^3),
V_R Volume of runners (dm^3),
ρ_M Specific weight of material (kg/dm^3),
Pr   Price of material ($\$/kg$).

Maintenance Costs

$$C_{MT} = \bar{h} \cdot C_W \qquad (4.14)$$

with
$\bar{h}$ Maintenance factor,
C_W Mold costs.

4.5   Cavity Layouts

4.5.1   General Requirements

After the number of cavities has been established, the cavities have to be placed in the mold as ingeniously as possible.

In modern injection molding machines the barrel is usually positioned in the central axis of the stationary platen. This establishes the position of the sprue. The cavities have to be arranged relative to the central sprue in such a way that the following conditions are met:

– All cavities should be filled at the same time with melt of the same temperature.
– The flow length should be short to keep scrap to a minimum.
– The distance from one cavity to another has to be sufficiently large to provide space for cooling lines and ejector pins and leave an adequate cross section to withstand the forces from injection pressure.
– The sum of all reactive forces should be in the center of gravity of the platen.

4.5.2 Presentation of Possible Solutions

Figures 4.24 and 4.19 present basic options for cavity layouts in a mold.

Circular layout 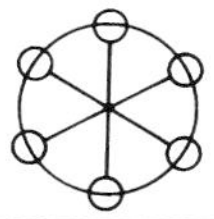	Advantages: Equal flow lengths to all cavities, easy demolding especially of parts requiring unscrewing device	Disadvantages: Only limited number of cavities can be accommodated
Layout in series 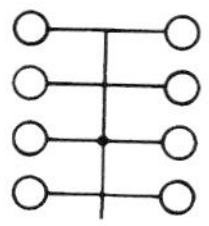	Advantages: Space for more cavities than with circular layout	Disadvantages: Unequal flow lengths to individual cavities, uniform filling possible only with corrected channel diameters (by using computer programs e.g. MOLDFLOW, CADMOULD etc.)
Symmetrical layout 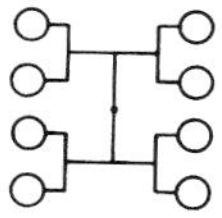	Advantages: Equal flow lengths to all cavities without gate correction	Disadvantages: Large runner volume, much scrap, rapid cooling of melt. Remedy: hot manifold or insulated runner

Figure 4.24 Comparison of cavity layouts [4.18]

4.5.3 Equilibrium of Forces in a Mold During Injection

Mold and clamping unit are loaded unevenly if the cavities are located eccentrically with respect to the central sprue. The mold can be forced open on one side. Flash and possible rupture of tie bars may occur as a consequence. Molds that have experienced flash once have a damaged sealing surface and will always produce flash again. Therefore, the first design law is the requirement that the resultant from all reactive forces (injection pressure) and the resultant from all clamping forces act in the center of the sprue. Figure 4.25 shows an eccentric and a centric runner system.

Figure 4.25 Centric and eccentric position of sprue and runner

 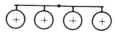

In complex molds the center of gravity has to be determined and the position of cavities in the mold established accordingly.

$$x_m = \frac{\sum (a_i x_i)}{\sum a_i}$$

(4.15)

a_i projected partial area.

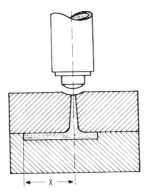

Figure 4.26   Determination of center of gravity [4.35]

$$x_m = \frac{\Sigma \, (a_i \cdot x_i)}{\Sigma \, a_i}$$

On the other hand, suitable mold design and remedial steps can also cause the resultant from the clamping forces to act in the center (Figures 4.27 and 4.28) [4.7, 4.35]. A balancing of forces with a compensator pin, however, increases the clamping forces.

4.5.4   Number of Parting Lines

Molding and runner are released in a plane through the parting line during mold opening. Thus all solidified plastic parts are ejected and the mold made ready for the next cycle.

A standard mold has one parting line. Molding and runner are demolded together. If the runner is to be automatically separated from the molding, as is frequently the case in multi-cavity molds or with multiple gating, then an additional parting line for the runner system is needed (three-plate mold) or a hot-runner mold (cold-runner mold for reactive materials) is used (exception: tunnel gate). Several parting lines are also necessary for stack molds.

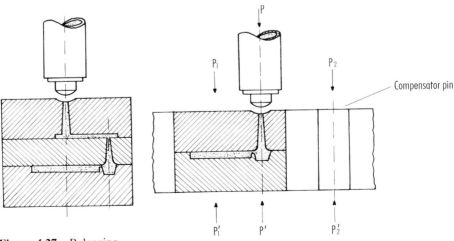

Figure 4.27   Balancing forces with floating plate (three-plate mold) [4.35]

Figure 4.28   Balancing with compensator pin [4.35]

For examples, refer to Section 4.4.1 (Classification of Molds).

Design Solutions

One parting line:
Standard mold,
Slide mold,
Split-cavity mold,
Mold with unscrewing device,
Hot-runner mold.

Several parting lines:
Three-plate mold,
Stack mold,
Insulated-runner mold.

Items Affecting Number of Parting Lines:

Part geometry,
Number of cavities n,
Runner system and gating,
Demolding system.

References

[4.1]   Johannaber, F.: Untersuchungen zum Fließverhalten thermoplastischer Formmassen beim Spritzgießen durch enge Düsen. Dissertation, Tech. University, Aachen, 1967.

[4.2]   Schröder, U.; Kaufmann, H.; Porath, U.: Spritzgießen von thermoplastischen Kunststoffen – Unterlagen für den theoretischen Unterricht. IKV, Verlag Wirtschaft und Bildung KG, Simmerath, 1976.

[4.3]   Menges, G.; Porath, U.; Thim, J.; Zielinski, J.: Lernprogramm Spritzgießen. IKV, Carl Hanser Verlag, Munich, 1980.

[4.4]   Mörwald, K.: Einblick in die Konstruktion von Spritzgießwerkzeugen. Garrels, Hamburg, 1965.

[4.5]   Amberg, J.: Konstruktion von Spritzgießwerkzeugen im Baukastensystem (Variantenkonstruktion). Unpublished report, IKV, Aachen, 1977.

[4.6]   Bangert, H.: Systematische Konstruktion von Spritzgießwerkzeugen und Rechnereinsatz. Dissertation, Tech. University, Aachen, 1981.

[4.7]   Menges, G.; Mohren, P.: Anleitung für den Bau von Spritzgießwerkzeugen. 2nd Ed., Carl Hanser Verlag, Munich, 1983.

[4.8]   Fertigungsplanung von Spritzgießwerkzeugen. Intermediate report about a DFG research EV 10/7, IKV/WLZ, Tech. University, Aachen, 1975.

[4.9]   Gastrow, H.: Der Spritzgieß-Werkzeugbau in 100 Beispielen. 3rd Ed., Carl Hanser Verlag, Munich, 1982.

[4.10] Catalog of Standards, Hasenclever & Co., Lüdenscheid.

[4.11] Jonas, R.; Schlüter, H.; Braches, L.; Thienel, P.; Bangert, H.; Schürmann, E.: Spritzgerechtes Formteil und optimales Werkzeug, Paper block X at the 9th Tech. Conference on Plastics, IKV, Aachen, March 8–10, 1978.

[4.12] Prospectus, Mechanica Generale, S. Paolo di Jesi, Italy, 1981.

[4.13] Leibfried, D.: Untersuchungen zum Werkzeugfüllvorgang beim Spritzgießen von thermoplastischen Kunststoffen. Dissertation, Tech. University, Aachen, 1971.

[4.141] Degalan-Formmassen für Spritzguß und Extrusion. Publication, Degussa, Hanau.

[4.15] Thienel, P.: Der Formfüllvorgang beim Spritzgießen. Dissertation, Tech. University, Aachen, 1977.

[4.16]   Göhing, U.: Ermittlung eines Algorithmus zur Bestimmung der techn.-wirtschaftlich optimalen Formnestzahl bei Thermoplastspritzgießwerkzeugen. Unpublished report, IKV, Aachen, 1976.

[4.17]   Koller, R.: Konstruktionsmethode für den Maschinen-, Geräte- und Apparatebau. Springer, Heidelberg, Berlin, New York, 1976.

[4.18]   Szibalski, M.; Meier, E.: Entwicklung einer qualitativen Methode für den Konstruktionsablauf bei Spritzgießwerkzeugen. Unpublished report, IKV, Aachen, 1976.

[4.19]   Drall, L.; Gemmer, H.: Berechnung der wirtschaftlichsten Formnestzahl bei Spritzgießwerkzeugen. Kunststoffe, 62 (1972), 3, pp. 158–165.

[4.20]   Gemmer, H.; Pröls, J.: Berechenbarkeit von Spritzgießwerkzeugen. VDI-Verlag, Düsseldorf, 1974.

[4.21]   Custodis, Th.: Auswahl der kostengünstigsten Spritzgießmaschinen für die Fertigung vorgegebener Produkte. Dissertation, Tech. University, Aachen, 1975.

[4.22]   Lichius, U.: Erarbeitung von Konzepten zur rechnerunterstützten Konstruktion von Spritzgießwerkzeugen und Erstellung einiger hierzu einsetzbarer Rechenprogramme. Unpublished report, IKV, Aachen, 1978.

[4.23]   Benfer, W.: Aufstellung eines Rechenprogrammes zur Ermittlung aller Hauptabmessungen eines Spritzgießwerkzeuges. Unpublished report, IKV, Aachen, 1977.

[4.24]   Schmidt, L.: Auslegung von Spritzgießwerkzeugen unter fließtechnischen Gesichtspunkten. Dissertation, Tech. University, Aachen, 1981.

[4.25]   Schläter, H.: Verfahren zur Abschätzung der Werkzeugkosten bei der Konstruktion von Spritzgießwerkzeugen. Dissertation, Tech. University, Aachen, 1981.

[4.26]   Krawanja, A.: Zeit- und Kostenplanung für die Herstellung von Spritzgießwerkzeugen. Unpublished graduation thesis at the Montan University, Leoben, Austria, 1976.

[4.27]   Haldenwanger, H. G.; Schäper, S.: Erfahrungen in der Rheologievorausberechnung von Kunststofformteilen. Paper, Annual VDI Conference: Plastics in the Automotive Industry, Mannheim, 1986.

[4.28]   Engelen, P.: Formteilauslegung mit CAD/CAM aufgezeigt an einem praktischen Beispiel. Lecture, VDI, Baden-Baden, Februar 1985.

[4.29]   Rehmert, W.: Behandlung von Umrüst- und Bemusterungskosten. Kunststoffe, 61 (1971), 6, pp. 441–443.

[4.30]   Burghoff, G.: Rüstzeitreduzierung in Spritzgießbetrieben. Dissertation, Tech. University, Aachen, 1983.

[4.31]   Verbal information from Revisa, Häggingen.

[4.32]   Verbal information from Netstal, Näfels.

[4.33]   Kalkulationsgrundsätze für die Berechnung von Spritzgießwerkzeugen. Fachverband Technische Teile in der GKV, Frankfurt.

[4.34]   Hüttner. H.-J.; Pistorius, D.; Rühmann, H.; Schürmann, E.: Kostensenkung durch Rüstzeitverkürzung beim Spritzgießen. Paper block XIII at the 9th Tech. Conference on Plastics, IKV, Aachen, March 8–10, 1978.

[4.35]   Morgue, M.: Modules d'injection pour Thermoplastiques. Officiel des Activités des Plastiques et du Caoutchouc, 14 (1967), pp. 269–276 and pp. 620–628.

5   Design of Runner Systems

5.1   Characterization of the Complete Runner System

The runner system accommodates the molten plastic material coming from the barrel and guides it into the mold cavity. Its configuration, dimensions and connection with the molded part affect the mold filling process and, therefore, largely the quality of the product. A design which is primarily based on economic viewpoints (rapid solidification and short cycles) is mostly incompatible with quality demands especially for technical parts.

A runner system usually consists of several components. This is particularly evident in multi-cavity molds. Figure 5.1 shows a runner system composed of

– sprue
– runners,
– gate.

Figure 5.1   Runner system [5.1]

The sprue bushing receives the plasticated material from the injection nozzle, which closes off the barrel and is pressed firmly against the sprue bushing. Frequently, a single cavity mold has only a sprue; the part is then said to be sprue-gated (Section 6.1). With multi-cavity molds, the sprue bushing feeds the melt into the runners. These are connected to the cavities via the gates.

The gate is an area of narrow cross-section in which flow is restricted. Its purposes are fourfold:

– to separate the molded part easily and cleanly from the runner system,
– to hold back the cooled skin that has formed on the cold walls of the runners (avoiding flash on the molded part),
– to heat the melt through shear before it enters the cavity,
– since the cross-section of the opening can be readily altered, the runner system can be balanced in such way that the melt enters each cavity at the same time and in the same condition.

5.2   Concept and Definition of Various Types of Runners

Depending on the temperature control, different types of runners may be distinguished:

– standard runner systems,
– hot-runner systems,
– cold-runner systems.

5.2.1   Standard Runner Systems

Standard runners are directly machined into the mold plates, which form the main parting line. The temperature is therefore the temperature of the mold. The melt remaining in the runner freezes and has to be demolded along with the molded part after each shot. In the case of thermoplastics, the frozen material can generally be recycled as regrind, whereas in the case of thermosets, it has limited scope for reuse and is unrecoverable material.

5.2.2   Hot-Runner Systems

Hot runners may be viewed as extended injection nozzles in the form of a block. Heat barriers isolate it from the cold mold. It contains the runner system consisting of central sprue bush, runners and gates or nozzles. The temperature of this block lies in the melting range of the thermoplastic melts. Hot-runners offer the following advantages:

– no loss of melt and thus less energy and work input,
– easier fully automatic operation,
– superior quality because melt can be transferred into the cavity at the optimum sites.

The disadvantages are:

– high costs,
– the risk of decomposition and production stoppages in the case of materials with low thermal resistance,
– thermal isolation from the hot-runner manifold block is problematic.

5.2.3   Cold-Runner Systems

Just as hot runners are used in molds for thermoplastics, cold runners are used in molds for reactive materials such as thermosets and rubber. Unlike the hot mold, which is kept at 160–180 °C, the cold runner must be kept at 80–120 °C in order that the material may not react prematurely in the runner. The advantages are the same as for thermoplastics, but there are additional difficulties:

– pressure consumption in cold runners is very high, a fact which makes the design more expensive,
– since the slightest temperature differences cause very large differences in viscosity, it is practically impossible to fulfill the requirement of introducing "the material into every cavity at the same time in the same condition".

For these reasons, specialty types only have established themselves for rubber and elastomers; cold runner molds are not used at all for thermoset molding compounds.

5.3   Demands on the Runner System

The dimensioning of a runner system is determined by a multitude of factors, which, in essence, result from the configuration of the molded part and the plastic material employed (Figure 5.2). The demands on quality and economics are listed in Figure 5.3.

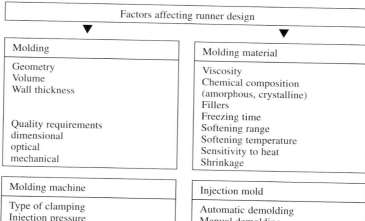

Factors affecting runner design	
Molding	**Molding material**
Geometry Volume Wall thickness Quality requirements dimensional optical mechanical	Viscosity Chemical composition (amorphous, crystalline) Fillers Freezing time Softening range Softening temperature Sensitivity to heat Shrinkage
Molding machine	**Injection mold**
Type of clamping Injection pressure Injection rate	Automatic demolding Manual demolding Temperature of runner system

Figure 5.2   Items which affect the design of a runner system [5.2]

Functions and demands	
1. Cavity filling with a minimum of knit lines	6. Length as short as technically feasible to keep losses in pressure, temperature and material small
2. Restrictions to flow as few as possible	7. Cross section so large that freezing time equals or slightly exceeds a little that of the molding. Only then can holding pressure remain effective until part is solid
3. Share of total weight as small as possible	8. Runner system should have little or no effect on cycle time
4. Ease of demolding	9. Place of gating at the thickest section of part
5. Appearance of part should remain unaffected	10. Location or design of gate so as to prevent jetting

Figure 5.3
Functions of and demands on the runner system [5.2]

5.4   Classification of Runner Systems

The design engineer can choose from a large number of runner systems to offer the optimum quality and economics to the user. These are:

I   Runners which remain with the molded part and have to be cut off afterwards.
II Runners which are automatically separated from the molded part and are demolded separately.
III Runners which are automatically separated from the molded part during demolding but remain in the mold.

This results in the classification shown in Figure 5.4.

Gating systems	
I	1. Sprue gate 2. Edge gate 3. Disk gate 4. Ring gate
II	5. Tunnel gate (submarine gate) 6. Pinpoint gate (in three-plate mold)
III	7. Pinpoint gate (with reversed sprue) 8. Runnerless gating 9. Runner for stack molds 10. Insulated runner 11. Hot manifold

Figure 5.4   Gating systems

In addition, there are several special types that will be discussed along with the various types of runner. To provide a first overview, the types of runners listed in Figure 5.4 and their characteristic features are summarized in Figure 5.5.

5.5   The Sprue

The melt enters the mold via a sprue which is generally machined in the sprue bushing. Together with the injection nozzle, which seals off the barrel, it must ensure a leakproof connection between the barrel and the mold during the injection process, which entails high mechanical load and thus is characterized by wear. The sprue bushing must therefore be replaceable. Except for hot manifolds, where planar contact surfaces (Figure 5.6) are frequently required, the sprue bushing of a normal manifold matches that shown in Figure 5.7. To be able to fulfill its functions, the following properties are required:

– wear resistance: therefore of made hardened steel,
– flexural fatigue strength: therefore a strong but not too large flange, and rounded edges,
– since the sprue always leaves a mark on the molded part: as small a diameter as possible,
– for a perfect seal, the orifices of the nozzles and bushing must be aligned. The diameter of the nozzle orifice (d_N) must be 1.5 mm smaller than that of the sprue bushing (d_S).

Type of gate		Characteristics
Sprue (gate)	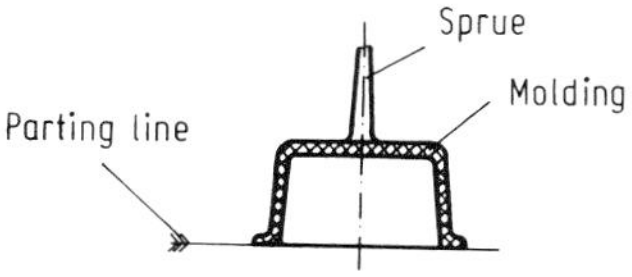	*Application:* for temperature-sensitive and high-viscous materials, high-quality parts and those with heavy sections *Advantages:* results in high quality and exact dimensions *Disadvantages:* postoperation for sprue removal, visible gate mark
Edge gate	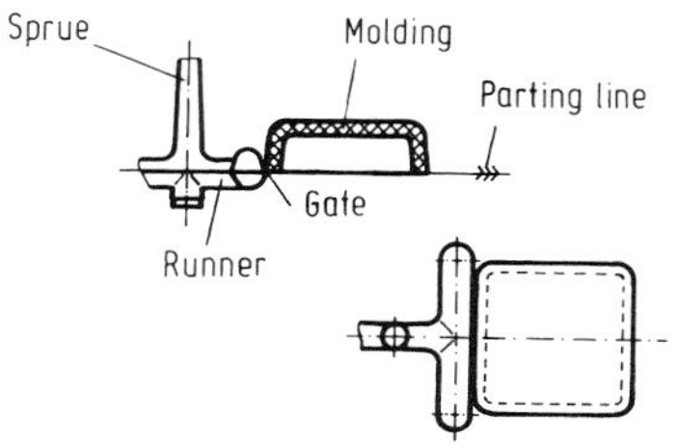	*Application:* for parts with large areas such as plates and strips *Advantages:* no knit lines, high quality, exact dimensions *Disadvantages:* postoperation for gate removal
Disk gate	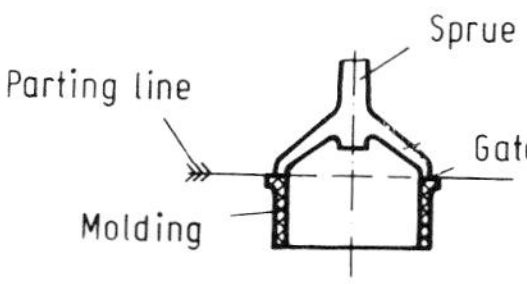	*Application:* for axially symmetrical parts with core mounted at one side only *Advantages:* no knit lines and no reduction in strength *Disadvantages:* postoperation for gate removal
Ring gate	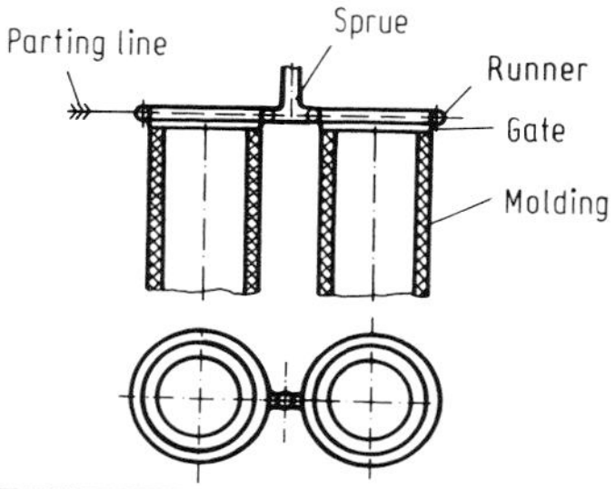	*Application:* for sleeve-like parts with core mounted at both sides *Advantages:* uniform wall thickness around circumference *Disadvantages:* slight knit line, postoperation for gate removal
Tunnel gate (submarine gate)	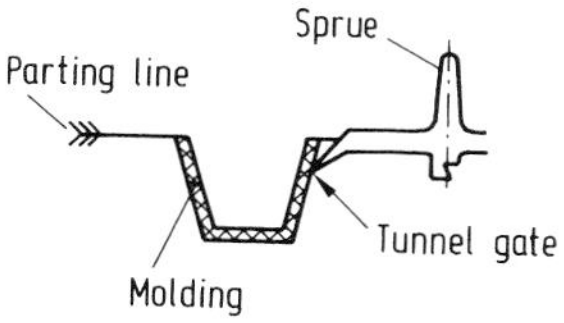	*Application:* primarily for smaller parts in multi-cavity molds and for elastic materials *Advantages:* automatic gate removal *Disadvantages:* for simple parts only because of high pressure loss

Figure 5.5   Summary of gate types [5.2 to 5.6]
(continued on next page)

Type of gate		Characteristics
Pinpoint gate (three-plate mold)	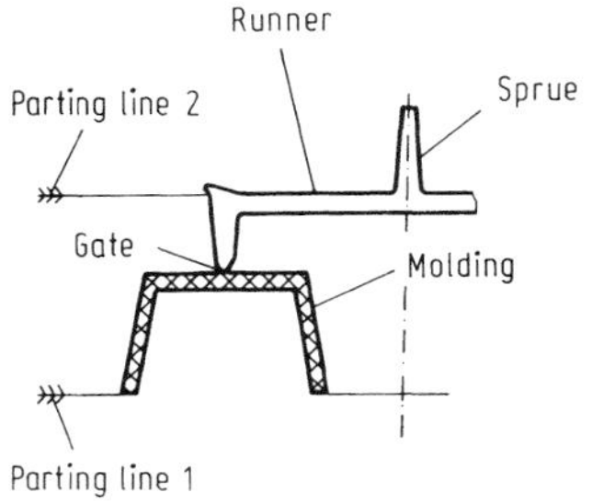 Runner / Parting line 2 / Sprue / Gate / Molding / Parting line 1	*Application:* for multi-cavity molds and center gating *Advantages:* automatic gate removal *Disadvantages:* large volume of scrap, higher mold costs
Pinpoint gate (with reversed sprue)	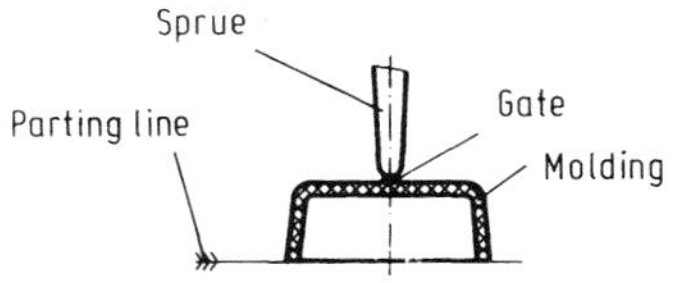 Sprue / Parting line / Gate / Molding	*Application:* for parts with automatic gate removal *Advantages:* no postoperation *Disadvantages:* preferably for thermally stable materials (PE, PS), limited use for others
Runnerless gating	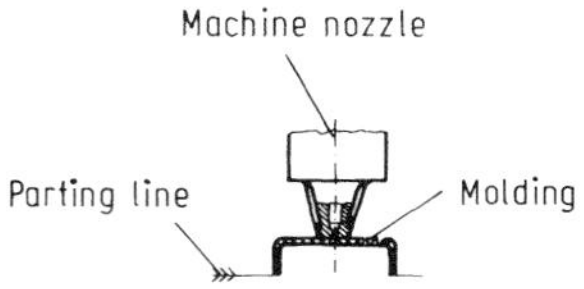 Machine nozzle / Parting line / Molding	*Application:* for thin-walled parts and rapid sequence of cycles *Advantages:* no loss of material for runner system *Disadvantages:* mark on part from nozzle
Gating of stack molds	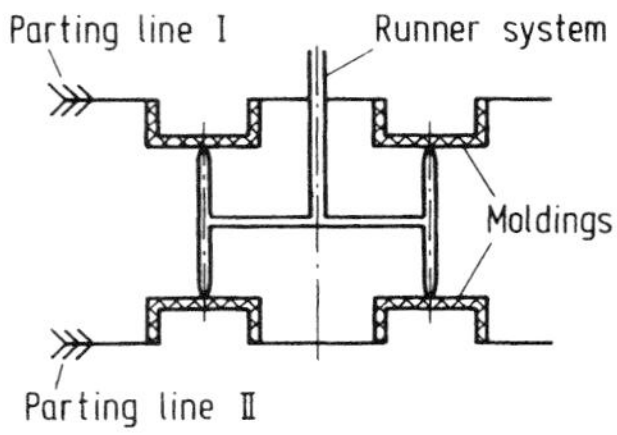 Parting line I / Runner system / Moldings / Parting line II	*Application:* flat and light-weight parts in multi-cavity molds *Advantages:* better utilization of machine's plasticating rate *Disadvantages:* large amount of scrap from voluminous runner system, higher mold costs *Note:* today generally used with hot manifold, thus no scrap but more expensive
Insulated runner molds	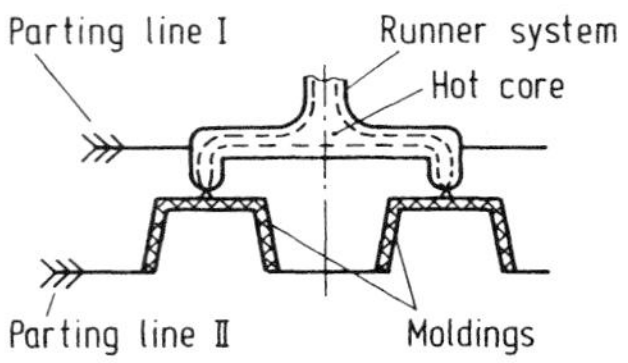 Parting line I / Runner system / Hot core / Parting line II / Moldings	*Application:* for materials with a large softening and melt temperature range and rapid sequence of cycles *Advantages:* automatic gate separation, material loss from runner only after shutdown *Disadvantages:* Danger of cold material getting into cavity after interruption

Figure 5.5 Summary of gate types (continued)
(continued on next page)

Type of gate		Characteristics
Hot manifold	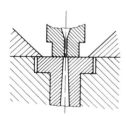	*Applications:* for high-quality, technical parts, independent of cycle time, also suitable for materials difficult to process *Advantages:* no material loss from runner system, automatic gate separation *Disadvantages:* expensive molds especially due to control equipment

Figure 5.5   Summary of gate types (continued)

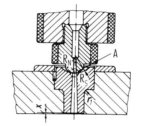

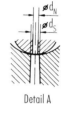

Detail A

Figure 5.6   Plane area of contact between machine nozzle and sprue bushing

Figure 5.7   Curved area of contact between machine nozzle and sprue bushing [5.3]

The radius of the spherical indentation in the sprue bushing (R_S) into which the tip of the nozzle extends, must be 1 mm greater than that of the nozzle tip R_N [5.7] (Figure 5.7).

Application of the following rules to the dimensions of the sprue (Figure 5.9) will ensure perfect quality and reliable operation:

– The diameter at the foot of the orifice should be roughly 1 mm greater than the gated molded part at its thickest point or greater than the diameter of the connecting runner. (This ensures that it freezes last and that the orifice remains open for the holding pressure.)
– The orifice must be tapered (> 1° and < 4°) and totally smooth, without furrows etc., around its circumference in order that the sprue may be pulled out of the orifice when the mold is opened. For this reason, it must not have any flash at its upper end (Figure 5.8).
– The lower orifice edge must be rounded to prevent the melt from pulling away from the wall to form a jet of material that would remain behind as a visible flaw on the surface of the molded part.

If these requirements are met, the sprue in single-cavity molds is pulled from the orifice and thus demolded by the molded part, which remains on the ejector-side of the mold half.

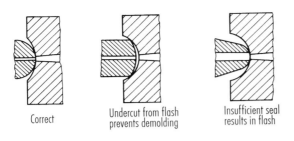

Correct

Undercut from flash
prevents demolding

Insufficient seal
results in flash

Figure 5.8   Correct and incorrect
design of areas of contact

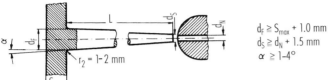

$d_F \geqq S_{max} + 1.0$ mm
$d_S \geqq d_N + 1.5$ mm
$\alpha \geqq 1\text{-}4°$

Figure 5.9   Guidelines for dimensioning sprues [5.8]

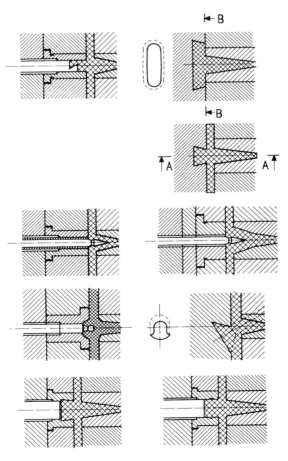

Figure 5.10   Design of sprue pullers
[5.9]

In multi-cavity molds, where the sprue serves only to feed the material to the runners, special demolding support is required. A sprue puller is installed opposite the sprue, the profiled tip of which acts as an undercut that grips the sprue (Figure 5.10). During ejection, the undercut releases the sprue, which can then drop out. This design also has the advantage of providing a cold-slug well.

Another, less common option for removing the sprue from the bushing is shown in Figure 5.11. The sprue bushing is spring-loaded. After the mold has been filled and the nozzle is retracted from the bushing, springs push back the bushing and loosen the sprue.

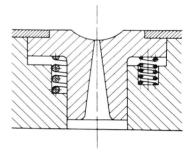

Figure 5.11   Spring loaded sprue bushing [5.7]
left side: Big spring = high force;
right side: Small spring = low force, therefore several circumferential springs

5.6   Design of Runners

Runners connect the sprue via the gate with the cavity. They have to distribute the material in such a way that melt in the same condition and under the same pressure fills all cavities at the same time.

The plasticated material enters the runners of a cooled mold with high velocity. Heat is rapidly removed from the material close to the walls by heat transfer, which then forms a skin. This provides a heat-insulating layer for the material flowing in the center of the channel. A hot, fluid core is formed, through which the plastic flows to the cavity. This hot core must be maintained until the molded part is completely solid; then the holding pressure can act fully to compensate for the volume contraction during solidification.

This requirement on the one hand and the wish for minimal pressure loss and maximum material savings on the other, determines the optimum geometry of the runner. The dimensions of the runner obviously depend on the maximum thickness of the molded part and the type of plastic being processed. The thicker the walls of the molded part, the larger the cross-section of the runner must be. As a rule, the cross-section must be roughly 1 mm larger than the molded parts are thick. A large cross-section promotes the filling process of the mold because the resistance to flow is smaller than in thin runners of the same length. It pays therefore to dimension the runner according to hydraulic laws. Section 5.9.7 explains how a runner system is optimized and balanced with computer assistance.

Figure 5.12 summarizes the factors affecting runner design. The objectives of a runner and the resulting demands can be taken from Figure 5.13. Figure 5.14 presents the most common cross-sections of runners and evaluates their performance.

Nomograms for a number of materials and their volumes or weights passing through the runner are presented in Figure 5.15. The data are empirical but the diameters of

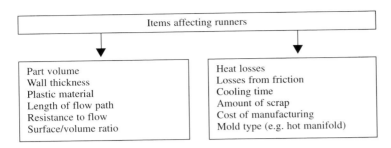

Figure 5.12
Factors which affect design and size of runners [5.2]

1. Conveying melt rapidly and unrestricted into cavity in the shortest way and with a minimum of heat and pressure loss.
2. Material must enter cavity (or cavities) at all gates at the same time under the same pressure and with the same temperature.
3. For reasons of material savings, cross-sections should be kept small although a larger cross-section may be more favorable for optimum cavity filling and maintaining adequate holding pressure. Larger cross-sections may increase cooling time.
4. The surface-over-volume ratio should be kept as small as feasible.

Figure 5.13
Functions of runners [5.2]

runners are to be determined as a function of their lengths with an acceptable pressure loss of less than 30 MPa.

The surface finish of a runner depends on the plastic to be molded. One can generally assume that it is of advantage not to polish a runner, so that the solid skin is better attached to the wall and not so easily swept along by the flowing material. With some plastics, however, runners have to be highly polished or even chrome plated in order to avoid flaws in the molded part. Critical plastics in this category are PVC, polycarbonate and polyacetal.

The cardinal demand for all mold cavities to be filled simultaneously with melt in the same condition is met very easily by making the flow paths identical. However, as shown in Figures 5.16–5.18, this can only be accomplished to a certain extent or at the expense of other drawbacks. This is why it has become standard practice to balance the distribution system by means of different runner or gate cross-sections.

5.7   Design of Gates

The gate connects the cavity (or molding) with the runner. It is usually the thinnest point of the whole system. Size and location are decided by considering various requirements (see Figure 5.19):

– it should be as small as possible so that material is heated but not damaged by shear,
– it must be easy to demold,
– it must permit automatic separation of the runners from the molded part, without leaving blemishes behind on the part.

Cross-sections for runners		
Circular cross-section 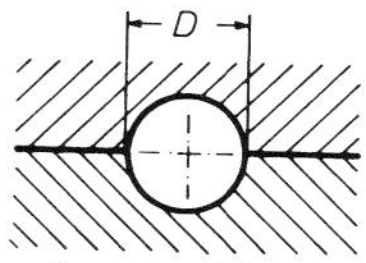 $D = s_{max} + 1.5$ mm	*Advantages:*	Smallest surface relative to cross-section, slowest cooling rate, low heat and frictional losses, center of channel freezes last therefore effective holding pressure
	Disadvantages:	Machining into both mold halves is difficult and expensive
Parabilic cross-section 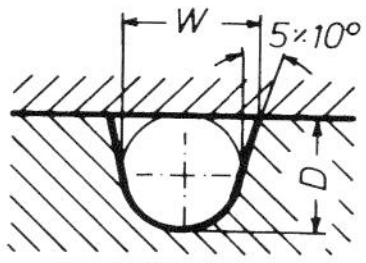 $W = 1.25 \cdot D$ $D = s_{max} + 1.5$ mm	*Advantages:*	Best approximation of circular cross-section, simpler machining in one mold half only (usually movable side for reasons of ejection)
	Disadvantages:	More heat losses and scrap compared with circular cross-section
Trapezoidal cross-section 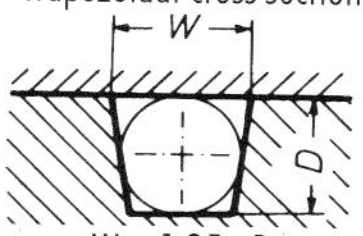 $W = 1.25 \cdot D$	Alternative to parabolic cross-section	
	Disadvantages:	More heat losses and scrap than parabolic cross-section
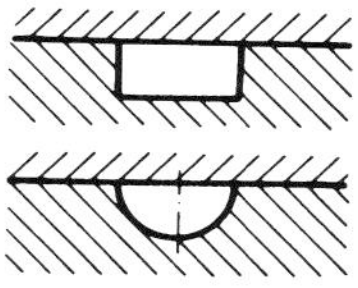	Unfavorable cross-sections have to be avoided	

Figure 5.14 Cross-sections for runners [5.2, 5.10–5.12]

The gate can be designed in various configurations. Thus, one distinguishes between a pinpoint and an edge gate. A special form is the sprue gate, which is identical with the sprue itself, as described in detail in Section 6.1.

In all gate types, except for the sprue gate, the gate is always the narrowest point in the gating system.

When flowing through narrow channels like a runner or gate, the material encounters a considerable resistance to flow. Part of the injection pressure is consumed and the temperature of the melt is noticeably raised. This is a desirable effect because

1. the melt entering the cavity becomes more fluid and reproduces the cavity better, and
2. the surrounding metal is heated up and the gate remains open longer for the holding pressure.

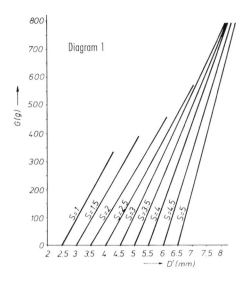

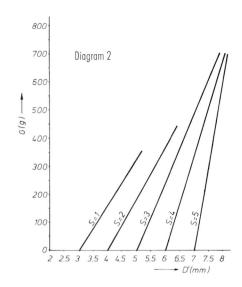

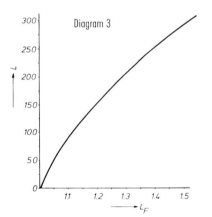

Figure 5.15   Guide lines for dimensioning cross-sections of runners [5.13]

Diagram 1 Applicable for PS, ABS, SAN, CAB.

Diagram 2 Applicable for PE, PP, PA, PC, POM.

Symbols:

S: Wall thickness of part (mm),

D': Diameter of sprue at its end (mm),

G: Weight of part (g),

L: Length of runner to one cavity (mm),

L_F: Correction factor.

Procedure (Diagram 3):

1. Determine G and S,

2. Take D' from diagram for material considered,

3. Determine L,

4. Take L_F from diagram 3,

5. Correct diameter or runner: $D = D' \times L_F$.

The optimum gate size that will not cause

1. thermal damage to the plastic or
2. too high a pressure loss

has to be determined by computation or experiment during a sample run. The runners can be balanced at the same time.

This is done in practice – this is generally necessary even if the design has been computed beforehand – as follows. The employee inspecting the mold changes the gates mechanically such that every cavity is filled uniformly at the same time with melt. This can be readily determined with consecutive short shots (Figures 5.20 and 5.21). In practice, it is accomplished by making the gates considerably smaller than necessary at first.

Circular layout 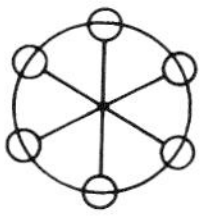	Advantages: Equal flow lengths to all cavities, easy demolding especially of parts requiring unscrewing device	Disadvantages: Only limited number of cavities can be accommodated
Layout in series 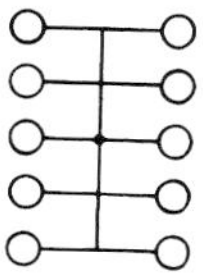	Advantages: Space for more cavities than with circular layout	Disadvantages: Unequal flow lengths to individual cavities, uniform filling possible only with corrected channel diameters (by using computer programs e.g. MOLDFLOW, CADMOULD etc.)
Symmetrical layout 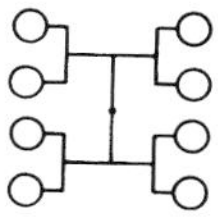	Advantages: Equal flow lengths to all cavities without gate correction	Disadvantages: Large runner volume, much scrap, rapid cooling of melt. Remedy: hot manifold or insulated runner

Figure 5.16   Cavity layouts with one parting line [5.2]

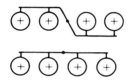

Figure 5.18 Centric (bottom) and eccentric (top) position of sprue and runner

Figure 5.17 Cavity layouts with one parting line

Number of cavities	Layout in series	Circular layout
1	⊙	⊙
2	 ○—•—○	 ○—•—○
3	—	
4	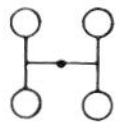	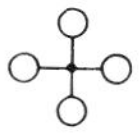
5	—	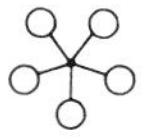
6	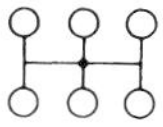	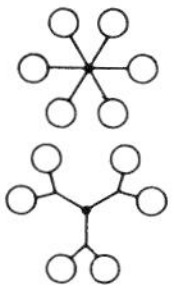
7	—	—
...	...	...

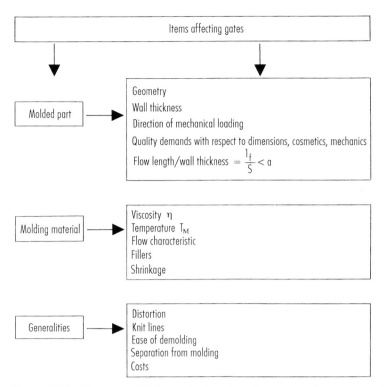

Figure 5.19   Factors which determine location, design, and size of gates [5.2]
a see Figure 4.10

Then they are enlarged during trial runs until all cavities are uniformly filled.

Figure 5.22 shows recommended locations and shapes of gates on the molded part. The gate can have a circular, semicircular or rectangular cross-section. The most favorable one is the rectangular gate. Easiest separation from the molded part is afforded by a semicircular one.

The gate is best connected to the runner as shown in Figure 5.22 (top). This does not, by itself, ensure the best flow characteristic into the cavity during filling. With some plastics, part of the frozen skin is swept into the cavity and causes blush marks (Figure 5.23). The plastic must not jet into the cavity either, but fill it uniformly beginning at the gate orifice. Jetting causes troublesome surface blemishes because the jetted material does not remelt in the material that follows. In noncritical cases a radius at the transition can already redress this effect.

Suggested dimensions for pinpoint and tunnel gates can be taken from Figure 5.24.

5.7.1   Position of the Gate at the Part

Since, with all homogeneous plastic materials, solidification of the melt in the cavities of the mold is an effect influenced by the heat of the mold and since thermal conduction is critically influenced by the wall thickness, the gate must always be positioned at the

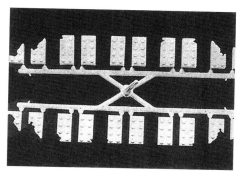

Figure 5.20 Irregular filling of cavities in a mold with imbalanced runner system [5.6]

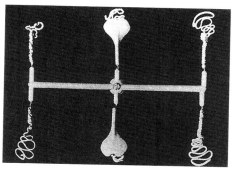

Figure 5.21 Filling of a mold with imbalanced runner system [5.6]

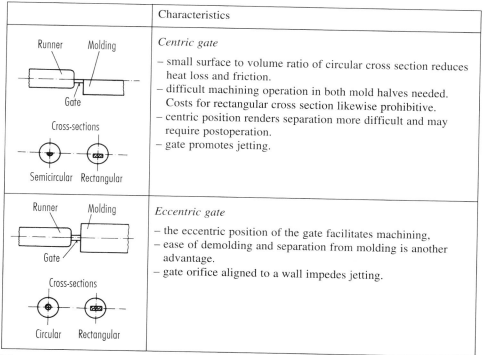

Figure 5.22 Cross-section of gates and their positions at the runners [5.2, 5.3, 5.10]

thickest cross-section. If the gate is not at the thickest section, voids and sink marks will be caused. They result from too little holding pressure because of premature freezing of the gate area.

(Processing of structural foam is an exception; with this technique the gate should be placed at the thinnest section. Filling is caused by the pressure of the developing gas, and the resistance to flow has to become smaller as filling progresses, to compensate for the diminishing gas pressure.)

Gate design	Characteristics
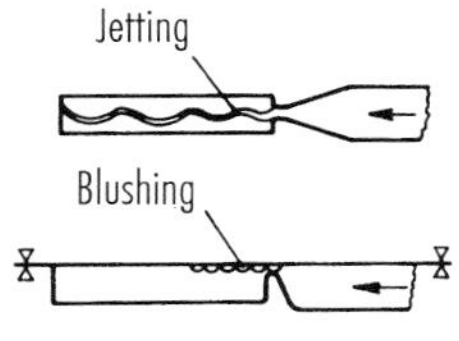 Jetting Blushing Poor gate design	Gate should be positioned in such a way that no jetting can occur causing troublesome marks; melt must impinge on wall or other obstacle. If gate is machined only into one mold half, cold "skin" may be carried into cavity. This also results in blush marks. Remedy: A special cold slug well accepts cold material.
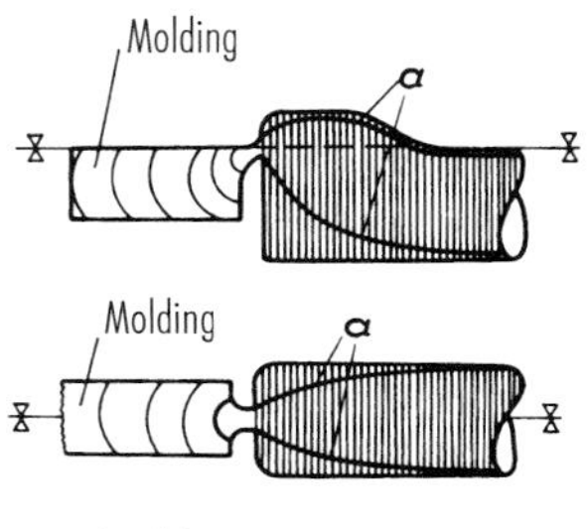 Molding Molding Good design practice	Centric location of gate with abrupt transition and rough walls prevents transport of cold surface layer. (a: indicates the boundaries of the hot, fluid core) Radius at transition causes laminar flow of melt into cavity and prevents jetting. Radii at transition make gate removal more difficult. They should, nevertheless, be preferred because of better flow conditions which result in higher quality with respect to dimensions and mechanical strength.

Figure 5.23   Guidelines for gate design [5.2, 5.12, 5.14, 5.15]

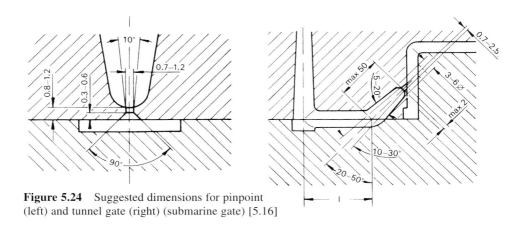

Figure 5.24   Suggested dimensions for pinpoint (left) and tunnel gate (right) (submarine gate) [5.16]

The position of the gate determines the direction of the material flow within the cavity. This causes so-called orientation, i.e. alignment of the molecules. Since the properties along and perpendicular to a molecule are very different, this also applies to many molded-part properties, e.g., the strength properties and shrinkage of moldings parallel to and perpendicular to the direction of flow. This effect, which is due to the orientation of the molecules, is all the more pronounced, the more the melt is sheared when it is freezing. The degree of orientation is therefore particularly high in thin-walled articles. The best values for tensile and impact strength are achieved in the direction of flow, while perpendicular to it, reduced toughness and increased tendency to stress cracking can be expected. Figures 5.25 to 5.27 exemplify the flow path of the melt for different gate positions and the effect on the strength of the molded part.

Before the mold is made, one has to clarify the type of loading and the direction of the principal stress. This is even more important with fiber-reinforced materials because the fibers should have the same direction as the maximum tensile stress in the molded part under load. Only in this direction do they sufficiently support the load.

In unreinforced high-viscosity materials, shrinkage always is a minimum in the direction of orientation (Figure 5.28). Such differential shrinkage can lead to distortion. This will be particularly extensive if, as in the case of fiber-reinforced materials, contraction in the fiber direction is suppressed and virtually only transverse shrinkage occurs.

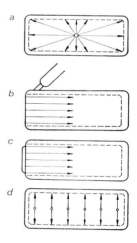

Figure 5.25 Flow path of melt with gates in various positions [5.11]
a Central sprue or pinpoint gating,
b Lateral standard gating causing desired turbulent flow,
c Edge gating,
d Multiple pinpoint gating

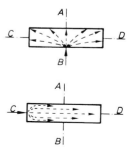

Figure 5.26 Molecular orientation perpendicular to flow of material with gate located at the long side. The mechanical strength in cross-section C–D is greater than in cross-section A–B [5.17]

Figure 5.27 Molecular orientation perpendicular to flow of material with gate located at the short side. The mechanical strength in cross-section A–B is greater than in cross-section C–D [5.17]

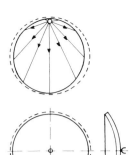

Figure 5.28   Effect of gate position on quality of a CAB molding [5.7]
top: Eccentric gating, shrinkage in the direction of flow is smaller than in transverse direction.
bottom: Centric gating results in concavity because of greater shrinkage in circumferential than in radial direction

Highly critical, too, is the occurrence of weld or knit lines where one flow of melt meets another and they are unable to penetrate each other. There are thus no molecules present that can absorb the forces at right angles to the direction of flow (Figure 5.29). Such lines are always optical defects and mechanically very weak in a fiber reinforced melt or in such materials which exhibit a liquid-crystalline structure. The further the weld lines are from the gate, the colder are the surfaces of the converging melt flows. They are thus all the more difficult to weld, i.e., they are the more critical weak points of the molded parts. This can be remedied by ensuring at later filling times or under holding pressure that the melt crosses them again at right angles. Modern gating techniques, such as cascade gates, can be used to obtain such effects.

On the other hand, in the case of parts that feature many flow obstructions, such as edge connectors (Figure 5.30), multiple pinpoint gating is perfectly adequate because, due to the short flow paths between two gates, the melt surfaces weld together well, i.e., the weld lines cannot form weak points. Figures 5.31 to 5.33 show further examples. Because separation is easy and can be automated, multiple pinpoint gates (Figure 5.31) are usually preferred over the otherwise better edge gates (Figure 5.32).

It is no longer a problem to get an idea during the design phase of how a certain gate position affects the quality of the molded part, because simulation software such as CADMOULD provide highly realistic results. But even simple graphical methods will convey an impression quickly (see Section 5.9: Flow Pattern Method).

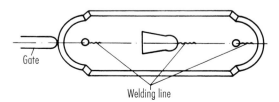

Figure 5.29   Knit lines behind holes or slots result in points of reduced strength [5.17]

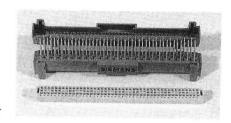

Figure 5.30 Edge connector

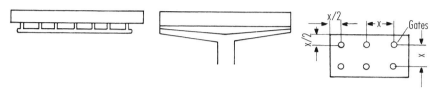

Figure 5.31 Multiple pinpoint gating

Figure 5.32 Edge gating

Figure 5.33 Principle of equal flow lengths

5.8 Runners and Gates for Reactive Materials

Minimizing the volume of the runners is particularly important for both elastomers and thermosets because this fraction of the material can only be recycled to a certain extent and generally needs to be disposed of. For economic reasons, however, multi-cavity molds with extensive runner systems are used on a large scale, with the result that cold-runner systems are gaining in importance as a means of reducing the material costs incurred. The design of the runner systems is basically the same as that of thermoplastic materials.

5.8.1 Elastomers

These materials are usually filled and so are highly viscous. They therefore use up almost all the injection pressure to overcome the resistance of the runner systems. Filling of the cavities, which often have large cross-sections, requires hardly any pressure. However, there is a risk of jetting. It is frequently thought that filling the mold by jetting rather than by frontal flow does not pose a problem since the curing process will largely eliminate the weakness of the weld lines. This is not correct. From a processing point of view, jetting is unfavorable because the molded part is not filled in a defined manner. For example, air can be introduced during injection and this will lower the quality (e.g., in burners). Furthermore, the stream of material formed can start to crosslink and this will lower the strength.

The high pressure consumption in the runner system mostly results in an opening of the mold in the parting line with extensive flashing (Figure 5.34). This creates very different filling conditions in the individual cavities that result in varying orientation and only partly filled, faulty molded parts. Furthermore, this flashing requires costly

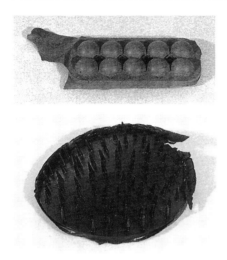

Figure 5.34   Elastomer molding with flash

machining and leads in the long term to destruction of the runner system. Only an adequately large and balanced runner system can remedy this. However, this is very problematic because the slightest temperature differences exert a considerable influence on the flow properties in the case of elastomers. To avoid an extensive gate system, recourse may be made to injection transfer molding (see Section 20.2), which is particularly suitable for small parts whose production requires no machining and generates little scrap.

5.8.2   Thermosets

The same systems in terms of arrangement, design and dimensions are used as described in Sections 5.6 and 5.7. Good results have even been achieved with tunnel gates, which allow the process to be automated. However, it is advisable to use inserts made of highly wear-resistant steels or those coated with appropriate hard materials for the gates when processing thermosets containing mineral powder or fibers, as these cause even more wear due to their low viscosity than do reinforced thermoplastics. In a series of tests, wear resistance was successfully bestowed on runners and cavities by chrome plating or other hard coatings.

5.8.3   Effect of Gate Position for Elastomers

The more complicated the part geometry, the more complex are the flow processes in the cavity. Although knit lines are welded well due to the crosslinking reaction, they still result in rejects in certain cases. Knit lines that always occur at the same place, and other obstacles to free flow, cause increased formation of deposits at these places. The same phenomenon can be observed at the end of a filling segment. The reason for this is the evaporation, to some degree, of low-molecular components such as waxes, oils and oligomers, which are trapped by the melt and condense again. This leads to a build-up of strongly adhering deposits causing mat spots on the surface of molded parts [5.52]. The position of the knit line also plays an important role, e.g. knit lines in sealing faces that would have a major adverse effect on the functionality of the molded part.

To ensure that the molded parts are of lasting high quality, care should be taken to avoid knit lines when designing molds.

5.8.4   Runners for Highly-Filled Melts

In special plastics processing methods, such as powder injection molding, up to 65 vol.-% filler may be added to the plastic material. The resultant change in rheological and thermodynamic properties of the mixture in the melt requires particular attention when designing and dimensioning the runner system in injection molds. Since the change in material behavior depends not only on the type of filler (e.g. metallic or ceramic powder) and its proportion in the mixture, but also on its macroscopic form (fiber, powder) and microscopic geometry (fiber length and diameter, surface texture, particle size and particle-size distribution), exact knowledge of the fillers used and their effect on the properties of the melt are crucial to the proper design of the runner system.

The melt viscosity of plastics increases markedly with the filler content, so that to completely fill the cavity much higher pressure is required than when unfilled thermoplastics are used. Since this leads to high wall shear stress and corresponding material load, this effect should be counteracted by keeping the flow resistance in the runner system low. In practice, this means that an extremely short runner system with large cross-section best meets the demands of highly filled melts.

Moreover, to an extent depending on the filler and filler content (see above), the shrinkage of a molding compound is much lower than that of unfilled plastics and so a greater draft is required for sprues to ensure better demolding. The choice of tool steel depends on the abrasiveness of the melt and the filler contained therein.

The particularly highly filled polymers used in powder injection molding often have no or very little melt elasticity (no memory effect), with the result that hardly any frontal flow occurs as the cavity is being filled. Jetting is counteracted at the design stage by positioning the gate such that the jet comes into contact with a cavity wall as it enters (e.g., side gate) or impinges on a flow restriction [5.18, 5.19]. For this reason, abrupt changes in wall thickness must be avoided as these can give rise to jetting.

When positioning the runner system, care must also be taken to avoid having knit lines in the molded part. Otherwise, low-filler areas form at the flow front, and if there is any orientation through the fibers, the knit line can suffer greatly from reduced strength and rigidity and potential fracture areas may be formed. The same problem occurs when plastic and filler separate, which can happen in areas of very high shear and under the influence of centrifugal forces. To suppress such demixing phenomena, sharp bends, corners and edges in the runner system must be avoided. Demixing can also be avoided by creating solid flow in the runner. This requires polishing the walls of the runners and the gate [5.20].

Unlike unfilled plastics, highly filled melts have a much higher thermal diffusivity which can be as much as 12 times that of unfilled plastics for high filler loads of ceramic or metallic particles. The use of a cooled runner system therefore leads to increased edge layer formation during the injection phase and to premature sealing of the gate. Consequently, due to the taper in the flow cross-section, the filling pressure requirement increases and the effective holding pressure time decreases. This often results in major quality problems for injection-molded parts because, on the one hand, a high filling pressure causes pronounced orientation in the molded part that in this form – and especially with filled materials – is often undesirable and troublesome. On the other hand, the maximum attainable flow-path lengths and the minimum part wall thickness are restricted by this. The gates and runners for molds that are intended for processing highly-filled polymers should therefore, also for these thermodynamic reasons, have a larger cross-section than is usual for thermoplastic molds.

An alternative to large-dimensioned gates, that also serves to counteract freezing effects, is hot-runner systems. These permit much longer, more selective influence to be exerted on the molded-part-formation process in the holding-pressure phase [5.21]. Since the formation of a frozen edge layer in the runner is suppressed, pressure losses are reduced. The disadvantage of this is the need for elaborate, thermal insulation of the hot-runner system toward the cavity with the risk of high orientation near the gate, where the material remains molten for a long time. This problem has been successfully eliminated in powder injection molding by using combinations of hot runners with short, freezing gates [5.22, 5.23]. The effect is that the areas of high orientation are pushed into the gate area to be later removed and therefore do not have any effect on the quality of the molded part.

5.9   Qualitative (Flow Pattern) and Quantitative Computation of the Filling Process of a Mold (Simulation Models) [5.24]

5.9.1   Introduction

It is often necessary to study the filling process of a finished mold in advance, that is during the conception of mold and molding. Examinations of this kind are generally summarized under the generic expression "rheological design" [5.25, 5.28] and make a qualitative and quantitative analysis of the later flow process possible. Qualitative analysis here is the composition of a flow pattern, which provides information concerning

– effective kind and position of gates,
– ease of filling individual sections,
– location of weld lines,
– location of likely air traps and
– directions of principal orientation.

Aids for theoretically composing a flow-pattern are the flow-pattern method [5.27 to 5.29] and calculation software for computers capable of graphics [5.29, 5.30].

The second step is the quantitative analysis. This is a series of calculations, which include the behavior of the material and assumed processing parameters. They determine mold filling data such as

– pressures,
– temperatures,
– shear rate,
– shear stresses, etc.

With the help of these calculations the effect of planned design features can be estimated, e.g.

– properties of the molded part,
– strength of weld lines,
– surface quality,
– damage to material,
– selection of material and machine,
– suitable range of processing, etc.

5.9.2   The Flow Pattern and its Significance

A flow pattern pictures the courses of the flow fronts in different areas of the cavity at various stages of the mold-filling process. The theoretical filling image corresponds with the production of short shots in a finished mold. Figures 5.35 to 5.37 demonstrate a comparison between a theoretical flow pattern and a series of short shots from a practical test.

Producing a flow pattern during part or mold design is beneficial because it makes it possible to recognize the location of weld lines and air trappings early on, that is before the mold is made.

If such problems are recognized, one can examine how the mold filling can be improved by

– a variation of position, kind and number of gates,
– a variation of the location of holes or different thicknesses of sections, which are demanded by design,
– introduction of facilities or restraints for the melt flow.

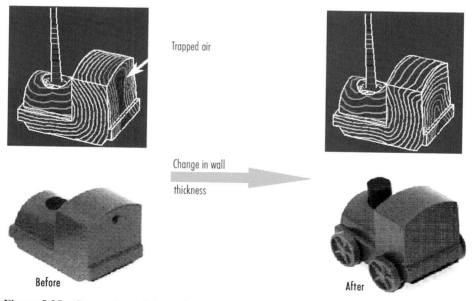

Figure 5.35   Comparison of theoretical flow pattern and injection trial

Figure 5.36   Series of short shots illustrating filling of box mold [5.24]

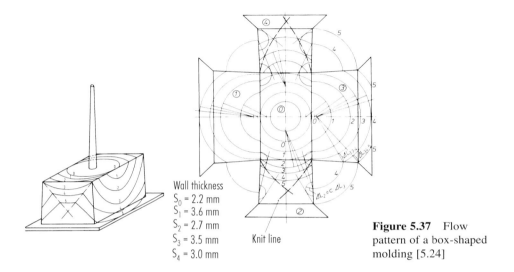

Wall thickness
$S_0 = 2.2$ mm
$S_1 = 3.6$ mm
$S_2 = 2.7$ mm
$S_3 = 3.5$ mm
$S_4 = 3.0$ mm

Knit line

Figure 5.37   Flow pattern of a box-shaped molding [5.24]

The production of a flow pattern is a pre-condition for the use of programs to compute pressure and temperature during the filling stage. Accurate computation requires a mental break down of the molded part or cavity into computable basic segments, which are established on the basis of the flow pattern.

5.9.3   Using the Flow Pattern for Preparing a Simulation of the Filling Process

For producing a flow pattern one starts with a plane presentation of the part or its development onto a plane.

In essence, three geometrical operations are necessary for such a development:

– cutting open a surface along an edge,
– turning a face around a fixed axis,
– stretching a curved surface (flatten it onto the plane of the paper).

The following considerations and simplifications result in developments, which favor the later production of a flow pattern:

– If possible, the actual part should be divided into subsections, which can be developed in a simple way (making the cut along an existing edge). The correlation in the developed presentation is then done by identifying the common connecting lines (cutting edges) or points (Figure 5.38).

– That face is the starting surface, on which the gate(s) is (are) placed or on which the longest flow paths can be expected.

– Areas, which cannot be directly included in the development of connected part sections (e.g. ribs), are folded separately onto the paper plane.

– Connection points of areas (ribs), which are presented in subdevelopments, have to be clearly identified (Figure 5.38).

– Another way of developing complex parts begins with a paper model, which is subsequently cut open.

– Conversely, a flow pattern can give a clear idea by cutting out the individual sections and joining the parts.

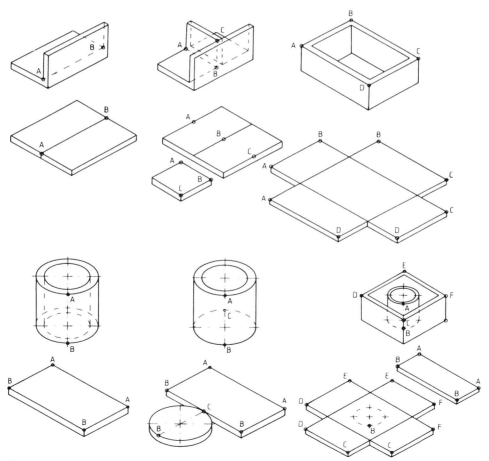

Figure 5.38   Examples for development on a plane [5.24]

5.9.4   Theoretical Basis for Producing a Flow Pattern

According to Hagen-Poiseuille the pressure demand to counter the resistance to flow in plate-like channels (W > H) is [5.31]:

$$\Delta p = 32 \; \varphi \bar{v}_F \; \frac{L \cdot \eta}{H^2 \cdot W} \qquad (5.1)$$

Wherein

Δp	Pressure consumption,
φ	Factor for real molds,
	For molds with W >> H $\rightarrow \varphi = 1.5$,
$\bar{v}_F$	Velocity of advancing flow front,
L	Length of flow path,
W	Width of segment,
H	Height of a segment,
η	Viscosity of fluid.

Since only the flow front is considered, it is permissible to assume that

a) all factors everywhere along the flow front are equal. This is the case as long as the whole range has the shape of a plate with uniform thickness;
b) the pressure is uniform, which automatically follows for the flow front;
c) the viscosity along the flow front is the same. This is the case as long as the melt along the flow front is of equal temperature and no larger differences in height exist, which would change the intrinsic viscosity, that is $H_1 / H_2 < 5$.

Always equally wide segments of the flow front are considered, that is $W_1 = W_2$. Thus, for two points of the flow front with different thickness H follows

$$v_{F_1} \frac{L_1}{H_1^2} = v_{F_2} \frac{L_2}{H_2^2} \qquad (5.2)$$

One introduces

$$v = \frac{\Delta L}{\Delta t}$$

and looks at the movement of the flow front in equal time steps so that $\Delta t = $ const.

then

$$\frac{\Delta L_1 L_1}{H_1^2} = \frac{\Delta L_2 L_2}{H_2^2} \qquad (5.3)$$

and because
$L_1 = \Delta L_1$ and $L_2 = \Delta L_2$
the result is

$$\frac{\Delta L_1}{H_1} = \frac{\Delta L_2}{H_2} \tag{5.4}$$

The advance of the flow front in equal time intervals corresponds with the ratio of thicknesses at the points considered.

Present experience demonstrates a very good agreement between flow pattern and practical results as long as one can assume that the flow front is uniformly supplied with melt. This holds true in all cases even for very different materials from very fluid reactive polyurethanes or caprolactams to elastomers and filled thermosets. If, however, narrow cross-sections, such as a living hinge, restrict the filling, then a deviation from the location of weld lines in the following area can be noticed. This, however, has never really interfered with the qualitative prediction of the location of weld lines. If this method, however, should serve to compute pressure and temperature in a simulation program (CADMOULD or MOLDFLOW) then the result has a considerable error in such cases.

5.9.5   Practical Procedure for Graphically Producing a Flow Pattern

5.9.5.1   Drawing the Flow Fronts

The model on which the method of the flow pattern is based, relies on the theory of wave propagation according to Huygens. It implies that every point of an "old" wave front (flow front) can be considered the starting point (center) of a so-called elementary wavelet (circular wave). The envelope of the new elementary wavelets is the new (next) wave front (flow front). The "new" flow front is the envelope of the new elementary wavelets, which expand in circular form from every point of the last flow front. The radius of every elementary wavelet is equal to the advance of the flow front Δl (Figures 5.39 and 5.40).

5.9.5.2   Radius Vectors for the Presentation of Shadow Regions

Areas of a part which are located in the "shadow" of openings cannot be directly reached by parallel or swelling flow from the gate. They are filled beginning from a flow front (Figure 5.41).

Vectors starting at the gate indicate these regions and offer points of support for producing the flow front (Figure 5.42).

The points P, where the vectors are tangent to the opening, are, as points of an "old" flow front, the origins of new elementary wavelets. They start the filling of the shadow regions (Figure 5.43 to 5.45).

With complex shapes of openings or barriers to flow it may be necessary to introduce more vectors during the process to completely cover the flow around them (Figure 5.46).

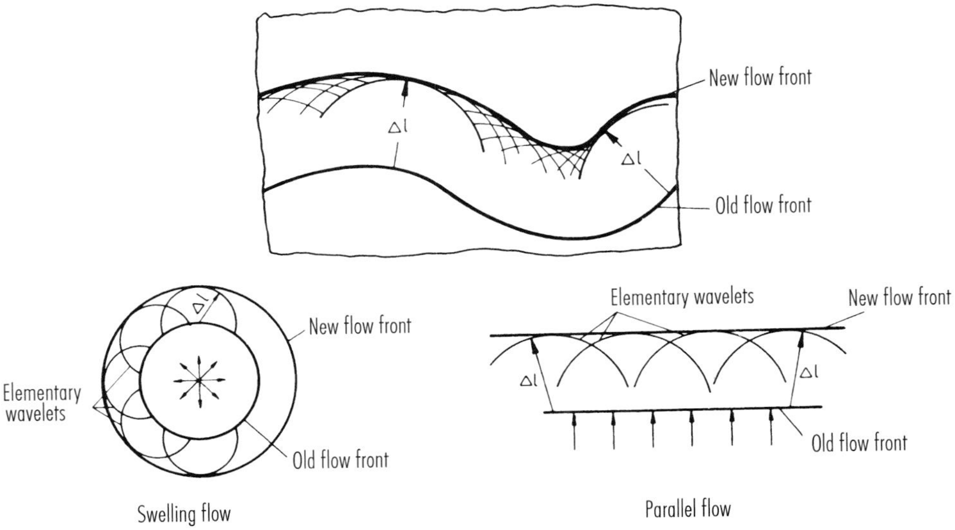

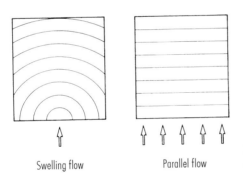

Figure 5.39   Methodology of designing a flow pattern [5.24]

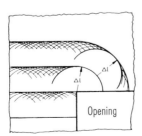

Swelling flow   Parallel flow

Figure 5.40   Application with pinpoint gate (left) and edge gate (right) [5.24]

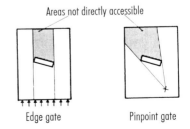

Figure 5.41   Design of a flow front with elementary wavelets behind an opening [5.24]

Figure 5.42   Vectors outline areas which are not directly accessible [5.24]

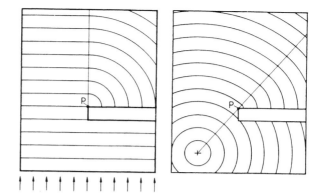

Figure 5.43   Design of a flow front behind flow barriers [5.24]
right: Edge gate,
left: Pinpoint gate

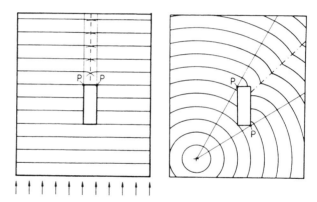

Figure 5.44   Design of flow front behind a rectangular opening [5.24]
right: Edge gate,
left: Pinpoint gate
+   gate
−·−   weld line

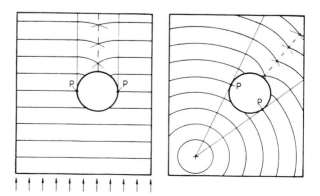

Figure 5.45   Design of flow front behind circular opening [5.24]
right: Edge gate,
left: Pinpoint gate

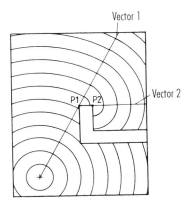

Figure 5.46   Design of flow front with several vectors [5.24] (pinpoint gate)

5.9.5.3 Areas with Differences in Thickness

A special benefit of the flow pattern method is the correct determination of the filling process even if there are differences in thickness (height). For a one-time step there is the relation

$$\frac{\Delta l}{H} = \text{const.} \tag{5.5}$$

This means that the ratio of advancement of the flow front Δl and its height H is the same in different regions of the cavity during the same time intervals Δt.

This relation becomes evident with center-gated plates, one with constant thickness, the other one with twice the thickness in one half (Figure 5.47).

The tangential design as an aid for producing a flow pattern is used if a continuous flow front is drawn in adjacent areas with different section thickness. The flow from a region with a thicker section into one with a thinner one is approximated by linear interpolation between known points of the "new" front. This method is an approximation of the central design, which will be discussed later.

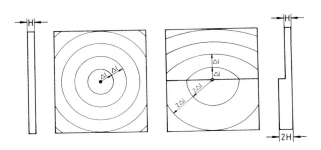

Figure 5.47   Flow front with varying wall thickness [5.24]

Various steps of this method:

1. The last flow front just touches the border of the region with a thicker section (II) at the point P (Figure 5.48).
2. First the continuation Δl_I in the old region is drawn. This results in the known points A and B of the new front, at the border of the region II. Starting at the point P of the old flow front the circle of an elementary wavelet is outlined in the region II. The radius results from the rule $\Delta l_{II} = \Delta l_I$ H_{II}/H_I (Figure 5.49).

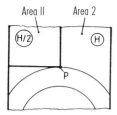

Area II   Area 2

Figure 5.48   Tangential design, step 1 [5.24]

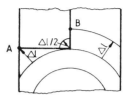

Figure 5.49   Tangential design, step 2 [5.24]

3. The tangents are drawn from the known point A and B of the "new" flow front to the circle of the swelling flow (Figure 5.50).
4. This generates the complete sequel of the new front (Figure 5.51).

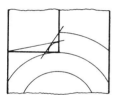

Figure 5.50   Tangential design, step 3 [5.24]

Figure 5.51   Tangential design, step 4 [5.24]

5. If the region II is thinner than the region I, the flow from I to II is pictured as described above. If the region II is thicker than region I the flow from I to II is considered by a second circle for a wavelet with tangents to it (Figure 5.52).

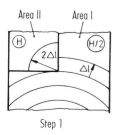

Area II   Area I

Step 1

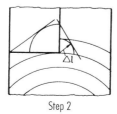

Step 2

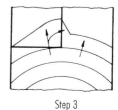

Step 3

Figure 5.52   Tangential design, $H_{II} > H_I$ [5.24]

The central design, another aid to produce a flow pattern, is a refinement of the tangential design.

Instead of a linear interpolation between known points, an interpolation of a circular arc is carried out. It should be primarily applied for

– large step sizes, or
– large differences in section thickness between adjacent regions.

Various steps of this method:

1. The last flow front just touches the border of a region with a different section thickness in the point P (Figure 5.53).
2. The new front is drawn with the advance Δl_I, in the old region. This results in known points A and B of the new flow front in the region II. Starting with point P of the old front the advance Δl_{II} is outlined in region II. It results from the rule $\Delta l_{II} = \Delta l_I \cdot H_{II}/H_I$. Thus another point C of the new flow front has been determined (Figure 5.54).
3. There are now three points of the new flow front A, B and C. The points A and C as well as B and C are connected by a straight line and their median verticals erected. Their intersection is the center M (Figure 5.55).
4. A circular arc through the points A, B and C represents the new flow front in the region II (Figure 5.56).

As a comparison the flow front according to the tangential design for the same example is shown (Figure 5.57).

Some more examples for both methods are presented in Figures 5.58 to 5.60. A combination of tangential and central design is also possible (Figure 5.61). The central design pictures the flow into region I and II and the tangential design the overflow from I to II (Figure 5.62).

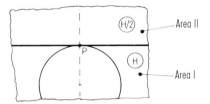

Figure 5.53   Central design, step 1 [5.24]

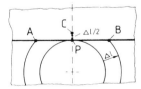

Figure 5.54   Central design, step 2 [5.24]

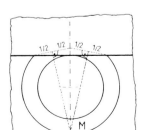

Figure 5.55 Central design, step 3 [5.24]

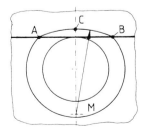

Figure 5.56 Central design, step 4 [5.24]

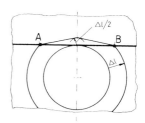

Figure 5.57 Flow front based on tangential design [5.24]

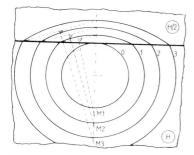

Figure 5.58 Determination of flow pattern with central design [5.24]

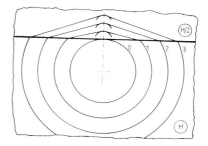

Figure 5.59 Determination of flow pattern with tangential design [5.24]

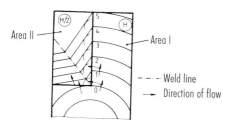

Figure 5.60 Determination of flow pattern with tangential design [5.24]

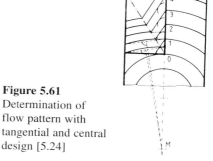

Figure 5.61 Determination of flow pattern with tangential and central design [5.24]

Here, tangential and central design can also be combined (Figure 5.63). The first one pictures the overflow from region II to I and the second one the flow into regions I and II.

5.9.5.4 Flow Patterns of Ribs

First, ribs are looked at which are thinner than the base (Figure 5.64). They are filled from the base and do not affect the flow pattern of the main body.

In Figure 5.64 venting should be provided in the corner C; the rib should not be machined as a closed pocket.

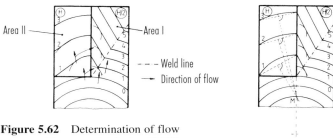

Figure 5.62   Determination of flow pattern with tangential design [5.24]

Figure 5.63
Determination of flow pattern with tangential and central design [5.24]

Another example is shown with Figure 5.65. Here provisions should be made for venting at the points A and B.

The filling of ribs which are thicker than the base is demonstrated with Figure 5.66. The advancement of melt in a thick rib affects the flow pattern of the base area. Venting should be provided in the corner C and at the ends of both weld lines.

5.9.5.5   Flow Pattern of Box-Shaped Moldings

In the development, connecting lines (cutting lines) of continuous regions are presented as curves, in special cases as straight lines. For the determination of a filling image the advances of the flow have to be picked off from one connecting line and transferred to the proper "opposite" connecting line to correctly determine an advance of flow or an overflow (Figure 5.67).

5.9.5.6   Analysis of Critical Areas

Knit lines are created by several merging melt flows. They occur, in any case, behind openings.

In a flow pattern, points of a knit line are recognized as "breaks" in the course of the flow front. Then the knit line results from a connecting line of these points.

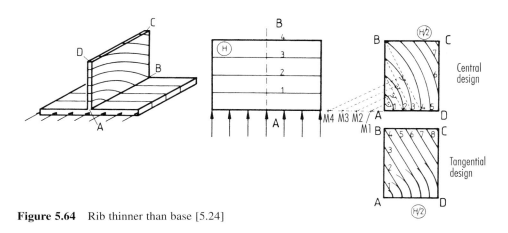

Figure 5.64   Rib thinner than base [5.24]

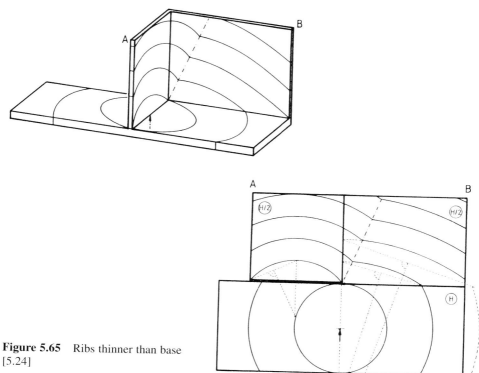

Figure 5.65   Ribs thinner than base [5.24]

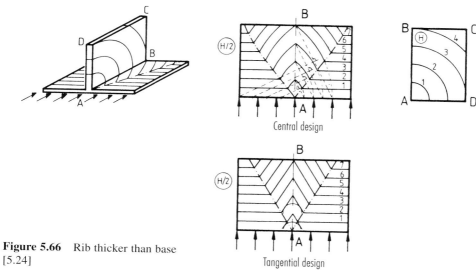

Figure 5.66   Rib thicker than base [5.24]

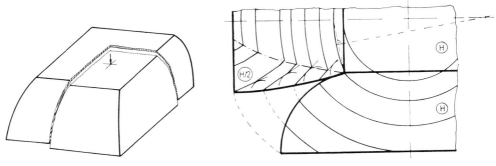

Figure 5.67   Junction in a box-shaped molding [5.24]

The smaller the angle of two merging flow fronts becomes, the more pronounced the knit line (and the reduction in quality) is (Figure 5.68).

Weld lines close to the gate are less critical than those far from it because they are created at relatively high temperatures and are possibly passed through by following melt. Weld lines far from the gate may present weak points because of cool melt and poor welding.

Air trapping occurs if flow fronts merge and the air cannot escape through a parting line or any other way (Figure 5.69). Besides incomplete filling, burned spots may be the consequence.

The faces at the parting lines should be carefully ground (grain size 240, but not finer).

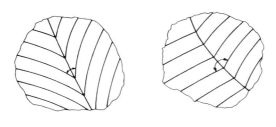

Figure 5.68   Weld lines [5.24]
left: Pronounced
right: Insignificant

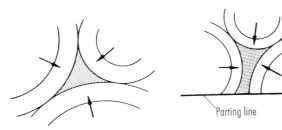

Figure 5.69   Air trapping [5.24]
left: Crucial
right: Not crucial

If the air cannot escape through an existing parting line, the problem can be redressed by:

– use of an additional parting line,
– relocation of ejector pins into areas of air trapping (air escapes through holes),
– venting pins,
– relocation of gate,
– modification of section thickness of part (see Chapter 7).

5.9.5.7   Final Comments

The following recommendations should provide help with complicated applications. They are based on frequent experience with flow patterns in practice.

– The *design with elementary wavelets* is well suited for the continuation of a flow front in regions of equal section thickness. It is always applicable and should be employed if difficulties arise with a different design.
– The *central design* is suggested if a region of different section thickness has to be filled. It is more accurate than the tangential design but also more work.
– The *tangential design* should be used if an overflow from a heavier section into a region with a thinner section perpendicular to the direction of flow takes place. Although the central design can be employed here, too, it takes much more time, in particular if for every flow front a new center has to be found.
– The maximum *"step width"* should be selected so long that a "problem point" is just reached.

Problem points are

– reaching of a region of different thickness,
– filling of a "shadow area" (radius vector),
– overflow from a region with heavier into one with thinner section (perpendicular to the direction of flow in the heavier section),
– merging of flow fronts (weld lines, air trapping).

5.9.6   Quantitative Analysis of Filling

A quantitative analysis is based on the concepts of fluid dynamics (rheology) and thermodynamics. It has to solve the fundamental equations for continuity, momentum, and energy. This is a matter of an interactive system of differential equations (Figure 5.70). For such a complex geometry as the cavity of an injection mold, no exact solution can be found, of course, so that some method of approximation has to be used. There are two possibilities:

a) The geometry is subdivided by a set of a finite number of nodal points equally spaced by small time intervals. Each point represents a finite difference approximation. With the resulting set of difference equations a final solution can be found [5.26, 5.27].
b) The geometry is divided into subregions called finite elements. An approximation with small time intervals is defined within each element according to known procedures and appropriate continuity conditions imposed on the inter-element boundaries (Figure 5.71) [5.9].

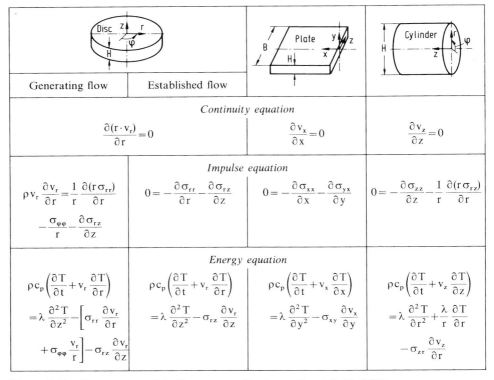

Figure 5.70   Computation of the flow process (basic equations) [5.26, 5.27]

Both methods call for the application of computers, which will be treated in detail in Chapter 14. They are suitable for other material melts or reactive liquids (RIM) as well.

5.9.7   Analytical Design of Runners and Gates

5.9.7.1   Rheological Principles [5.32]

A general flow is completely described by the conservation laws of mass, momentum and energy and by a rheological and thermodynamic equation of state. The rheological equation of state, also called the law of materials, describes the relation between the flow rate field and the resultant stress field. This accounts for all the flow properties of the polymer concerned. The describing, explaining and measuring of the flow properties of materials is a key subject of the science of deformation and flow of matter, known as *rheology* [5.33]. The principles of rheology will be introduced in this chapter in as far as they are necessary for the design of gates and runners from an engineer's point of view. Polymer melts do not exhibit purely viscous behavior, but possess a not inconsiderable amount of elasticity. Their properties, therefore, lie between those of an ideal fluid and an ideal (Hookean) solid. They are said to exhibit *viscoelastic behavior* or *visco-*

Generating flow	Established flow	Plate	Cylinder

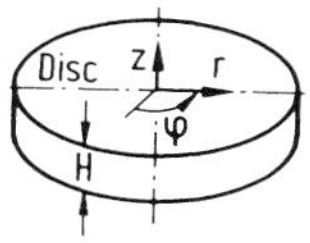

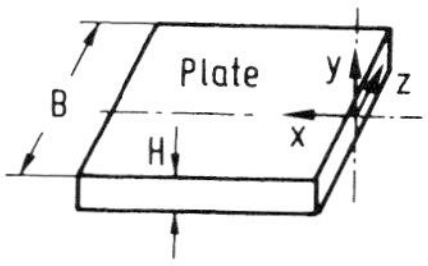

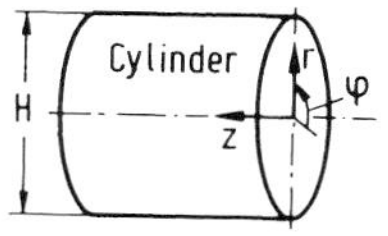

Velocity profile

$$v_r = \int_0^{H/2} \frac{1}{\eta} \left\{ \int_0^z \left[\left(-\frac{\partial p}{\partial r} - \frac{v_r}{r} \right) \cdot \frac{1,5\,\eta_0}{r} - \rho \cdot v_r \right] dz \right\} dz$$

$$v_r = \left(-\frac{\partial p}{\partial r} \right) \cdot \int_z^{H/2} \frac{1}{\eta} z\, dz$$

$$v_x = \left(-\frac{\partial p}{\partial x} \right) \cdot \int_y^{H/2} \frac{1}{\eta} y\, dy$$

$$v_z = \frac{1}{2} \left(-\frac{\partial p}{\partial z} \right) \cdot \int_r^{R} \frac{1}{\eta} r\, dr$$

Mean velocity

$$\bar{v}_r = \frac{2}{H} \int_0^{H/2} v_r\, dz$$

$$\bar{v}_r = \frac{2}{H} \left(-\frac{\partial p}{\partial r} \right) \cdot \int_0^{H/2} \frac{1}{\eta} z^2\, dz$$

$$\bar{v}_x = \frac{2}{H} \left(-\frac{\partial p}{\partial x} \right) \cdot \int_0^{H/2} \frac{1}{\eta} y^2\, dy$$

$$\bar{v}_z = \frac{1}{2R^2} \left(-\frac{\partial p}{\partial z} \right) \cdot \int_0^{R} \frac{1}{\eta} r^3\, dr$$

Shear rate

$$-\frac{\partial v_r}{\partial z} = \frac{1}{\eta} \int_0^z \left[\left(-\frac{\partial p}{\partial r} \right) - \frac{v_r}{r} \left\{ \frac{1,5\,\eta_0}{r} - \rho v_r \right\} \right] dz$$

$$-\frac{\partial v_r}{\partial z} = \left(-\frac{\partial p}{\partial r} \right) \cdot \frac{1}{\eta} \cdot z$$

$$-\frac{\partial v_x}{\partial y} = \left(-\frac{\partial p}{\partial x} \right) \cdot \frac{1}{\eta} \cdot y$$

$$-\frac{\partial v_z}{\partial r} = \frac{1}{2} \left(-\frac{\partial p}{\partial z} \right) \cdot \frac{1}{\eta} \cdot r$$

Pressure gradient implicit in velocity profile

$$\left(-\frac{\partial p}{\partial r} \right) = \frac{\dot{V}}{4\pi r \int_0^{H/2} \frac{1}{\eta} z^2\, dz}$$

$$\left(-\frac{\partial p}{\partial x} \right) = \frac{\dot{V}}{2B \int_0^{H/2} \frac{1}{\eta} y^2\, dy}$$

$$\left(-\frac{\partial p}{\partial z} \right) = \frac{2\dot{V}}{\pi \int_0^{R} \frac{1}{\eta} r^3\, dr}$$

$$\dot{V} = 2 \cdot \pi \cdot r \cdot H \cdot \bar{v}_r$$

$$\dot{V} = B \cdot H \cdot \bar{v}_x$$

$$\dot{V} = \pi \cdot R^2 \cdot \bar{v}_z$$

Figure 5.71 Computation of the flow process (definitive equations) [5.26, 5.27]

elasticity. It is therefore common to distinguish between material data pertaining to purely viscous behavior and material data pertaining to elastic behavior when describing rheology.

Viscous Melt Properties

In the flow processes of the kind that occur in injection molds, the melt is mainly sheared. This so-called shear flow is due to adhesion of the polymer melts to the surfaces of the mold halves shaping them (Stokes' adhesion). The result is a change in flow rate across the channel cross-section that is described by the shear rate:

$$\dot{\gamma} = -\frac{dv}{dy} \tag{5.6}$$

v   Flow rate,
y   Direction of shear.

Under stationary shear flow, shear stress τ occurs between the fluid layers. In the simplest case of a *Newtonian fluid*, the shear stress τ is proportional to the shear rate $\dot{\gamma}$, yielding:

$$\tau = \eta \cdot \dot{\gamma} \tag{5.7}$$

The proportionality factor is called the dynamic shear viscosity or simply *viscosity*.
    It has the unit Pa s. The viscosity is a measure of the internal flow resistance of a sheared fluid.
    In general, polymer melts do not exhibit Newtonian behavior. Their viscosity is not constant but instead depends on the shear rate. By analogy with the equation for Newtonian fluids, Equation (5.7), the flow law is as follows:

$$\tau = \eta(\dot{\gamma}) \cdot \dot{\gamma} \tag{5.8}$$

and

$$\eta(\dot{\gamma}) = \frac{\tau}{\dot{\gamma}} \neq \text{const.} \tag{5.9}$$

Note:
The viscosity of many polymers exhibits a more or less pronounced time dependency (thixotropy, rheopexy, lag in viscosity under sudden shear or strain [5.33–5.35]). However, this time dependency is usually ignored when injection molds are being designed because usually the appropriate data are not available. And yet, this supposition can cause errors, albeit small ones, in calculations. The time dependency will be ignored in the following discussion.

Viscosity and Flow Function
A double logarithmic plot of viscosity as a function of shear rate (at constant temperature) for polymers has the basic shape shown in Figure 5.72. It can be seen that the viscosity remains constant at low shear rates but declines as the shear rate increases.
    Behavior of the kind that the viscosity decreases with the shear rate is termed *structural viscosity* or *pseudoplasticity*. The constant viscosity for low shear rates is

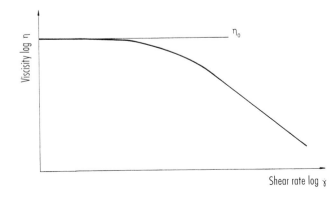

Figure 5.72   Viscosity as a function of shear rate in the form of a viscosity curve

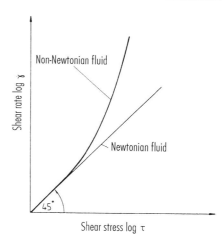

Figure 5.73   Shear rate as a function of shear stress in the form of a flow curve

called the Newtonian viscosity, lower Newtonian intrinsic viscosity or when $\dot{\gamma} = 0$ as *zero viscosity 0.*

Not only can the viscosity be plotted as a function of the shear rate to yield a *viscosity curve*, but the shear stress can be plotted against the shear rate (also on a double logarithmic scale) to yield a *flow curve* (Figure 5.73).

In a Newtonian fluid, the shear rate is proportional to the shear stress. A double logarithmic plot of the pairs of values yields a straight line with a slope of 1, i.e., the angle between the abscissa and the flow curve is 45°. Any deviation by the flow curve from this slope is therefore a direct indicator of non-Newtonian flow behavior.

A pseudoplastic fluid would have a slope greater than 1 on this plot, i.e., the shear rate increases progressively with increase in shear stress.

Mathematical Description of Pseudoplastic Melt Behavior
Various mathematical models have been developed to describe the viscosity and flow curves; they differ in the amount of mathematics involved on the one hand and in their adaptability to concrete experimental data and thus accuracy on the other. Overviews are provided in [5.33, 5.36]. The models most commonly employed for thermoplastics and elastomers are briefly discussed below. In the following chapters, reference will be made exclusively to the models described here.

Power Law of Ostwald and de Waele [5.37, 5.38]
When the flow curves of different polymers are plotted on a double logarithmic scale, the resultant curves have two approximately linear ranges separated by a transition range (Figure 5.74). In many cases, we are often only dealing with one of the two ranges and so we only need one function, namely

$$\dot{\gamma} = \Phi \cdot \tau^{m} \tag{5.10}$$

to describe it mathematically. Equation (5.10) is called the power law of *Ostwald* and *de Waele*, where the exponent m is the flow exponential and Φ is the fluidity. The flow exponent m characterizes the flow capability of a material and its deviation from non-Newtonian behavior. The following applies:

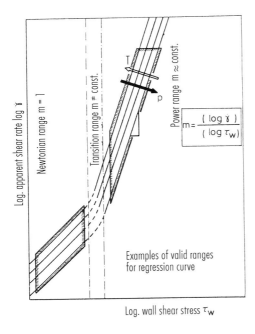

Figure 5.74   Approximation of the flow curve by a power law [5.32]

$$m = \frac{\Delta \lg \dot{\gamma}}{\Delta \lg \tau} \tag{5.11}$$

m is the slope of the flow curve in the range under consideration on a double logarithmic plot (Figure 5.74).

For polymer melts, m usually has a value in the range 1 to 6; for the shear-rate range of 10^0 to 10^4 s^{-1}, this range is mostly 1 to 4. When m = 1, $\Phi = 1/\eta$, i.e. the liquid is Newtonian.

Since

$$\eta = \tau / \dot{\gamma}$$

Equation (5.10) can be rearranged to yield the viscosity function:

$$\eta = \Phi^{-1} \cdot \tau^{1-m} = \Phi^{-\frac{1}{m}} \cdot \dot{\gamma}^{\left(\frac{1}{m}-1\right)} \tag{5.12}$$

Since

$$k = \Phi^{-\frac{1}{m}}$$

$$n = \tfrac{1}{m}$$

we arrive at the usual form of the viscosity function

$$\eta = k \cdot \dot{\gamma}^{n-1} \tag{5.13}$$

k is called the consistency factor. It indicates the viscosity at a shear rate of 1/s. The viscosity exponent n is equal to 1 for Newtonian behavior and is in the range 0.7 and 0.2

for most polymers. It describes the slope of the viscosity curves in the range under consideration (Figure 5.74).

The power law has a very simple mathematical construction and therefore allows almost all simple flow problems to be analyzed that can be resolved by the Newtonian model. The disadvantage of this approach is that, on vanishing shear rate, the viscosity value becomes infinitely large when the power law of Ostwald and de Waele is used, with the result that the roughly shear-rate-independent Newtonian range is not described. A further disadvantage is that the flow exponent m is included in the dimension of the fluidity.

In general, the power model is only accurate enough to describe a flow or viscosity curve over a certain shear-rate range. For a given accuracy, the size of this range depends on the curvature of the curve.

Where the power law is to be used to describe a large range of a flow curve, the curve can be divided into segments. Φ and m have to be determined for each segment [5.39].

Standard collections of rheological data [5.40, 5.41] are available that list values for Φ and m for various shear-rate ranges.

Carreau Model [5.41–5.43]
A model that is increasingly being used in practical injection-mold design is the triple parameter model of *Carreau*:

$$\eta(\dot{\gamma}) = \frac{A}{(1 + B \cdot \dot{\gamma})^C} \tag{5.14}$$

$[A]$ = Pa s, $[B]$ = s and $[C]$ = –

Here, A is the zero viscosity, B is the so-called reciprocal transition rate and C is the slope of the viscosity curves in the pseudoplastic range for $\dot{\gamma} \to \infty$ (Figure 5.75). This Carreau model has the advantage that it correctly reflects the actual material behavior over a broader shear-rate range than that afforded by the power law, and that it also yields meaningful viscosity values for $\dot{\gamma} \to 0$.

Moreover the dependency of the pressure loss on the volumetric flow rate can be calculated for a cylindrical gate and a plane-parallel gap in analytically closed form [5.41, 5.43]. This model can therefore also be used for making rough calculations with a calculator. This applies particularly when easier-to-handle approximate solutions are to be used instead of the analytically correct solution [5.41, 5.43].

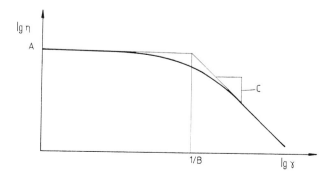

Figure 5.75 Approximation of the viscosity curve by the Carreau law [5.32]

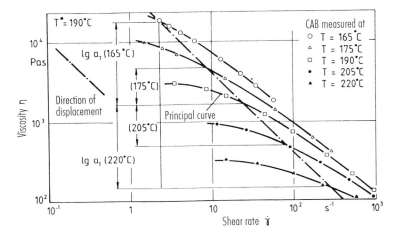

Figure 5.76
Viscosity function for CAB at different temperatures [5.32]

Temperature Influence

A double logarithmic plot of the viscosity values for the same polymer melt at different temperatures (Figure 5.76) reveals the following information:

– First, the temperature has a much greater influence on the viscosity at low shear rates than at high rates, particularly in the zero viscosity range.
– Second, although the position of the viscosity curves in the diagram varies according to temperature, their shape remains the same.

It can be shown for almost all polymer melts (so-called thermo-rheologically simple fluids [5.44]) that the viscosity curves can be converted into a single temperature-invariant master curve. The temperature displacement factor can be calculated in various ways [5.41] of which the most common, namely the WLF equation, will be presented here. With the aid of this equation, which was developed by Williams, Landel and Ferry [5.45], a viscosity curve at any arbitrary known reference temperature T_0 may be displaced to the desired temperature T as follows:

$$\lg \frac{\eta(T)}{\eta(T_0)} = \lg \ \alpha_T = \frac{C_1(T_0 - T_S)}{C_2(T_0 - T_S)} - \frac{C_1(T - T_S)}{C_2(T - T_S)} \tag{5.15}$$

where T_0 is the reference temperature and T_S the standard temperature.
For $T_S \approx T_E + 50\,°C$ [5.45], where T_E is the softening temperature, $C_1 = -8.86$ and $C_2 = 101.6$.

5.9.7.2 Determining Viscous Flow Behavior under Shear with the Aid of a Capillary Viscometer

The capillary viscometer is one of the most widespread analytical instruments. It works on the principle that the liquid or melt under investigation is forced through a capillary with an annular cross-section (but a circular annular slot cross-section is also possible) under the hydrostatic pressure of its own weight or also by external excess pressure. The flow-through rate is measured; it is dependent on the viscosity.

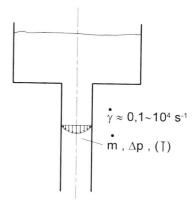

$\dot{\gamma} \approx 0{,}1{\sim}10^4 \ s^{-1}$

$\dot{m} \ , \ \Delta p \ , \ (T)$

Figure 5.77 Principle of the capillary rheometer [5.32]

The isothermal flow of Newtonian liquids in capillaries is described by the Hagen-Poiseuille law. This assumes that the flow is laminar and stationary, and that adhesion to walls occurs.

This yields a flow-through of

$$Q = \frac{\pi \cdot R^4 \cdot \Delta p}{8 \ \pi \ L}$$

(5.16)

and a wall shear stress of

$$y_w = \frac{4 \ Q}{\pi \ R^3}$$

(5.17)

Since these relations only hold for Newtonian liquids, a flow test performed on non-Newtonian materials yields an apparent shear rate D_s, which is lower at the wall (Figure 5.78) than the true shear rate.

To obtain the true shear rate for non-Newtonian liquids, a correction called the Rabinowitsch correction, is applied. This correction is based on the assumption that a power law, such as that of Ostwald and de Waele, describes the flow behavior of the melt.

If the throughput Q is known, the apparent shear rate D_s and the wall shear τ_w can be determined. Since the desired flow function must be fulfilled both at the tube wall and in the interior, the apparent shear rate can be linked to the true rate to yield the following relationship for the annular capillary:

$$y = \frac{D_s}{4}\left(3 + \frac{d(\log D_s)}{d(\log \tau_w)}\right)$$

(5.18)

And for the slot capillary:

$$y = \frac{D_s}{3}\left(2 + \frac{d(\log D_s)}{d(\log \tau_w)}\right)$$

(5.19)

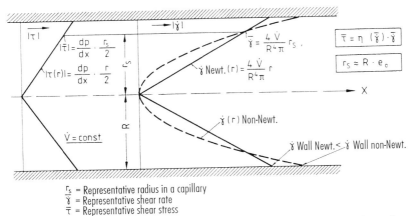

r_s = Representative radius in a capillary
$\dot{\gamma}$ = Representative shear rate
$\bar{\tau}$ = Representative shear stress

Figure 5.78   Shear rate distribution for a Newtonian and a non-Newtonian substance in laminar pipe flow [5.32]

Another possibility for pseudoplastic substances consists in determining so-called "representative variables". The aim here is no longer to look at the wall but rather to consider the point at which the two shear rate profiles overlap (Figure 5.78). For capillaries, this representative spot is always given with sufficient accuracy by $r_s = R \cdot e_0$, where $e_0 \approx 0.815$. For a slot-shaped geometry, $r_s = R \cdot e_\square$, where $e_\square \approx 0.772$. It is thus possible to describe the flow condition of non-Newtonian materials by the laws for Newtonian behavior. Given the same throughput Q, it is easy to calculate the representative shear rate.

Figure 5.79 shows the basic structure of a capillary viscometer. The material to be investigated is melted in a heated cylinder and forced through a nozzle at a defined rate by a piston to produce the measurement. The measured loss in pressure and the preset volumetric flow rate are used for calculating the viscosity. Performing the measurement on different nozzle lengths allows the flow-in pressure loss in the dead-water area at the nozzle inlet to be eliminated with the aid of the Bagley correction.

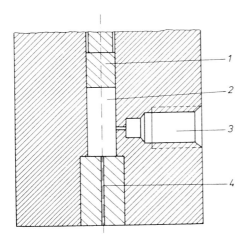

Figure 5.79   Principle of the capillary viscometer
1 Piston, 2 Cylinder, 3 Pressure sensor bore, 4 Capillary [5.32]

5.9.7.3   Elongational Viscosity

Aside from undergoing shear, the material is also subject to elongation, particularly where extreme cross-sectional changes occur, such as between the runner and the gate. Like the behavior of viscous materials under shear, viscous elongational flow can be described mathematically with the aid of a viscosity function.

The pressure losses that arise due to elongation are generally determined separately from those due to shear and are then superposed on these to form an overall pressure loss [5.46, 5.47]. Models for calculating the volumetric flow rate-pressure characteristic at changes in cross-section in flow channels have been developed by [5.48]. Formulas for calculating the additional pressure loss due to elongation in transfer molds are presented in [5.49] for the following mold areas: pressure losses in the transfer chamber and in the transition area from the transfer chamber to the gate nozzles, pressure losses in the conical nozzles and in the gates. The elongational viscosity is described by a power law and the relation between the elongational and shear viscosity for biaxial loading of the compound at low elongation rates is given by

$$\eta_D = 6\,\eta_S \tag{5.20}$$

The derived relations for the various mold areas are discussed in detail in [5.50, 5.51]. In [5.52] they are also used for designing elastomer injection molds with cold-runner systems.

To determine the elongational viscosity function is very time-consuming, a fact that explains why this material property has not yet been taken into account when injection molds are being designed.

5.9.7.4   Simple Equations for Calculating Loss of Pressure in Gates and Runners

Tables 5.1 and 5.2 list the equations for calculating the pressure drop as a function of length as well as the shear rates for various geometries of the intended runner and gate, both for Newtonian and pseudoplastic material behavior. These relations are based on the following suppositions:

– isothermal flow (all flowing mass particles at the same temperature),
– stationary flow (no temporal change in flow profile),
– laminar flow (Re < 2100),
– incompressible liquid (constant density),
– no allowance for flow-in and flow-out effects,
– wall adhesion.

By way of example, let us assume that the simple hot-runner system shown in Figure 5.80 has to be designed.

The gate diameter D_{G2} of part 2 is to be adapted such that the two parts are filled under the same conditions (simultaneous entry by the melt into the mold halves; end of flow paths reached simultaneously).

The material to be used is characterized by the following data:
Power law: consistency factor: $k = 1.14 \cdot 10^4$ kg/(ms^{2-n}); flow exponent: $n = 0.6$

$$\eta = k \cdot a_T \cdot \dot\gamma^{n-1} \tag{5.21}$$

Table 5.1   Pressure drop $\Delta p/L$ [5.34]

Geometry	Newtonian $(\tau = \eta \cdot \dot{\gamma})$	Intrinsic viscosity $\left(\tau^m = \frac{1}{\phi}\cdot\dot{\gamma}\right)$	Intrinsic viscosity, repres. quantities
Circular (tube)	$\dfrac{\Delta p}{L} = \dfrac{8\eta \cdot \dot{V}}{\pi R^4}$	$\dfrac{\Delta p}{L} = \left[\dfrac{2^m(m+3)\cdot \dot{V}}{\phi \pi R^{m+3}}\right]^{\frac{1}{m}}$	$\dfrac{\Delta p}{L} = \dfrac{8\bar{\eta}\cdot \dot{V}}{\pi R^4} = \dfrac{8\bar{\eta}\cdot \bar{\dot{\gamma}}}{\pi R} = \dfrac{8\bar{\eta}\cdot \bar{v}_z}{R^2}$
Annular aperture	$\dfrac{\Delta p}{L} = \dfrac{12\eta \cdot \dot{V}}{\pi D H^3}$; $H \ll D$	$\dfrac{\Delta p}{L} = \left[\dfrac{2^{m+1}(m+2)\dot{V}}{\phi \pi D H^{m+2}}\right]^{\frac{1}{m}}$	$\dfrac{\Delta p}{L} = \dfrac{12\bar{\eta}\cdot \dot{V}}{\pi D H^3}$ $\dfrac{\Delta p}{L} = \dfrac{8\bar{\eta}\cdot \dot{V}}{\pi(R_a^2 - R_i^2)\cdot \bar{R}}$ $\bar{R} = R_a\left(1 + k^2 + \dfrac{1-k^2}{\ln k}\right)^{\frac{1}{2}}$; $k = \dfrac{R_i}{R_a}$
Rectangular aperture	$\dfrac{\Delta p}{L} = \dfrac{12\eta \cdot \dot{V}}{B\cdot H^3}$ (B ≫ H) $\dfrac{\Delta p}{L} = \dfrac{12\eta \cdot \dot{V}}{B\cdot H^3 \cdot fp}$ (B/H ≤ 20)	$\dfrac{\Delta p}{L} = \left[\dfrac{2^{m+1}(m+2)\dot{V}}{\phi B\cdot H^{m+2}}\right]^{\frac{1}{m}}$	$\dfrac{\Delta p}{L} = \dfrac{12\bar{\eta}\cdot \dot{V}}{B H^3} = \dfrac{12\bar{\eta}\bar{v}}{H^2}$
Irregular cross-section (general)	a) $\dfrac{\Delta p}{L} = \dfrac{12\eta \cdot \dot{V}}{B\cdot H^3\cdot fp}$ (Approximation!) b) $\dfrac{\Delta p}{L} = \dfrac{2\eta U^2 \cdot \dot{V}}{A^3}$	$\dfrac{\Delta p}{L} = \left[\dfrac{(m+3)U^{m+1}\dot{V}}{2\phi A^{m+2}}\right]^{\frac{1}{m}}$ (Approximation!)	— ($\bar{\eta}$ not defined with general validity)

Table 5.2   Shear rates $\dot\gamma$ Shear rate or the mold wall $\dot\gamma_w$ [5.32]

Geometry	Newtonian $(\tau = \eta \cdot \dot\gamma)$	Non-Newtonian $\left(\tau^m = \dfrac{1}{\varnothing}\dot\gamma\right)$
Circular (tube)	$\dot\gamma = \dfrac{4\bar{v}_z \cdot r}{R^2} = \dfrac{4 \cdot \dot{V} \cdot r}{R^4 \pi}$ $\dot\gamma_w = \dfrac{4 \cdot \dot{V}}{\pi R^3}$	$\dot\gamma = (m+3)\dfrac{\bar{v}_z}{R}\left(\dfrac{r}{R}\right)^m$ $\dot\gamma_w = \dfrac{(m+3)\cdot \dot{V}}{\pi R^3}$
Annular aperture	$\dot\gamma_w = \dfrac{6 \cdot \dot{V}}{\pi D H^2}$ $\dot\gamma_w = \dfrac{6 \cdot \dot{V}}{\pi (R_a + R_i)(R_a - R_i)^2}$	$\dot\gamma_w = \dfrac{2(m+2)V}{\pi D H^2}$ $\dot\gamma_w = \dfrac{2(m+2)\dot{V}}{\pi (R_a + R_i)(R_a - R_i)}$
Rectangular aperture	$\dot\gamma_w = \dfrac{6 \cdot \dot{V}}{B \cdot H^2}$	$\dot\gamma = (m+2)\dfrac{2\bar{v}_z}{H}\left(\dfrac{2x}{H}\right)^m$ $\dot\gamma_w = \dfrac{2(m+2)\dot{V}}{BH^2}$
Conical aperture	$\dot\gamma = \dfrac{4\dot{V}}{\pi r^3}\left[\dfrac{1-\left(\dfrac{r_0}{R_0}\right)^3}{3-\left(\dfrac{R_0}{r_0}-1\right)}\right]^{\frac{3}{4}}$	
Irregular cross-section	a) $\dot\gamma = \dfrac{2 \cdot \dot{V} \cdot U}{A^2}$ b) $R_{eq} = \sqrt{\dfrac{A}{\pi}}$ $\dot\gamma_w = \dfrac{4 \cdot \dot{V}}{\pi R_{eq}^3}$ (Approximation)	

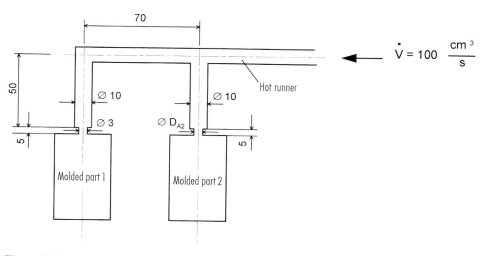

Figure 5.80   Simple hot runner system

Dimensions of the hot-runner system:

Hot runner as far as molded part 2: L_{HR2} = 50 mm; D_{HR} = 10 mm; L_{G2} = 5 mm; D_{G2} = ?

Hot runner as far as molded part 1: L_{HR1} = 120 mm; D_{HR} = 10 mm; L_{G1} = 5mm; D_{G1} = 3 mm

For simultaneous filling of the parts, the pressure on both flow paths must be equal:

$$\Delta p_1 = \Delta p_2$$

$$\frac{\eta_{HR} \cdot L_{HR1}}{D^4_{HR}} + \frac{\eta_{G1} \cdot L_{G1}}{D^4_{G1}} = \frac{\eta_{HR} \cdot L_{HR2}}{D^4_{HR}} + \frac{\eta_{G2} \cdot L_{G2}}{D^4_{G2}} \tag{5.22}$$

For the various viscosities, the following applies:

$$\eta_i = k \cdot a_T \cdot \dot{\gamma}_i^{n-1} \tag{5.23}$$

Given the representative shear rate:

$$\dot{\gamma}_{i, rep} = e_o \cdot \frac{32 \cdot \frac{\dot{V}}{2}}{D_i^3} \tag{5.24}$$

equation 5.22 becomes:

$$\eta_i = k \cdot a_T \cdot (e_o \cdot 16\dot{V})^{n-1} \cdot \frac{1}{D_i^{3(n-1)}} \tag{5.25}$$

This is substituted into equation (5.21) to solve for D_{G2}. When the numerical values have been inserted, D_{G2} computes to 2.61 mm. This is a plausible result since the much greater resistance generated in this gate, which has a smaller diameter than gate 1, compensates the shorter flow path to molded part 2.

5.10 Special Phenomena Associated with Multiple Gating

To discover particular phenomena of materials, a mold was developed [5.52] based on the theoretical considerations above that would demonstrate dependency on an operating level. The runner system shown in Figure 5.81 was chosen. It has equal flow paths and equal gate cross-sections that would guarantee simultaneous onset of filling and a uniform filling pattern.

Trials on various materials (HDPE, LDPE, POM) confirmed the assumptions again. Very different pictures of the filling process were obtained with PA 6 and even more so with glass-reinforced nylon 12. A difference in the onset of filling could not be noticed but the throughput through the outside gates was considerably larger than through the inside gates (Figure 5.81). This phenomenon was more pronounced at lower material temperatures (< 250 °C) and at high injection speeds. However, at high material temperatures (approx. 300 °C) and extremely low injection speeds, no advancing in the sense just described was noticeable. This effect is best illustrated by a change of colors during molding to observe the origin of the melt in the individual cavities. It can be clearly seen, as the low Reynolds numbers (Re < 20) would suggest, that completely laminar flow is maintained even in the gate. This means that, above all, a flow path close to the wall remains there. At the point of branching, the core material penetrates against the opposite wall and lingers there afterwards (Figure 5.82) [5.53].

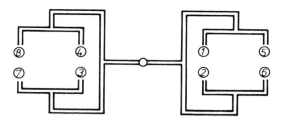

Figure 5.81   Runner system [5.25]

The reason for the advancing of melt could therefore be due to the dependency of melt viscosity on shearing time in the various zones. Those portions of melt which are close to the runner wall during the filling process (melt for the outside cavities) have experienced a larger shearing time integral than the melt which fills the inside cavities. This is confirmed by the simultaneous onset of filling already mentioned because the gate-filling phase differs from the mold-filling phase. A considerable part of the melt which is sheared in the runner during gate filling remains in the runner to fill the gate next to the runner wall. As a result, the design of a runner system demands that:

– The flow of melt must be divided up symmetrically (Figure 5.82).
– Two branches must not be placed in the same plane.

From this example (glass-reinforced nylon 12), it is evident that the initially derived methods of estimation are sufficient if the consequences of the flow path study are taken into consideration.

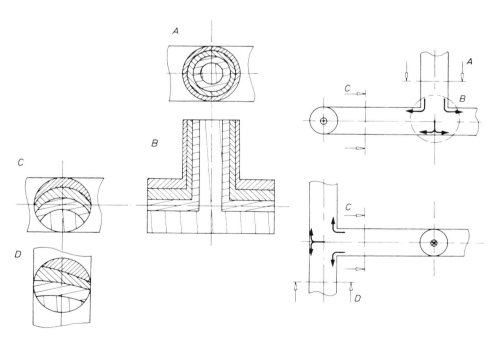

Figure 5.82   Melt distribution in a runner system [5.25]

It is, however, more practical today to employ one of the commercial simulation programs (CADMOULD, MOLDFLOW, C-MOLD, etc.) to balance a runner system. This subject is treated in Chapter 14.

5.11   Design of Gates and Runners for Crosslinking Compounds

5.11.1 Elastomers

5.11.1.1 Calculation of Filling Process

The design of gates and runners, as in thermoplastics processing, is performed under rheological and thermal aspects. In addition, however, allowance must be made for the reaction kinetics of the material as a function of prevailing temperatures.

Rheological mold design has long been based on simulation programs such as CADMOULD [5.54], RUBBERSOFT [5.55] and FILCALC [5.56]. The flow behavior is described by the Carreau or power laws (see also Chapter 14).

To guarantee dependable mold filling, scorching must not occur during the injection phase. The injection time available to the processor for filling the mold and during which scorching has not occurred is called the incubation time. It is material and temperature dependent. The occurrence of scorching is described mathematically by the scorch index. This is the ratio of elapsed (injection) time and incubation time and can assume values from "0" to "1". As soon as the value "1" is attained, the incubation period has elapsed and actual crosslinking begins. Consequently, scorch indices below "1" are required for dependable mold filling. Figure 5.83 shows the scorch index distribution for

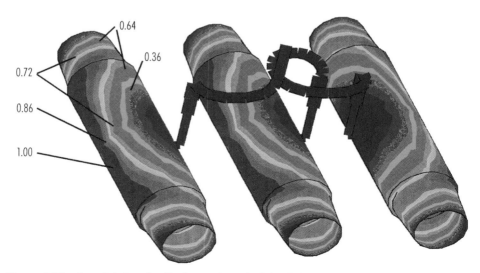

Figure 5.83   Scorch index distribution at the end of the filling phase

a 3-cavity mold for a medium-voltage bushing at the end of the filling phase. Scorching can therefore be expected during the injection phase for the chosen parameter setting.

It is suggested in [5.57] that so-called processing windows be established for frequently used materials. These windows are based on tests performed on plate molds and results from rheometers and plastometers. With this, scorch is also registered, which plays an important part in the case of elastomers. However, creating them involves a relatively high input.

The following section explains how different materials and mold geometries influence processing window qualitatively. After this, several calculated processing windows are discussed.

5.11.1.2 Effect of Processing Characteristics on the Basis of Processing Windows

The processing properties of an elastomer may be characterized by its

– viscosity,
– incubation time (t_i time) and
– curing velocity (difference between t_{90} and t_i).

Of greatest interest during processing is the question of how different material types or different lots in the process behave and how potential processing problems can be eliminated. These questions can be discussed with the aid of a processing window. Figure 5.84 demonstrates how the position of the

– injection-time line,
– injection limit,
– holding-time line, and
– heating-time line

vary according to material (or lot).

Higher viscosity causes higher pressure losses in the mold. Higher pressure losses lead to very different effects [5.58] depending on the injection-speed control of the injection-molding machine. With a closed-loop control, the injection speed is decreased continuously. With an open-loop control, in contrast, the speed is decreased more or less irregularly. How low the speed may become depends on the dimension of the hydraulic drive. If the machine reaches its pressure limit (maximum injection pressure) an injection speed results which corresponds with the equilibrium between injection pressure and pressure loss [5.58].

Therefore higher pressure losses (due to a higher viscosity) have a very different effect. If there is still an adequate pressure reserve, the injection time remains constant (closed-loop control) or becomes a little longer (open-loop control). At this time, however, the injection pressure rises. This means another, higher injection pressure counts for the injection-time line. Because of the higher pressure loss, more energy is dissipated during injection and the temperature of the material increases. The curing process proceeds more rapidly and the incubation phase is shortened. This shifts the injection limit to smaller time periods (Figure 5.84, effect a). Consequently, increasing viscosity narrows the possible filling-time range (distance between injection-time line and injection limit).

If the machine is operated at its pressure limit, the injection time is increased by a material with higher viscosity. This means that the injection-time line is shifted to the

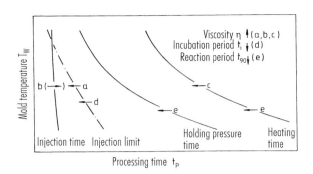

Figure 5.84   Effect of material properties in a processing window (qualitative response) a–e effect (see text) [5.53]

right (Figure 5.84, effect b, arrow in parentheses). Then the injection-time line always represents the maximum injection pressure of the machine. Although the same amount of energy is dissipated during injection (equal pressure loss), filling lasts longer and the material is heated up, therefore, by thermal conduction. The higher melt temperature may again result in shorter incubation time and a narrowing of the possible filling range.

The effects described here also cause the holding-pressure and heating phases to start with higher material temperatures, of course. The higher melt temperature at the onset of the holding-pressure phase has almost no effect in the gate area. The material rapidly attains the mold temperature in the (usually) small cross-section of the gate anyway and, with this, the maximum curing velocity possible. The holding-pressure time (distance between injection line and holding-pressure time) is hardly shortened at all as a result. In heavy sections, however, the higher material temperature (at the end of the injection process) shortens the curing time, that is, the distance between the injection-time and heating-time lines (Figure 5.84, effect c). The decisive factors for this are faster equalization of temperature and, therefore, higher curing velocity.

In contrast, a shorter incubation time only shifts the injection limit to the left (Figure 5.84, effect d). Storage of a material mixture increases the viscosity and shortens the incubation period at the same time. These two effects reinforce each other and reduce the maximum filling time.

Higher vulcanization rates of a material shorten only the holding-pressure and heating time (Figure 5.84, effect e). The corresponding lines also move to the left. However, the processes that occur in the injection phase are not affected.

Certain processing problems can often be attributed to different causes or resolved by different measures. A processing window can help to choose the most effective measure.

5.11.1.3 Criticism and Examples Concerning the Processing-Window Model

The calculation of a processing window does not consider the feeding process, although it can affect the course of the process if the conditions are unfavorable. Such conditions are

– a feeding time longer than the heating time, and
– attaining high material temperature during feeding.

In the first case, the processing time or the cycle, respectively, is increased and the calculated window pictures a processing time that is too short.

In the case of extreme feeding conditions, the material temperature can become so high that part of the incubation phase is already used up before injection. The assumption for the calculation of the processing window that injection starts with a scorch index of zero becomes erroneous.

From processing windows, as shown in Figure 5.85 for two molds and two different gates, one can pick up the changes of processing conditions and obtain help in designing molds and, above all, gates. The complete processing window for molds C and D is presented in Figure 5.85. For the two graphs, top and center, holding-pressure and heating times are calculated from the 50 MPa-injection-time line. One can recognize that the flat runner shifts the heating-time line slightly towards shorter time periods. With an injection limit for a scorch index of 10%, mold D allows a maximum mold temperature of only 164 °C. This results in a significantly increased minimum processing time for mold D compared to C. The latter allows a mold temperature of 200 °C because of the longer distance between injection-time line and injection limit. Thus the processing time in mold C can be reduced to almost half of that of mold D (78 s against 138 s). Only a higher injection pressure of 70 MPa increases the distance from the injection-time line to the injection limit in mold D to the extent that a temperature of 200 °C is also possible (Figure 5.85, bottom). With this step, minimum processing time can be reduced to 68 s.

Figure 5.86 demonstrates the effect of different heights of the runner h_r during the injection phase (mold A: h_r = 6 mm; mold B: h_r = 4 mm). The cavity depth is 6 mm in both molds.

With equal injection pressure, the injection line shifts to double the time because of the increased pressure loss in the runner of mold B. The position of the injection limit remains about the same in both molds. Figure 5.87 compares the effect of different cavity depths h_C during the injection phase (mold A: h_C = 6 mm; mold C: h_C = 12 mm). The height of the runner h_r is the same in both molds (h_r = 6 mm).

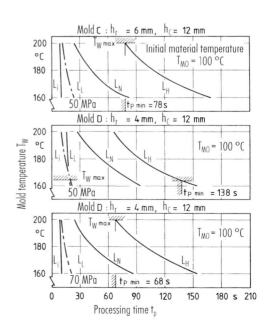

Figure 5.85 Processing window for two runner heights h_r. Test molds C and D, plot of injection time L_i, of injection limit L_L. (S_1 = 10%), of holding pressure L_N, (X_r = 30%) and of heating time L_H, (X_C = 80%); material: NBR [5.53]

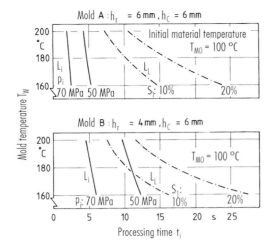

Figure 5.86   Processing window for two runner heights h_r.
Test molds A and B, plot of injection time L_i for two injection pressures p_i and of injection limit L_L for different scorch indices S_1; material: NBR [5.53]

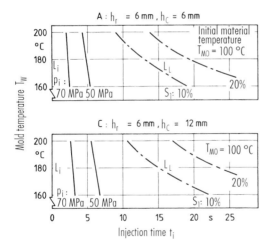

Figure 5.87   Processing window for two cavity heights h_C.
Test molds A and C, plot of injection time L_i for different injection pressures p_i and of injection limit L_L for different scorch indices S_1; material: NBR [5.53]

5.11.2 Thermosets

5.11.2.1 Flow-Curing Behavior of Thermosets

Calculating the pressure loss in the case of thermosets is difficult. One reason for this is additional flow effects in the form of elongational flow losses and uncompacted flow front areas. Another reason is, it is difficult to determine the viscosity function since the high filler content and the rapid crosslinking reaction make it almost impossible to conduct a measurement with the aid of a high-pressure capillary viscometer.

In practice, therefore, generally simple test procedures are used for characterizing the material. For example, test specimens are compression molded with constant molding compound weight under defined molding pressure (rods, slabs, tumblers, etc.). The resultant thickness or length of the test specimen then acts as a measure of the complex flow behavior. However, a low-viscosity but highly-reactive compound may lead to the same results as a high-pressure, less-reactive compound. A separate description of the flow or curing behavior is possible however, in principle by recording different curves of flow resistance over time (e.g. Brabender Plastograph or Kanavec viscometer). Admittedly, the demands on the material in these tests are not in the same order of magnitude as in injection molding, so it is difficult to translate the data obtained.

A method for estimating pressure losses is presented in [5.59]. Starting from a mechanical analogy, formulas are developed for calculating pressure losses in straight channels, at cross-section transition and in turnarounds. To be sure, the use of a trial mold to determine the material constants contained in the formulas is relatively laborious, but it pays off if the molding compound rarely changes.

The following formula is to be used for calculating the pressure loss:

$$\Delta p = R \cdot \overline{v}_F \cdot \frac{C \cdot L}{A} \tag{5.26}$$

Where

R is the resistance to flow, specific to the material,
$\overline{v}_F$ is the medium flow velocity of cross-section,
C is the circumference of the flow channel,
L is the length of the flow channel,
A is the cross-sectional area of the flow channel.

The experiments are now conducted with molds and the results recorded as shown in Figure 5.88.

The evaluation of the test results with the help of Equation (5.25) can be presented in a double-logarithmic system of coordinates (Figure 5.88).

$$R = \frac{\Delta p \cdot A}{\overline{v}_F \cdot C \cdot L} \tag{5.27}$$

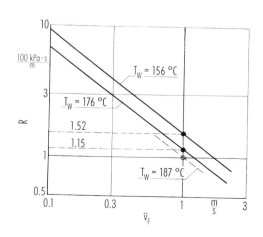

Figure 5.88   Flow resistance in a phenolic resin [5.59]

There is a clear connection with the resistance to flow R. An effect of the channel geometry cannot be seen anymore. It is completely included in the geometry coefficient CL/A.

The function R is a function of the velocity and invariant with respect to geometry. According to these preconditions a computation of arbitrary channel geometries is possible. The mathematical relations for different geometric shapes are summarized in Table 5.3. More detailed information is to be found in [5.26, 5.59–5.63].

Table 5.3   Pressure calculation

Geometry	Pressure drop	Characteristics
Straight channels	$$\Delta p = R_0 \frac{\overline{v}_F^{a+1}}{v_0^a} \cdot \frac{C \cdot L}{A}$$	V_0 = 1 m/s a = tan α, Slope of the curves in the double-logarithmic plot R = f(v_F) R_0 = f(v_0)
Elbows	$$\Delta p = \frac{R_0}{v_0^a} \cdot \left[\overline{v}_F\right]^{a+1} \cdot \frac{CL}{A} \cdot (R_{E_0} \cdot \psi^c + 1)$$	ψ = r_K/B, Geometry function ψ_0 = 1 c = tan β, Slope of the curves in the double-logarithmic plot R_E = f(ψ) R_{E0} = f(ψ_0)
Cross-sectional transitions	$$\Delta p_A = \Delta p_{A_0} \cdot \left[\frac{\overline{v}_F}{v_0}\right]^f \cdot \left[\frac{\varphi}{\varphi_0}\right]^g$$	φ = A_1/A_2 -1, Cross-sectional ratio ϕ_0 = 1 Δp_{A_0} = f(v_0,ϕ_0) f = tan γ, Slope of the curves in the double-logarithmic plot Δp_{A0} = f(v) g = tan δ, Slope of the curves in the double-logarithmic plot Δp_{A_0} = f(φ)

References

[5.1]   Pye, R.G.E.: Injection Mould Design (for Thermoplastics). Ilitte Books Ltd., London, 1968.
[5.2]   Szibalski, M.; Meier, E.: Entwicklung einer quantitativen Methode für den Konstruktionsablauf bei Spritzgießwerkzeugen. Unpublished report, IKV, Aachen, 1976.
[5.3]   Möhrwald, K.: Einblick in die Konstruktion von Spritzgießwerzeugen. Verlag Brunke Garrels, Hamburg, 1965.

[5.4]   Spritzgießtechnik von Vestolen. Publication, Chemische Werke Hüls AG, Marl.

[5.5]   Christoffers, K.-E.: Formteilauslegung, verarbeitungsgerecht. Das Spritzgußteil. VDI-Verlag, Düsseldorf, 1980.

[5.6]   Kunststoff-Verarbeitung im Gespräch, 1: Spritzgießen. Publication, BASF, Ludwigshafen, 1979.

[5.7]   Morgue, M.: Moules d'injection pour Thermoplastiques. Officiel des Activities des Plastiques et du Caoutchouc, 14 (1967), pp. 269–276 and 14 (1967), pp. 620–628.

[5.8]   Kegelanguß, Schirmanguß, Bandanguß. Technical Information, 4.2.1, BASF, Ludwigs-hafen/Rh., 1985.

[5.9]   Zawistowski, H.; Frenkler D.: Konstrukcja form wtryskowych do tworzyw termoplastycznych. Wydawnictwo Naukowo-Techniczne, Warszawa, 1984.

[5.10] Menges, G.; Mohren, P.: Anleitung für den Bau von Spritzgießwerkzeugen. Carl Hanser Verlag, Munich, 1974.

[5.11] Spritzgießen von Thermoplasten. Publication, Farbwerke Hoechst AG, Frankfurt/M., 1971.

[5.12] Schmid, A.: Leitsätze, Angüsse, Anschnitte, Lehrgangshandbuch Spritzgießwerkzeuge. VDI-Bildungswerk, Düsseldorf, December 1971.

[5.13] Stank, H.-D.: Anforderungen an den Anguß, seine Aufgaben, Anordnung am Spritzgußteil. Anguß- und Anschnittprobleme beim Spritzgießen. Ingenieurwissen, VDI-Verlag, Düsseldorf, 1975.

[5.14] Speil, Th.: Fertigungsgenauigkeit und Herstellung von Kunststoffpräzisionsteilen, Lehrgangshandbuch Spritzgießwerkzeuge. VDI-Bildungswerk, Düsseldorf, December 1977.

[5.15] Cechacek, J.: Problematik der Werkzeugkonstruktion. Plaste und Kautschuk, 22 (1975), 2, p. 183.

[5.16] Crastin-Sortiment, Eigenschaften, Verarbeitung. Company brochure, Ciba-Geigy AG, Basel, August 1977.

[5.17] Gestaltung von Spritzgußteilen aus thermoplastischen Kunststoffen. VDI-Richtlinie, 2006, VDI-Verlag GmbH, Düsseldorf, 1970.

[5.18] Appel, O.: Übertragbarkeit von Auslegungsregeln für Spritzgießwerkzeuge auf Werkzeuge für die Pulvermetall Spritzgießtechnik, Unpublished report, IKV, RWTH, Aachen, 1988.

[5.19] German, R. M.: Powder Injection Molding, Metal Powder Industries Federation, Princeton, New Jersey, 1990.

[5.20] Greim, J.: Keramische Formmassen – Anwendungen und Verarbeiten durch Spritzgießen, Neue Werkstoffe und Verfahren beim Spritzgießen, VDI-Verlag, Düsseldorf, 1990.

[5.21] Bielzer, R.: Ermittlung von Kriterien zur systematischen Binderauswahl beim Pulverspritzgießen und Übertragung des Verfahrens auf PTFE, Dissertation, RWTH, Aachen, 1992.

[5.22] Ricking, T.: Analyse der thermischen Vorgänge in einem Heißkanalwerkzeug zur Herstellung keramischer Bauteile, Unpublished report, IKV, RWTH, Aachen, 1996.

[5.23] König, K.: Entwicklung statistischer Prozeßmodelle für den Keramikspritzguß, Unpublished report, IKV, RWTH, Aachen, 1997.

[5.24] Menges, G.; Schmidt, Th. W; Hoven-Nievelstein, W. B.: Handbuch zur Berechnung von Spritzgießwerk zeugen. Verlag Kunststoff-Information, Bad Homburg, 1985.

[5.25] Schürmann, E.: Abschätzmethoden für die Auslegung von Spritzgießwerkzeugen. Dissertation, RWTH, Aachen, 1979.

[5.26] Schmidt, L.: Auslegung von Spritzgießwerkzeugen unter fließtechnischen Gesichts-punkten. Dissertation, RWTH, Aachen, 1981.

[5.27] Lichius, U.: Rechnerunterstützte Konstruktion von Werkzeugen zum Spritzgießen von thermoplastischen Kunststoffen. Dissertation, RWTH, Aachen, 1983.

[5.28] Bangert, H.: Systematische Konstruktion von Spritzgießwerkzeugen und Rechnereinsatz. Dissertation, RWTH, Aachen, 1981.

[5.29] Lichius, U.; Bangert, H.: Eine einfache Methode zur Vorausbestimmung des Fließfrontver-laufs beim Spritzgießen von Thermoplasten. Plastverarbeiter, 31 (1980), 11, pp. 671–676.

[5.30] Schacht, Th.: Unpublished report, IKV, Aachen, 1984.

[5.31] Thienel, P.: Der Formfüllvorgang beim Spritzgießen von Thermoplasten. Dissertation, RWTH, Aachen, 1977.

[5.32] Michaeli, W.: Extrusionswerkzeuge für Kunststoffe und Kautschuk. Carl Hanser Verlag, Munich, 1991.
[5.33] Pahl, M.: Praktische Rheologie der Kunststoffschmelzen und Lösungen. VDI-Verlag, Düsseldorf, 1982.
[5.34] Meissner, J.: Rheologisches Verhalten von Kunststoffschmelzen und Lösungen. In: Praktische Rheologie der Kunststoffe. VDI-Verlag, Düsseldorf, 1978.
[5.35] Gleißle, W.: Kurzzeitmessungen zur Ermittlung der Fließeigenschaften von Kunststoffen bis zu höchsten Schergeschwindigkeiten. In: Praktische Rheologie der Kunststoffe. VDI-Verlag, Düsseldorf, 1978.
[5.36] Bird, R. B.; Armstrong, R. C.; Hassager, O.: Dynamics of Polymeric Liquids. Vol. 1: Fluid Mechanics. Wiley, New York, 1977.
[5.37] Ostwald, W.: Über die Geschwindigkeitsfunktion der Viskosität disperser Systeme. Kolloid-Z., 36 (1925), pp. 99–117.
[5.38] Waele, A. de: J. Oil Colour Chem. Assoc., 6 (1923), p. 33.
[5.39] Schulze-Kadelbach, R.; Thienel, E.; Michaeli, W.; Haberstroh, E.; Dierkes, A.; Wortberg, J.; Wäbken, G.: Praktische Stoffdaten für die Verarbeitung von Plastomeren. Conference volume. Plastics Conference, IKV, Aachen, 1978.
[5.40] Kenndaten für die Verarbeitung thermoplastischer Kunststoffe. VDMA (ed.). Vol. 2: Rheologie, Carl Hanser Verlag, Munich, 1982.
[5.41] Kenndaten für die Verarbeitung thermoplastischer Kunststoffe. VDMA (ed.). Vol. 4: Rheologie, Carl Hanser Verlag, Munich, 1986.
[5.42] Carreau, P. J.: Rheological equations from Molecular Network Theories. Ph. D. Thesis. University of Wisconsin, 1968.
[5.43] Geiger, K.; Kühnle, H.: Analytische Berechnung einfacher Scherströmungen aufgrund eines Fließgesetzes vom Carreauschen Typ. Rheol. Acta, 23 (1984), pp. 355–367.
[5.44] Laun, H. M.: Rheologie von Kunststoffschmelzen mit unterschiedlichem molekularem Aufbau. Kautschuk, Gummi, Kunststoffe, 40 (1987), 6, pp. 554–562.
[5.45] Williams, M. L. et. al.: The Temperature Dependence of Relaxation Mechanism in Amorphous Polymers and other Glassforming Liquids. J. Am. Chem. Soc., 77 (1955), 7, pp. 3701–3706.
[5.46] Cogswell, F. N.: Polymer Melt Rheology. A Guide for Industrial Practice. George Godwin Ltd. Halted Press, 1981.
[5.47] Kemper, W.: Kriterien und Systematik für die theologische Auslegung von Spritzgießwerkzeugen. Dissertation, RWTH, Aachen, 1982.
[5.48] Benfer, W.: Rechnergestützte Auslegung von Werkzeugen für Elastomere. Dissertation, RWTH, Aachen, 1986.
[5.49] Schneider, W.: Spritzpressen technischer Gummiformteile. Dissertation, RWTH, Aachen, 1985.
[5.50] Pötsch, G.: Berechnungsverfahren zur Abschätzung der Druckverluste von Transfermoldingwerkzeugen. Unpublished report, IKV, Aachen, 1985.
[5.51] Pötsch, G.: Dehndruckverluste in Transfermoldingwerkzeugen. Unpublished report, IKV, Aachen, 1986.
[5.52] Maaß, R.: Die Anwendung der statistischen Versuchsmethodik zur Auslegung von Spritzgießwerkzeugen mit Kaltkanalsystern. Dissertation, RWTH, Aachen, 1995.
[5.53] Hengesbach, H. A.; Egli, L.; Schürmann, E.; Neuenschwander, S.: Injection tests at Bühler, Uzwil, 1978.
[5.54] Programmsystem CADMOULD. IKV, Aachen, 1996.
[5.55] Krehwinkel, T.; Schneider, C.: RUBBERSOFT – ein Programmsystem für Elastomerverarbeiter. Kautschuk, Gummi, Kunststoffe, 41 (1988), p. 953 ff.
[5.56] Usermanual FILCALC, Version 5.0. RAPRA Technology Ltd., Shawbury, 1997.
[5.57] Schneider, Ch.: Das Verarbeitungsverhalten von Elastomeren im Spritzgießprozeß. Dissertation, RWTH, Aachen, 1986.
[5.58] Rörick, W.: Prozeßregelung im Thermoplast-Spritzgießbetrieb. Dissertation, RWTH, Aachen, 1979.
[5.59] Paar, M.: Auslegung von Spritzgießwerkzeugen für vernetzende Formmassen. Dissertation, RWTH, Aachen, 1984.
[5.60] Kemper, W.: Kriterien und Systematik für die theologische Auslegung von Spritzgießwerkzeugen. Dissertation, RWTH, Aachen, 1982.

[5.61] Feichtenbeiner, H.: Auslegung eines Spritzgießwerzeuges mit 4fach Kaltkanalverteiler für Elastomerformteile. Unpublished report, IKV, Aachen, 1983.

[5.62] Busch, E.: Fließtechnische Auslegung von Sprtzgießwerkzeugen für Duromere. Unpublished report, IKV, Aachen, 1983.

[5.63] Andermann, H.: Abschätzmethode für Druckverluste an Querschnittsübergängen in Angußsystemen. Unpublished report, IKV, Aachen, 1983.

6   Design of Gates

6.1   The Sprue Gate

The sprue gate is the simplest and oldest kind of gate. It has a circular cross-section, is slightly tapered, and merges with its largest cross-section into the part.

The sprue gate should always be placed at the thickest section of the molded part. Provided proper size, the holding pressure can thus remain effective during the entire time the molded part solidifies, and the volume contraction during cooling is compensated by additional material forced into the cavity. No formation of voids or sink marks can occur. The diameter of the sprue gate depends on the location at the molded part. It has to be a little larger than the section thickness of the molded part so that the melt in the sprue solidifies last. The following holds (Figure 5.9):

$$d_F \geqq S_{max} + 1.0 \text{ (mm)}. \tag{6.1}$$

It should not be thicker, though, because it then the melt solidifies too late and extends the cooling time unnecessarily.

To demold the sprue without trouble it should taper off towards the orifice on the side of the nozzle. The taper is

$$\alpha \geqq 1\text{–}4°. \tag{6.2}$$

American standard sprue bushings have a uniform taper of 1/2 inch per foot, which is equivalent to about 2.4°.

The orifice towards the nozzle has to be wider than the corresponding orifice of the nozzle. Therefore

$$d_A \geqq d_D + 1.5 \text{ mm} \tag{6.3}$$

(Refer to Figure 5.9 for explanation of symbols)

If these requirements are not met, undercuts at the upper end are formed (Figure 5.8).

Very long sprues, that is if the mold platens are very thick, call for a check on the taper. Possibly another nozzle has to be used in the injection molding machine.

To a large degree the release properties of the sprue also depend on the surface finish of the tapered hole. Scores from grinding or finishing perpendicular to the direction of demolding have to be avoided by all means. Material would stick in such scores and prevent the demolding. As a rule the interior of sprue bushings is highly polished.

A radius r_2 (Figure 5.9) at the base of the sprue is recommended to create a sharp notch between sprue and molding and to permit the material to swell into the mold during injection.

To its disadvantage, the sprue always has to be machined off. Even with the most careful postoperation, this spot remains visible. This is annoying in some cases, and one could try to position the sprue at a location that will be covered after assembly of the article. Since this is often impractical, the sprue can be provided with a turnaround so

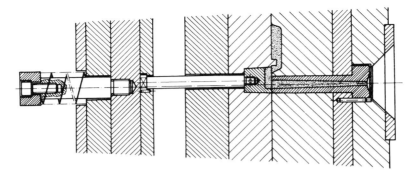

Figure 6.1   Sprue with turnaround [6.1] (also called "overlap gate")

that it reaches the molded part from the inside or at a point not noticeable later on (Figure 6.1). The additional advantage of such redirected sprues is the prevention of jetting. The material hits the opposite wall first and begins to fill the cavity from there [6.2]. Machining as a way of sprue removal is also needed here.

Another interesting variant of a sprue gate is shown in Figure 6.2 It is a curved sprue, which permits lateral gating of the part. It is used to achieve a balanced position of the molded part in the mold, which is now loaded in the center. This is only possible, however, for certain materials, such as thermoplastic elastomers.

6.2   The Edge or Fan Gate

An edge gate is primarily used for molding parts with large surfaces and thin walls. It has the following advantages:

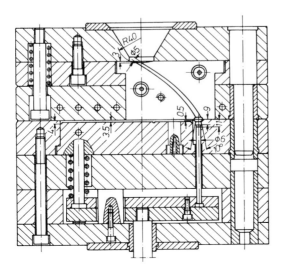

Figure 6.2   Curved sprue [6.3]

– parallel orientation across the whole width (important for optical parts),
– in each case uniform shrinkage in the direction of flow and transverse (important for crystalline materials),
– no inconvenient gate mark on the surface.

The material leaving the sprue first enters an extended distributor channel, which connects the cavity through a narrow land with the runner system (Figure 6.3). The narrow cross-section of the land acts as a throttle during mold filling. Thus, the channel is filled with melt before the material can enter the cavity through the land. Such a throttle has to be modified in its width if the viscosity changes considerably.

The distributor channel has usually a circular cross-section. The relationship of Figure 6.3 generally determines its dimensions. They are comparable with the corresponding dimensions of a ring gate, of which it may be considered a variant.

Besides the circular channel, a fishtail-shaped channel is sometimes met (Figure 6.4). This shape requires more work and consumes more material, but it results in excellent part quality due to a parallel flow of the plastic into the cavity.

Dimensioning was mostly done empirically so far. Today it can be accomplished with the help of rheological software packages such as CADMOULD, MOLDFLOW, etc. (see Chapter 14).

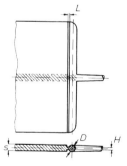

Figure 6.3 Edge gate with circular distributor channel [6.1, 6.4]
D = s to 4/3 s + k,
k = 2 mm for short flow lengths and thick sections,
k = 4 mm for long flow lengths and thin sections,
L = (0.5 to 2.0) mm,
H = (0.2 to 0.7) s.

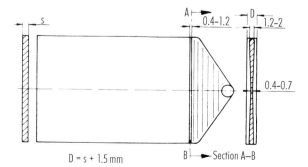

Figure 6.4 Edge gate with adjusted cross section resulting in uniform speed of flow front [6.5]

6.3   The Disk Gate

The disk gate allows the uniform filling of the whole cross-section of cylindrical, sleeve-like moldings, which need a mounting of the core at both ends. The disk can be of a plane circular shape (Figure 6.8) or a cone usually with 90° taper ("umbrella" gate) (Figure 6.5) and distributes the melt uniformly onto the larger diameter of the molded part. This has the advantage that knit lines are eliminated. They would be inevitable if the parts were gated at one or several points. Besides this, a possible distortion can be avoided. With proper dimensions there is no risk of a core shifting from one-sided loading either. As a rule of thumb, the ratio between the length of the core and its diameter should be smaller than

$$\frac{L_{core}}{D_{core}} < \frac{5}{1} \tag{6.4}$$

[6.5] (see also Chapter 11: Shifting of Cores).

If the core is longer, it has to be supported on the injection side to prevent shifting caused by a pressure differential in the entering melt. In such cases a ring gate should be employed (Section 6.4). A design like the one in Figure 6.6 is poor because it results again in knit lines with all their shortcomings.

The "umbrella" gate can be connected to the part in two different ways; either directly (Figure 6.5) or with a land (Figure 6.7). Which kind is selected depends primarily on the wall thickness of the molded part.

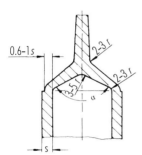

Figure 6.5   Disk gate [6.5] 90° taper

There is another type of umbrella gate known as a disk gate [6.5, 6.6]. A disk gate permits the molding of cylindrical parts with undercuts in a simple mold without slides or split cavities (Figure 6.8, left).

6.4   The Ring Gate

A ring gate is employed for cylindrical parts, which require the core to be supported at both ends because of its length.

The melt passes through the sprue first into an annular channel, which is connected with the part by a land (Figure 6.9). The land with its narrow cross-section acts as a throttle during filling. Thus, first the annular gate is filled with material, which then

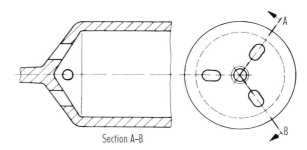

Figure 6.6 Conical disk gate
with openings for core support
[6.5]

Section A–B

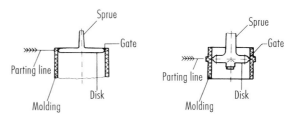

Figure 6.7 Disk gate

Figure 6.8 Disk gates [6.5, 6.6]

enters the cavity through the land. Although there is a weld line in the ring gate, its effect is compensated by the restriction in the land and it is not visible, or only slightly visible.

The special advantage of this gate lies in the feasibility of supporting the core at both ends. This permits the molding of relatively long cylindrical parts (length-over-diameter ratio greater than 5/1) with equal wall thickness. The ring gate is also utilized for cylindrical parts in multi-cavity molds (Figure 6.9). Although similar in design, a disk gate does not permit this or a core support at both ends.

The dimensions of a ring gate depend on the types of plastics to be molded, the weight and dimensions of the molded part, and the flow length. Figure 6.10 presents the data for channels with circular cross-section generally found in the literature.

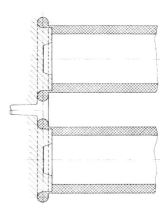

Figure 6.9 Sleeves with ring gates and interlocks for
core support [6.1]

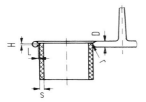

Figure 6.10   Ring gate with circular cross-section [6.4, 6.5]
D = s + 1.5 mm to 4/3 s + k,
L = 0.5 to 1.5 mm,
H = 2/3 s to 1 to 2 mm,
r = 0.2 s,
k = 2 mm for short flow lengths and thick sections,
k = 4 mm for long flow lengths and thick sections.

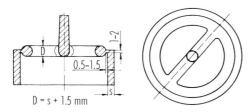

Figure 6.11   Internal ring gate [6.5]

The gates in Figures 6.9 and 6.10 are called external ring gates in the literature [6.5]. Consequently, a design according to Figure 6.11 is called internal ring gate. It exhibits the adverse feature of two weld lines, is more expensive to machine, and complicates the core support at both ends.

A design variation of the common ring gate can be found in the literature. Since it is basically the usual ring gate with only a relocated land (Figure 6.12), a separate designation for this does not seem to be justified.

6.5   The Tunnel Gate (Submarine Gate)

The tunnel gate is primarily used in multi-cavity molds for the production of small parts which can be gated laterally. It is considered the only self-separating gating system with one parting line, which can be operated automatically.

Part and runner are in the same plane through the parting line. The runners are carried to a point close to the cavities where they are angled. They end with a tapered hole, which is connected with the cavities through the land. The tunnel-like hole which is milled into the cavity wall in an oblique angle forms a sharp edge between cavity and tunnel. This edge shears off the part from the runner system [6.7].

There are two design options for the tunnel (Figures 6.13a and 6.13b). The tunnel hole can be pointed or shaped like a truncated cone. In the first case the transition to the molded part is punctate, in the second it is elliptical. The latter form freezes more slowly

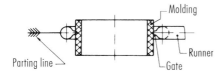

Figure 6.12   External ring gate (rim gate)
[6.6]

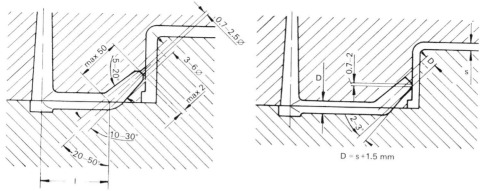

Figure 6.13a Tunnel gate with pointed tapered tunnel [6.5]

Figure 6.13b Tunnel gate with truncated tapered tunnel [6.5]

and permits longer holding pressure time. Machining is especially inexpensive because it can be done with an end-mill cutter in one pass.

For ejection, part and runner system must be kept in the movable mold half. This can be done by means of undercuts at the part and the runner system. If an undercut at the part is inconvenient, a mold temperature differential may keep the molded part on the core in the movable mold half as can be done with cup-shaped parts.

The system works troublefree if ductile materials are processed. With brittle materials there is the risk of breaking the runner since it is inevitably bent during mold opening. It is recommended therefore, to make the runner system heavier so that it remains warmer and hence softer and more elastic at the time of ejection.

In the designs presented so far, the part was gated laterally on the outside. The tunnel is machined into the stationary mold half and the molded part is separated from the runner during mold opening. With the design of Figure 6.14 the part, a cylindrical cover,

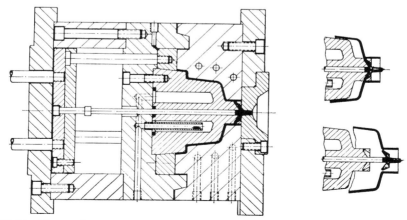

Figure 6.14   Mold with tunnel gates for molding covers [6.8]

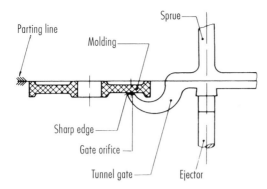

Figure 6.15   Curved tunnel gate [6.6]

is gated on the inside. The tunnel is machined into the core in the movable mold half. The separation of gate and part occurs after the mold is opened by the movement of the ejector system. The curved tunnel gate (Figure 6.15) functions according to the same system.

6.6   The Pinpoint Gate in Three-Platen Molds

In a three-platen mold, part and gate are associated with two different parting lines. The stationary and the movable mold half are separated by a floating platen, which provides for a second parting line during the opening movement of the mold (Figure 6.16). Figures 6.17 and 6.18 show the gate area in detail.

This system is primarily employed in multi-cavity molds for parts that should be gated in the center without undue marks and post-operation. This is particularly the case with cylindrical parts where a lateral gate would shift the core and cause distortion.

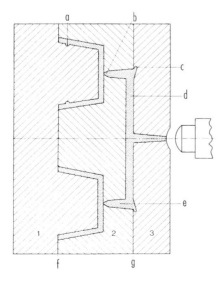

Figure 6.16   Three plate mold [6.9]
1 Movable mold half, 2 Floating plate,
3 Stationary mold half,
a Undercut in core, b Gate, c Undercut,
d Runner, e Sprue core, f Parting line 1,
g Parting line 2.

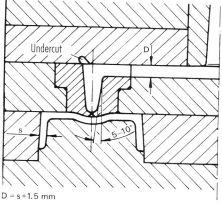

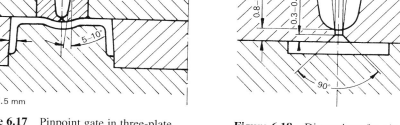

Figure 6.17   Pinpoint gate in three-plate mold [6.5]

Figure 6.18   Dimensions for pin point gate [6.6]

Thin-walled parts with large surface areas are also molded in such a way in single cavity molds. Multiple gating (Figure 6.19) is feasible, too, if the flow length-over-thickness ratio should call for this solution. In this case special attention has to be paid to knit lines as well as to venting.

The opening movement of a three-platen mold and the ejection procedure separate part and runner system including the gate. Thus, this mold provides a self-separating,

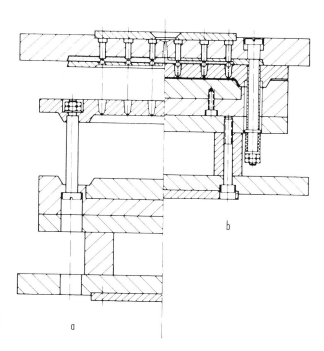

Figure 6.19   Three plate mold for multiple gating in series [6.10] a Open, b Closed.

automatic operation. The mold is opened first at one and then at the other parting line, thus separating moldings and runner system.

6.7   Reversed Sprue with Pinpoint Gate

The reversed sprue is frequently enlarged to a "pocket" machined into the stationary mold half. It is connected with the cavity by a gate channel with reversed taper.

During operation the sprue is sealed by the machine nozzle and fully filled with plastic during the first shot. With short cycle times the material in the sprue remains fluid, and the next shot can penetrate it. The nozzle, of course, cannot be retracted each time.

The principle of operation of a reversed-sprue gate is demonstrated in Figure 6.20. The hot core in the center, through which fresh material is shot, is insulated by the frozen plastic at the wall of the sprue bushing. Air gaps along the circumference of the bushing obstruct heat transfer from the hot bushing to the cooled mold. The solution shown in Figure 6.20 functions reliably if materials have a large softening range such as LDPE, and the molding sequence does not fall short of 4 to 5 shots per minute [6.11].

If these shorter cycle times are impractical, additional heat has to be supplied to the sprue bushing. This can be done rather simply by a nozzle extension made of a material with high thermal conductivity. Such materials are preferably copper and its alloys. The design is presented in Figure 6.21. The tip of the nozzle is intentionally kept smaller than the inside of the sprue bushing. With the first shot the gap is filled with plastic, which protects the tip from heat loss to the cool mold later on.

Major dimensions for a reversed-sprue design can be taken from Figure 6.22.

The gate diameter like that of all other gates depends on the section thickness of the part and the processed plastic material and is independent of the system. One can generally state that smaller cross-sections facilitate the break-off. Therefore, as high a melt temperature as possible is used in order to keep the gate as small as possible.

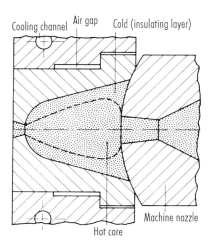

Figure 6.20   Bushing for reversed sprue [6.9]

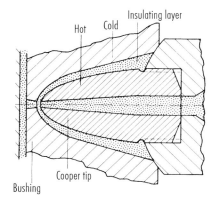

Figure 6.21   Reversed sprue heated by nozzle point [6.9]

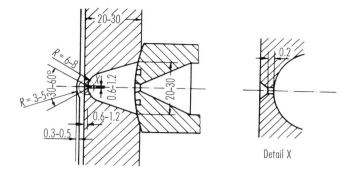

Figure 6.22 Reversed sprue with pinpoint gate and wall thickening opposite gate for better distribution of material [6.11] right: Detail X (Dimensions in mm)

A tapered end of the pinpoint gate is needed, even with its short length of 0.6 to 1.2 mm, so that the little plug of frozen plastic is easily removed during demolding and the orifice opened for the next shot.

Some plastics (polystyrene) have a tendency to form strings under those conditions. In such cases a small gate is better than a large one. Large gates promote stringing and impede demolding.

It is practical to equip the nozzle with small undercuts (Figure 6.22), which help in pulling a solidified sprue out of the bushing. The sprue can then be knocked off manually or with a special device (Figure 6.23).

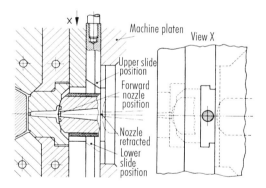

Figure 6.23 Sprue strike-off slide in a guide plate between mold and machine platen [6.12]

A more elegant way of removing the sprue from the bushing is shown in Figure 6.24. The reversed sprue is pneumatically ejected. An undercut holds the sprue until the nozzle has been retracted from the mold. Then an annular piston is moved towards the nozzle by compressed air. In this example it moves a distance of about 5 mm. After a stroke of 3 mm the air impinges on the flange of the sprue and blows it off [6.12].

6.8 Runnerless Molding

For runnerless molding the nozzle is extended forward to the molded part. The material is injected through a pinpoint gate. Figure 6.25 presents a nozzle for runnerless molding.

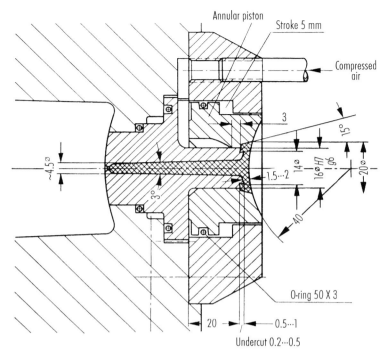

Figure 6.24   Reversed sprue with pinpoint gate and pneumatic sprue ejector [6.12] Dimensions in mm

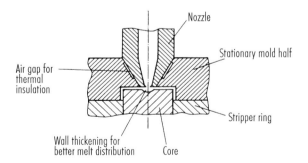

Figure 6.25   Sprueless gating

The face of the nozzle is part of the cavity surface. This causes pronounced gate marks (mat appearance and rippled surface) of course. Therefore, one has to keep the nozzle as small as possible. It is suggested that a diameter of 6 to 12 mm not be exceeded. Because the nozzle is in contact with the cooler mold during injection- and holding-pressure time, this process is applicable only for producing thin-walled parts with a rapid sequence of cycles. This sequence should not be less than 3 shots per minute to avoid a freezing of the nozzle, which is only heated by conduction. The applicability of this procedure is limited and it is used for inexpensive packaging items.

The principle is successfully employed when the material is further distributed through runners as in a three-platen mold.

6.9   Molds with Insulated Runners

Properly designed insulated runners, i.e., with thermally controlled gate, offer several advantages over hot runners. These are:

– Thanks to the lack of dead spots and to the smooth channel, insulated runners are dependable, provided that fairly well stabilized materials are used. But all common thermoplastic materials nowadays meet this condition.
– Since the thermal insulation arises itself through melt deposited at the wall, the temperature distribution of the melt will always be very uniform.
– Insulated runners are always economical if constant operation with uniform cycles is guaranteed. It is not suitable, however, for extended interruptions.
– The higher the throughput, i.e., the greater the shot weight at normal wall thickness, the more dependable are insulated runners.
– Because insulated runners are very easy and quick to clean, they are particularly recommended when frequent color changes have to be made or when recycled material is used for which it cannot be guaranteed that entrained impurities will not lead to blockage or unclean, patchy surfaces.
– Properly designed insulated runners are both cheaper to buy and to maintain than hot runners.

A distinguishing feature of a well designed insulated runner is that it has minimal heat loss. This means that thermal equilibrium will be reached pretty quickly with low energy input on startup or after interruptions. Good design requires the following measures:

– good insulation effect through thick, outer insulation (generous channel cross-section),
– an isolated air gap (a chimney effect must not occur in the air gap),
– minimal contact areas between channel block and mold,
– carefully calculated installation of cartridge heaters in the channel block to compensate for losses at critical points during long cycle times.

It is always advisable – and absolutely vital for heat-sensitive plastics such as POM, PC, PBT, etc. – that the gate area be carefully designed. Neither the critical shear rate may be exceeded nor may material that is too cold be transported into the mold. Furthermore, material that is too hot must not remain there to decompose. The following measures will produce an ideal temperature profile in the gate area:

An internally heated needle (Figure 6.26) serving as the energy supply element in the transition area to the cavity must have a temperature profile well adjusted to the plastic for processing. This means that the tip of the needle must keep the melt precisely at its ideal processing temperature, while it must not overheat the melt along its shaft, and in the area of the guide bushing the temperature of the plastic should just about be that of freezing.

For some years now, three standard types of tried and proven modules have been available in two sizes for materials such as PS, ABS, PC, PE, PP, PA, POM and PBT (see Figures 6.27 and 6.28) (e.g., supplied by KBC System, Bellanger, 1271 Givirns, Switzerland):

– for gate diameters in the range: 0.6 to 2.5 mm: MIDI,
– for gate diameters in the range 2.0 to 5 mm: MAXI.

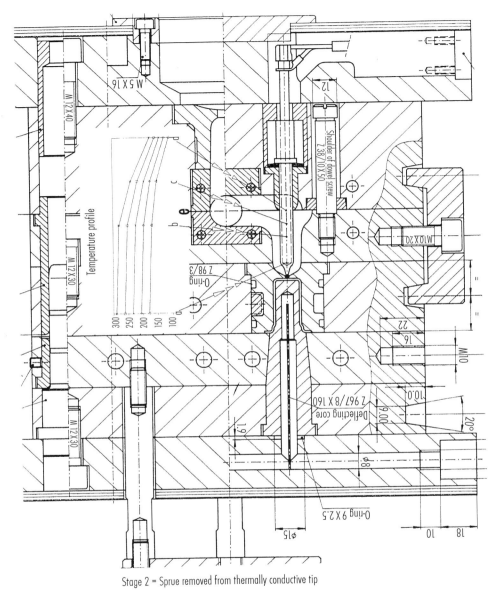

Stage 2 = Sprue removed from thermally conductive tip

Figure 6.26   Insulated runner mold with internally heated needle

Application Areas

PE moldings weighing 0.15 g can still be produced at a rate of 8 shots/minute with insulated runners, although the heat input into the system is correspondingly low for small shot weights. In these cases, more energy must be fed to the runner by means of cartridge heaters. Nevertheless, the insulated runners require barely one fourth of the

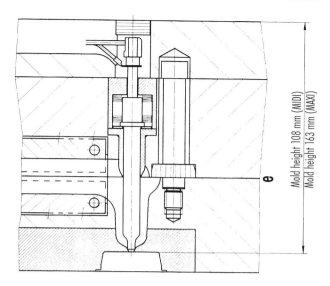

Figure 6.27 Internally heated needles

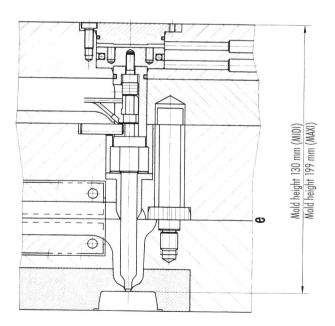

Figure 6.28 Internally heated needles

electric energy required by hot runners. It may generally be assumed that the size and weight of the producible molded parts are governed only by the rheological limits of the plastic melts used, i.e. the shear rate at the gate.

Practical Experience Gained with Insulated Runners

Thanks to its simple construction, clear functionality and self-sealing capability, the insulated runner is easy to operate. There are few practiced operatives who consider the freezing of the insulated runner during protracted production breaks to be a serious disadvantage. Quite the opposite is true. They appreciate the fact that the second parting line is easy and quick to open by simply moving two retaining clamps and that the frozen material can be removed in one movement (Figure 6.29). The mold is then ready for production again after two to three cycles. This is quite advantageous because, when disruptions occur in the case of hot runners, these are by far more complicated to dismantle and clean. Furthermore, protracted disruptions with hot runners cause problems because the material degrades if the heating is not turned off. An insulated runner can be completely cleaned within a few minutes, whereas production has to be stopped for hours when this happens to hot runners.

6.10   Temperature-Controlled Runner Systems – Hot Runners

Runner systems in conventional molds have the same temperature level as the rest of the mold because they are in the same mold block. If, however, the runner system is located in a special manifold that is heated to the temperature of the melt, all the advantages listed below accrue. Runner manifolds heated to melt temperature have the task of distributing the melt as far as the gates without damage. They are used for all injection-molded thermoplastics as well as for crosslinking plastics, such as elastomers and thermosets.

In the case of thermoplastics, these manifolds are usually referred to as the hot-runner system, the hot manifold, or simply as hot runners. For crosslinking plastics, they are known as cold runners.

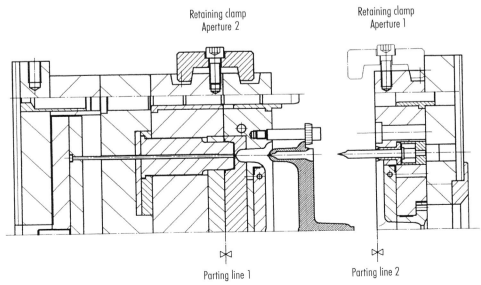

Figure 6.29   Retaining clamps make insulated runners easier to clean

6.10.1 Hot-Runner Systems

Hot-runner systems have more or less become established for highly-automated production of molded thermoplastic parts that are produced in large numbers. The decision to use them is almost always based on economics, i.e. production size. Quality considerations, which played a major role in the past, are very rare now because thermoplastics employed today are almost all so stable that they can be processed without difficulty with hot-runner systems that have been adapted accordingly.

Hot-runner systems are available as standard units and it is hardly worthwhile having them made. The relevant suppliers offer not only proven parts but also complete systems tailored to specific needs. The choice of individual parts is large.

Table 6.1 Hot runner systems suppliers in North America (selection) (see also Table 17.2)

D-M-E Company	Madison Heights, MI/USA
Dynisco HotRunners	Gloucester, MA/USA
Eurotool	Gloucester, MA/USA
Ewicon Hotrunner Systems	East Dundee, IL/USA
Gunther Hot Runner Systems	Buffalo Grove, IL/USA
Hasco-Internorm	Chatsworth, CA/USA
Husky	Bolton, Ontario/Canada
Incoe	Troy, MI/USA
Manner International	Tucker, GA/USA
Mold-Masters	Georgetown, Ontario/Canada
Thermodyne HotRunner Systems	Beverly, MA/USA

6.10.1.1 Economic Advantages and Disadvantages of Hot-Runner Systems

Economic Advantages:
- **Savings in materials** and costs for regrind.
- **Shorter cycles;** cooling time no longer determined by the slowly solidifying runners; no nozzle retraction required.
- **Machines can be smaller** because the shot volume – around the runners – is reduced, and the clamping forces are smaller because the runners do not generate reactive forces since the blocks and the manifold block are closed.

Economic Disadvantages:
- Much more complicated and considerably more expensive.
- More work involved in running the mold for the first time.
- More susceptible to breakdowns, higher maintenance costs (leakage, failure of heating elements, and wear caused by filled materials).

Technological Advantages:
- Process can be automated (demolding) because runners do not need to be demolded.
- **Gates at the best position;** thanks to uniform, precisely controlled cooling of the gate system, long flow paths are possible.
- **Pressure losses minimized,** since the diameter of the runners is not restricted.
- **Artificial balancing of the gate system;** balancing can be performed during running production by means of temperature control or special mechanical system (e.g. adjustment of the gap in a ring-shaped die or use of plates in flow channel). (Natural balancing is better!)

– **Selective influencing of mold filling;** needle valve nozzles and selective actuation of them pave the way for new technology (cascade gate system: avoidance of flow lines, in-mold decoration).
– **Shorter opening stroke** needed compared with competing, conventional three-platen molds.
– **Longer holding pressure,** which leads to less shrinkage.
Technological Disadvantages:
– Risk of thermal damage to **sensitive** materials because of long flow paths and dwell times, especially on long cycles.
– Elaborate temperature control required because non-uniform temperature control would cause different melt temperatures and thus non-uniform filling.

6.10.1.2 Hot Runners for Various Applications and New Possibilities

Figure 6.30 shows the basic possibilities that are available.

Hot-runner systems are almost always used when large series have to be made in highly automated production. However, they also permit new technological variants based on the possibility of positioning the gates so as to yield the best quality molded parts. They are primarily connected to needle valve nozzles, which are actuated with precise timing.

Cascade gating (Figure 6.31): needle valve nozzles that – depending on the filling – are opened and closed so that the flow front is always fed by the last nozzle to have been passed [6.14, 6.15].

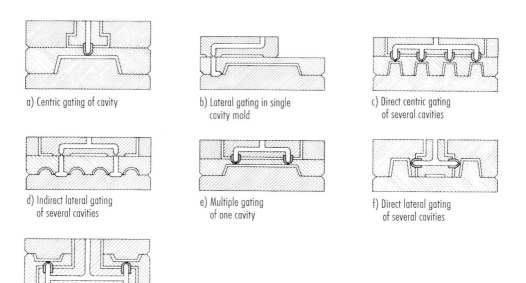

a) Centric gating of cavity

b) Lateral gating in single cavity mold

c) Direct centric gating of several cavities

d) Indirect lateral gating of several cavities

e) Multiple gating of one cavity

f) Direct lateral gating of several cavities

g) Hot manifold for stack molds

Figure 6.30 a–g   Modes of melt transport in hot manifolds [6.13]

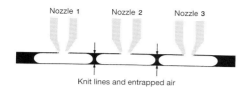

Knit line with entrapped air at the confluence of two melt fronts in conventional injection molding

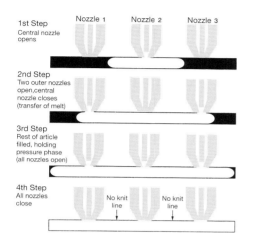

Cascade control over the needle valve nozzles yields a uniform melt front without knitlines in the molded part (the central nozzle shown here also has a needle valve)

Figure 6.31
Cascade injection
[6.15]

This allows:

- **Avoidance of weld lines** (e.g. requirement for vehicle body exterior parts). These large-surface parts require gates. This would normally give rise to weld lines. The cascade gating technique pushes the flow front forward in relays, whereby each nozzle opens only after the front has just passed it and the previous nozzle closes at the same time.
- **In-mold decoration** (integrated lamination with textiles or film) has become possible because the lower pressures no longer displace the inserted textile, and so no folds or other flaws occur. This method works on the principle of avoiding weld lines.
- **Multi-cavity mold** with cavities of different geometry and volume. Also known as family molds because parts of different volume that belong together are produced simultaneously in one mold by one shot.
- **Since injection pressure and holding pressure** may be actuated independently of each other, opening and closing can be adjusted to the conditions of each cavity.
- **Controlled volume balancing** means that a weld line can be shifted into a non-critical area of the molded part.
- **Stack molds**, i.e. doubling or quadrupling of production in the same time scale thanks to two or more mold platens and parting lines.

6.10.1.3 Design of a Hot-Runner System and its Components

Hot-runner molds are ambitious systems in a technological sense that involve high technical and financial outlay for meeting their main function of conveying melt to the gate without damage to the material. Such a design is demonstrated in Figure 6.32.

Typical hot runner system

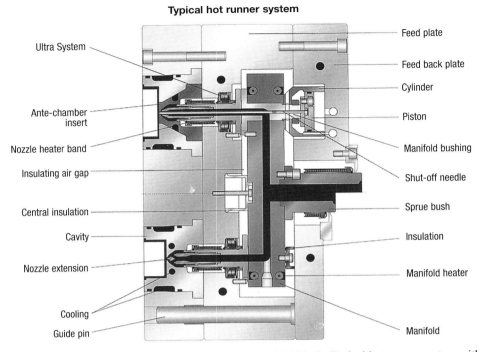

Ultra System

Ante-chamber insert

Nozzle heater band

Insulating air gap

Central insulation

Cavity

Nozzle extension

Cooling

Guide pin

Feed plate

Feed back plate

Cylinder

Piston

Manifold bushing

Shut-off needle

Sprue bush

Insulation

Manifold heater

Manifold

Figure 6.32   View through an externally heated manifold block. Typical hot runner system with two different gate nozzles. Top: A needle valve nozzle with pneumatic actuation; bottom: an open nozzle point for a small mold mark, of the kind used for thermoplastics. Manifold is heated with tubular heaters [6.16].
(Husky)

Hot runners are classified according as they are heated:

– insulated-runner systems (see Section 6.9) and
– genuine hot-runner systems.

The latter can be further sub-classified according to the type of heating (see Figure 6.35 [6.17]):

– internal heating, and
– external heating.

Heating is basically performed electrically by cartridge heaters, heating rods, band heaters, heating pipes and coils, etc. To ensure uniform flow and distribution of the melt, usually a relatively elaborate control system comprising several heating circuits and an appropriate number of sensors is needed. The operating voltage is usually 220 to 240 V, but small nozzles frequently have a low voltage of 5 V, and also 15 V and 24 V operating voltage.

Externally/Internally Heated Systems
The two possibilities are shown schematically in Figure 6.33, while Figure 6.34 shows the flow conditions and the resultant temperature distributions in the melt for both types

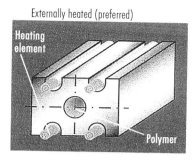
Externally heated (preferred)
Heating element
Polymer

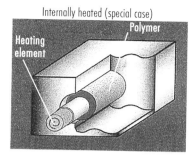
Internally heated (special case)
Polymer
Heating element

Figure 6.33
Cross-sections of the flow-channel in the manifold
Source: DuPont [6.17]

of heating. For the sake of completeness, it should be mentioned that this distinction between internal and external heating applies only to the manifold blocks because it is common practice to heat, for instance, the blocks externally and the nozzles internally.

The major advantages and disadvantages of the two types are immediately apparent from Figure 6.34.

Externally Heated System:
Advantage:
Large flow channels cause low flow rate and uniform temperature distribution.

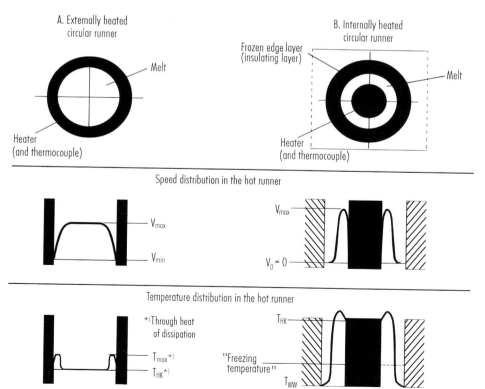

A. Externally heated circular runner — Melt — Heater (and thermocouple)

B. Internally heated circular runner — Frozen edge layer (insulating layer) — Melt — Heater (and thermocouple)

Speed distribution in the hot runner

V_{max}, V_{min}

V_{max}, $V_0 = 0$

Temperature distribution in the hot runner

*) Through heat of dissipation
T_{max}*), T_{HK}*)

T_{HK}, "Freezing temperature", T_{WW}

Figure 6.34   Hot runner systems. Comparison of internally and externally heated systems [6.18]

Disadvantage:
The temperatures required for external heating have to be very much higher (see Figure 6.35 [6.19] for PA 66). Here, the mold temperature is approximately 100 °C and the manifold temperature is at least 270 °C; this means there is a temperature difference of approximately 170 °C from the mold block, which means:

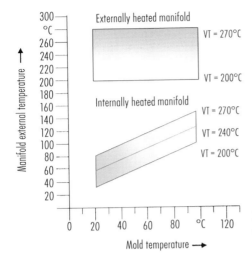

Figure 6.35   External temperatures of manifold systems as a function of mold temperature [6.19]

– special measures required for fixing the hot-runner nozzles to the gates because of the considerable thermal expansion,
– risk of disruption if this is not adequately resolved,
– higher heating power (over 500 W per 100 mm line for a typical cross-section measuring 40 · 7 mm²),
– insulation from the mold block,
– large, unsupported areas and therefore, with large-surface molds, risk of bowing of the mold platen on the feed side if this has not been designed thick enough and thus, as a direct consequence, the mold becomes very heavy.

Internally Heated System
A frozen layer of plastic forms on the inner surface of the channel and functions as an insulation layer.

– The heat requirement of the system is much lower (roughly 55 W per 100 mm length of inside tube).
– The temperature differences between mold and manifold blocks are negligible; therefore measures that would have been necessary for large heat expansion are not needed.
– The hot manifold of an internally heated system is a compact block that is bolted tightly to the mold. Consequently, the mold is very rigid and no measures are required for centering the nozzles and gates. This also allows the plate on the machine side to be manufactured as one block consisting of fixed mold with in-built manifold and corresponding rigidity [6.20] (Figure 6.36).

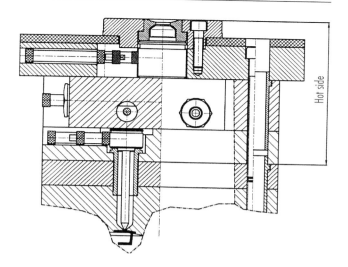

Hot side

Figure 6.36 Cross-section
through a mold with hot side
[6.20]

The melt volume is small and so the dwell times of the flowing melt are short. On the
other hand, the flow rates are very much greater and this can damage the material.

It is not advisable to use internally heated systems for sensitive materials.

When deciding on a certain system, advice can be obtained from suppliers. All of the
major ones supply more than one system [6.19, 6.21].

6.10.1.3.1 Sprue Bushing

The sprue bushing serves to transfer the melt from the machine into the manifold. In
order to satisfy the basic requirement of uniform melt temperature, this spot must also
be carefully heated and must therefore generally be fitted with its own heating circuit and
temperature sensors. If the temperature in this area is too low for thermoplastics sensitive
to high temperatures, there may be complaints about the surface quality of the finished
parts because there may be a temperature difference of 20 to 30 °C in the melt on account
of the large lengths of sprue bushings of 30 to 50 mm [6.21]. They must therefore be
heated.

Since the plastic melt is shot through the hot runner into the injection mold under high
pressure, a high nozzle contact pressure is necessary in order to achieve a permanent and
melt-tight connection to the hot runner. Naturally the same conditions apply here as for
any other sprue bushing. Since, with hot runners, the distance between machine nozzle
and mold is often large – e.g., if clamping systems are required on the feed side in the
mold – extended, heated nozzles are required in such cases (Figure 6.37).

Since there are no temperature differences between machine and manifold, it is not
necessary to detach the machine nozzle from the sprue bushing. So-called extended
nozzles and extended bushings have become commonplace (Figure 6.38) because they
ensure that no melt escapes either into the cavity or out of the bushing and also that
decompression can be readily performed.

Decompression is an established method of preventing melt drooling from a hot
runner gate into the empty cavity after demolding, thereby leading to lower quality and
disrupting operations. It is generally performed by retracting the screw in the cylinder
but may also be effected by retracting the extended nozzle in the extended bushing.

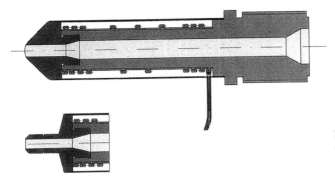

Figure 6.37   Machine nozzle with integrated heater [6.22]

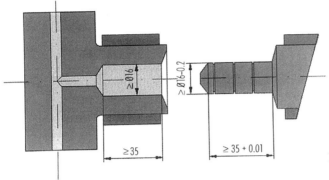

Figure 6.38   Dipping nozzle (extended) [6.22]

Nozzles and bushings are available as standard parts and it is not worthwhile having them made.

6.10.1.3.2 Melt Filters

As a result of blockages in the hot runners, particularly in the narrow cross-sections of the gate nozzles, which are caused by melt that is not totally clean, it is very common to install filters nowadays (Figure 6.39). Roßbach [6.23] always recommends this precaution, not just when virgin material is being processed or when the machines have a clamping force of less than 5000 kN (larger machines have molds whose gates are so large that common impurities do not become trapped). In all cases, actually, it is necessary to know the pressure losses in order to be able to estimate whether mold filling will still be accomplished without error. The pressure loss is usually < 30% of the standard pressure of a nozzle without filter.

A filter cannot be installed on the mold if decompression is employed. In this case, the filter should be installed in the nozzle of the machine as shown in Figure 6.40.

6.10.1.3.3 Manifold Blocks

6.10.1.3.3.1 Single-Cavity Molds

There are several reasons for installing a heated sprue in the case of single-cavity molds, e.g., when a prototype has to be produced under exactly the same conditions as parts

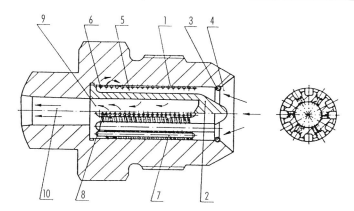

1 Location holes,
2 Filter insert,
3 Locking ring,
4 Transition to nozzle of
   injection molding
   machine,
5 Feed channel,
6 Tangential filter groove,
7 Intermediate channel,
8 Radial filter holes,
9 Collecting channel,
10 Die orifice

Figure 6.39   Filter insert with radial holes and tangential grooves [6.23]

from a later series to be made in a multi-cavity mold. Only in such cases is the same holding pressure and thus the same shrinkage adjustable. Figure 6.41 shows a needle valve nozzle and a nozzle with thermal valve for simple applications.

6.10.1.3.4   *Manifold Beams*

6.10.1.3.4.1 *Multi-Cavity Molds*
The melt is fed from the screw bushing via the runners to the gate nozzles. With identical cavities, natural balancing is preferred, i.e., the cross-sections and distances to every sprue bushing have the same dimensions (see Section 5.6). However, as discussed in Section 5.6, it is possible, with the same means, to compensate for different lengths by changing the channel cross-sections, i.e., to balance artificially. As already briefly mentioned, apart from needle valve nozzles, there are other mechanical or thermal (usually more simple) ways of controlling the flow rate to the various cavities.

   In contrast to internally heated manifolds, with externally heated manifolds, manifold beams are used instead of manifold blocks (Figure 6.42). This is so enough space remains for installing the support pillars, which have to prevent unpermissible bending of the platen on the fixed mold half when the cavities are being filled.

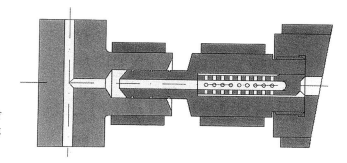

Figure 6.40   Pressure relief
with an dipping nozzle using
a melt filter

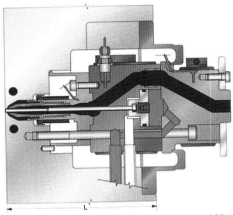

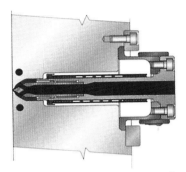

Needle valve for simple applications with length L of 80 to 155 mm

Gating with a thermal shut-off nozzle is the most common way of eliminating the cold sprue

Figure 6.41   Hot runner for simple (single-cavity) molds. Left, with needle valve; right, with thermal closure [6.16]
(Husky)

Melt flow
Flow channels on the same plane should be equally long and have the same diameter in order to ensure that the melt undergoes the same drop in pressure and experiences the same shear on its way from the machine to all cavities.

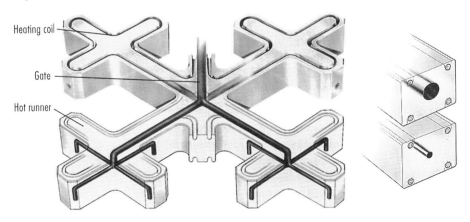

Optimum flow channel contours
Each application imposes specific demands on molded part weight, filling time, material type and processing conditions. Flow studies ensure that hot runner systems are optimally designed. Smaller channel diameters increase shear and pressure drop to the benefits of faster color changes and shorter dwell times. Larger diameters are chosen for shear-sensitive polymers and applications involving pressure restrictions.

Figure 6.42   Manifold block for feeding 16 gate nozzles [6.16]
(Husky)

The melt runners should naturally be as smooth as possible in order that no melt may get trapped. In addition, the design of all turnarounds must promote flow, i.e. large radii are required, sharp corners are forbidden. In the less expensive runners, the channels are bored and honed. For the corners, turnaround pieces are required that fit into the channel (see Figure 6.43). They are held in place by special sealing elements. There is no hiding the fact that these channels can be better cleaned.

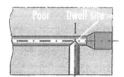 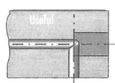 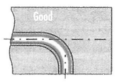

Figure 6.43
Turnarounds in the manifold [6.17]
Source: DuPont

Details on heating hot runners are provided in Section 6.10.1.6.

In order to minimize the number of heating circuits and controls and to be able to utilize failsafe, inexpensive tubular heaters, various hot runner system manufacturers offer manifold beams with heat-conduction tubes (see Chapter 17). These failsafe, maintenance-free tube-like bodies ensure uniform heat distribution even at those points where a heat gradient is present, such as in spacers, centering pieces and mounting pieces. This results in a relatively inexpensive, failsafe and, when properly designed, virtually isothermal hot manifold.

The bores are generally chosen such that acceptable flow rates are obtained on the one hand and tolerably long dwell times on the other. Diameters of 6 to 8 mm are chosen for medium throughputs.

There have also been trials [6.24] to bolt together the manifold from high-pressure hydraulic pipes and fittings. They are then surrounded with a band heater and insulated individually. Particular advantages are:

– the mass to be heated up is very much smaller than in manifold beams,
– thermal expansion is easily compensated by bending the tubes,
– more space is available for the supporting columns of the mold platens and these can be distributed better,
– easy to clean and disassemble,
– inexpensive.

A good example is the production of multi-component moldings with a hot runner system that consists of such tubes bolted together because the two requisite distribution systems would take up a great deal of space if they were made from manifold blocks. Separate temperature control is also easier to ensure.

Insulation of the external heated runners, in as far as the rigidity of the mold platens allows this, are usually of an air pocket with spacers consisting of poorly conducting metal, e.g., titanium and ceramic (Figure 6.32).

6.10.1.4 Nozzles for Hot-Runner Molds

The nozzle forms the connection between hot manifold and cavity. The essential requirements imposed are:

– Transport of as homogeneous and isothermal a melt as possible to the mold.

Figure 6.44   Hot runner system for a car fender [6.16].
Hot runner systems for injection molding of large automotive parts such as bodywork components and fender trim require injection on the moving mold half and "Class A" surfaces. Encapsulated, premounted, and prewired manifold systems are available for large molds whose core or cavity takes up the entire mold half. This simplifies installation and maintenance. Pre-mounted hot runner system with five nozzles, two of which are parallel, for injection at the rear of the fender trim.

– Thermal separation between hot manifold and cooled mold. The mold should not experience an undue temperature rise in the gate area (dull, wavy regions) and the gate should not cool to the extent that it freezes.
– Clean, reproducible separation between the fluid content of the runner and the solidifying part during demolding (no forming of strings and no drooling).

It can be seen that, relative to normal molds, the demands imposed on the nozzles have undergone little change. However, a large number of new variants have come into existence.

The advantages of the various types of nozzles may be described as follows:

Open Nozzles (Figure 6.45): Offer flow advantages and are used in conventional molds where such requirements have to be met. They are also used for filled, abrasive molding compounds on account of their relatively high insensitivity. Finally, there are sometimes spatial reasons for resorting to these gates, which require a certain amount of machining for removing the sprue.

Nozzles with Tips (Figure 6.46): The tips are hot due to the very good thermal conduction of their mounting, e.g. in the nozzle platen, because they must carry the heat into the melt at the gate that is at risk of freezing. They are, therefore made of highly conducting materials, usually copper or copper-beryllium. They thereby, and function as flow aids. It is particularly important for the sprue to tear off cleanly, which is precisely why these nozzles come in a variety of designs to suit the material for processing. This applies particularly to hot-edge nozzles. Very high-quality nozzles feature soldered-in heating wires that are controlled by their own control loop, which utilizes a dedicated

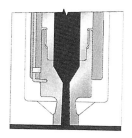
a)

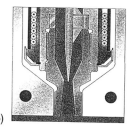
b)

Figure 6.45 Hot runner gate nozzle with sprue (indirect gating), particularly recommended for abrasive melts [6.16] (Husky)

a) Thermal seal (TS) nozzles:
Tapered gate, open nozzle tip, large, free gate
b) Thermal seal nozzle with torpedo:
Thermal seal nozzles require a balance of conditions in the gate area in order that the material may tear off readily. The additional torpedo extends the processing window by minimizing the consequences of cycle interruptions and possible forming of strings. Thermal seal nozzles are ideal for the gate of cold runners or gating onto molded-part surfaces when a small sprue is not problematic. The nozzles have an extended tip that forms a part of the shape-giving cavity surface and whose contours can be adapted. The design also simplifies the installation of the nozzle tip. A negative nozzle taper ensures that the material tears off at the tip of the cone. The corresponding height of the sprue depends on the gate diameter and the plastic being processed. The swappable, thermal seal nozzles of hardened steel are suitable for a wide range of amorphous and crystalline polymers and offer long service lives, even when abrasive materials are used.

sensor installed there. Many of these nozzles do not have pinpoint gates but rather ring gates as, due to similar or sometimes superior optical design, the flow speed is much smaller than in the pinpoint gate on account of the relatively large surface area. They, therefore, come in a variety of designs to suit the material for processing.

Needle Valve Nozzles (Figure 6.47): These are increasingly being used where injection is performed segment-wise, e.g., with a cascade gate. Actuation is usually performed pneumatically, but there are hydraulically actuated nozzles available. The latter are mainly used for large molds since they require less space. Hydraulically actuated nozzles still suffer from the reputation of leaking at precisely the wrong moment.

Whereas hot runners may be heated with 220 to 220 V, the small, narrow, and closely arranged nozzles have necessitated the development of 5 V, 15 V and 24 V heaters. Due to their close spatial arrangement of down to 11 mm, wiring of the individually heated, loop-controlled nozzles presents a problem [6.21]. In all cases that do not require the narrowest temperatures, indirect heating is preferred; it is maintenance-free and less expensive. For this reason, the heat-conducting elements, which are enveloped by the melt, are made of highly heat-conducting materials (usually copper-beryllium) or else heat pipes are used.

More details of the various nozzles are to be found in the text accompanying the diagrams.

A particular problem of externally heated distributors is sealing off of the nozzles against the mold. A good solution to this problem seems to be that afforded by Husky, called ultra-sealing technology. The seal is effected by disk springs and is described in Figure 6.48.

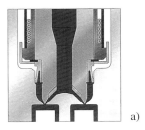

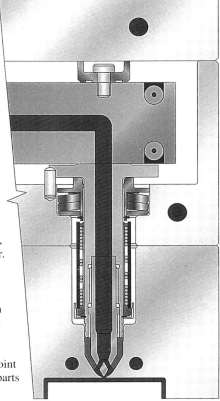

b)

a)

Figure 6.46   Pinpoint gates for hot runner gate
nozzles with tips or torpedo and tunnel distributor
for side gate [6.16]
(Husky)
Pinpoint gate:
A hot-tip (HT) or pinpoint gate is used when a
small gate sprue is not problematic. Its height
depends on several factors: gate diameter and land,
cooling in the gate area, type and grade of polymer.
Most materials are suitable for pinpoint gates. The
maximum gate diameter is usually 3 mm.
The needle valve is recommended for larger gate
diameters. Since the quality of the gate depends on
controlled hot-cold transition of the material in the
gate, the design of the cooling system in the gate
region is critical.
To realize gate distances less than 26 mm, multi-point
gates (MPs) may be used. These allow up to four parts
to be gated in a common cavity block, and this
reduces the size of the mold and the investment costs.
a) MP nozzles. These allow up to four parts to be
gated via the same nozzle housing. The possible distances between the gates range from 7 mm
to 30 mm.
b) HT nozzles. An exchangeable insulating cap reduces the amount of insulating plastic film that
coats the nozzle tip. This speeds up color change and enables heat-sensitive plastics to be
processed.

6.10.1.5   Data Concerning the Design of Hot Runner Manifolds

Although hot runner manifolds are rarely made in-house nowadays, some dimensional
data are provided below.

6.10.1.5.1   Manifold Beams
The material should be a C 60 or higher-grade steel. The diameters of the channels may
be chosen from Table 6.2. When shot weights are low and the channels are shorter than
200 mm, the shot weights alone determine the diameters in this table. If the channels are
longer, the channel diameters must be enlarged in order to reduce pressure losses and
thus to keep shear heating to a minimum.

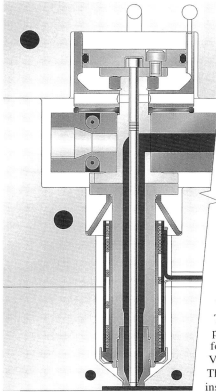

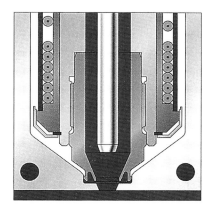

The pneumatic needle valve allows controlled injection via the hot runner nozzles in a programmed sequence. It is therefore also suitable for "family" molds in which molded parts of different weight are injection molded, and larger parts are filled sequentially.

VX nozzle:
This swappable nozzle of hardened tool steel forms part of the shape-giving cavity surface. This makes for easier maintenance work in the gate area.

VG nozzle:
The nozzle tip features particularly good thermal insulation – ideal for amorphous polymers.

Figure 6.47   Hot-runner gate nozzles with needle valve. Left: for semicrystalline plastics; right: amorphous plastics [6.16]
(Husky)

The turnarounds would be made of corner pieces with fits of, e.g., H 7, n 6 and mounted with sealing plugs. The turnarounds naturally would have to be secured against twisting; no undercuts must form in the channel (compare Figure 6.43).

In in-house production, the manifold would be made of high-pressure pipes and fittings (see Figure 6.49) or manifold beams.

The robust tubular heaters would normally be used for the heating elements. They are inserted into milled grooves with a thermally conducting cement (Figure 6.49). The grooves should approximate the isotherms that can be determined and printed out with the aid of an appropriate heat-calculation program.

For insulation purposes, an air gap of 3 to 5 mm is left all around the manifold. The insulation can be improved by inserting crumpled aluminum foil. Spacers can be made of titanium.

Table 6.2   Guidelines for dimensioning channels in hot runner molds [6.13, 6.28]

Channel diameter (mm)	Channel length (mm)	Shot weight/cavity (g)
5		Up to approx. 25
6		50
8		100
6 to 8	Up to 200	
8 to 10	200 to 400	
10 to 14	Over 400	

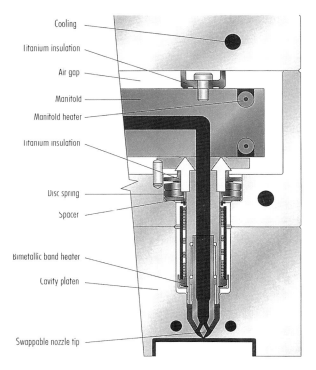
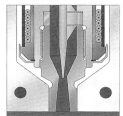

Figure 6.48   Hot runner gate nozzle with the Husky patented sealing system featuring disc springs [6.16]
(Husky)

The patented ultra-sealing system facilitates hot runner operation. The design prevents potential damage by cold-start leakage or the failure of overheated components. A disc spring unit presses the nozzle housing during assembly against the hot runner manifold, thereby bringing the preliminary load to bear that is necessary for dependably sealing the system while the temperature is still below the flow temperature of the material. While the manifold is warming up, the disc springs absorb the thermal expansion, even in the case of excessive overheating temperatures. The wide processing window of ± 100 °C allows the same hot runner to process a number of different plastics using the same channel dimensions and gates.

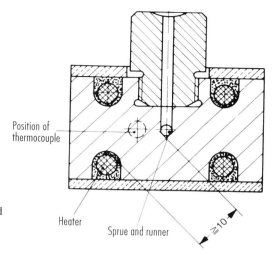

Figure 6.49   Cross-section of on manifold where the heating elements and the temperature sensors are installed [6.26]

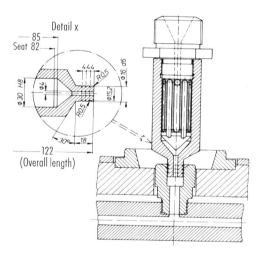

Figure 6.50   Sprue bushing, pressure-relief design with filter [6.25]

6.10.1.5.2 Nozzle Design

The free channel diameter must match that of the channels in the nozzle. The gate diameters, on the other hand, should be chosen on the basis of Table 6.3. They depend on the weight of the individual molded parts and roughly correspond to those of normal molds. The risk of degradation through excessive shear rates tends to be lower with hot runner manifolds than with pinpoint gates in conventional molds because the melt here flows into the gates at a higher temperature. Moreover, there no the need to heat up the melt prior to entry into the mold; this means that the diameters or free cross-sections can be made somewhat smaller. They must be small enough for sprue puller gates, so that pull-off does not present any problem; this behavior differs from molding compound to molding compound and is also dependent on the temperature.

Table 6.3   Guide values for dimensioning pinpoint gates [6.29]

Shot weight (g)	Pinpoint gate ø (mm)	Shot weight (g)	Pinpoint gate ø (mm)
to 10	0.4 to 0.8	40 to 150	1.2 to 2.5
10 to 20	0.8 to 1.2	150 to 300	1.5 to 2.6
20 to 40	1.0 to 1.8	300 to 500	1.8 to 2 8

It is therefore advisable, when having a hot runner made in-house, to use appropriate software (e.g. CADMOULD) to calculate both its rheological and its thermal behavior. Clues about the thermal performance to be installed are provided in Section 6.10.1.6.1. This information can be resorted to, however, it the power output is to be measured very accurately, it may also be calculated with the aid of a thermal design program (e.g. from CADMOULD). However, 25 to 30% must be added on to the result in order to cover mainly radiation losses.

All nozzles must be fitted with a thermocouple and their heating system must have its own control loop. This is the only way to ensure that the nozzles can be synchronized. Controllers with a PIDD structure are best [6.27]. The controllers should be connected to the machine control such that the temperatures are automatically adjusted to lower levels during breaks in operation or longer stoppages in order that no degradation, or even decomposition, may occur in the manifold area.

6.10.1.5.3   Notes on Operating Hot Runners

When heating hot runners with external heaters, it is advisable not to cool the molds themselves at first. Even better is to keep them as warm as possible with hot water, instead of with the cooling water, in order that the manifolds may attain their set values faster.

Color changes can take a very long time and be expensive on material. For medium-sized to large molds, between 50 and 100 shots must be allowed for. It is therefore best to avoid color changes if at all possible but, where this cannot be helped, to clean the hot runner prior to using the next color. This is relatively easily accomplished in drilled channels in the manifold by removing the stoppers and then heating until the plastic remaining in the channels melts at the edges so that the rest can be pushed out. Insulated runner manifolds definitely have an advantage in this respect.

6.10.1.6   Heating of Hot Runner Systems

6.10.1.6.1   Heating of Nozzles

There are three ways to heat nozzles in hot manifolds. One distinguishes:

– indirectly heated nozzles,
– internally heated nozzles,
– externally heated nozzles.

With indirectly heated nozzles heat is conducted from the manifold through heat-conducting nozzles or probes to the gate. To control the temperature of the individual nozzles independently of one another, the corresponding sections of the manifold have to be heated separately. This is usually done with paired heater cartridges along the runner in the nozzle area. Indirect heating of nozzles has the disadvantage that for small temperature changes at the gate, required for proper filling or smooth gate separation, a far greater change of the manifold temperature is needed. This leads inevitably to changes in the melt temperature in the runner, too. This undesirable change in melt temperature can produce an adverse effect on the quality of the parts. It is better to control the nozzle temperature independently of the manifold. This can be done with directly heated nozzles.

For *internally heated nozzles*, diameter and length of cartridge heaters are determined by the dimensions of the nozzle. One should strive for a cartridge diameter as large as possible to have a low watt density.

Table 6.4 lists recommended watt densities according to [6.28]. Cartridges with a length of more than 75 mm should have an apportioned power output. A suitable variation in the winding provides more heat at the generally cooler end and less in the center, which is normally too hot.

Table 6.4   Dimensioning of cartridge heaters [6.28]

Cartridge ″	Length (mm)	Watt density (W/cm²)
¼	30	35
	75	23
³⁄₈	30	27
	200	13
½	50	20
	200	13

Hot-manifold nozzles with external heating are heated by band heaters, tubular heater cartridges or helical tubular heaters. Because of the large size but low power output of 4 W/cm², the use of band heaters is rather limited.

6.10.1.6.2 Heating of Manifolds

Hot manifolds with indirectly heated nozzles are heated with cartridge heaters. They permit heating of the individual nozzle areas separately, in contrast to tubular heaters, which are discussed later on. The cartridges are arranged on both sides of the runners. The distance from the runner is about equal to the cartridge diameter. The positioning in longitudinal direction has to be optimized by measuring the temperature distribution.

Tubular heaters can be recommended for manifolds with directly heated nozzles. These sturdy heating elements make a very uniform heating of manifolds possible; the probability of failure is small. The tubing is bent and inserted into milled grooves along the manifold and around nozzles from top and bottom. The grooves are milled with a slightly excessive dimension, e.g., 8.6 mm for an 8.2 mm heater diameter. When the tubing is inserted, it is embedded with heat-conducting cement and covered with steel sheet. The distance of the heaters from the runner should be somewhat larger than the tubing diameter.

The most important elements for heating of the hot runners are summarized in Figure 6.51. Their use depends primarily on the requisite heating power and space considerations. The maximum heating power in the smallest space is attained with high-performance heater cartridges. However, the problems grow as the Watt density increases. Aside from the high failure rate, there is the risk of local overheating of the hot runner or its elements. For this and control reasons, the heating elements should not

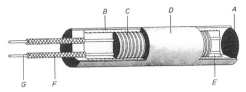

a) High density heater cartridge, Watt density 10 to 130 W/cm²:
A Bottom welded airtight, B Insulator: highly compressed, pure magnesium oxide, C Filament, D Shell, E Ceramic body, F Glass fiber insulation, G Temperature resistant

b) Tubular heater, Watt density up to about 30 W/cm²

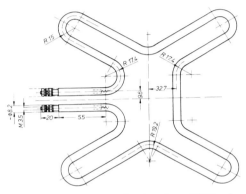

c) Tubular heater, Watt density about 8 W/cm²

d) Helical tubular heater

Figure 6.51   Heating elements for hot manifolds [6.29, 6.30]

be oversized. The Watt density should not exceed 20 W/cm³, where possible. The most important precondition for acceptable service life of the heating cartridges is good thermal transfer to the heated object. For this, the requisite roughed fit demanded by the heater cartridge manufacturers must be observed strictly. Nevertheless, replacement of heater cartridges will remain unavoidable, and so simple assembly is crucial.

Insufficiently insulated hot-runner molds lose energy from radiation. With reflector sheets of aluminum mounted between manifold and platens, energy savings of up to 35% can be achieved [6.31].

6.10.1.6.3 Computing of Power Output

The power output to be installed can be calculated with the equation:

$$P = \frac{m \cdot c \cdot \Delta T}{t \cdot \eta_{tot}}$$

(6.5)

m   Mass of the manifold (kg),

c   Specific heat of steel = 0.48 kJ/(kg · K),
ΔT   Temperature differential between desired melt temperature and manifold temperature at the onset of heating,
t   Heating-up time (s),
η_{tot} Total efficiency (electric-thermal) (ca. 0.4 to 0.7, mostly 0.6).

6.10.1.6.4 Temperature Control in Hot Manifolds

Hot-runner molds are extremely sensitive to temperature variations in nozzle and gate area. Even a temperature change of a few degrees can result in rejects. Exact temperature control is, therefore, an important precondition for a well functioning and automatically operating hot-runner mold. In principle, each nozzle should be controlled separately, because only then can the melt flow through each nozzle be influenced individually.

The control of the manifold itself is less critical. One measuring and control point is sufficient for smaller manifolds with tubular heaters. Thus, a four-cavity mold with directly heated nozzles requires at least 5 temperature-control circuits.

6.10.1.6.5 Placement of Thermocouples

There are two critical places in the nozzle area. One is the gate, the temperature of which is important for ease of flow and holding pressure; the other one is the point of greatest heat output, usually the middle of the cartridge heater where the material is in danger of thermally degrading. The best compromise is measuring the temperature between these two points. A proven method for externally heated nozzles is presented in Figure 6.52. Heaters with built-in thermocouples are often used for heated probes. Then the thermocouple should be at the end of the cartridge close to the tip of the probe. If the probe is sufficiently thick, miniature thermocouples of 0.8 mm diameter can be brought to the tip of the probe in a small groove.

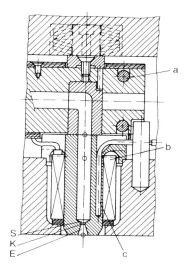

Figure 6.52   Heated nozzles for indirect gating [6.13]
S Restriction slit, K Cross section constriction at the nozzle outlet, E Expansion part, a Tubular heater, b Enclosed cylindrical heater, c Temperature sensor

Similar considerations apply to the manifold. Thermocouples should never be installed at the relatively cool ends of the manifold. This could pose the risk of overheating in the center. They should be located between the runner and the hottest spot of the cartridge. It is also obvious that the vicinity of a spacer or dowel would give a wrong temperature reading. With tubular heaters the thermocouple is positioned in the area of highest temperature, that is in the center close to the sprue bushing. For good reproducibility all thermocouples should be securely installed in the mold because thermocouples and kind of mounting can cause a considerable error in measuring. Only secured thermocouples ensure error-free read-out when the mold is put to use again.

With externally heated blocks, an installed output of 0.002 W/mm^3 volume of the manifold is expected. The heating elements are usually tubular heaters and panel heaters. The latter have the advantage of being more suitable for molds that require highly accurate matching of the temperatures across several heating loops. However, they are less robust than tubular heaters.

6.10.2 Cold Runners

When injection-molding crosslinking plastics, the same design criteria with regard to the gating system may be applied as are used for injection molding thermoplastics. However, there is the disadvantage that, aside from the molded part, the molding compound also fully crosslinks in the runner system of hot runner molds and, unlike thermoplastics, cannot be remelted and returned to the process.

These material costs of fully crosslinked runner systems, which do not contribute to added value, are the most important reason for fitting out injection molds with cold runner systems. Admittedly, these do incur higher mold costs, so that cold runner molds are only worth while for large series in which the mold costs do not constitute a major factor in production costs [6.33].

6.10.2.1 Cold-Runner Systems for Elastomer Injection Molds

The task of the cold runner system is to keep the melt at a temperature at which scorching of the elastomer will be reliably prevented. The thermal separation of the cold runner from the heated cavity saves on materials and produces other advantages [6.34–6.36] that are of interest in the context of greater productivity and higher molded part quality, as well as greater degrees of automation. Examples are [6.37]:

– longer service lives, since there is no damage caused by flash residues,
– low thermal loading during the injection phase,
– reduction in heating time through higher mold temperature,
– easier automation,
– greater design freedom in rheological dimensioning and balancing the system.

In the simplest case, in which only one cavity is directly gated, the cold runner is the extension of the machine nozzle as far as the cavity. It is more common to have a runner system for several cavities.

The basic design of a cold runner shown in Figure 6.53 consists of the following modules: manifold block, nozzles, and temperature control with insulation. The manifold block contains the runners, the turnarounds, and the branch points. It comes in various designs, each with advantages and disadvantages.

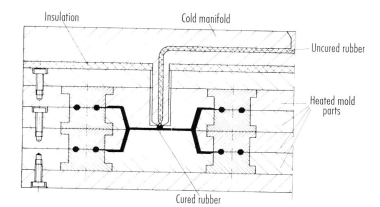

Insulation   Cold manifold

Uncured rubber

Heated mold parts

Cured rubber

Figure 6.53 Simple cold-runner design [6.38]

The nozzles connect the manifold block to the mold. They either lead direct to the molded part or to a submanifold which in turn supplies several cavities. The simplest type of nozzle is the uncooled one. However, it should only be used if the nozzles do not extend far into the cavity and a lifting cold runner block can be used [6.40] (see Figure 6.56).

If a molding is to be directly gated with a cold-runner nozzle, a more elaborate thermal separation is required. The cooled nozzles of the mold in Figure 6.54 for molding small bearings extend into the cavity area. The separation point of the gate is closely located to the molded part by a ceramic insert (Figure 6.55), which impedes heat transfer from the hot stationary mold platen into the cold runner. The gate separation is always in the transition range between cured and uncured elastomer [6.41].

Thermal separation can also be obtained by leaving the cold runner in contact with the hot mold for certain time periods only. The cold manifold is here, even in its movements, an independent component (Figure 6.56). The molded part in this 20-cavity cold-runner

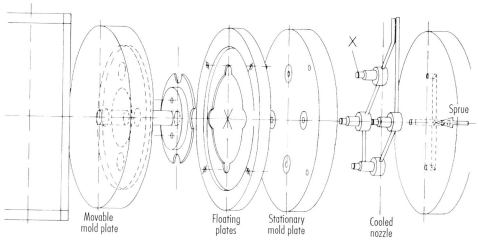

Movable mold plate   Floating plates   Stationary mold plate   Cooled nozzle   Sprue

Figure 6.54 Cold-runner mold for the production of bearings [6.41]

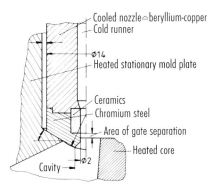

Figure 6.55   Design of the gate area of a cold runner [6.41]

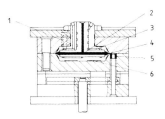

Figure 6.56   Cold-runner mold for elastomers [6.44]
1 Lifting device, 2 Cold manifold,
3 Cooling lines, 4 Pinpoint gate,
5 Molding, 6 Slots for air passage

mold are gated sideways without scrap. The cold-runner manifold is clamped in the parting line and is lifted off the hot mold parts during the mold-opening phase [6.42].

Another design solution starts with the idea that a contact between cold runner and hot mold is only needed as long as pressure can be transmitted, that is, until the gate is cured. Then a lifting of the cold runner at the end of the compression stage results in considerable technological advantages because the thermal separation is achieved in an almost ideal manner [6.43].

A corresponding mold is presented in Figure 6.57. The cold runner has the shape of a nozzle and is the immediate extension of the injection unit. Lifting of the cold runner is

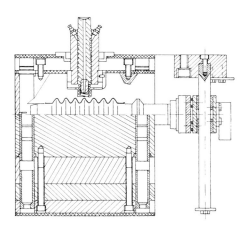

Figure 6.57   Cold-runner mold for molding of folding bellows [6.44]

caused by a spring, which lifts the cold-runner nozzle from the mold after the machine nozzle has been retracted. Now the mold can be heated without any heat exchange between the cold runner and the mold.

Figure 6.58 demonstrates the effect of the lift-off on the temperature development in the nozzle during one molding cycle. In the case of a lifting nozzle, one can clearly see how the temperature rises because of the heat flow from the mold into the nozzle. It drops back to its original level immediately after the nozzle is lifted off. This temperature development is not critical for the material in the nozzle. The contacting nozzle progresses and the nozzle is finally clogged [6.44].

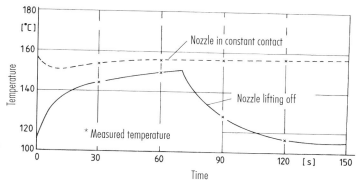

Figure 6.58 Change in melt temperature of a cold runner nozzle over one cycle [6.44]

This mold concept also permits multi-cavity molds to be designed. The special feature of the mold in Figure 6.59 is that it has two parting lines. The first parting line serves the conventional demolding. If during production an interruption occurs, e.g. by cured material in the nozzle, throwing of a locking bracket opens a second parting line and with it a plane of maintenance. The nozzles can be taken from the opened mold and purged. If the curing has progressed into the cold runner, it can be completely disassembled.

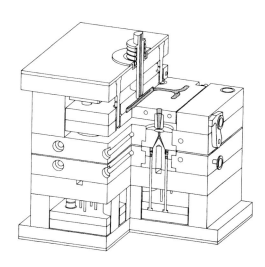

Figure 6.59 Eight-cavity cold-runner mold with curing disk gates [6.43, 6.44]

When designing cold runners, the following criteria should be observed to ensure optimum functionality [6.37]:

– minimal pressure loss: the lower the pressure consumption in the cold runner, the more pressure is available for the actual mold filling and the lower are the buoyancy forces that can lead to flash,
– no dead-water areas at turnarounds and branches,
– simultaneous filling of all mold cavities,
– low dwell times of the molding compound in the runner, to prevent scorching,
– adequate thermal separation of cold runner and mold for attaining adequate crosslinking in the gate area and avoiding scorching in the main runner,
– mechanical loading of the cold runner nozzles during transmission of the machine force in moveable cold runners,
– if interruptions in production occur and the material crosslinks in the runner, it should be easy to clean the runner.

These criteria should not be seen as being distinct from the design of the molded part, the performance of the injection molding machine or the mixture for processing. Successful use of the cold runner technique also necessitates appropriate training of the employees in order that they may be able to employ it competently. A detailed presentation of the advantages and disadvantages of the cold runner technique is contained in [6.33].

A special variant of the cold runner injection molding is the temperature-controlled transfer chamber used in injection transfer molding (ITM). ITM came into existence by applying transfer molding to an injection molding machine. The transfer chamber is filled with rubber from the injection mold unit via a runner in the transfer plunger. Appropriate heat-control keeps the transfer chamber at the plasticating temperature in order that the elastomer will not crosslink there. Figure 6.62 is a schematic diagram of the individual phases of the complete ITM process, which are described in the table below.

The duration of the various phases varies with the molded part and elastomer. In the manufacture of rubber-metal components, a further process phase may be needed for

Table 6.5 Sequence of processes in ITM shown in Figure 6.60

Name of phase	Description of process
Phase 1: Close Phase 2: Injection	Closing of all mold platens; space in front of screw (1) is filled Elastomer injected into cooled transfer chamber (2)
Phase 3: Transfer	Filling of cavities (3) via the runners (4) in the insulating plate (5) through closing of the transfer chamber
Phase 4: Heating	Crosslinking of the molded parts and the sprue slug through introduction of heat via the heating platen (6)
Phase 5: Opening	Opening of the mold platens; automatic separation of the sprue slug from the molded part; plasticating in the space in front of the screw for the next shot
Phase 6: Demolding	Removal of molded parts; removal of gate slug, e.g. with the aid of a brushing device; perhaps blowing off of mold platens and introduction of release agent

installing the inserts. The use of a transfer chamber heated to the plasticating temperature results in virtually scrap-free production, since only the sprue slugs in the short runners between the transfer chamber and the cavity area crosslink. The process is mainly used to manufacture a high number of small molded elastomer parts in one mold.

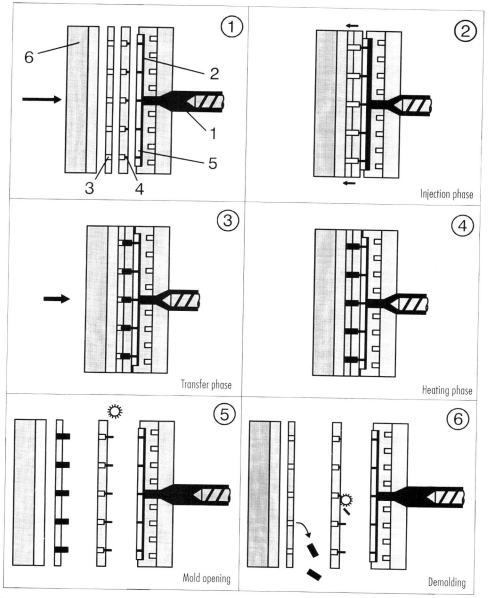

Figure 6.60   Schematic representation of phases in the ITM process

6.10.2.2 Cold-Runner Molds for Thermosets

Cold-runner molds are also employed for processing thermosets. Here one has to differentiate the kind of material to be processed. Based on lot to lot deviations, processing problems may occur with polycondensates. Experience has led to the limitation of today's runner systems on temperature-controlled sprue bushings or, in some cases, controlled machine nozzles for polycondensates.

With heat-controlled sprue bushings, a distinction has to be drawn between those with and those without a fixed pull-off point. The former have the advantage that a predefined pull-off point exists from a purely geometric point of view in the form of a cross-sectional constriction. The disadvantage is that greater pressure is needed for flow-through and thus there is greater stress on the material. Studies have shown that a uniform pull-off point is also attained in those sprue bushings without cross-sectional constriction, especially when the cycle time is very constant [6.45].

The use of the cold-runner technique for polymerized materials is wide-spread, particularly for wet polyester resins because of their low viscosity and the resulting low injection pressure for filling a mold [6.46].

A particular mold design is the use of cold runners in the so-called cassette technique (Figure 6.61).

The mold in Figure 6.61 is equivalent in its design to a two-platen mold with tunnel gate. The cold runner is formed by a temperature-controlled manifold (medium: water), which is vertically mounted to the stationary mold platen. A substantial advantage of this design is based on the ease of assembly or disassembly at interruptions or the end of a production run. The cold runner can be uncovered inside the machine in the open mold and subsequently cleaned. Mold costs are higher by 20 to 25% if compared with a

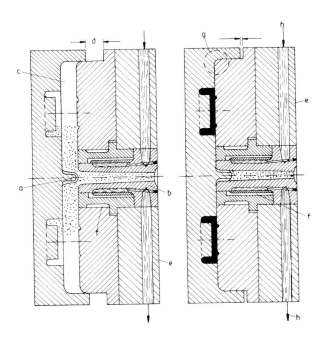

Figure 6.61 Common pocket mold for processing thermosets by coining [6.48]
left side: During injection,
right side: Closed,
a Distributor, b Sprue bushing,
c Common filling space,
d Coming gap, e Insulating sheet,
f Air space, g Closing shoulder,
h Heat exchanger

conventional mold and have to be compensated by material savings. Thus, material losses in a corresponding eight-cavity mold could be reduced from 12 to 3% [6.47].

In cases where multiple gating is needed for certain moldings (e.g. headlamp reflectors), production without cold-runner cassettes is often not conceivable [6.46].

Cold-runner technique for thermosets is also used in the so-called common-pocket process (Figure 6.62). A combination with this process is the RIC technique (Runnerless Injection Compression), which reduces scrap to a minimum in a simple way. At the same time flashing is diminished. The plasticated material flows through a temperature-controlled runner into the slightly opened mold and is distributed there. The material is pushed into the cavities and formed by the clamping motion of the mold. The material distributor penetrates the tapered sprue bushing and closes it against the parting line. Temperature control keeps the material in the runner fluid and ready for the next shot [6.48].

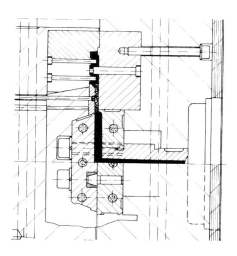

Figure 6.62   Cold runner mold
Bucher/Mueller system with tunnel gate
[6.47]

6.11   Special Mold Concepts

6.11.1 Stack Molds

A special mold design has come into use, the stack mold, for molding shallow, small parts in large quantities such as tape cassettes. Here, cavities are located in two or more planes corresponding to two parting lines and are filled at the same time (Figure 6.63). A molding machine with an exceptionally long opening stroke is needed. An increase in productivity of 100% as one might expect from doubling the number of cavities cannot be realized because of the time needed for the longer opening and closing strokes. The increase in productivity is about 80% [6.49]. The clamping force should be 15% higher than for a standard mold [6.49].

Hot manifolds are now employed exclusively. A stack mold with two parting lines has three main components, a stationary and a movable mold half, and a middle section. It contains the runner system (Figure 6.64).

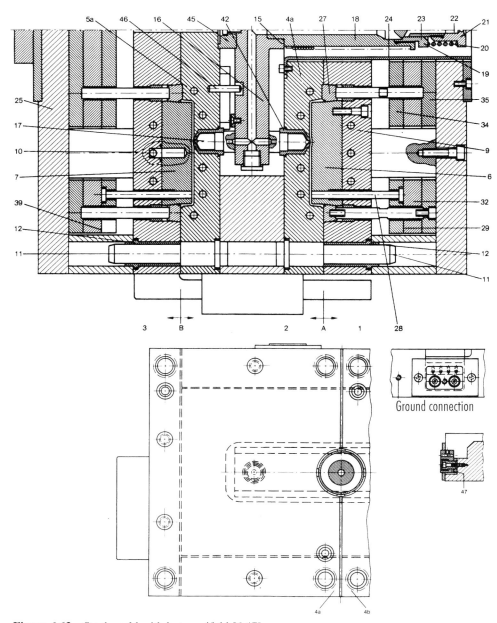

Figure 6.63   Stack mold with hot manifold [6.47]
A, B Parting lines; 1–3 Leader for center plate; 4a–4b Mold plates; 7–8 Cores; 9–10 Mold
plates; 11 Leader for mold plates; 12 Leader bushing; 15 Heater for sprue; 16 Hot manifold;
17 Sprue to molded part; 18 Mold plate, as per 9; 19–23 Central sprue to machine as extended
nozzle; 24 Sprue; 27 Stripper ring; 28 Ejector pins; 29–35 Ejector system; 39 Retainers;
42 Heated nozzle; 45–46 Interlocks

The mold section mounted on the movable platen and the center section are moved in the direction of the machine axis during demolding. With this, the extension is removed from the nozzle. The extension has to be sufficiently long that no leakage material can drop onto the leader pins and stick there during mold opening. This would impede their proper functioning [6.53, 6.54]. For this reason many stack molds are operated today with telescopic extensions and without nozzle retraction. While the outer section on the clamping side is mounted on the movable machine platen and moves positively with it during mold opening and closing, special elements are necessary to guide and control the movement of the center section. Because of the frequently large size of the molds utilizing the whole platen area, center sections are attached to the tie bars or are guided by means of guide bars with guide shoes [6.53, 6.54] (see Figure 6.64).

Today the motion is primarily produced by toggles or sometimes by racks (Figure 6.65). Previously systems were employed which used separate hydraulic cylinders for moving the center section.

Toggle and rack control open at both parting lines smoothly and simultaneously. Toggle controls also offers the option of using, within a certain range, opening strokes of different lengths. This allows the molding of parts with one height in one stack and parts with a different height in the other one. The curves of the opening path can be

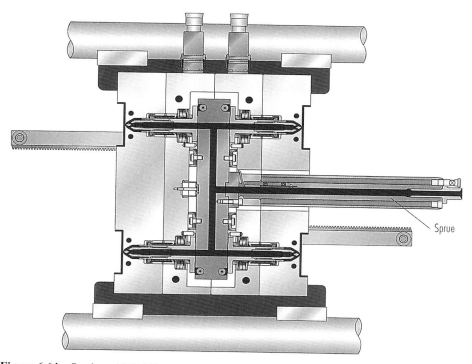

Figure 6.64   Stack mold [6.16]
Hot runner system for stack mold manifold and sprues and gates molds. The sprue is normally mounted at the level of the mold center and feeds the melt into the middle of the hot runner manifold. From there, the melt is distributed uniformly to all cavities of both mold daylights.

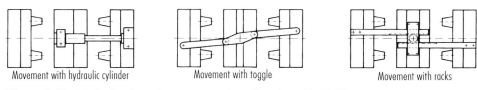

Figure 6.65   Methods of moving center section of stack molds [6.55]

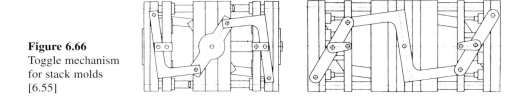

Figure 6.66
Toggle mechanism
for stack molds
[6.55]

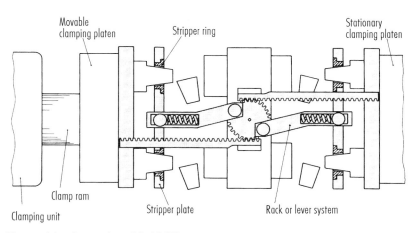

Figure 6.67   Ejector drive for stack molds [6.55]

adjusted within a wide range depending on pivotal point and toggle geometry. At the same time ejection is actuated by the same elements that move the center section. Various kinds of toggle control are shown with Figure 6.66. The rack control in Figure 6.67 is less rigid and permits a gentle start and build-up of demolding forces because of springs in the pulling rods connected to the crank drive.

6.11.2 Molds for Multicomponent Injection Molding

There are a large number of multicomponent injection molding techniques, in terms of processes and of names, which are explained in Table 6.6 [6.56, 6.57].

6.11.2.1 Combination Molds

Two-component combination injection molding in which two melts are introduced into the cavity in succession via separate gating systems requires special mold techniques since those areas of the mold that become filled by the second melt must be blocked off when the first material is injected, in order that it does not penetrate into those areas.

Table 6.6 Definition of several multicomponent injection molding processes

Process name	Definition
Multicomponent injection molding	All injection molding methods in which two or more materials are processed
Composite injection molding	Several melts are injected via several gate systems into the cavity in succession
2-Color injection molding	As above, but using one material in different colors
Multicolor injection molding	Same as 2-color injection molding, but using more than 2 colors
2-Component sandwich injection molding	Two melts are injected in succession through a gate system, to form a core and outer layer
Bi-injection	Two melts are injected simultaneously via two gating systems into the cavity

This separation has allowed the development of two-component combination injection-molded parts, such as housings with integrated seals.

The separation may be effected in either of two ways: by the rotating mold systems shown in Figure 6.68 and by the non-rotating core-back technique shown in Figure 6.69 [6.57–6.59].

Molds with Rotating Mold Platen or Rotating Mold Half

A rotating mold has several gating stations and different cavities. For a two-colored part, the first colored section is created by injection at the first mold position. After sufficient time has elapsed for the melt to cool, the mold opens and the mold-part section turns 180° into the second position. The mold closes to form the second cavity into which the second color or another material is injected via a second injection position. In the first mold position, meanwhile, the first molded-part section is being created again. In a similar fashion, three-colored parts can be made using three injection and mold positions and rotations of 120°. The mold is rotated either by means of a standard rotary platform that can be attached to the machine, irrespective of the mold, or by means of a rotary device integrated into the mold that allows a rotary plate to operate. The advantage of the standard rotary platform is its universal method of use, and in the smaller and less expensive design of the molds used. Usually the mold platen on the ejector side is designed to be the rotating side since rotation of the nozzle-side mold platen is more complicated in terms of gating system and rotating system. These molds require high precision mold making but are dependable in operation and do not require any elaborate melt feed [6.57, 6.60]. Typical applications are car tail light covers [6.60], three-colored keyboards [6.61], and the vent flaps of the Golf motorcar [6.62].

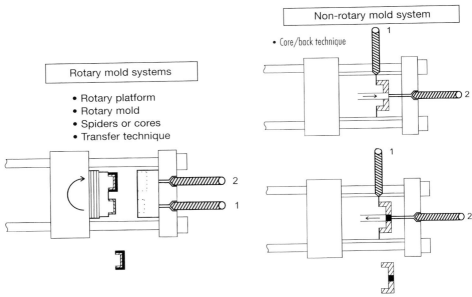

Figure 6.68   Rotary mold systems for composite injection molding

Figure 6.69   Core-back technique

Molds with Rotary Cores or Spiders
In this technique, only part of the ejector- or nozzle-side cavity with injected pre-molding is rotated (Figure 6.70). Both mold platens remain in position.

Molds with Transfer or Insert Technique
After the pre-molding is made in the first cavity, it is transferred by a handling device or by hand into the second cavity and molded to produce the final part with a second material. The term transfer technique is also used to describe using a different machine for molding to produce the final part. Generally, these molds are preferred to rotary molds for economic reasons because the complicated rotary device can be dispensed with, and usually more cavities can be accommodated on the mold platen. Furthermore, thermal separation of the pre-molding and final-molding positions is easier to accomplish (particularly important for thermoplastic-thermoset laminates). Disadvantages are the need for precise centering of the pre-moldings [6.57].

Molds with Retractable Slides and Cores (Core-Back Molds)
With comparatively low mold costs, it is possible to produce multicolor or multi-component injection molded parts in one mold without the need for opening the machine in between and further transport of a molded part by means of the core-back technique. The cavity spaces for the second material are first closed by movable inserts or cores and are opened only after the first material has been injected. The components can be arranged beside, above, or inside each other. This method does not suit material pairs that will not join or bond to each other since it is not possible to produce effective undercuts for interlocking with the injection partner. Furthermore, injection in these molds can only be carried out sequentially and not in parallel as in other methods. This results in longer

cycle times [6.57]. Separate temperature control of the cores or inserts is beneficial since the temperature of the impact surface onto which the second melt is injected can be controlled more accurately [6.64].

In combination injection molding, the rotary mold systems often employ hot runners for the pre-molding so as to yield a gateless pre-molding, since the gate interferes during

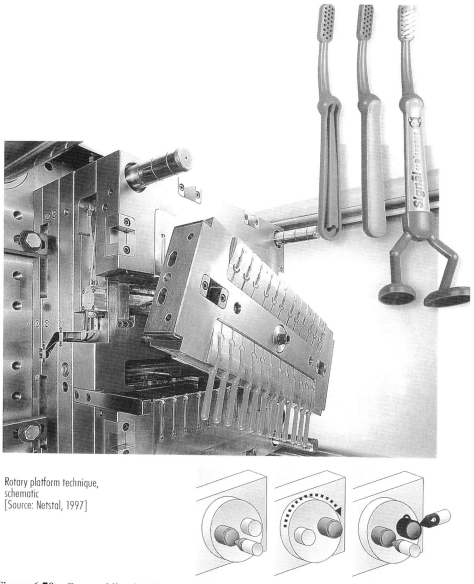

Rotary platform technique,
schematic
[Source: Netstal, 1997]

Figure 6.70   Overmolding by the rotary technique. Here: toothbrush made of two components [6.63]

rotation or transfer [6.64, 6.65] and would otherwise have to be removed prior to transfer.

The choice of method for a particular molded part must be established individually from technical and, economic aspects for every application. It must be remembered, however, that rotary mold systems are generally more expensive because of the need for two cavities and from the machine point of view, need a large distance between tie bars in order to be rotatable. Rotary molds do, however, offer greater design freedom and the possibility of thermal separation of the kind required for the manufacture of rubber-thermoplastic combinations (e.g. PA/SLR).

6.11.2.2 Two-Component Sandwich Injection Molds

In contrast to combination injection molding, sandwich molding theoretically does not require a special mold technology and may be performed with standard injection molds. Two melts are injected through a joint gating system into the cavity, to form a core and an outer skin. The melts meet in an adapter between the nozzle peaks of the injection units and the sprue bushing of the mold. It should be noted that all deviations from rotationally symmetrical molded-part geometry with central gating cause non-uniform core material distribution.

6.11.2.3 Bi-Injection Molds

In this injection molding method, two different melt components are fed into the cavity simultaneously through different gating systems. The weld line is affected by the positions of injection and wall thicknesses in the mold as well as by the injection parameters of the two components [6.68].

References

[6.1]   Kegelanguß, Schirmanguß, Ringanguß, Bandanguß. Technical Information, 4.2.1, BASF, Ludwigshafen/Rh., 1969.

[6.2]   Spritzgießen von Thermoplasten. Publication, Farbwerke Hoechst AG, Frankfurt/M., 1971.

[6.3]   Sowa, H.: Wirtschaftlicher fertigen durch verbesserte Angußsysteme. Plastverarbeiter, 29 (1978), 11, pp. 587–590.

[6.4]   Kohlhepp, K. G.; Mohnberg, J.: Spritzgießen von Formteilen hoher Präzision, dargestellt am Beispiel von Polyacetal. Kunststoff-Berater, 10 (1974), pp. 577–584.

[6.5]   Crastin-Sortiment, Eigenschaften, Verarbeitung. Publication, Ciba-Geigy, Basel, August 1977.

[6.6]   Christoffers, K. E.: Formteilgestaltung, verarbeitungsgerecht. Das Spritzgußteil. VDI-Verlag, Düsseldorf, 1980.

[6.7]   Tunnelanguß, Abreiß-Punkt-Anguß. Technical Information, 4.2.3, BASF, Ludwigshafen/Rh., 1969.

[6.8]   Thonemann, O. E.: Anguß- und Anschnitt-Technik für die wirtschaftliche Herstellung von Spritzgußteilen aus Makrolon. Plastverarbeiter, 14 (1963), 9, pp. 509–524.

[6.9]   Kunststoffverarbeitung im Gespräch, 1: Spritzgießen. Publication, BASF, Ludwigshafen/Rh., 1979.

[6.10] Spritzguß-Hostalen PP. Handbook. Farbwerke Hoechst AG, Frankfurt/M., 1980.

[6.11] Vorkammer-Punktanguß-Isolierverteiler. Technical Information, 4.2.4, BASF, Ludwigshafen/Rh., 1969.

[6.12] Durethan BK. Tech. Ringbuch der Farbenfabriken, Bayer AG, Leverkusen, 1967.

[6.13] Goldbach, H.: Heißkanal-Werkzeuge für die Verarbeitung technischer Thermoplaste (ABS, PA, PBT, PC). Plastverarbeiter, 29 (1978), pp. 677–682, and 30 (1979), pp. 591–598.

[6.14] Gauler, K.: Heißkanalsysteme mit Ventilanschnitten. Kunsstoffe, 87 (1997), 3, pp. 338–340.

[6.15] Homes, W.: Kaskadenspritzgießen vermeidet Bindenähte. Kunststoffe, 86 (1996), 9, pp. 1269–1272.

[6.16] Husky: Husky-Heißkanalsysteme. Werkschrift der Husky Injection Moulding Systems Ltd., Bolton, Canada, D-62507 Wiesbaden-Igstadt, Phone: +49 611 950 850.

[6.17] Leidig, K.; Poppe, E. A.; Schinner, K.: Technische Kunststoffe: Die Top Ten der Spritzgießprobleme. Schwierigkeiten mit Heißkanälen (8), DuPont de Nemours GmbH, D-61343, Bad Homburg.

[6.18] Eiden, G.: Werkzeuge für die Herstellung von Präzisionsteilen aus technischen Thermoplasten. Reprint of 8th Tooling Conference at Würzburg: Der Spritzgießformenbau im internationalen Wettbewerb. September 24–25, 1997.

[6.19] Braun, P.: Innenbeheizte Heißkanalsysteme. Kunststoffe, 87 (1997), 9, pp. 1184–1886.

[6.20] Heiße Seiten. Plastverarbeiter, 48 (1997), 10, pp. 96–98.

[6.21] Braun, P.: Universelles Heißkanalsystem. Osterr. Kunststoffzeitschrift, 28 (1997), 5/6, pp. 98–102.

[6.22] Sander, W.: Homogene Schmelze-Temperaturverteilung mit Standard-Elementen. Plastverarbeiter, 42 (1991), 12, pp. 55–59.

[6.23] Roßbach, R.: Schmelzefilter für den Spritzgießer. Kunststoffe, 85 (1995), 2, pp. 193–195.

[6.24] Zimmermann, W.; Hack, K.: Einfaches Heißkanal Schmelzeleitsystem für Spritzgießwerkzeuge. Plastverarbeiter, 44 (1993), 4, pp. 18–22.

[6.25] Konstruktionszeichnung. H. Weidmann AG, Raperswil/Schweiz.

[6.26] Heißkanalsystem – Indirekt beheizter Wärmeleittorpedo C.2.1. Technische Kunststoffe, Berechnen – Gestalten – Anwenden. Publication, Hoechst AG, Frankfurt/M., October 1979.

[6.27] Mit Microelektronik Temperaturen genau regeln. Plastverarbeiter, 41 (1990), 5, pp. 78–80.

[6.28] Der Heißkanal. Publication, Plastic Service GmbH, Mannheim.

[6.29] Catalog. Firma Hotset, Lüdenscheid.

[6.30] Publication. Türk und Hillinger, Zuttlingen.

[6.31] Wärmeverluste an Heißkanalblöcken – Reflektorbleche schaffen Abhilfe. Plastverarbeiter, 37 (1986), 4, pp. 122–123.

[6.32] Schauf, D.: Angußloses Entformen von Spritzgußteilen. In: Das Spritzgießwerkzeug. VDI-Verlag, Düsseldorf, 1980.

[6.33] Krehwinkel, T.; Schneider, Ch.: Kaltkanaltechnik-pro und contra. In: Spritzgießen und Extrudieren von Elastomeren, VDI-Gesellschaft Kunststofftechnik, VDI-Verlag, Düsseldorf, 1996.

[6.34] Cottancin, G.: Gummispritzformen mit dem Kaltkanalverfahren. Kautschuk Gummi Kunststoffe, 33 (1980), 10, pp. 839–841.

[6.35] Kaltkanaltechnologie beim Elastomerspritzgießen. Publication, Klöckner Ferromatic Desma, Werk Achim, 1983.

[6.36] Schneider, Ch.: Das Verarbeitungsverhalten von Elastomeren im Spritzgießprozeß. Dissertation, RWTH, Aachen, 1987.

[6.37] Maaß, R.: Die Anwendung der statistischen Versuchsmethodik zur Auslegung von Spritzgießwerkzeugen mit Kaltkanalsystem. Dissertation, RWTH, Aachen, 1995.

[6.38] Cottancin, G.: Gummispritzformen für das Kaltkanalverfahren. Gummi, Asbest, Kunststoffe, 33 (1980), 9, pp. 624–633.

[6.39] Bode, M.: Spritzgießen von Gummiformteilen. In: Spritzgießen von Gummiformteilen. VDI-Verlag, Düsseldorf, 1988, pp. 1–33.

[6.40] Cottancin, G.: Gummispritzpressen mit dem Kaltkanalverfahren. Gummi, Asbest, Kunststoffe, 33 (1980), 9, pp. 624–633.

[6.41] Lommel, H.: Einflüsse der Prozeßgrößenverläufe un der Werkzeuggestaltung auf die Qualität von Elastomerformteilen. Unpublished report, IKV, 1984.

[6.42] Holm, D.: Aufbau von Werkzeugen für Spritzgießmaschinen. In: Spritzgießen von Elastomeren. VDI-Verlag, Düsseldorf, 1978.

[6.43] Benfer, W.: Rechnergestützte Auslegung von Spritzgießwerkzeugen für Elastomere. Dissertation, RWTH Aachen, 1985.

[6.44] Menges, G.; Barth, P.: Erarbeitung systematischer Konstruktionshilfen zur Auslegung von Kaltkanalwerkzeugen. AIF-Abschlußbericht, IKV, Aachen, 1987.

[6.45] Kloubert, T.: Reduzierung von ausgehärtetem Angußmaterial beim Spritzgießen von rieselfähigen Duroplasten – Kaltkanaleinsatz und Partikelrecycling. Dissertation, RWTH, Aachen, 1996.

[6.46] Niemann, K.: Kaltkanaltechnik-Stand und Einsatzmöglichkeiten bei Duroplasten. Angußminimiertes Spritzgießen, SKZ, Würzburg, June 23–24, 1987.

[6.47] Gluckau, K.: Wirtschaftliche Verarbeitung von Duroplasten auf Spritzgießmaschinen in Kaltkanalwerkzeugen. Plastverarbeiter, 31 (1980), 8, pp. 467–469.

[6.48] Braun, U.; Danne, W.; Schönthaler, W.: Angußloses Spritzprägen in der Duroplastverarbeitung. Kunststoffe, 77 (1987), 1, pp. 27–29.

[6.49] Hotz, A.: Mehr-Etagen-Spritzgießwerkzeuge. Plastverarbeiter, 29 (1978), 4, pp. 185–188.

[6.50] Hartmann, W.: Spritzgießwerkzeuge in Etagenbauweise mit "Thermoplay"-Heißkanaldüsen für Verpackungs-Unterteile aus Polystyrol. Plastverarbeiter, 32 (1981), 5, pp. 600–605.

[6.51] Moslo, E. P.: Runnerless-Moulding. SPE-Journal, 11 (1955), pp. 26–36.

[6.521 Moslo, E. P.: Runnerless-Moulding. New York, 1960.

[6.531 Lindner E.; Hartmann, W.: Spritzgießwerkzeuge in Etagenbauweise. Plastverarbeiter, 28 (1977), 7, pp. 351–353.

[6.54] Schwaninger, W.: Etagenwerkzeuge insbesondere als Alternative zum Schnelläufer. Der Spritzgießprozeß, Ingenieurwissen, VDI-Verlag, Düsseldorf, 1979.

[6.55] Publication, Husky.

[6.56] Eckardt, H.: Mehrkomponentenspritzgießen – Neue Werkstoffe und Verfahren beim Spritzgießen. VDI-Verlag, Düsseldorf, 1990, pp. 149–194.

[6.57] Langenfeld, M.: Werkzeugtechnik zur Herstellung von Mehrkomponenten-Spritzgießteilen. Mehr Farben – Mehr Materialien – Mehr Komponenten – Spritzgießtechnik, SKZ, Würzburg, May 1992, pp. 83–102.

[6.58] Eckardt, H.: Verarbeitung von TPE auf Spritzgießmaschinen und Möglichkeiten der Anwendung des Mehrkomponenten-Spritzgießens. Thermoplastische Elastomere im Aufwärtstrend, SKZ, Würzburg, October 1989, pp. 95–123.

[6.59] Kraft, H.: "Erweiterte Anwendungsbereiche für das Mehrkomponenten-Spritzgießen". In: Kunststoffe, 83 (1993), 6, pp. 429–433.

[6.60] Bodeving, C.: "Horizontalmaschine mit Drehtisch". In: Kunststoffe, 85 (1995), 9, pp. 1244–1254.

[6.61] "Beweglich aus dem Spritzgießwerkzeug". In: Plastverarbeiter, 36 (1985), 4.

[6.62] "Vollbewegliche Lüfterklappen bei jedem Spritzzyklus". In: Plastverarbeiter, 44 (1993), 2.

[6.63] Netstal News. No. 31, Netstal-Maschinen AG, Näfels/Schweiz, April 1997.

[6.64] "Heißkanaltechnik beim Mehrkomponenten-Spritzgießen". In: Plastverarbeiter, 45 (1994), 10.

[6.65] Krauss, R.: Spritzgießtechnik für die Mehrfarbentechnik. In: Konstruieren von Spritzgießwerkzeugen, VDI-Verlag, Düsseldorf, 1987, pp. 6.1–6.18.

[6.66] Mehr-Rohstoff-Technologie. Publication. Battenfeld, Meinerzhagen, September 1996.

[6.67] "Combimelt-Technologie für Mehrfarben und Mehrkomponentenspritzguß". In: KGK Kautschuk, Gummi, Kunststoffe, 49 (1996), 7–8.

[6.68] Johannaber F.; Konejung, K.: Mehrkomponenten-Technik beim Spritzgießen. Mehr Farben – Mehr Materialien – Mehr Komponenten – Spritzgießtechnik, SKZ, Würzburg, May 1992, pp. 7–20.

7   Venting of Molds

During mold filling the melt has to displace the air which is contained in the cavity. If this cannot be done, the air can prevent a complete filling of the cavity. Besides this the air may become so hot from compression that it burns the surrounding material. The molding compounds may decompose, outgas or form a corrosive residue on cavity walls. This effect can occasionally be noticed in poorly vented molds at knit lines or in corners or flanges opposite the gate.

Burns usually appear as dark discolorations in the molded part and render it useless. If this residue is not carefully removed again and again, it may cause irreparable damage to the mold from corrosion and abrasion. Table 7.1 [7.1] summarizes the major consequences of inadequate mold venting.

Table 7.1   Consequences of inadequate mold venting [7.1]

For the molded part	For the mold	For injection molding
Burn marks due to **diesel effect**	**Abrasion** (leaching) through combustion residues in the combustion gas → diesel effect	**Irregular processes** through blockage of venting channels
Structural defects/surface defects through detachment of the polymer from a structured mold wall	**Corrosion** by aggressive gases → diesel effect	**Longer cycle** times due to increased back pressure in the cavity
Overpacking due to injection pressure set too high when vents clogged	**Mold coated** by combustion residues in the combustion gas → diesel effect	**Short service life** of machine due to higher loading
Displacement of weld lines due to changes in vents (e.g. through increasing clogging of venting elements)	Mold exposed to direct heat due to strong air heating during compression (diesel effect) ⇒ **hardening of outer layer** (varying with steel grade)	Escaping gases during composting of the polymer (diesel effect), may be **harmful to health**, depending on material
Entrapped air (voids)	**Increased cleaning** of venting elements	**Longer setup time** through higher scrap rate
Incomplete mold filling	Higher **repair** and **maintenance costs**	**Greater need for pressure** due to increased back pressure in the cavity
Reduction in strength, especially at weld lines		Injection molding machine has **higher energy requirements**

For systematization of the subject "mold venting", a distinction is drawn between "passive" venting, where the air escapes from the cavity due to the pressure of the incoming melt, and "active" venting where a pressure gradient is created to artificially remove the air.

7.1   Passive Venting

Most molds do not need special design features for venting because air has sufficient possibilities to escape along ejector pins or at the parting line. This is particularly true if a certain roughness is provided at the parting line such as planing with a coarse-grained grinding wheel (240 grain size). This assumes filling the cavity in such a way that the air can "flow" towards the parting line. A grinding direction radial to the cavity has been successfully tested (Figure 7.1). Air pockets must not be created.

Figure 7.1   Special ground sections in sealing surface [7.2 to 7.4]

The configuration of the part, its position in the mold, and its gating have a considerable effect on venting. This can be best demonstrated with some examples. The tumbler in Figure 7.2 is gated at the bottom. The confined air is pushed towards the parting line and can escape there. Special provisions are not necessary. The design of Figure 7.3 is different. The tumbler is gated on the side. During mold filling, the melt flows around the core first, closes the parting line and then rises slowly pushing the air ahead into the bottom of the tumbler. Here the air is compressed and overheated. To prevent this, special design steps are necessary. The same holds true for the mold in Figure 7.4a. The enclosed air in the rib cannot escape because the melt first crosses the base of the rib.

Figure 7.2   Tumbler gated at bottom, favorable gate location

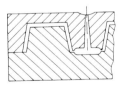

Figure 7.3   Tumbler laterally gated, poor gate location for venting

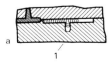 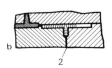

Figure 7.4   Part with rib [7.3]:
a Air cannot escape from rib section and is trapped (1),
b Remedial measure: Additional parting line (2)

In both cases adding a joint can provide venting of the cavity. The tumbler mold (Figure 7.3) can obtain an additional joint if the bottom piece is made separately (Figure 7.5) or a cylindrical insert is used, which lets the air escape. A mark on the bottom of the tumbler, of course, cannot be avoided. If it is a nuisance, it can be converted into a decorative line [7.5].

Figure 7.5   Laterally gated cup. Mold with additional parting line: Venting through parting line (left) or venting pin (right)

In the case of a rib (Figure 7.4a), an additional joint for venting is obtained by dividing the forming inserts into two pieces (Figure 7.4b).

For the presented solution it was assumed that the air can escape through joining faces, but this is feasible only if the faces have sufficient roughness and the injection process is adequately slow to allow the air to escape. This solution fails for molding thin-walled parts with very short injection times. Here, special venting channels become necessary.

In the case of the center-gated tumbler (Figure 7.2), a solution is found by machining an annular channel into the parting-line plane into which the air can escape via one or more venting gaps during injection, and then from the mold through a venting channel. Dimensions for these channels can be taken from Figures 7.6 and 7.7.

A logical development of this approach is that of continuous venting [7.6, 7.8] or peripheral venting [7.9]. The annular gap is not interrupted by flanges but is continuous. Penetration by the melt is prevented by adjusting the gap width so as to just prevent ingress of melt. Usual gap widths vary from 10 µm to 20 µm in accordance with the polymer employed [7.8].

Figure 7.6   Cup mold with annular channel for venting [7.6]

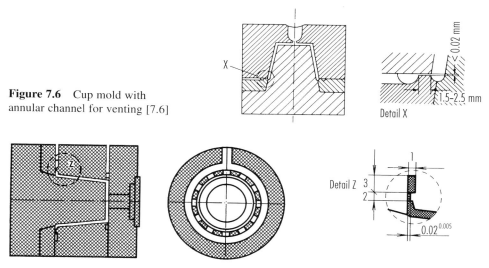

Figure 7.7   Mold venting through venting gaps and annular channel [7.7]

Figures 7.8 and 7.9 demonstrate another option for venting molds with large surface areas by using a set of lamellae. One has to bear in mind, though, that these venting elements leave marks on the molded part and may interfere with cooling lines.

Packs of lamellae are also of advantage if multiple gating is needed and an exact location of knit lines cannot be determined. Reference is made here to Section 5.9 where it was demonstrated how knit lines and locations of possible air trapping can be

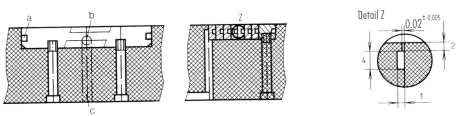

Figure 7.8   Venting with a set of lamellae [7.7]:
a Spring, b Venting channel through lamellae, c Connection of venting channel with the outside

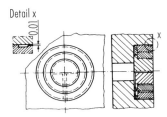

Figure 7.9   Venting with sleeves [7.10]

predetermined with the help of the filling image method. The safest way to rule determine such potentially harmful points is to perform a computer simulation (see Chapter 14).

Porous inserts of, e.g. sintered metals that open out into free space, have not proved suitable because they more or less clog rapidly, the rate depending on the molding compound being processed [7.7]. If the sintered metal opens into a cooling channel, however, completely new perspectives open up for the use of sintered metal mold inserts. In a manner analogous to the water-jet pump, the existing cooling water circuit can provide active venting of the cavity [7.9, 7.11, 7.12]. The water from the circuit draws the air out of the cavity via the sintered metal inserts, without water entering the cavity.

If ejector pins are located in an area where air may be trapped, they can usually be used for venting. Venting can be facilitated by enlarging the ejector pin hole (Figure 7.10). This solution offers an additional advantage. If needed, compressed air can be blown into the hole to support demolding. In addition, the gaps are cleaned by the movement of the pins.

Frequently, so-called venting pins are employed. They can be grooved (Figure 7.12) or kept 0.02 to 0.05 mm smaller in diameter (material dependent) than the receiving hole for a length of 3 mm (Figure 7.11) [7.10]. A venting channel follows, in which the air can expand and from where it reaches the outside through an axial groove.

A design according to Figure 7.13 is called "self-cleaning venting pin with ejector function" in the literature. The definite advantage is based on the precise centering of the ejector, which ensures a defined venting gap [7.7].

In multi-cavity molds or in molds with multiple gating, venting should already start in the runners so that air there cannot get into the cavity. This prevents extensive degradation of the material by burning and the gate system can be ground and reused without major loss of quality. As one can see from Figure 7.14, the same rules for venting channels apply here as for the venting at the end of the flow path.

There is finally the question left about the size of the venting gap. To avoid flashing, a certain gap width, which is plastic-specific, cannot be exceeded after mold clamping.

The critical gap width for specific plastics is as follows [7.10, 7.13–7.15]:

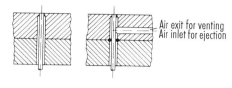

Figure 7.10 Design of holes for ejector pins permitting improved venting of cavity [7.6]: Enlarged hole about 3 mm below the cavity wall surface

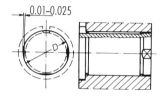

Figure 7.11 Fluted venting pin [7.6]

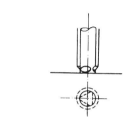

Figure 7.12 Venting pin [7.6]

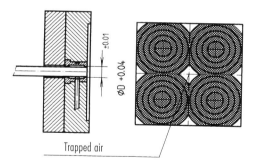

Figure 7.13 Self-cleansing venting pin [7.7]

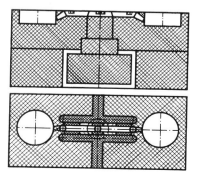

Figure 7.14 Venting of runners [7.7]

Crystalline thermoplastics: 0.015 mm,
PP, PA, GF-PA, POM, PE
amorphous thermoplastics: 0.03 mm,
PS, ABS, PC, PMMA
for extremely fluid materials   0.003 mm.

If the venting gap is considered a rectangular diaphragm, and one assumes the validity of the laws of dynamics of gases, then, with the volume of part and runners and the injection time, the flow rate can be determined, which has to flow across the venting gap to vent the mold [7.16, 7.17]:

$$\dot{V} = \frac{V_M + V_R}{t_I} \qquad (7.1)$$

$\dot{V}$ Flow rate,
V_M volume of molding,
V_R volume of runner system,
t_I injection time.

If the flow rate is equated to the admittance of the assumed rectangular diaphragm (venting gap), the width of the gap can be calculated from the equation:

$$L = A \sqrt{\frac{T_K}{293}} 2 \cdot 10^{-2} \qquad (7.2)$$

$$A = \frac{L}{\sqrt{\dfrac{T_K}{293}}} \cdot 50 \qquad (7.3)$$

$$\sqrt{\frac{T_K}{293}} = \frac{L}{A} \cdot 50 \qquad (7.4)$$

$L =$ V [m³/s],
$A =$ b · h cross section of gap [cm²],
$b =$ width of gap [cm],
$h =$ height of gap [cm],
$T_K =$ temperature of air [K].

For a molded part with a total volume (part + runner volume) of 10 cm³ which is produced with an injection time of 0.2 s Equation (7.2) results in a gap width of 12.5 mm if one assumes that a gap height of 0.02 mm does not yet cause flashing and the air temperature is 293 °K (20 °C). Of course this venting gap has to be located where air trapping can be expected and not some place else. Several points of air trapping are anticipated, then the mold has to be equipped with several vents with the sum of their cross sections at least equal to the predetermined cross section.

7.2 Active Venting

Aside from active venting via the cooling water circuit, which was already mentioned above, partial or complete evacuation of the cavity prior to the injection process is possible. This type of venting is used in the injection molding of microstructures, since conventional venting gaps are too large for the extremely low-viscosity plastic melts used, there, and would clog [7.18]. In thermoset and elastomer processing, there are applications that require evacuation of the cavity, primarily to improve the accuracy of reproduction and the molded part quality [7.19].

The structure of a vacuum system is shown schematically in Figure 7.15 [7.20]. The circuit diagram also contains a vacuum accumulator. This is connected in series if the evacuation of large volume parts is to be accelerated; furthermore, the power consumption of the vacuum pump can then be reduced overall.

Evacuation of the molds is only efficient, however, if the complete mold is sealed off. Due to the many moving parts on the mold, such as slides, ejectors, etc., this is extremely complicated and virtually impossible to achieve. The mold is instead surrounded with a

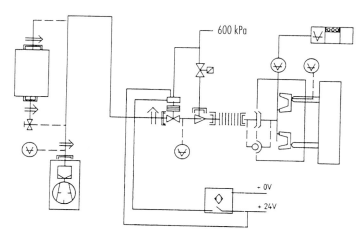

Circuit symbols as set out in DIN 28401

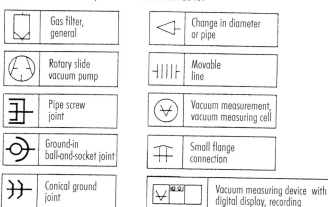

Gas filter, general	Change in diameter or pipe
Rotary slide vacuum pump	Movable line
Pipe screw joint	Vacuum measurement, vacuum measuring cell
Ground-in ball-and-socket joint	Small flange connection
Conical ground joint	Vacuum measuring device with digital display, recording

Figure 7.15 Circuit for a vacuum unit [7.20]

closed jacket or box that has just one parting line. This type of construction for a microinjection mold is shown schematically in Figure 7.16. To an extent depending on the type of demolding, this design requires either no other or very few moving parts projecting out of the vacuum space; these, however, can be readily sealed [7.21].

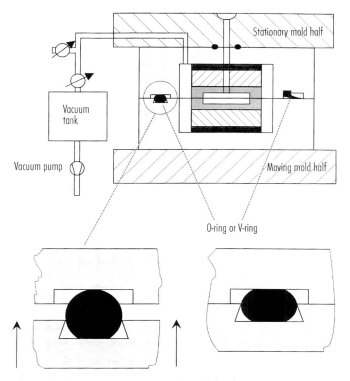

Figure 7.16   Evacuation of the mincroinjection mold [7.1]

7.3   Venting of Gas Counter-Pressure Injection Molds

Structural foam parts frequently have a rough surface as the decomposition of the blowing agent starts immediately with injection into the mold. This can be avoided by suppressing the foaming of the blowing-agent-containing melt through injecting against a pressure that has been generated in the mold cavity prior to injection. This counter-pressure must be precisely as large as the blowing agent pressure. The mold also has to be sealed (Figure 7.17). This may be done with the aid of heat-resistant seals (O-rings). To fill the mold with gas it is best to arrange a collection channel around the mold cavity that is connected via gaps to the cavity (Figure 7.18). These gaps and the collection channel allow the gas to escape again as the mold is being filled [7.22–7.24]. To keep the counter-pressure constant during injection and also to ensure venting, the collection channel is connected to magnetically-controlled pressure-control valves.

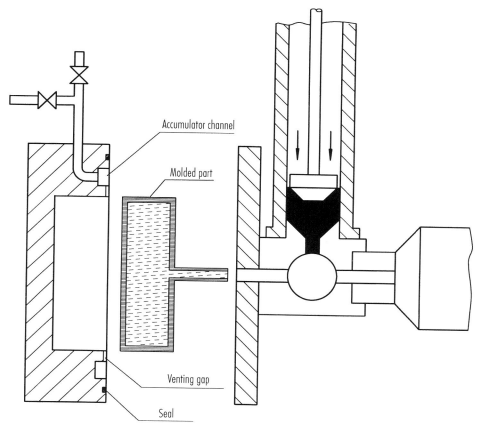

Figure 7.17   Schematic diagram of the gas counterpressure process [7.21]

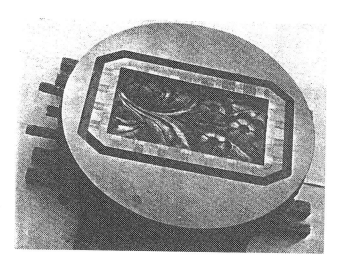

Figure 7.18   Mold for injection molding a decorative wood-carving imitation under gas counterpressure. The mold cavity wall consists of electrolytically deposited hard nickel shell

The dimensions of the gap required for filling and venting the mold are obtained with the aid of Equation (7.1, passive venting).

If complete venting of the mold is not possible, e.g. into blind holes, spring-actuated venting pins may be used (Figure 7.19). The required amount of pressure is readily adjusted manually via the pre-tension of the spring.

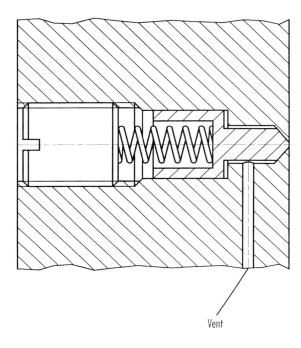

Vent

Figure 7.19   Schematic diagram of a venting pin

References

[7.1]   Notz, F.: Entlüftung von Spritzgießwerkzeugen. Plastverarbeiter, 45 (1994), 11, pp. 88–94.
[7.2]   Weyer, G.: Automatische Herstellung von Elastomerartikeln im Spritzgießverfahren. Dissertation, Tech. University, Aachen, 1987.
[7.3]   DE PS 1 198 987 (1961) Jurgeleit, H. F.
[7.4]   DE PS 1231 878 (1964) Jurgeleit, H. F.
[7.5]   Stoeckhert, K.: Werkzeugbau für die Kunststoffverarbeitung. 3rd Ed., Carl Hanser Verlag, Munich, 1979.
[7.6]   Giragosian, S. E.: Continous mold venting. Mod. Plast., 44 (1966), 11, pp. 122–124.
[7.7]   Sander, W.: Formverschmutzung (Formbelag)-verschleiß und Korrosion bei Thermoplastwerkzeugen. Paper presented at the 2nd Tooling Conference at Würzburg, October 4–5, 1988.
[7.8]   Rees, H.: Mold Engineering. Carl Hanser Verlag, Munich, 1995.
[7.91   Allen, P.: A non-traditional approach to mold cooling and venting. SPE Injection Molding Div. Conference, Columbus, OH, Oktober 20–22, 1981, pp. 71–77, Confer. 831.
[7.10] Hartmann, W.: Entlüften des Formhohlraums. Paper at the VDI Conference, Nürnberg, December 6–7, 1978.
[7.11] Smith, B.: Venting is vital. British Plastics and Rubber, May 1986, p. 22.

[7.12]   Water-line venting saves the job. Plastics World, 46 (1988), 3, pp. 27–28.

[7.13]   Ufrecht, M.: Die Werkzeugbelastung beim Überspritzen. Unpublished report, IKV, Aachen, 1978.

[7.14]   Huyjmans, H.; Packbier, K.; Schürmann, E.: Trial run with a two-cavity mold with 12 gates at NWM, s'Hertogenbosch, 1978.

[7.15]   Stitz, S.; Schürmann, E.: Measurements of deformations of injection molds at H. Weidmann, Switzerland, 1976.

[7.16]   Wutz, M.; Hermann, A.; Walcher, W.: Theorie und Praxis der Vakuumtechnik. Vieweg, Braunschweig, Wiesbaden, 1986.

[7.17]   Speuser, G.: Evakuierung von Spritzgießwerkzeugen für die Elastomerverarbeitung. Unpublished report, IKV, Aachen 1987.

[7.18]   Rogalla, A.: Analyse des Spritzgießens mikrostrukturierter Bauteile aus Thermoplasten. Dissertation, RWTH, Aachen, 1997.

[7.19]   Meiertoberens, U.; Herschbach, Ch.; Maaß, R.: Verbesserte Technologien für die Elastomerverarbeitung. Gummi, Fasern, Kunststoffe, 47 (1994), 10, pp. 642–649.

[7.20]   Michaeli, W.; Weyer G.; Speuser G.; Kretzschmar, G.: Entlüftung von Formnestern beim Spritzgießen von Elastomeren. Kautschuk + Gummi, Kunststoffe, 44 (1991), 12, pp. 1146–1153.

[7.21]   Winterkemper, A.: Entwicklung eines Werkzeugkonzeptes für das Spritzgießen von Mikrostrukturen. Unpublished report, IKV, Aachen.

[7.22]   Semerdjiev, S.; Popov, N.: Probleme des Gasgegendruck-Spritzgießens von thermoplastischen Strukturschaumstoffen. Kunststoffberater, 4/1978, pp. 198–201.

[7.23]   Eckardt, E.: Schaumspritzgießverfahren – Theorie und Praxis. Kunststoffberater, 1983, 1/2, pp. 26–32.

[7.24]   Semerdjiev, S.; Piperov, N.; Popov, N.; Mateev, E.: Das Gasgegendruck-Spritzgieß-verfahren zum Herstellen von thermoplastischen Strukturschaumteilen. Kunststoffe, 64 (1974), 1, pp. 13–15.

8 The Heat Exchange System [8.1, 8.2]

The velocity of the heat exchange between the injected plastic and the mold is a decisive factor in the economical performance of an injection mold. Heat has to be taken away from the thermoplastic material until a stable state has been reached, which permits demolding. The time needed to accomplish this is called cooling time. The amount of heat to be carried off depends on the temperature of the melt, the demolding temperature, and the specific heat of the plastic material.

For thermosets and elastomers, heat has to be supplied to the injected material to initiate curing.

Primarily, the cooling of thermoplastics will be discussed here in detail. To remove the heat from the molding the mold is supplied with a system of cooling channels through which a coolant is pumped. The quality of a molding depends very much on an always constant temperature profile, cycle after cycle. The efficiency of production is very much affected by the mold as an effective heat exchanger (Figure 8.1). The mold

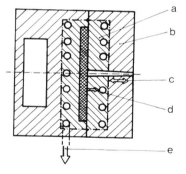

Figure 8.1 Heat flow in an injection mold [8.1]
a Region of cooling,
b Region of cooling or heating,
c $\dot{Q}_E$ = Heat exchange with environment,
d $\dot{Q}_P$ = Heat exchange with molding,
e $\dot{Q}_C$ = Heat exchange through coolant

has to be heated or cooled depending on the temperature of the outside mold surface and that of the environment. Heat removal from the molding and heat exchange with the outside can be treated separately and then superimposed for the cooling channel region. If the heat loss through the mold faces outweighs the heat to be removed from the molding, the mold has to be heated in accordance with the temperature differential. This heating is only a protection for shielding the cooling area against the outside. Cooling the molding remains in the foreground. The above mentioned relationships, however, remain applicable for all kinds of molds for thermoplastics as well as for thermosets. The latter case includes heat supply for curing. Thus the term heat exchange can be applied under all conditions.

8.1   Cooling Time

Cooling begins with the mold filling, which occurs during the time t_I. The major amount of heat, however, is exchanged during the cooling time t_C. This is the time until the mold opens and the molding is ejected. The design of the cooling system must depend on that section of the part that has to be cooled for the longest time period, until it has reached the permissible demolding temperature T_E.

The heat exchange between plastic and coolant takes place through thermal conduction in the mold. Thermal conduction is described by the Fourier differential equation. Because moldings are primarily of a two-dimensional nature and heat is only removed in one direction, the direction of their thickness, a one-dimensional computation is sufficient. (Solutions for one dimensional heat exchange in the form of approximations have been compiled by [8.3, 8.4] for a length-over-wall-thickness ratio L/s > 10.) Elastomers, however, may have very different shapes and Figure 8.14 presents, therefore, all conceivable geometries.

In the case of one-dimensional heat flow, the Fourier differential equation can be reduced to:

$$\frac{\partial T}{\partial t} = a \frac{\partial^2 T}{\partial x^2}$$

(8.1)

with $a = \dfrac{k}{\rho \cdot c_p}$ = thermal diffusivity.

In these and the following equations:
a   Thermal diffusivity,
a_{eff} Effective thermal diffusivity,
t   Time,
t_C Cooling time,
s   Wall thickness,
x   Distance,
ρ   Density,
k   Thermal conductivity,
c_P Specific heat capacity,
T_E Demolding temperature,
$\overline{T}_E$ Mean demolding temperature,
$\hat{T}_E$ Maximum demolding temperature,
T_M Melt temperature,
T_W Cavity-wall temperature,
$\overline{T}_W$ Average cavity-wall temperature,
θ   Cooling rate,
Fo   Fourier number.

Assuming that, immediately after injection, the melt temperature in the cavity has a uniform constant value of $T_M \neq f(x)$, the temperature of the cavity wall jumps abruptly to the constant value $T_W \neq f(t_C)$ and remains constant, then according to [8.3]

$$\frac{T_E - T_W}{T_M - T_W} = \frac{4}{\pi} e^{-\frac{a \cdot \pi^2}{s^2} \cdot t} \cdot \sin\frac{\pi \cdot x}{s} \tag{8.2}$$

is a solution of the differential Equation if only the first term of the rapidly converging series

$$T_E - T_W = \frac{4}{\pi}(T_M - T_W) \cdot \sum_{n=0}^{\infty} \frac{1}{2n+1} \cdot e^{-\frac{a(2n+1)^2 \pi^2 t}{s^2}} \cdot \sin\frac{(2n+1)\pi \cdot x}{s} \tag{8.3}$$

is considered.
Hence

$$\frac{\overline{T}_E - T_W}{T_M - T_W} = \frac{8}{\pi^2} \cdot e^{-\frac{a \cdot \pi^2}{s^2} \cdot t} \tag{8.4}$$

or resolved with respect to the cooling time:

$$\frac{s^2}{\pi^2 \cdot a} \ln\left(\frac{8}{\pi^2} \frac{T_M - T_W}{\overline{T}_E - T_W}\right) = t_C \tag{8.5}$$

If this equation is rearranged to

$$\frac{t_C \cdot a}{s^2} = \frac{1}{\pi^2} \cdot \ln\left(\frac{8}{\pi^2} \cdot \frac{T_M - T_W}{\overline{T}_E - T_W}\right) \tag{8.6}$$

the dimensionless representation of the cooling process (Figure 8.2) for the average part temperature is obtained

$$\frac{\overline{T}_E - T_W}{T_M - T_W} = \theta \tag{8.7}$$

It is called the excess temperature ratio and can be interpreted as cooling rate.

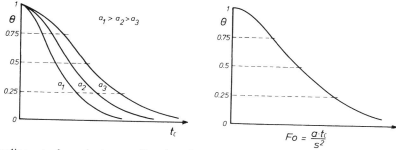

Figure 8.2   Cooling rate dependent on cooling time (left) and Fourier number (right) [8.1]

$$\frac{t_c \cdot a}{s^2} = \text{Fo is the dimensionless Fourier number} \qquad (8.8)$$

According to Equation (8.6) the cooling rate θ is only a function of the Fourier number:

$$\theta = f(\text{Fo}). \qquad (8.9a)$$

If the term $\dfrac{t_c \cdot a}{s^2} = \text{const.}$, the cooling rate is always the same.

Instead, the average temperature the calculation can also be based on the maximum temperature in the center of the molding (Figure 8.3). Then the equation for the dimensionless cooling rate is:

$$\frac{\hat{T}_E - T_W}{T_M - T_W} = \hat{\theta} \qquad (8.9b)$$

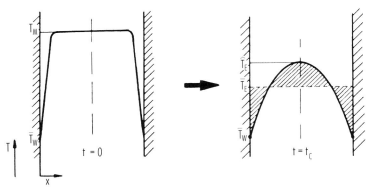

Figure 8.3   Temperature plot in molding [8.2]
T_M Temperature of material,
$\overline{T}_W$ Average temperature of cavity wall,
$\hat{T}_E$ Temperature at demolding, center of molding,
$\overline{T}_E$ Temperature at demolding, integral mean value,
t_C Cooling time

The different patterns of cooling rates can be presented dimensionless by a single curve (Figure 8.2). Although injection molding does not exactly meet the required conditions, the cooling time can be calculated with adequate precision as experience confirms.

As far as injection molding of thermoplastics is concerned, investigations [8.5] have demonstrated that demolding usually takes place at the same dimensionless temperature, that is with the same cooling rate $\hat{\theta} = 0.25$ based on the maximum temperature in the center or $\overline{\theta} = 0.16$ based on the average temperature of the molding. Therefore, it was possible to come up with a mean value for the thermal diffusivity a, the effective thermal diffusivity a_{eff}. The thermal diffusivity proper for crystalline materials is presented by an unsteady function.

8.2 Thermal Diffusivity of Several Important Materials

Figure 8.4 presents the effective thermal diffusivity of unfilled materials with a cooling rate of $\hat{\theta} = 0.25$. Figure 8.5 depicts the change in thermal diffusivity dependent on the cooling rate with polystyrene as an example. This information should permit a possible conversion into other cooling rates.

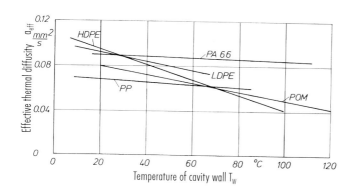

Figure 8.4 Effective mean thermal diffusivity of crystalline molding materials [8.6]

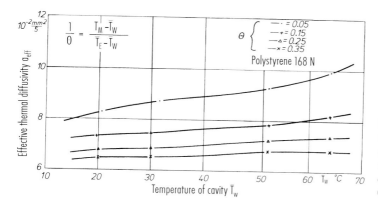

Figure 8.5 Effective mean thermal diffusivity versus mean temperature of cavity wall $\bar{T}_W$ with θ as parameter [8.1]

The thermal diffusivity of filled materials changes in accordance with the replaced volume [8.7]. Figure 8.6 shows the effective thermal diffusivity of polyethylene with various quartz contents (percent by weight) as a function of the cooling rate.

Criteria such as shrinkage, distortion and residual stresses are unimportant in structural foam parts for all practical purposes. The cooling time is solely determined by the outer skin, which has to have sufficient rigidity for demolding. Otherwise, remaining pressure from the blowing agent causes swelling of the part after release from the retaining cavity. Independent of the thickness of structural foam parts, the cooling rate can be taken

$$\hat{\theta} = 0.18 \text{ to } 0.22 \text{ (Figure 8.7)}$$

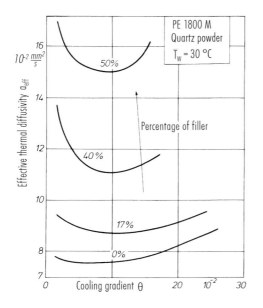

Figure 8.6   Effective thermal diffusivity of polyethylene filled with quartz powder [8.1]

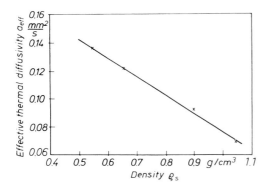

Figure 8.7   Effective thermal diffusivity dependent on density of structural foam [8.1]
(Styrofoam parts 4–8 mm thick, cooling rate θ = 0.2)

8.2.1   Thermal Diffusivity of Elastomers

For elastomers, the heat of reaction can be neglected because of its small magnitude. One can calculate and proceed like one does with thermoplastics.

Due to a high content of carbon black the thermal diffusivity is shifted to higher values similar to filled polyethylene (Figure 8.6):

$$a_{eff} \sim 1 \text{ to } 2 \text{ mm}^2/s$$

8.2.2   Thermal Diffusivity of Thermosets

Thermosets can develop a considerable higher heat of reaction. The amount of released heat depends on the degree of cross-linking and the percentage of reacting volume of the polymer. High contents of filling materials have a dampening effect. Therefore, no data can be provided. They can be obtained from the raw-material producer or by determining them with a differential calorimeter.

How much heat of reaction has to be expected can be measured with a reacting molding by plotting the increase in temperature versus the time, as is shown in Figure 8.8. The area of the "hump" is an assessment of the exothermic heat of reaction of this molding. With a hump area of small size compared with the total area under the temperature curve, one can disregard its share.

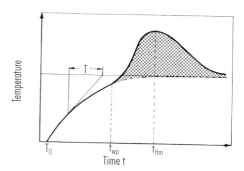

Figure 8.8 Characteristic temperature development of a reactive material [8.8]

8.3 Computation of Cooling Time of Thermoplastics

8.3.1 Estimation

Since cooling of all materials is physically similar, one can often estimate the cooling time with the simple correlation:

$$t_C = c_C \cdot s^2 \tag{8.10}$$

For unfilled thermoplastics
c_C = 2 to 3 [s/mm²],
t_C Cooling time,
s Wall thickness.

8.3.2 Computation of Cooling Time with Nomograms

With the help of mean thermal diffusivities a_{eff}, nomograms can be drawn, which allow for an especially simple and fast determination of the cooling time.

The cooling time t_C is plotted against the cavity-wall temperature T_W for a number of constant demolding temperatures $\hat{T}_E$ and various wall thicknesses s. The presented cooling-time dependence is valid for plane moldings (plates without edge effect) with symmetrical cooling (Figures 8.9 and 8.10).

Besides the diagram presentation, nomograms (Figure 8.11) can be used which are derived from the following equation (valid for plates):

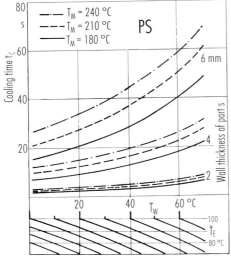

Figure 8.9   Cooling time diagram (PS) [8.1]

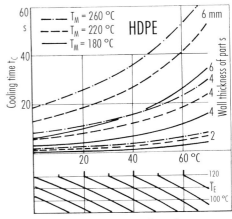

Figure 8.10   Cooling time diagram (HDPE) [8.1]

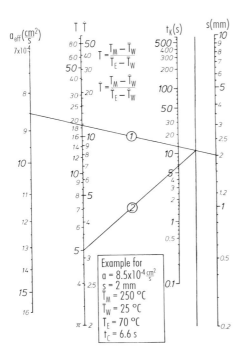

Figure 8.11   Nomogram for computation of cooling time [8.1]

$$t_C = \frac{s^2}{a_{eff} \pi^2} \ln \left(\frac{8}{\pi^2} \frac{T_M - T_W}{\overline{T_E} - T_W} \right). \tag{8.11}$$

The following correlation is valid for cylindrical parts:

$$t_C = \frac{R^2}{a_{eff} \cdot 5 \cdot 8} \ln \left(0.7 \cdot \frac{T_M - T_W}{\overline{T_E} - T_W} \right). \tag{8.12}$$

Reference data for melt, wall and demolding temperature as well as the average density between melt and demolding temperature can be found in Table 8.1.

Table 8.1 Material data [8.12]

Material	Melt temperature (°C)	Wall temperature (°C)	Demolding temperature (°C)	Average density (g/cm³)
ABS	200–270	50–80	60–100	1.03
HDPE	200–300	40–60	60–110	0.82
LDPE	170–245	20–60	50–90	0.79
PA 6	235–275	60–95	70–110	1.05
PA 6.6	260–300	60–90	80–140	1.05
PBTP	230–270	30–90	80–140	1.05
PC	270–320	85–120	90–140	1.14
PMMA	180–260	10–80	70–110	1.14
POM	190–230	40–120	90–150	1.3
PP	200–300	20–100	60–100	0.83
PS	160–280	10–80	60–100	1.01
PVC rigid	150–210	20–70	60–100	1.35
PVC soft	120–190	20–55	60–100	1.23
SAN	200–270	40–80	60–110	1.05

8.3.3 Cooling Time with Asymmetrical Wall Temperatures

If there are asymmetrical cooling conditions from different wall temperatures in the cavity, the cooling time can be estimated in the same manner by using a corrected part thickness [8.9]. The asymmetrical temperature distribution is converted to a symmetrical one by the thickness complement s' (Figure 8.12). The following estimate results from a correlation, which is discussed in [8.9]:

$$s' \approx \frac{2s}{\dfrac{\dot{q}_2}{\dot{q}_1} + 1} \qquad \dot{q}_2 \leq \dot{q}_1 \tag{8.13}$$

$\dot{q}$ = Heat flux density.
For $\dot{q}_2 = 0$ (one-sided cooling) s' = 2s; the cooling time is four times that of two-sided cooling.

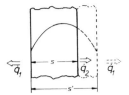

Figure 8.12   Illustration of corrected part thickness [8.1]

The cavity-wall temperatures determine the different heat-flux densities, which in turn provide the corrected wall thickness. The thickness finally allows the cooling time to be estimated.

8.3.4   Cooling Time for Other Geometries

Besides flat moldings, almost any number of combinations from plates, cylinders, cubes, etc. can be found in practice. The correlation between cooling rate and Fourier number has already been demonstrated with a plane plate as an example. This relationship can also be shown for other geometrical forms such as cylinder, sphere, and cube. Figure 8.13 presents this correlation for the cooling rate θ in the center of a body according to [8.10 and 8.11]. This also permits calculating or estimating other configurations. The necessary formulae are summarized in Figure 8.14.

For practical computation, additional simplification is possible. The cooling rate θ can be expressed by the ratio of the average part temperature on demolding $\overline{T}_E$ over the melt temperature $T_M = T_0$ and plotted versus the Fourier number for cylinder and plate (Figure 8.15).

To determine the cooling characteristic of a part that can be represented by a combination of a cylinder and a plate (cylinder with finite length) or by three intersecting infinite plates (body with rectangular faces), the following simple law [8.10] can be applied:

$$\theta_1 (Fo_1) \cdot \theta_2 (Fo_2) = \theta_{1,2} \tag{8.14}$$

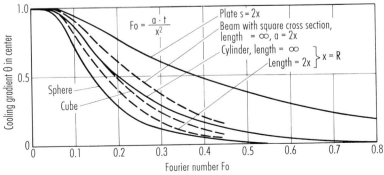

Figure 8.13   Temperature in center of molding if surface temperature is constant [8.10]

Geometry	Boundary condition	Equation
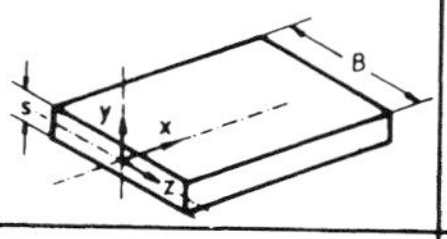 (plate)	Plate $\dot{Q}_z = 0$ $\dot{Q}_x = 0$	$t_c = \dfrac{s^2}{\pi^2 \cdot a} \cdot \ln\left(\dfrac{8}{\pi^2} \cdot \dfrac{T_M - \bar{T}_W}{\bar{T}_E - \bar{T}_W}\right)$ $t_c = \dfrac{s^2}{\pi^2 \cdot a} \cdot \ln\left(\dfrac{4}{\pi} \cdot \dfrac{T_M - \bar{T}_W}{\hat{T}_E - \bar{T}_W}\right)$
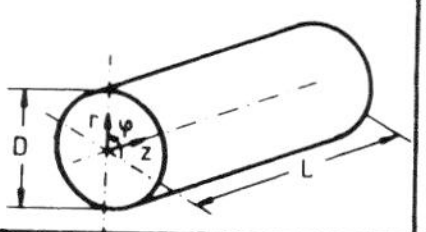 (cylinder)	Cylinder $\dot{Q}_\varphi = 0$ $\dot{Q}_z = 0$ $L \gg d$	$t_c = \dfrac{D^2}{23.14 \cdot a} \cdot \ln\left(0.692 \cdot \dfrac{T_M - \bar{T}_W}{\bar{T}_E - \bar{T}_W}\right)$ $t_c = \dfrac{D^2}{23.14 \cdot a} \cdot \ln\left(1.602 \cdot \dfrac{T_M - \bar{T}_W}{\hat{T}_E - \bar{T}_W}\right)$
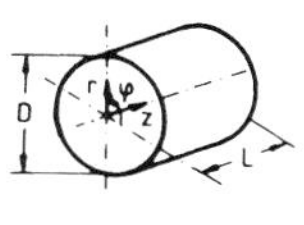 (cylinder)	Cylinder $\dot{Q}_\varphi = 0$ $L \sim d$	$t_c = \dfrac{1}{\left(\dfrac{23.14}{D^2} + \dfrac{\pi^2}{L}\right) \cdot a} \cdot \ln\left(0.561 \cdot \dfrac{T_M - \bar{T}_W}{\bar{T}_E - \bar{T}_W}\right)$ $t_c = \dfrac{1}{\left(\dfrac{23.14}{D^2} + \dfrac{\pi^2}{L}\right) \cdot a} \cdot \ln\left(2.04 \cdot \dfrac{T_M - \bar{T}_W}{\hat{T}_E - \bar{T}_W}\right)$
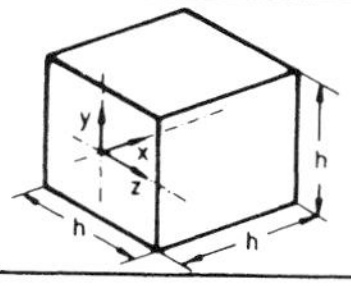 (cube)	Cube	$t_c = \dfrac{h^2}{3 \cdot \pi^2 \cdot a} \cdot \ln\left(0.533 \cdot \dfrac{T_M - \bar{T}_W}{\bar{T}_E - \bar{T}_W}\right)$ $t_c = \dfrac{h^2}{3 \cdot \pi^2 \cdot a} \cdot \ln\left(2.064 \cdot \dfrac{T_M - \bar{T}_W}{\hat{T}_E - \bar{T}_W}\right)$
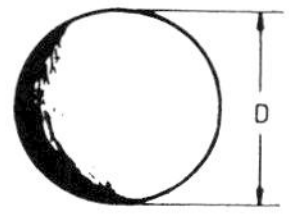 (sphere)	Sphere	$t_c = \dfrac{D^2}{4 \cdot \pi^2 \cdot a} \cdot \ln\left(2 \cdot \dfrac{T_M - \bar{T}_W}{\hat{T}_E - \bar{T}_W}\right)$
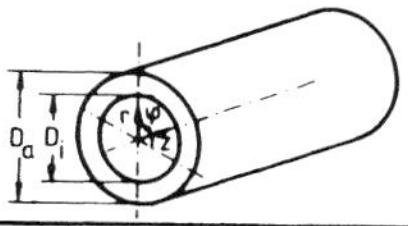 (hollow cylinder)	Hollow cylinder $\dot{Q}_\varphi, \dot{Q}_z = 0$ $r < D_i/2$: $\dot{Q}_r = 0$	same as plate with $s = D_a - D_i$
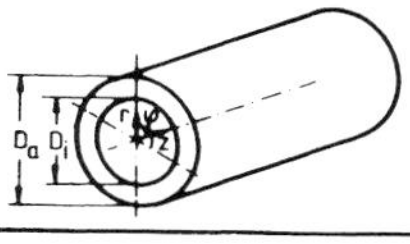 (hollow cylinder)	Hollow cylinder $\dot{Q}_\varphi, \dot{Q}_z = 0$	same as plate with $s = (D_a - D_i)/2$

Figure 8.14 Equations for the computation of cooling time [8.12]

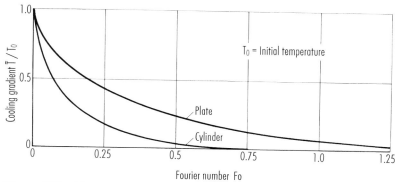

Figure 8.15 Mean temperature if surface temperature is constant [8.1]

Multiplication of the cooling rates θ_1 and θ_2 for the corresponding basic geometric elements and for the respective Fourier number results in the cooling rate for the combined geometry. Thus, the average or the maximum temperature (in the center) of a cylinder of finite length at a defined time can be computed. Because a cylinder of finite length is formed by a cylinder of infinite length and a plate with a thickness equal to the length of the cylinder, the corresponding cooling rates can be taken from Figure 8.13 or 8.15 by using the appropriate Fourier numbers (plate and cylinder). After multiplication, the result is the cooling rate of the cylinder of finite length. Thus, the boundary effect of ribbing, cutouts, studs, etc. on cooling time can be estimated in a simple way. The presented correlations permit the computation of every conceivable cooling process of moldings with sufficient accuracy.

Example:
How long is the cooling time of the cylindrical part of Figure 8.16?

Figure 8.16   Cylindrical part

The cooling rate results from the multiplication of the cooling rate of a plate (thickness s = 13 mm = 2x) with the cooling rate of a cylinder (diameter D = 15 mm = 2R).

$$\theta = \theta_{plate} \cdot \theta_{cyl} = \frac{\hat{T}_E - T_W}{T_M - T_W}$$

Material: PMMA,
Cavity temperature: 40 °C,
Melt temperature: 220 °C,
Max. demolding temperature: 120 °C,
Thermal diffusivity: 0.07 mm²/s,

$$\theta = \frac{120° - 40°}{220° - 40°} = 0.44$$

$$Fo = \frac{a \cdot t}{x^2}, \quad Fo_{cyl} = \frac{a \cdot t \cdot 4}{D^2}, \quad Fo_{plate} = \frac{a \cdot t \cdot 4}{s^2}$$

$Fo_{cyl} = 0.00124\ t$
$Fo_{plate} = 0.00166\ t.$

The corresponding cooling rates θ_p and θ_c for various cooling times t can be taken from Figure 8.13. The result of the multiplication is the cooling rate of the molding after the time t.

t(s)	80	100	120	140	160
Fo_c	0.099	0.124	0.149	0.173	0.198
Fo_p	0.133	0.166	0.199	0.232	0.265
θ_c	0.860	0.810	0.700	0.610	0.510
θ_p	0.900	0.850	0.800	0.700	0.670
θ	0.770	0.690	0.560	0.427	0.340
$\hat{T}_E(°C)$	178	164	141	117	101

The table above shows that after ca. 135 seconds the temperature in the center of the molding has dropped below the demolding temperature of 120 °C (after 140 s already 117 °C).

$\hat{T}_E$ Max. demolding temperature,
Fo_c Fourier number – cylinder,
Fo_p Fourier number – plate,
θ_c Cooling rate – cylinder,
θ_p Cooling rate – plate.

8.4 Heat Flux and Heat-Exchange Capacity

8.4.1 Heat Flux

8.4.1.1 Thermoplastics

A mold for processing thermoplastics has to extract fast and uniformly, so much heat from the melt injected into the cavity as is possible to render the molding sufficiently rigid to be demolded.

During this process, heat flows from the molding to the walls of the cavities.

To calculate this heat flux and design the heat-exchange system the total amount of heat to be carried into the mold has to first be determined. It is calculated from the enthalpy difference Δh between injection and demolding (Figure 8.17).

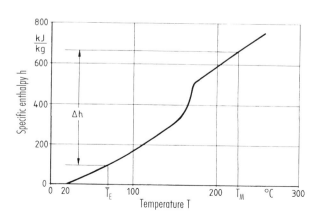

Figure 8.17 Enthalpy plot of polypropylene [8.2]
Δ h Enthalpy difference

The variation of the specific enthalpy of amorphous and crystalline thermoplastics can be described by a function of the following form:

$$h_{(T)} = C_1 + C_2 \cdot T + C_3 \cdot \exp(C_4 \cdot T - C_5) \qquad \text{for } T < C_8,$$
$$h_{(T)} = C_6 \cdot T + C_7 \qquad\qquad\qquad\qquad\qquad \text{for } T > C_8. \tag{8.15}$$

The enthalpy difference related to the mass can, with the average density and the volume, be converted to the amount of heat, which has to be extracted from the molding and conveyed to the mold during the cooling stage.

Because the heat flow in the mold is considered quasi-steady, the amount of heat is distributed over the whole cycle time and results in the heat flux from the molding to the mold:

$$\dot{Q}_{KS} = \Delta h \cdot \frac{m_{KS}}{t_C} \tag{8.16a}$$

$$\dot{Q}_{KS} = \frac{\Delta h \cdot \rho_{KS} \cdot V}{t_C} \tag{8.16b}$$

Δh Enthalpy difference,
ρ_{KS} Average density between injection and demolding temperature,
m_{KS} Mass injected into mold.
t_C Cooling time

In the range of quasi-steady operation, heat flux that is supplied to the mold (counted as positive) and heat flux that is removed from the mold (counted negative) have to be in equilibrium. Therefore, one can strike a heat flux balance, which has to take into account the following heat flows (Figure 8.18):

$\dot{Q}_{KS}$ Heat flux from molding (Equation 8.16),
$\dot{Q}_E$ Heat exchange with environment,
$\dot{Q}_{AD}$ Additional heat flux (e.g., from hot runner),
$\dot{Q}_C$ Heat exchange with coolant.

Then the heat flux balance is:

$$\dot{Q}_{KS} + \dot{Q}_E + \dot{Q}_{AD} + \dot{Q}_C = 0 \tag{8.17}$$

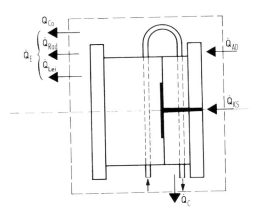

Figure 8.18   Heat flow assessment in an injection mold [8.2]

With this, the necessary heat exchange with the coolant can be computed after the heat exchange with the environment and any additional heat flux have been estimated.

The heat exchange with the environment can be divided into different kinds of heat transport [8.13]:

$\dot{Q}_{Co}$ Heat exchange by convection at the side faces of the mold

$$\dot{Q}_{Co} = A_s \cdot \alpha_A \cdot (T_{Mo} - T_E) \tag{8.18}$$

A_s Area of mold side faces,
α_A Coefficient of heat transfer to air (in slight motion: $\alpha \sim 8$ W/m^2 K).

$\dot{Q}_{Rad}$ Heat flux from radiation at the side faces of the mold

$$\dot{Q}_{Rad} = A_S \cdot \varepsilon \cdot C_R \cdot \left[\left(\frac{T_{abs.\ Mo}}{100} \right)^4 - \left(\frac{T_{abs.\ E}}{100} \right)^4 \right] \tag{8.19}$$

C_R Constant of radiation 5.77 W/(m^2 K^4),
ε Emissivity,
for steel: polished   = 0.1,
            clean   = 0.25,
            slightly rusted = 0.6,
            heavily rusted = 0.85.

$\dot{Q}_C$ Heat flux from conduction into machine platens

This portion can be calculated with a factor of proportionality h (analogous to coefficient of heat transfer) [8.14]:
h = for carbon steel ~ 100 W/(m^2 K),
     low-alloy steel ~ 100 W/(m^2 K),
     high-alloy steel ~ 80 W/(m^2 K),

$$\dot{Q}_C = A_{Cl} \cdot h \cdot (T_{Mo} - T_E) \tag{8.20}$$

A_{Cl} Area of clamping faces of mold.

Thus, the heat exchange with the environment is:

$$\dot{Q}_E = \dot{Q}_{Co} + \dot{Q}_{Rad} + \dot{Q}_C. \tag{8.21}$$

With these equations and an estimate of the mold dimensions and the temperature of its faces the heat exchange with the environment can be computed. Heat flux balances can also be established for individual mold segments if the heat flow across the borders of the segments is negligibly smaller or can be accounted for by an additional heat flow. If larger mold areas are divided into smaller segments to determine the heat flux, then this heat flux can be considered by a heat-flow ratio [8.15]:

$$C_q = \frac{\dot{Q}_E}{\dot{Q}_{KS}} \tag{8.22}$$

In addition, the heat-flow ratio makes a characterization of the operating range of the heat-exchange possible (Figure 8.19).

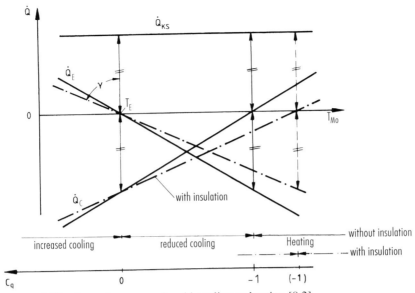

Figure 8.19   Operating range of mold cooling or heating [8.2]

$C_q > 0$: A thermoplastic molding supplies heat to the mold ($\dot{Q}_{KS} > 0$). In this case additional heat from the environment is supplied to the cooler external mold faces ($\dot{Q}_E > 0$). The heat-exchange system has to be designed for increased cooling. Insulating the mold lowers the demand on the efficiency of the system.

$-1 < C_q < 0$: Part of the heat flow from the molding is transferred into the environment ($\dot{Q}_E < 0$). Thus, only reduced cooling is required from the heat-exchange system. A heat-exchange system is basically not needed for simple parts in the case of $C_q = -1$; for other values of C_q a modification of the cycle to $t'_C = t_C/-C_q$ results in this point of operation. However, this would make the mold dependent on the temperature of the environment and exclude a control of the heat exchange system. A uniform cooling could not be maintained.

$C_q < -1$: The heat transfer to the environment is larger than the transfer from the molding because of a high external mold temperature. The heat-exchange system has to be designed as a heating system to avoid a lowering of the cavity-wall temperature. Insulation reduces the demand on the efficiency of the heat-exchange system.

If insulation is employed, it must be considered in the computation of the heat exchange with the environment. It does not only reduce energy costs for increased cooling or heating but also lowers the dependency of thermal processes on the temperature of the environment (Figure 8.19).

A problem remains the unknown external mold temperature T_{Mo}.

It can be estimated or determined by iteration as follows:

$T_{Mo} = T_E$ That is, the heat exchange with the environment is neglected (permissible only with small unheated molds).

$T_{Mo} = T_C$ (Temperature of coolant). Results in highest heat flux into the environment.

$T_{Mo} = T_{CC}$ (Temperature of cooling channel). Transpires in the course of the computation.

With, the average distance $\bar{l}$ cooling channel and external mold surface one can calculate

$$T_{Mo} = T_{CC} + \frac{\dot{Q}_E \cdot \bar{l}}{(A_S + A_{Cl})k_M} \tag{8.23}$$

It is

A_S Area of external mold surface,

A_{Cl} Clamping area of mold,

k_M Thermal conductivity of mold material

$\bar{l}$ Distance between cooling channel and external mold surface.

Since T_{CC} is still unknown, one estimates the heat flow in a first step with 2 (temperature of coolant). When in the course of further calculation the temperature of the cooling channel has been found, it can be inserted to improve the accuracy of the calculation.

8.4.1.2 Reactive Materials [8.16]

8.4.1.2.1 Thermosets

Thermosets set free considerable amounts of heat from reaction. They cannot be disregarded. The simplest way to determine them is a DSC analysis (Figure 8.20).

This analysis provides the possibility of quantitatively correlating heat of reaction with temperature and time. The relationship can be clearly demonstrated with exothermic reactions (phenolic resin). The presented graph shows the energy needed to raise the temperature with the desired speed (constant heating rate of 10 °C if not noted other-wise). The integral under the curve (shown dark in Figure 8.20) is a good approximation of the heat of reaction, the distance between the two curves, and the heat effect. If sufficiently high temperatures are attained, there is no heat effect any more, and one can assume a complete curing. Then the total heat of reaction corresponds to a degree of cross-linking of 100%. This method is so reliable that different degrees of pre-curing can be determined with phenolic resin [8.17]. If a specimen presents incomplete curing either by too short a testing time or too low a temperature, a second pass shows a clearly smaller peak, ands its area corresponds with the residual curing. For this reason, the DSC analysis is a suitable procedure for the thermal characterization of a reactive material and its kinetics of reaction.

The dashed line in Figure 8.20 is, strictly taken, a curved line as it would be obtained with a completely cured material. The used plotter programs, however, depend on straight base lines. The considered temperature range has been determined by preliminary testing.

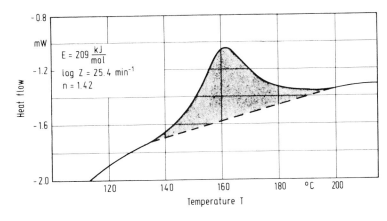

Figure 8.20 DSC plot [8.16] (DSC = Differential Scanning Calorimetry); phenolic resin

Kinetics of the Curing Reaction

Because of the large number of curing reactions occurring, an exact description of the kinetics of the reactions is very complex. This is equally true for rubber as it is for resins.

One can look at the whole curing process as one single reaction, although there are several reactions, which run partly parallel and partly consecutively. It can be described by a reaction-kinetic expression. Several of such expressions are found in the literature [8.18 to 8.22].

A simple expression of a reaction of the n-th order is sufficient as velocity Equation [8.17].

$$\frac{dc}{dt} = K_{(T)} \cdot (1 - c)^n \tag{8.24}$$

c   Share of cross-links (= degree of curing),
dc/dt   Velocity of reaction,
$K_{(T)}$ Velocity constant (temperature dependent),
n   Formal order of reaction (temperature independent).

By integrating Equation (8.24) a definite equation is obtained for the time t of reaction

$$t = \frac{1}{(n - 1) \cdot K_{(T)}} \cdot \left(\frac{1}{(1 - c)^{n-1}} - 1 \right) \tag{8.25}$$

The temperature dependence of the velocity constant K is described by an Arrhenius expression:

$$K_{(T)} = Z \cdot \exp\left(-\frac{E_a}{R \cdot T} \right) \tag{8.26}$$

with
Z   Maximal possible velocity constant,
E_a Energy of activation,

R   Gas constant (8.23 J/mol K),
T   Temperature [K].

By entering into Equation (8.25) one obtains a definite equation for the reaction time as a function of the degree of curing and the temperature T. The magnitudes of Z, E_a and n are typical for the respective material and have to be found by experiment.

$$t = \frac{1}{(n-1) \cdot Z \cdot e^{(-E_a/R \cdot T)}} \cdot \left(\frac{1}{(1-c)^{n-1}} - 1 \right)$$ (8.27)

Some authors [8.18, 8.19, 8.22] use a more general form of the kinetic expression; the calculation is very complicated and is only marginally more accurate.

Determination of the Reaction Parameters

The DSC analysis provides a practical combination of data and the test can be well transferred to real events occurring in the mold. The specimen is heated up at a constant rate. The supplied energy is plotted and reduced by the amount which is needed for a reference specimen [8.23, 8.24]. Figure 8.21 presents a typical plot for a phenolic resin. The shaded area corresponds with the heat of reaction during curing.

The peak in the range from 120 to 190 °C pictures the reaction of cross-linking. According to [8.25] the rising side of the peak is evaluated. The area enclosed by peak and base line is determined by integration (Figure 8.20). The total area is equivalent to a 100% degree of curing. The degree of curing reached at a certain point in time is established by the corresponding area, for which the actual temperature is the upper limit of integration. The parameters n, E_a and Z can be found by consecutive linear regression. How well the found parameters match can be checked by an isothermal test [8.17]. For this test a temperature has to be established, which results in a ca. 50% curing after about 50 minutes. This can be tested in a second pass.
The correct data in this example are:
E_a = 203 kJ/kmol,
log Z = 24.5 l/min,
n   = 1.34.

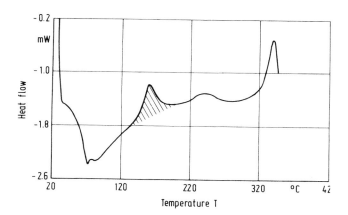

Figure 8.21   DSC plot
for a phenolic resin [8.16]

If these data are inserted into Equation (8.27), the curing time is obtained as a function of temperature and conversion corresponding to the degree of curing.

If the temperature is plotted against the time with the degree of curing as parameters, Figure 8.22 is obtained. A general heating rate can be estimated.

Phase Diagram of Curing

Knowledge of the necessary residence time in the mold is of decisive significance for the user of reactive materials. On the one side, sufficient curing should be secured for reasons of quality; on the other side, one aims at a time period as short as possible for economical reasons. To estimate the curing time correctly, the advancement of curing has to be connected with the temperature development of the molding.

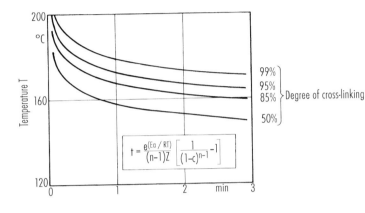

$$t = \frac{e^{(Ea/RT)}}{(n-1)Z}\left[\frac{1}{(1-c)^{n-1}} - 1\right]$$

Figure 8.22   Degree of cross-linking as a function of time and temperature for a phenolic resin [8.16]

The energy use provides the differential equation for the temperature field of a plane plate.

$$\frac{\partial T}{\partial t} = a \cdot \frac{\partial^2 T}{\partial t^2} \tag{8.28}$$

This equation can be transformed into a differential equation, which presents the temperature field of the equation [8.28]. Figure 8.23 pictures the temperature distribution in a part of 5 mm thickness. Coordinate $x = 0$ stands for the edge of the molding, $x = 2.5$ for its center.

With this, one is able to specify the temperature as a function of time for every point in the wall of the molding. Therefore it is reasonable to include the advancement of curing in this consideration because it is a function of time and temperature as well. Thus, basically the degree of curing can be calculated for every location within the wall of the molding and for every point in time. There are some simplifying assumptions, though. The reaction parameters were determined at a low heating rate, so that they are a good approximation for the isothermal case, too. Variations during testing and different evaluation procedures result in a relatively large scattering of the parameters, which, however, balance out to a great extent.

If one looks at definite locations within the section of a molding, one can plot the temperature as a function of time in the coordinate system of Figure 8.22. The

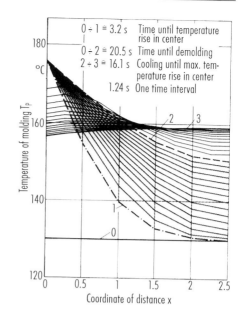

Figure 8.23 Temperature development in a molding [8.16]

temperature development at the edge and in the center is of interest. Figure 8.24 shows a combination of temperature development and degree of curing.

The value of this certainly rather rough presentation should not be seen in a precise calculation of the degree of curing, but in a method of estimating the tendency of all important parameters and their efficient connection with one another. Cycle time and mold temperature can be rather correctly established in advance.

Effect of Heat of Reaction

Especially with materials which undergo a distinct exothermic reaction, the question arises of how much does the heat of reaction affect the mold temperature. A theoretical solution to the equation of thermal conduction is not possible. The specific thermal capacity of a specimen can be determined relatively precise with a DSC analysis, however, if one assumes a constant thermal conductivity and density.

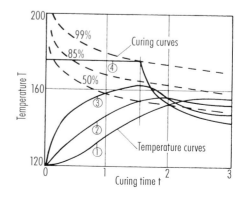

Figure 8.24 Diagram presenting temperature, time, and rate of cross-linking for phenolic resin [8.16]
Part thickness 10 mm (c_p = const.)
Distance from cavity wall:
① 5 mm,
② 4 mm,
③ 2.5 mm,
④ 0 mm

This establishes the heat flux which is needed to heat up a specimen at a certain heating rate. In the diagram (Figure 8.25) this is presented as the difference between a so-called base line and the plot from the specimen (dash-dotted lines). The base line represents the thermal losses of the specimens: radiation and convection. The curve of the specimen has to be corrected by the appropriate heat of reaction [8.29], because no specific thermal capacity is defined for a reaction in progress. If this correction is omitted, the following equation is the result:

$$\dot{Q}_{meas.} - \dot{Q}_{base} = m \cdot c_{p\,eff} \cdot \frac{\Delta T}{\Delta t} \tag{8.29}$$

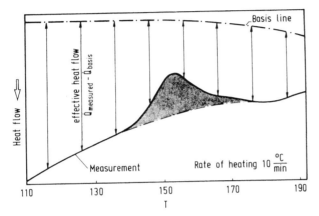

Figure 8.25   Specific heat capacity of a DSC analysis [8.16] (Phenolic resin)

An "effective thermal capacity" $c_{p\,eff}$ can be established with this correlation for a constant heating rate and the known weight of the sample m. With this effective quantity the effect of the heat of reaction is introduced into the temperature function. The effective specific thermal capacity is a temperature-dependent quantity (Figure 8.26). To introduce this quantity into the computation an approximation of the function in Figure 8.26 was proposed according to [8.28, 8.30] by using equations of a straight line by sections.

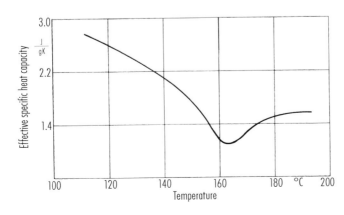

Figure 8.26 $c_{p\,eff}$ as a function of temperature [8.16]

If this effect is taken into account, the result is a more rapid temperature rise, which is especially noticeable in heavy sections. The temperature function attains higher temperatures for the center of the part faster with this more accurate description. The necessary degree of curing is reached sooner. The residence time in the mold is estimated more accurate and can be set shorter. Similar results are reported by [8.31].

Elastomers

The calculation corresponds to that for thermosets, however, when designing the mold one can neglect the reaction heat for noncritical materials. The thermal diffusivities of some typical elastomers are listed in Table 8.2.

Material	Hardness Shore A	Thermal-diffusivity a mm²/s
NBR-1	40	0.147
NBR-2	70	0.145
NR-1	45	0.093
NR-2	60	0.122
ACM	75	0.218
CR	68	0.188

Table 8.2   Thermal diffusivity a of some elastomers [8.32]

In practice, however, pinpoint gates in multi-cavity molds are frequently employed. Then precuring (also called scorching) may occur. To consider this phenomenon it is suggested to establish so-called windows of operation [8.32, 8.33].

8.5   Analytical, Thermal Calculation of the Heat-Exchange System Based on the Specific Heat Flux (Overall Design)

A simple analytical calculation (rough design) can be used as a first approximate value for thermal design of the cooling system if one does not wish to rely totally on experience. The molded part is simply regarded as a slab. Short calculation times produce results for temperatures, heat fluxes, and cooling geometry that form a good starting value for the numerical calculation.

In practical cases, it has been shown for molds up to approximately 100 °C that the errors are negligible in the case of just two-dimensional calculation. There are good reasons for this since one is on the safe side with a two-dimensional calculation and the calculation necessarily contains a series of further assumptions, such as coolant temperature, heat-transfer coefficient in the cooling channels, which, unfortunately, lie on the more unreliable side. It may therefore be assumed that the error given by two-dimensional calculation compensate for the other errors in terms of providing a more reliable design.

These principles may be used to design the cooling system. It must be borne in mind that: the calculation is based on a two-dimensional viewpoint. This simplification can naturally lead to errors of the outer edges and corners of the mold, if for example no insulation is present at high mold temperatures.

8.5.1   Analytical Thermal Calculation

The analytical thermal calculation can be subdivided into separate steps (Figure 8.27). The necessary time to cool down a molding from melt to demolding temperature with a given cavity-wall temperature is established in the cooling-time computation. This can be done with equations for a variety of configurations (Section 8.3.4) for different sections of moldings, which have to be rigid enough to be demolded and ejected at the end of the cooling time. The longest time found with this calculation is decisive for further proceedings.

	Design steps		Criteria
1	Computation of cooling time	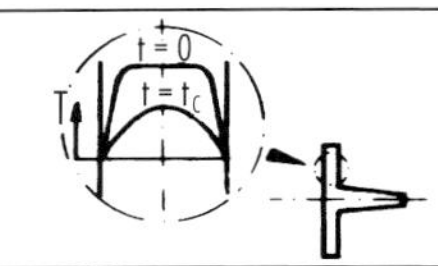	Minimum cooling time down to demolding temperature
2	Balance of heat flow	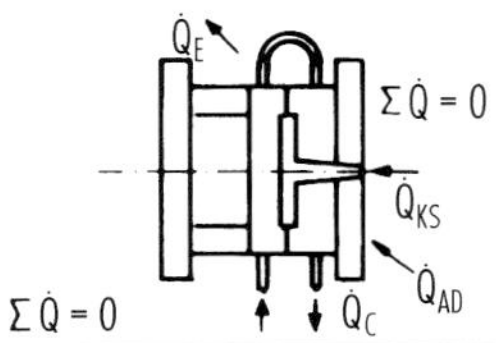	Required heat flow through coolant
3	Flow rate of coolant	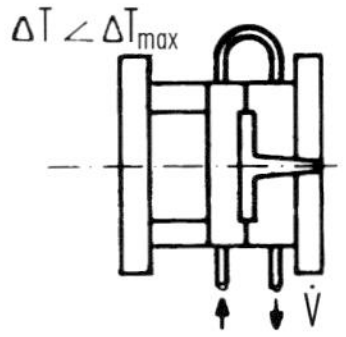	Uniform temperature along cooling line
4	Diameter of cooling line	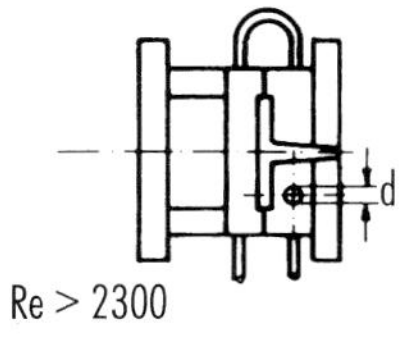	Turbulent flow
5	Position of cooling line	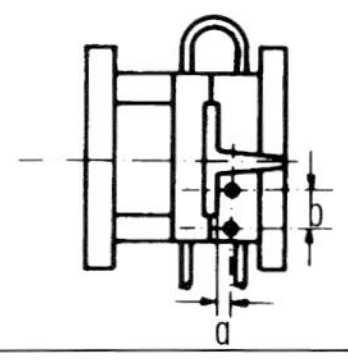	Heat flow uniformity
6	Computation of pressure drop	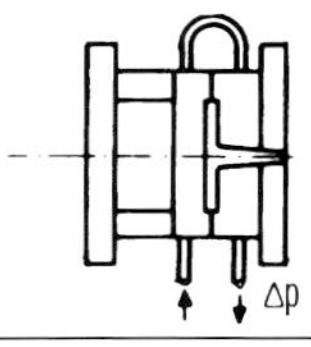	Selection of heat exchanger Modification of diameter or flow rate

Figure 8.27   Analytic computation of the cooling system [8.2]

With the heat-flux balance, which has to be taken in by the coolant is computed. This calls for consideration of additional heat input, heat exchange with the environment, and eventual insulation. The heat exchange with the environment is found with the estimated outside dimensions of the mold and the temperature of the mold surface. An approximation for the latter is the temperature of the coolant, which is assumed for the time being.

The heat-flux balance does not only provide information about the operating range of the heat-exchange system, but also indicates design problems along the way. High heat flux, which has to be carried away by the coolant and which occurs particularly with thin, large moldings of crystalline material, requires a high flow rate of the coolant and results in a high pressure drop in the cooling system. Then the use of several cooling circuits can offer an advantage. A low heat flux to be taken up by the coolant may result in a small flow rate of the coolant in channels with common diameter and with this in a laminar flow. For this reason, a higher flow rate should be accomplished than that resulting from the criterion of temperature difference between coolant entrance and exit. After the flow rate of the coolant has been calculated, the requirement of a turbulent flow sets an upper limit for the diameter.

The position of the cooling channels to one another as well as their distance from the surface of the molding result from the calculated heat flux including compliance with the limits of homogeneity (Figure 8.28). The calculation can be done with a number of preconditions:

– specification of cooling error and computation of distances,
– specification of distance from molding surface,
– specification of distances among cooling channels,
– specification of whole length of cooling line,
– specification of both channel distances and computation of required flow rate of coolant. In this case another heat-flux balance becomes necessary.

Tabulating the points of the functions presented with Figure 8.28 is an additional beneficial option. Such a table makes it possible to fit the position of individual cooling channels into the mold design without relocating parting lines or ejector pins.

For the analytical calculation, the configuration of the molding is simplified by a plate with the same volume and surface as the molding. This allows segmenting the internal

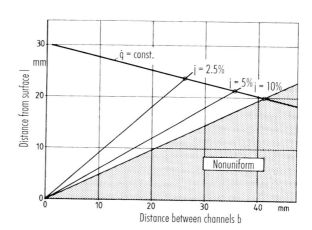

Figure 8.28   Position of cooling line and temperature uniformity [8.2]

area of the mold with each segment having a width equal to the calculated distance between the cooling channels.

With the position of the cooling channels one can determine the whole length of the cooling system, the length of supply lines, the number of corners, elbows and connectors and enter them into the drawing. This information permits the computation of the pressure loss and the necessary capacity of heat exchangers. In the same way one can calculate the minimum diameter of the cooling channel or the maximum flow rate of the coolant for a given permissible pressure drop.

One has to consider that a large part of the pressure drop does not occur solely in the cooling line in the mold, but also in connecting lines, connectors and turns, which must also be taken into account.

8.5.1.1 Calculating the Cooling Time

With the aid of the equations presented in Section 8.3.4 (Figure 8.14), it is possible to calculate the necessary cooling time. To ensure a dependable design, it is always better to use the critical (i.e., thickest) cross-section of the part for the calculation.

8.5.1.2 Heat Flux Balance

The efficiency of the heat-exchange system of a mold depends on the amount of heat that can be extracted from the plastic material in the cavity at a certain cavity-wall temperature in the shortest time span possible. Therefore it is the task of the heat-exchange system to ensure the desired temperature of the cavity wall (see also Section 8.4).

In contrast to [8.34, 8.35] one can apparently do without a transfer coefficient between cavity wall and material if the wall temperature is defined as the contact temperature between plastic and cavity wall, which, as shown below, has to be calculated. This holds for as long as the molded part is still lying against the cavity wall.

The contact temperature is the maximum temperature during the cycle that establishes itself when the molding compound is in contact with the wall. It is expressed as:

$$T_{C_{max}} = \frac{b_P \cdot T_{P_0} + b_W \cdot T_{W_0}}{b_P + b_W} \qquad b = \sqrt{\rho \cdot k \cdot c} \qquad (8.30)$$

$T_{C_{max}}$ Contact temperature,
T_{P_0} Material temperature before contact,
T_{W_0} Cavity-wall temperature before contact,
b Heat penetrability,
k Thermal conductivity,
ρ Density,
c Specific heat capacity.

Depending on time and cycle, the temperature of the cavity surface varies between contact and minimum temperature, which is determined by the temperature of the coolant. The average of both provides sufficient accuracy for calculating cooling time and necessary heat-exchange capacity. Therefore this average temperature represents the cavity-wall temperature $\overline{T}_W$.

$$\overline{T}_W = \frac{T_{W_{max}} + T_{W_{min}}}{2} = \frac{T_{W_{max}} - \overline{T}_{Cool}}{2} \tag{8.31}$$

In a steady state the cavity surface with the temperature $\overline{T}_W$ acts as a heat sink for the molding and a heat source for the mold. The resulting heat flux can be determined as follows:

Quantity of heat from the molding:

$$Q_M = \Delta h \cdot A_M \cdot s \cdot \rho \tag{8.32}$$

Heat flux from the molding into each mold half:

$$\dot{Q}_M = \frac{\Delta h \cdot A_M \cdot s \cdot \rho}{2 \cdot t_c} \tag{8.33}$$

Δh Enthalpy difference (Figure 8.17),
ρ Density of melt,
$\dot{Q}_M$ Heat flux from melt,
s Wall thickness,
A_M Surface area of molding,
t_C Cooling time $= C_C \cdot s^2$,
t_c Cycle time $t_C + t_n$ (t_n is the nonproductive time for opening, closing and ejecting).

with

$$C_C = \frac{1}{\alpha \cdot \pi^2} \cdot \ln\left(\frac{8}{\pi^2} \cdot \frac{1}{\theta}\right) = \frac{t_K}{s^2} \tag{8.34}$$

Heat-flux density from molding:

$$\dot{q} = \frac{\Delta h \cdot A_M \cdot s \cdot \rho}{A_M \cdot t_c} \tag{8.35}$$

For the first estimate, the nonproductive time (the period of time during which no molding compound is in the cavity) is ignored initially.

When the nonproductive time is ignored, the equation can be simplified further:

$$\dot{q} = \frac{\Delta h \cdot \rho \cdot s}{K' \cdot s^2} = \frac{\Delta h \cdot \rho}{K'} \cdot \frac{1}{s} \tag{8.36}$$

If one now combines $\dfrac{\Delta h \cdot \rho}{K'} = K$ into a single material constant, then

$$\dot{q} = K \cdot \frac{1}{s} \tag{8.37}$$

Within the usual range of processing temperature the heat-flux density only depends on the material. It is obvious that it has to increase with decreasing wall thickness to ensure the required cavity temperature. Besides this, the heat-flux density, in future called specific heat flux, represents a specific material characterization if the generally small machine effect is neglected. The machine effect becomes apparent by the extra time needed for the completion of movements (ejection, etc.). The heat flow from the molding to the mold is interrupted with the demolding process; the coolant, however, continues to cool the mold during this secondary time, which reduces the specific heat flux accordingly.

Figures 8.29 and 8.30 present the specific heat flux for PS and HDPE as a function of secondary machine time. The maximum occurs with a wall thickness for which cooling and secondary time are equal. The specific heat flux as a function of cavity temperature is shown in Figures 8.31 and 8.32 for a number of wall thicknesses. The specific heat flux can facilitate the selection of the right mold concept already in the planning stage and also presents a basis for the design of the heat-exchange system of the mold.

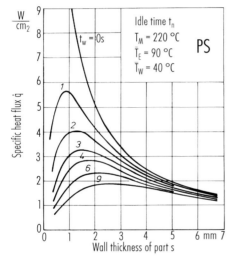

Figure 8.29   Specific heat flux $\dot{q}$ PS as a function of wall thickness [8.1]

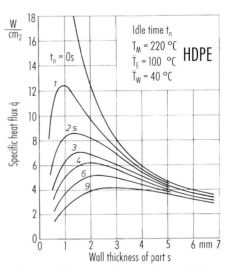

Figure 8.30   Specific heat flux $\dot{q}$ HDPE as a function of wall thickness [8.1]

8.5.1.3   Coolant Throughput

For the determination of the coolant throughput, the specific heat flux $\dot{q}$ (Figures 8.31 and 8.32) is the central design parameter. Multiplying the specific heat flux by the corresponding surface area of the molding A_M results in the amount of heat which has to be removed by the coolant during one cycle, which belongs to A_M. The necessary flow rate of the coolant $\dot{V}_C$ is determined by the permissible rise in coolant temperature ΔT_C.

The difference in coolant temperature ΔT should not exceed 3 to 5 °C in order to ensure adequate homogeneous cooling along the cooling circuit. The maximum permissible temperature difference may be used to calculate the minimum necessary coolant throughput ΔT_{max}.

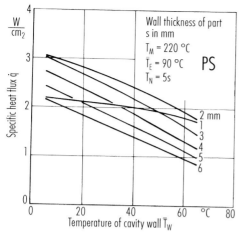

Figure 8.31 Specific heat flux q̇ PS as a function of temperature of cavity wall [8.1]

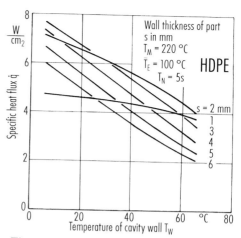

Figure 8.32 Specific heat flux q̇ HDPE as a function of temperature of cavity wall [8.1]

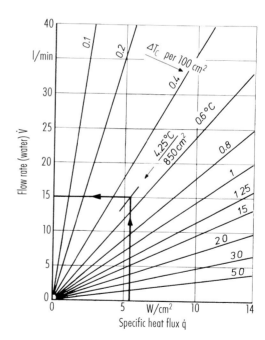

Figure 8.33 Computation of flow rate of cooling water (molding surface = 100 cm²) [8.1]

$$\dot{V}_C = \frac{\dot{q} \cdot A_M}{\rho_C \cdot c_C \cdot \Delta T_C}$$

(8.39)

ρ_C Density of coolant,
c_C Specific heat of coolant.

8.5.1.4 Temperature of the Cooling Channel

If a pressure drop Δp and common channel geometry are given, the necessary flow rate for the coolant (which is predetermined by the pressure generated by the heating unit with a deduction for pressure losses in the feed lines to the mold) can only be ensured by the channel diameter d_C to be selected.

$$d_C = \sqrt[4]{\frac{\rho_C \cdot \dot{V}_C^2 \cdot 16}{\Delta p \cdot 2 \cdot \pi^2 \cdot 3600} \cdot \left(f_C \cdot \frac{L_C}{d_C} + n_C \cdot K_C \right)} \tag{8.39}$$

f_C Friction factor in pipes (= 0.05 in Figure 8.34),
L_C Channel length in cm (= 200 in Figure 8.34),
n_C Number of turns (= 10 in Figure 8.34),
K_C Resistance coefficient (= 1.9 in Figure 8.34),
$\dot{V}_C$ Flow rate of coolant, l/min,
d_C Diameter of cooling channel, cm,
Δp Pressure drop, 10^2 kPa.

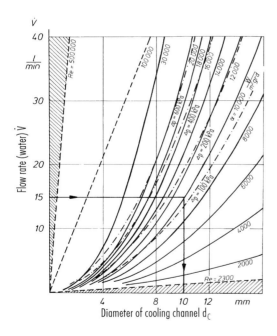

Figure 8.34 Determination of cooling channel diameter d_C and coefficient of heat transfer [8.1]

Since the final cooling-channel arrangement is not yet established, the number of turns and the length-over-diameter ratio L_C/d_C have to be estimated for the time being. To avoid iteration, a rather small pressure drop should be chosen. Because it represents a minimum requirement, variations of line pressure, length of supply lines and degree of contamination (from deposits and corrosion) have to be generously considered (corrosion during storage of the mold can be minimized by drying with compressed air at the end of operations and proper storage).

Measurements on molds have demonstrated [8.9] that the friction factor in pipes without deposits can be described according to Blasius' law:

$$f_C = \frac{0.3164}{Re^{1/4}} \tag{8.40}$$

With few deposits (after about 100 working hours = usual operating range) one can estimate $f_C \approx 0.04$ [8.9].
Resistance coefficients for sharp and rounded bends were reported as follows:
90° sharp bend   K = 1.9 [8.9],
90° round bend   K = 0.4 [8.9].

(Literature [8.36, 8.37] contains values for the 90° sharp bend of between 1.13 and 1.9 and for the 90° round bend of between 0.4 and 0.9).
   The coefficient of heat transfer can be calculated in combination with Nusselt, Reynolds and Prandtl numbers [8.38]:

$$\alpha = \frac{k_C}{d_h} \cdot \left\{ 0.0235 \cdot \left[Re^{0.8} - 230 \right] \cdot \left[1.8 \cdot PR^{0.3} - 0.8 \right] \right\} \cdot K_f, \tag{8.41}$$

d_h Hydraulic diameter (4 x cross section/circumference),
k_C Thermal conductivity of coolant,
K_f Correction factor.

The dimensionless parameters compute to:

$$Re = \frac{4 \cdot \dot{V}}{\pi \cdot \upsilon \cdot d_h} \tag{8.42}$$

$$Pr = \frac{\upsilon \cdot \rho \cdot c_p}{k_C} \tag{8.43}$$

The equation for the heat transfer coefficient α applies to the range of turbulent flow with a considerably better heat transfer than with laminar flow (Re < 2300). Another disadvantage of the range of laminar flow is the transition into turbulent flow behind corners and changes in cross-sectional areas, which causes a better heat transfer for a certain length. The demand for a turbulent flow results in an upper limit for the diameter:

$$d < \frac{4 \cdot \dot{V}}{\pi \cdot 2300 \cdot v} \tag{8.44}$$

Common diameters for cooling channels in most applications are far below this limit. Arrival of laminar flow in water of 20 °C can only be assumed with a flow rate of less than 2 l/min.
   With the channel diameter determined, the coefficient of heat transfer can be computed with the flow rate of the coolant for $2300 < Re < 10^6$, $0.7 < Pr < 500$, and $L_C \gg d_C$:

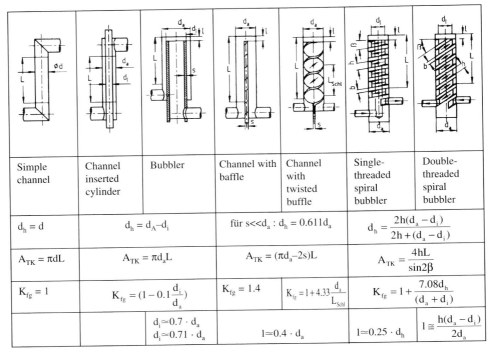

Simple channel	Channel inserted cylinder	Bubbler	Channel with baffle	Channel with twisted buffle	Single-threaded spiral bubbler	Double-threaded spiral bubbler
$d_h = d$	$d_h = d_A - d_i$		für s<<d_a : $d_h = 0.611 d_a$		$d_h = \dfrac{2h(d_a - d_i)}{2h + (d_a - d_i)}$	
$A_{TK} = \pi d L$	$A_{TK} = \pi d_a L$		$A_{TK} = (\pi d_a - 2s)L$		$A_{TK} = \dfrac{4hL}{\sin 2\beta}$	
$K_{fg} = 1$	$K_{fg} = (1 - 0.1\dfrac{d_i}{d_a})$		$K_{fg} = 1.4$	$K_{fg} = 1 + 4.33\dfrac{d_a}{L_{Schl}}$	$K_{fg} = 1 + \dfrac{7.08 d_h}{(d_a + d_i)}$	
	$d_i \approx 0.7 \cdot d_a$ $d_i \approx 0.71 \cdot d_a$		$l \approx 0.4 \cdot d_a$		$l \approx 0.25 \cdot d_h$	$l \cong \dfrac{h(d_a - d_i)}{2d_a}$

Figure 8.35   Cooling elements [8.2]

$$\alpha = \left(0.037 \left(\frac{\dot{V}_C \cdot 4 \cdot 1000}{d_C \cdot \pi \cdot v_C \cdot 60} \right)^{0.75} - 180 \right) \cdot \mathrm{Pr}^{0.42} \cdot \frac{k_C}{a_C} \qquad (8.45)$$

v_C Kinematic viscosity of coolant [cm²/s],
d_C Cooling channel diameter [cm],
Pr   Prandtl number = v/a,
k_C Thermal conductivity of coolant [W/cm K].

The dependency of the coefficient of heat transfer on flow rates (expressed in Figure 8.34 by Δp), temperature and channel diameter is demonstrated for several coolants in Figures 8.36, 8.37 and 8.38. It is evident that water is the most effective coolant.

8.5.1.5   Position of the Cooling Channels

The distance between two cooling channels in a mold is based on the relationship given in Figure 8.39.

For the dimensions Figure 8.40 can be used. The example is based on an average temperature of coolant of $\overline{T}_C = 20\ °C$ and an average mold temperature of $\overline{T}_W = 62\ °C$.

The ratio A_C/A_M is presented with Figure 8.41 as a function of the distance of the channels between each other and their diameters.

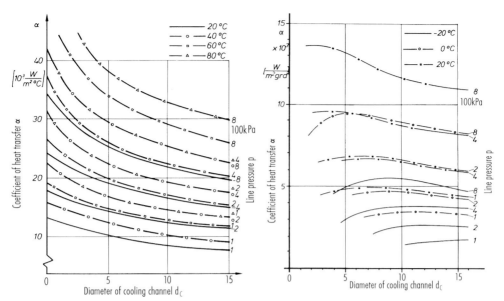

Figure 8.36   Coefficient of heat transfer α, coolant: water [8.1]

Figure 8.37   Coefficient of heat transfer α, coolant: brine (20%) [8.1]

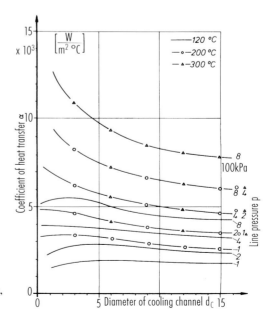

Figure 8.38   Coefficient of heat transfer α, coolant: oil (Marlotherm S) [8.1]

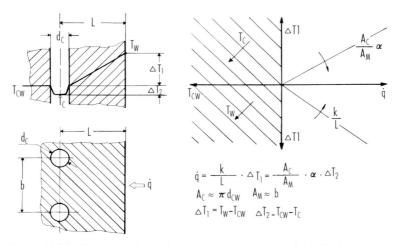

Figure 8.39   Picture of thermal reactions (steady conduction) [8.1]
(The ratio A_C/A_p is presented in Figure 8.41)
A_p Surface of molding, A_C Surface of cooling channel

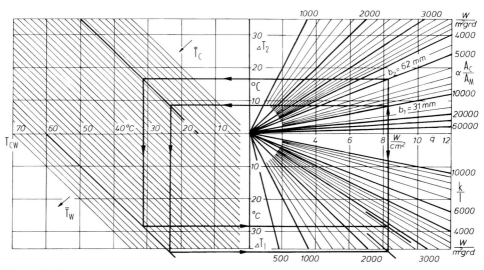

Figure 8.40   Computation of heat-transfer data [8.1]

$$\dot{q} = \frac{k_W}{L} \cdot \Delta T_1 \qquad\qquad (8.46a)$$

$$\Delta T_1 = \overline{T}_W - T_{CW} \qquad\qquad (8.46b)$$

$\overline{T}_W$ Average temperature of cavity wall,
T_{CW} Temperature of cooling channel wall.

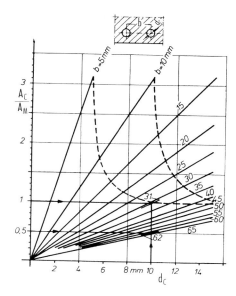

Figure 8.41   Cooling-channel layout [8.1]

The same specific heat flux is transferred into the coolant. This results, depending on the coefficient of heat transfer, in the temperature differential ΔT_2 (convective heat transfer).

$$\dot{q} = \alpha \cdot \Delta T_2 \tag{8.47}$$

$$\Delta T_2 = T_{CW} - T_C \tag{8.48}$$

T_C Temperature of coolant.

The ratio between the areas of the cooling-channel wall A_{CW} and the corresponding molding surface A_M affects the specific heat flux $\dot{q}$ like α the coefficient of heat transfer. Hence:

$$\dot{q} = \frac{A_{CW}}{A_M} \cdot \alpha \cdot \Delta T_2 \tag{8.49}$$

The heat fluxes $\dot{q}$ transferred by thermal conduction and convection must naturally be equal, and the temperature differences ΔT_1 and ΔT_2 have the temperature of the cooling-channel wall T_{CW} as common parameter. In an alignment chart with four quadrants, appropriate common abscissa can, therefore, be selected (Figure 8.39). With the exception of $\dot{q}$ and $\overline{T}_W$, which are interdependent (Figures 8.31 and 8.32), all other parameters can be arbitrarily varied. With this, the required closed solution is made possible. There has to be a functional correlation between the cooling-line distances l and b, if the specific heat flux is constant. The distance l can be taken with adequate accuracy from the thermal conductance k_M/l with the thermal conductivity of the mold material k_M (Figure 8.42), if one remains in the range of validity b of 2 to 5 d_C.

   The goal of thermal design, aside from attaining the necessary heat flux for maintaining the predetermined mold wall temperature, is uniform cooling of the

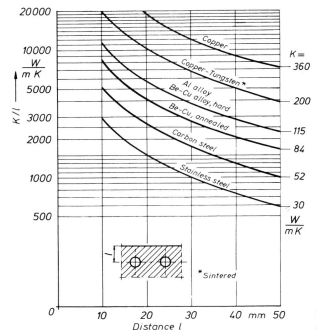

Figure 8.42   Distance between cavity wall and cooling line [8.1]

molding. There are different combinations of l (distance from molding to cooling line) and b (distance between cooling lines) that produce the same temperature difference:

– small values for b (many cooling lines) require large distances l,
– large values for b (few cooling lines) require small distances.

These two demands are equivalent in terms of thermal efficiency. The uniformity of the cooling is, however, different and smaller in the second case. To maximize homogeneity (uniform surface temperature) at the part surface, a small cooling error j is desirable. The cooling error, j, which is defined by local differences in heat flux, can be used as a measure of uniformity (Figures 8.43 and 8.44)

$$j = \frac{q_{max} - q_{min}}{\dot{q}} \; [\%] \tag{8.50}$$

$$j = 2.4 \cdot Bi^{0.22} \cdot \left(\frac{b}{l}\right)^{2.8 \cdot k} \tag{8.51}$$

with

j   = Cooling error,
Bi = Biot number $= \dfrac{\alpha \cdot d_C}{k_M}$
k   = ln b/l.

Range of validity for distances of cooling channels:
l = 1 to 5 d$_C$,
b = 2 to 5 d$_C$.

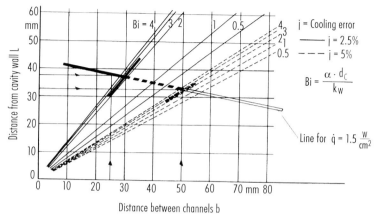

Figure 8.43 Feasible cooling line layout [8.1]

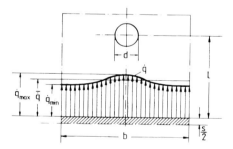

Figure 8.44 Distribution of heat flow in segment [8.1]

For crystalline plastics, the cooling error should be a maximum 2.5–5% and for amorphous plastics 5–10% in order to avoid inhomogeneous molded part properties (e.g., waviness, differences in gloss) [8.1].

Distance b arises from the predetermined ratio of cooling channel to molded part area (e.g., A$_C$/A$_M$ = 0.5 and 1), which establishes the line of equal cooling efficiency. The line of equal cooling efficiency offers the designer a large number of alternative channel arrangements for avoiding bolts, ejector pins, etc.

8.5.1.6 Design of Cooling Circuit

8.5.1.6.1 Flow Rate of Coolant
The temperature of the coolant changes between entrance and exit in accordance with the heat flux.

$$\dot{Q}_C = \dot{V} \cdot \rho \cdot c_p \cdot \Delta T$$

(8.52)

$$\Delta T = T_{in} - T_{out} \tag{8.53}$$

The average temperature difference should not exceed a maximum of 3 to 5 °C to ensure a uniform heat exchange over the whole length of the cooling line. A required minimum flow rate $\dot{V}$ can be calculated from the permissible maximum temperature difference. However, the rate also depends on the arrangement of the cooling elements. With an arrangement in series the permissible temperature difference applies to the sum of heat fluxes from all segments; with parallel arrangement it applies to each segment. Parallel arrangement results in a lower flow rate and smaller pressure drop. However, parallel arrangement calls for an adjustment of flow rates with throttles [8.1] and a constant monitoring during production; for this reason it is not recommended.

8.5.1.6.2 Pressure Drop
The flow through the heat-exchange system causes pressure drops, which are an additional criterion for a controlled design of heat-exchange systems and a boundary condition for the heat exchanger.

If the pressure drop is higher than the capacity of the heat exchanger, then the necessary flow rate and, with this, the permissible temperature difference between coolant entrance and exit, cannot be met. The consequences are nonuniform cooling of the molding and heterogeneous properties and distortion of the molding. For calculating the pressure drop, different causes have to be considered:

– pressure drop from the length of the cooling element,
– pressure drop from turnabouts, corners and elbows,
– pressure drop from spiral flow,
– pressure drop from changes in cross-sectional area,
– pressure drop in connectors,
– pressure drop from connecting lines.

The total pressure drop is the sum of all items. The equations used to compute the pressure drop [8.1, 8.15, 8.40, 8.41, 8.42] are too extensive to be listed here because of all the effects they include. However, with a bit of practical experience, they can readily be estimated with sufficient accuracy.

From the total pressure drop and the heat flux to the coolant one can conclude the capacity of the heat exchanger:

$$P = \Delta p \cdot \dot{V} + |\dot{Q}_C| \tag{8.54}$$

Where
P   = Pumping efficiency of the heating unit,
Δp = Pressure loss,
$\dot{V}$ = Volumetric flow,
$\dot{Q}_C$ = Heating efficiency of the coolant.

8.6   Numerical Computation for Thermal Design of Molded Parts

Through the use of simulation programs and thanks to the processing power of modern computers, it is possible to calculate the temperature range in the injection mold.

Numerical procedures are used for this, so that the Fourier differential equation for heat conduction can be solved without the simplifications presented in Section 8.1.

$$\rho c_p \frac{\partial T}{\partial t} = \frac{\partial}{\partial x}\left(k\frac{\partial T}{\partial x}\right) + \frac{\partial}{\partial y}\left(k\frac{\partial T}{\partial y}\right) + \frac{\partial}{\partial z}\left(k\frac{\partial T}{\partial z}\right) \tag{8.55}$$

Differential methods and nowadays preferably finite element programs are used for this. Since there is a great deal of work involved in the compiling the computational net for the three-dimensional calculation, two-dimensional programs are very widespread. They usually supply enough information for the designer and so are also presented here.

8.6.1 Two-Dimensional Computation

In mold design, it is often necessary to optimize cooling at certain critical points, such as corners or rib bases. There is no need to perform a computation for the whole mold, and anyway, such a computation would unnecessarily extend the processing time. It is sufficient in this case to analyze the critical area. Two-dimensional computation is well suited to this.

In a two-dimensional computation, a section of the point under consideration is taken through the mold. When selecting the section, it is important that as little heat as possible is dissipated vertically to the section plane. Because this heat flow is not allowed for, it would reduce the accuracy and informativeness of the study.

The mold section under consideration is then overlaid with a computational net with which the numerical computation is performed. Various material combinations, starting temperatures, thermal boundary conditions, and process settings can be taken into account.

The results of the computation are the temporal temperature curves in the section plane. It sometimes makes sense therefore to perform the computation for several cycles in order to be able to analyze start-up processes and to capture the temperature distribution throughout the mold.

In this computational method, it is advantageous that the processing time is short and the net generation is relatively simple. For critical part areas, such as corners, rib bases and abrupt changes in wall thickness, results can be obtained relatively quickly.

8.6.2 Three-Dimensional Computation

If the temperature ranges for the entire mold and the quantities of heat to be dissipated via the cooling channels are to be analyzed, there is no getting round a three-dimensional computation. To this end, the entire mold along with all cooling channels must be simulated.

There are two computational philosophies available for the computation. There are programs that see the mold as being infinitely large. In them, the position and the number of the cooling channels alone decide on the temperature conditions in the mold. Heat flow to the environment is ignored [8.43]. For the computation, only the molded part and the cooling channels need to be modeled.

If the influence of mold inserts and heat exchange with the environment is to be considered, this approach is unsuitable and the entire mold has to be simulated. The outlay on modeling and the processing time increase accordingly. However, the results are then all the more precise.

The advantages of 3D computation over analytical computations lie [8.43]

– in solving in several directions, even for complex geometries and heat flow,
– in more accurate simulation of the cooling conditions,
– in intelligible results (color plots),
– in rapid "playing through" of variants (processing conditions, cooling channel arrangements),
– in good coupling to computation modules for the filling and holding phase as well as to shrinkage and distortion programs.

A further effect that can only be taken into account with a 3D computation is the influence of the parting line on the mold wall temperature distribution. This will be explained below with an example. At the parting line, heat conduction is much poorer relative to the bulk material. This exerts an effect, particularly in the case of differently cooled mold halves, on the exchanged heat flux q̇. Figure 8.45 shows the results obtained with and without parting line influence. It may be clearly seen that the colder, lower mold half without parting line influence is heated in the edge zone. At the cavity edge there is a temperature minimum. If the slight insulating effect of the parting line is taken into account, there will be a temperature maximum taken instead at this point.

The computation shows that in critical cases – molds that are operated at high temperatures – large, non-permissible temperature differences may establish themselves. It is often, therefore, expedient to carry out such computational analyses.

8.6.3   Simple Estimation of the Heat Flow at Critical Points

Corners of moldings, especially with their differences in surface, areas, have high cooling rates on the outside and a low rate inside the corner (Figure 8.46). Immediately after injection, the melt solidifies on the surface and the temperature maximum is in the center of a section. With progressing solidification more melt solidifies on the outside

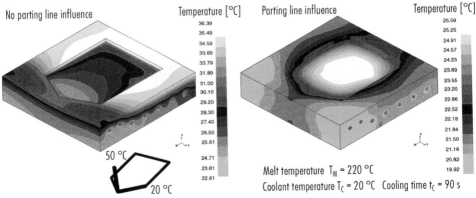

Figure 8.45   Influence of parting line on the mold wall temperature

than on the inside of the corner because the heat-exchange areas are of different size and more heat is dissipated on the outside than on the inside. Figure 8.46 demonstrates that the remaining melt moves from the center towards the inside. At the end of the cooling time, the melt which solidifies last is close to the internal surface.

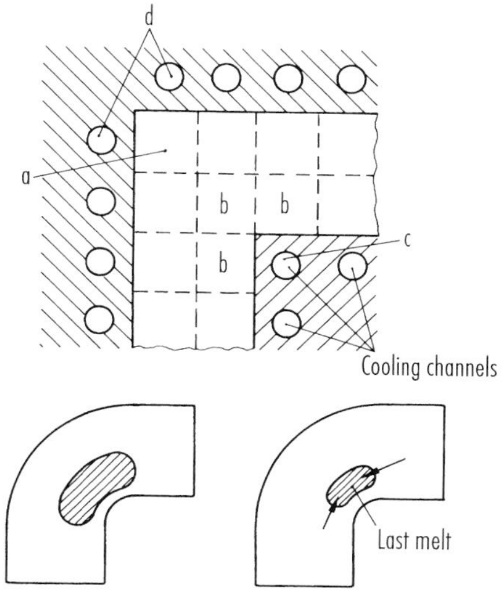

Figure 8.46   Freezing of melt in a corner [8.1]
The drawing at the top shows that the farthest square a on the convex side is affected by two cooling channels d. On the concave side three squares b are affected by only one cooling channel c. Consequently melt close to the concave side will solidify last

A material deficit during solidification of the last melt is generated because the shrinkage cannot be compensated by melt supplied by the holding pressure. Tensile stresses are created accordingly. These stresses are counterbalanced by the rigidity of the mold. After demolding, the external forces have ceased and the formation of a stress equilibrium in the part causes warpage or deformation. Besides this, voids and sink marks and even spontaneous cracking may occur. Deformation can be eliminated, however, if the remaining melt, and with it the forces of shrinkage, are kept in the plane of symmetry. Then an equilibrium of forces through-out the cross section is generated if the last material solidifies in the center.

8.6.4   Empirical Correction for Cooling a Corner

One draws the corner of the part and the planned cooling channels on an enlarged scale. Then the cross section of the corner is divided into rectangles of equal size with one side equal to half the thickness of the section (s/2); the other one equal to the distance between two cooling channels. Thus, the area is pictured, which is cooled by one cooling channel (cooling segment). By comparing areas and adjustment, one hole at the corner is either eliminated or the holes are shifted in such a way that equal cooling surfaces (ratio of holes to rectangles) are generated (Figure 8.46).

8.7   Practical Design of Cooling Systems

8.7.1   Heat-Exchange Systems for Cores and Parts with Circular Cross-Section

Adapting the specific heat flux to requirements and ensuring it in all areas of a molding, particularly in critical sections, may cause considerable difficulties.

A slender core is a characteristic example for mold parts which are accessible only with difficulties. Because of unawareness of the serious consequences (increased cooling time) or for reasons of manufacturing, such cores are often left without any particular cooling. Cooling occurs only from the mold base through the core mount. With decreasing secondary time and, consequently, reduced time for core cooling between ejection and injection, heating-up of cores without separate cooling is unavoidable. Core temperatures of the magnitude of the demolding temperature are definitely possible. If intense cooling of the core base is feasible, then an undesirable temperature gradient from the tip of the core to the base is the result. A high temperature differential between core wall and coolant impairs the dynamic characteristics, which are important for start-up and leads to high time constants, this means a long time until the temperatures of the mold level out to a constant value. (The basic correlations for describing the dynamics are presented in [8.44 to 8.46].)

Because of the already mentioned increase in cycle time, an uncooled core can result in parts of inferior quality and even fully interrupt a production. This becomes particularly apparent with cores having a square or rectangular cross section. With uncooled cores, sink marks or distorted sides can hardly be avoided. Therefore, provisions for cooling of cores should always be made. To do so, the following options are available dependent on the diameter or width of the core (Figure 8.47).

If diameter or width are minor, only air cooling is feasible most of the time. Air is blown from the outside during mold opening or flows through a central hole from the inside. This procedure, of course, does not permit maintaining exact mold temperatures (Figure 8.47a).

A better cooling of slender cores is accomplished by using inserts made of materials with high thermal conductivity, such as copper, beryllium-copper, or high-strength sintered copper-tungsten materials (Figure 8.47b). Such inserts are press-fitted into the core and extend with their base, which has a cross section as large as it is feasible, into a cooling channel.

The most effective cooling of slender cores is achieved with bubblers. An inlet tube conveys the coolant into a blind hole in the core. The diameters of both have to be adjusted in such a way that the resistance to flow in both cross sections is equal. The condition for this is ID/OD = 0.5. The smallest realizable tubing so far are hypodermic needles with an OD of 1.5 mm. To guarantee flawless operation in this case, the purity of the coolant has to meet special demands. Bubblers are commercially available and are usually screwed into the core (Figure 8.47d). Up to a diameter of 4 mm the tubing should be beveled at the end to enlarge the cross section of the outlet (Figure 8.47c).

Bubblers can be used not only for core cooling but also for flat mold sections, which cannot be equipped with drilled or milled channels.

A special bubbler has been developed for cooling rotating cores in unscrewing molds (Figure 8.47e).

It is frequently suggested to separate inlet and return flow in a core hole with a baffle (Figure 8.47f). This method provides maximum cross sections for the coolant but it is difficult to mount the divider exactly in the center. The cooling effect and with it the temperature distribution on one side may differ from those of the other side. This disadvantage of an otherwise economical solution, as far as manufacturing is concerned, can be eliminated if the metal sheet forming the baffle is twisted. This "cooling coil" is self-centering. It conveys the coolant to the tip and back in the form of a helix and makes for a very uniform temperature distribution (Figure 8.47g).

Further logical developments of baffles are one or double-flighted spiral cores (Figure 8.47h).

A more recent, elegant solution uses a so-called heat pipe (Figure 8.47i). This is a closed cylindrical pipe filled with a liquid heat conductor, the composition of which depends on the temperature of use. It has an evaporation zone where the liquid evaporates through heat and a condensation zone where the vapor is condensed again. The center zone serves the adiabatic heat transfer. Heat pipes have to be fitted very accurately to keep the resistance between pipe and mold to a minimum. They have to be cooled at their base as described for inserts of highly conductive metals (Figure 8.47b). Heat pipes are commercially available from 3 mm upward. They can be nickel-coated and then immediately employed as cores.

For core diameters of 40 mm and larger a positive transport of coolant has to be ensured. This can be done with inserts in which the coolant reaches the tip of the core through a central hole and is led through a spiral to its circumference, and between core and insert helically to the outlet (Figure 8.47j). This design weakens the core only insignificantly.

Cooling of cylindrical cores and other circular parts should be done with a double helix (Figure 8.47k). The coolant flows to the tip in one helix and returns in the other one. For design reasons, the wall thickness of the core should be at least 3 mm in this case. For thinner walls another solution is offered with Figure 8.47l. The heat is removed here by a beryllium-copper cylinder intensely cooled at its base.

Another way of cooling poorly accessible mold areas (narrow cores) is not to use conventional mold steels for the cavity but rather to use instead a microporous material (TOOLVAC©), through which liquid gas, usually CO_2, flows (Figure 8.47m). The gas expands in the special material, thereby absorbing heat energy via the pore surface and transports it via the evacuation channels out of the mold [8.53, 8.54].

In the CONTURA© system [8.54, 8.55], the mold core is separated such that at a certain distance close to the mold wall cooling channels may be milled so as, on the one hand, to increase the surface area available for heat exchange and, on the other, to allow the cooling channel system to follow the mold wall contour at a close distance (8.47n). In this case, a more uniform temperature distribution in the core ensures better mold reproduction of the part as well as shorter cooling times. The use of a suitable joining method (high-temperature soldering under vacuum) joins all section lines together again.

If there are several cores in a mold to be cooled simultaneously, solutions are demonstrated with Figure 8.48 and 8.49. They represent a cooling layout in series or parallel.

With cooling in series the individual cores are supplied with coolant one after the other. Since the temperature of the coolant increases and the temperature differential between molding and coolant decreases with the increasing flow length of the coolant, a uniform cooling of cores and thus of moldings is not provided. With such a system in a multi-cavity mold the quality of all parts will not be the same. To avoid this shortcoming, parallel cooling is employed.

With parallel cooling the individual cores are supplied with coolant from a main channel. Another collecting channel removes the coolant. Thus, each core is fed with

Figure	Diameter or width of core	Characteristic	Design
a	$\geqq 3$ mm	Heat removal by air from the outside when mold is open; continuous cooling only feasible if part has openings. Cooling of closed mold achieved with sucked-in water	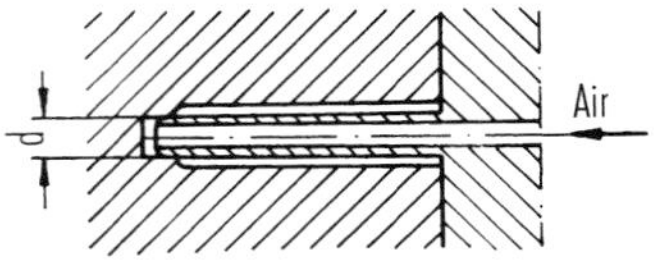
b	$\geqq 5$ mm	Heat-conducting copper is connected to cooling line. Base of insert should be enlarged	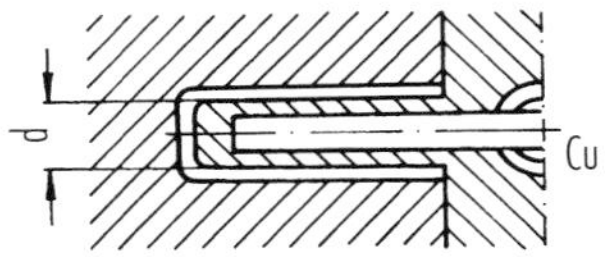
c	$\geqq 8$ mm	Bubbler with beveled tip (4 mm)	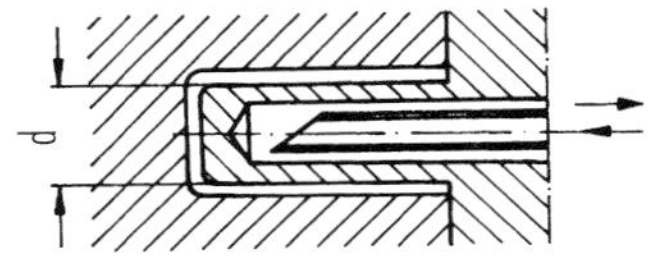
d		ID/OD = 0.5	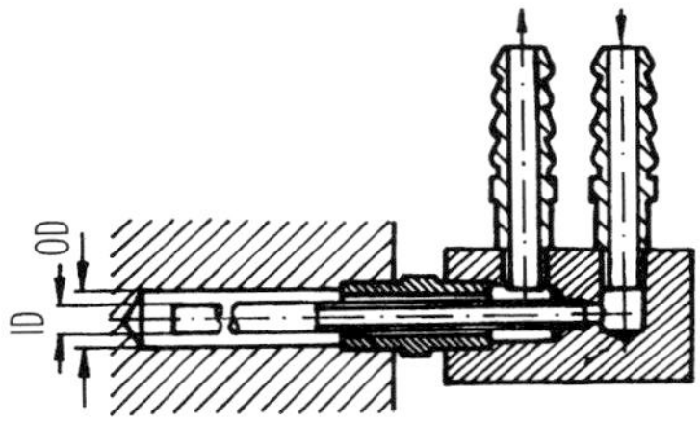
e		Bubbler for rotating cores	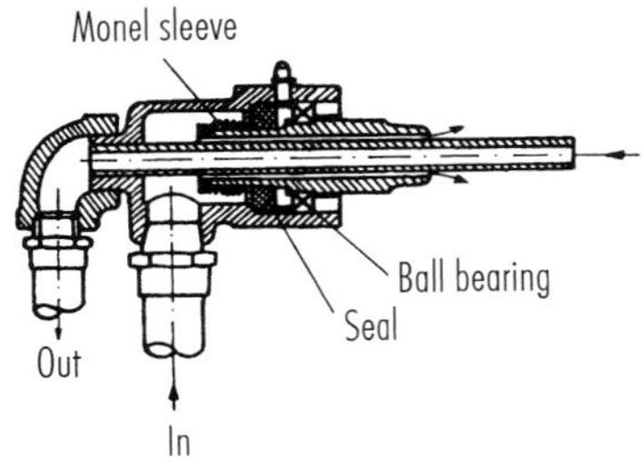
f		Baffle	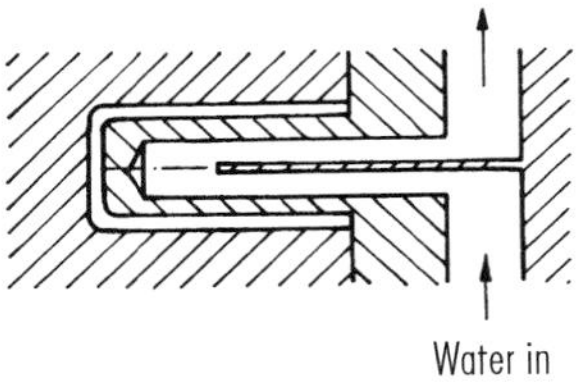

Figure 8.47   Core cooling techniques [8.47 to 8.55] (continued on next page)

Figure	Diameter or width of core	Characteristic	Design
g		Twisted baffle	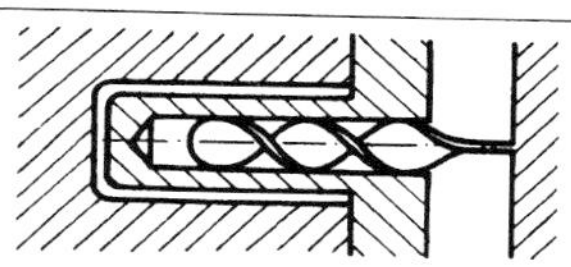
h		Spiral core, single and double spiral Loose fit Diameter 12–50 mm (refer also to "Standards")	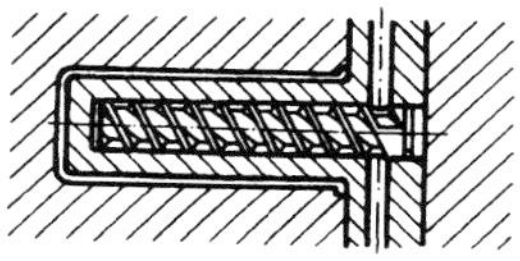
i		Thermal pin (heat pipe) from 3 mm dia, installation with tamp rings or silver or copper compound	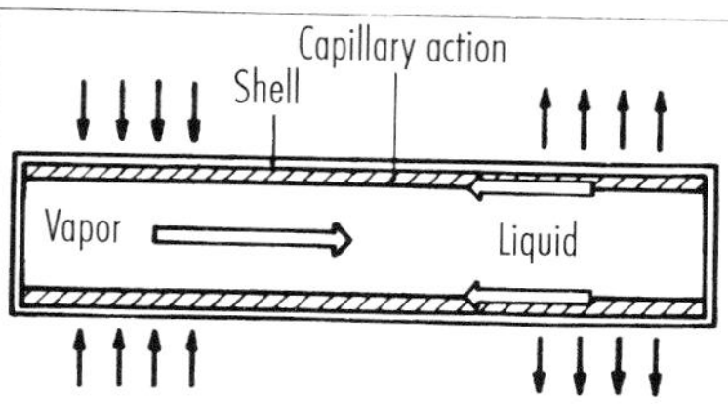
j	$\geqq 40$ mm	Helical cooling channel	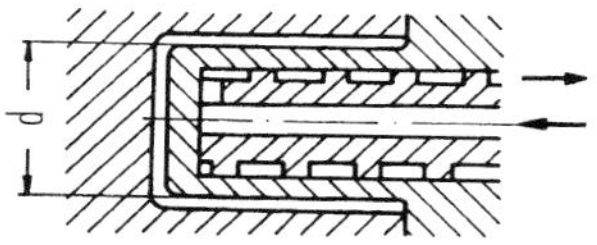
k	Internal core $S \geqq 4$ mm	Double helix and bubbler	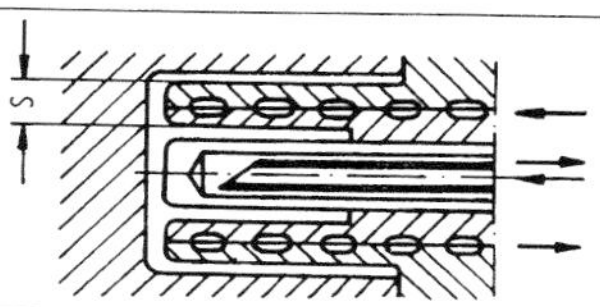
l		Molding; b⁺ Be-Cu sleeve, thickness ≤ 3 mm; b Steel, thickness > 3 mm; c Helical cooling channel, d Welded stainless steel part	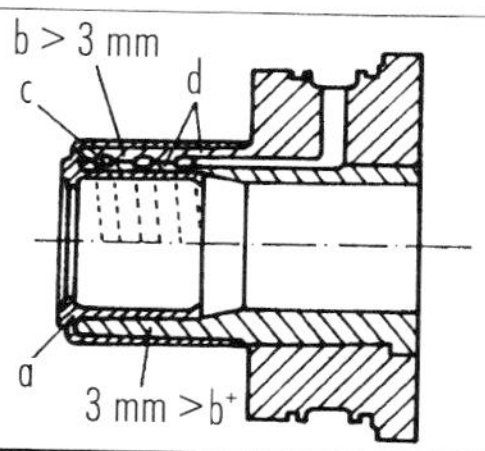
m		a Microporous material b Capillary tube for CO_2 feed	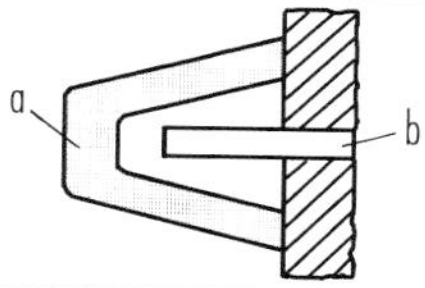
n		Slicing of core Milling of modified heating channels Joining of core	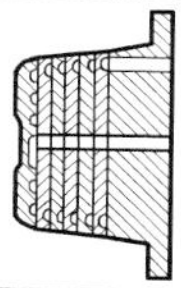

Figure 8.47 (continued) Core cooling techniques [8.47 to 8.55]

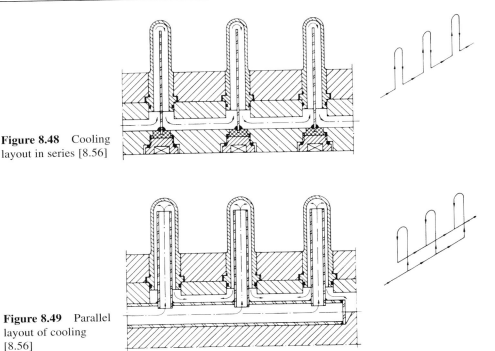

Figure 8.48   Cooling layout in series [8.56]

Figure 8.49   Parallel layout of cooling [8.56]

coolant of the same temperature. This provides for a uniform cooling [8.56] if, in addition, one sees to it that the coolant volume is equally divided.

As a more elegant, although more costly way of cooling, each core could be equipped with a bubbler (Figure 8.47d) separately supplied with coolant.

All these cooling systems are well suited for cooling parts with circular cross section.

The helical design in single- or double-flighted form can be used equally well for cores or for cavities.

8.7.2   Cooling Systems for Flat Parts

One has to distinguish between circular and angular parts here. For circular parts the system presented with Figure 8.50 has been successfully used in practice. The coolant flows from the center (opposite the gate) to the edge of the part in a spiral pattern. This offers the advantage of the largest temperature differential between molding and coolant at the hottest spot. The temperature of the coolant increases as it flows through the spiral, while the melt has already cooled down to some degree because of the length of its flow. Thus the temperature differential is getting smaller, and less heat is removed. This results in a rather uniform cooling. The uniformity is improved even more if a second spiral is machined into the mold, parallel to the first one, for the return flow of the coolant. This system is expensive to make but produces high-quality and particularly distortion-free parts. It has been used for molding precision gears and compact discs [8.57].

Of course both mold halves must be equipped with this cooling system for molding high-quality parts.

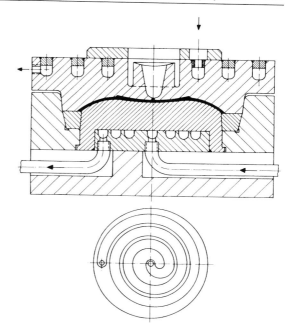

Figure 8.50   Cooling line in spiral design [8.56]

For economic reasons, molds for circular parts have frequently straight, through-going cooling channels. This cannot, of course, produce a uniform temperature distribution (Figures 8.51 and 8.47). Consequently distortion of the part may occur.

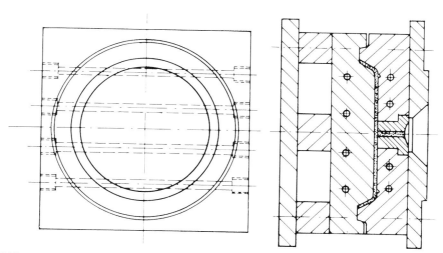

Figure 8.51   Straight cooling channels. Poor design for circular parts [8.58]

Straight cooling lines should only be used, at best, in molds for rectangular parts. Drilling straight through the mold plate is most cost effective [8.51]. The ends are plugged and the coolant is positively directed into cross bores by diverting plugs and rods (Figure 8.53).

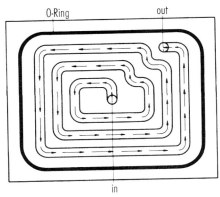

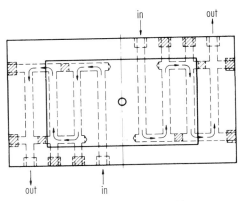

Figure 8.52   Cooling line layout in spiral form for rectangular parts [8.59]

Figure 8.53   Rectangular part with center gating [8.59]

Considerably more expensive is the cooling system presented in Figure 8.52. The cooling channel is milled into the plate conveying the coolant in form of a spiral from the center towards the edge. This system is justified only for central gating because of its costs. Another cooling system for centrally gated, rectangular parts is equally effective but less costly (Figure 8.53). The system consists of blind holes drilled into the mold plate.

If the part is gated at the side, the coolant can, of course, also be supplied from the side (Figure 8.54).

High-quality parts from multi-cavity molds can be produced if the same cooling conditions are ensured for each cavity, that is, each cavity has to be cooled separately. This can be done by arranging several cooling circuits parallel as shown in Figure 8.55, however, equal flow rates are not guaranteed by this design. This always needs additional control.

All these systems presented so far for cooling flat parts can also be used for box-shaped parts after being appropriately modified. The location of the gate determines the more practical layout of the cooling lines, either in series or parallel.

As an example for a parallel layout the core cooling of a mold for refrigerator boxes is presented with Figure 8.56. This system can only be made cost effectively by drilling blind or through-going holes. Plugs or welding has to be used to achieve positive flow of the coolant. This may result in weakened or otherwise hazardous spots. Plugs may cause marks in transparent parts. Welding may distort the core to such an extent that even a finishing machining cannot compensate for the dimensional deviation.

It is suggested, therefore, to cool rectangular cores with the same systems as circular ones in accordance with Figure 8.47 using parallel or series layouts (Figure 8.57).

8.7.3   Sealing of Cooling Systems

Plugging and welding to close cooling-line ends as well as sealing the system with a plate on top of it (Figure 8.52) are problematic. There is always the danger that a slight bending of the plates has already caused the channels not to be sealed any more against

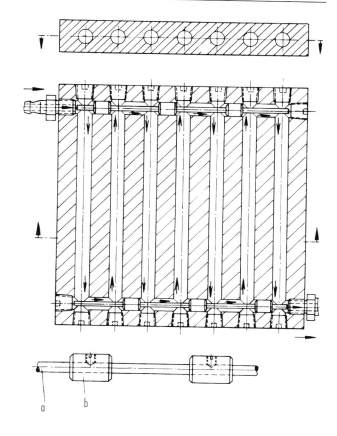

Figure 8.54   Straight cooling channels for rectangular parts gated laterally [8.56] a Rod, b Diverting plug

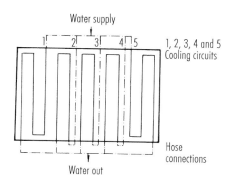

Figure 8.55   Parallel layout of several cooling circuits for a large surface [8.60]

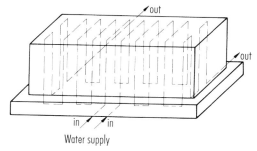

Figure 8.56   Parallel layout of core cooling for box mold [8.59, 8.60]

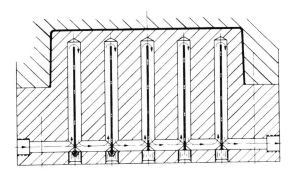

Figure 8.57   Cooling circuit for core of a box mold [8.60]

one another or against the outside. Even a "short-cut" between channels is already a defect because it creates uncooled regions where no coolant flows. Thus, the plates have to be bolted in adequately small intervals.

Another problem are holes for ejector pins, etc. They have to be carefully and individually sealed, e.g., by O-rings or by applying pasty sealants. Sealants are applied to the cleaned surface with a roller, or continuously squeezed from a tube and cured between the matching faces at room temperature and under exclusion of air. Such products seal gaps up to 0.15 mm. They are temperature resistant in the range from –55 to 200 °C.

To facilitate disassembly, O-rings are used considerably more often for sealing the cooling systems. Depending on the mold temperature, they can be made of synthetic or natural rubber, and of silicone or fluoro rubber. The groove which accommodates the O-ring, should be of such a size as to cause a deformation of 10% of the ring after assembly. Figure 8.58 shows O-rings for sealing a core cooling in parallel layout [8.56].

One uses according to temperature

– below 20 °C: O-rings of synthetic rubber,
– above 20 °C: O-rings of silicone or fluoro rubber,
– above 120 °C: Copper-asbestos.

8.7.4   Dynamic Mold Cooling

In the injection molding of thermoplastics there are specialty applications in which the requirements imposed on cooling not only concentrate on rapid cooling of the part but also require brief or local heating. In other words, the mold is heated to e.g. the temperature of the molten plastic prior to injection. When the filling phase is finished, the part is cooled to the demolding temperature. This is known as dynamic or variothermal mold cooling.

Examples of such applications are low-stress and low-oriented injection molding of precision optical parts [8.61]. The hot cavity walls permit relaxation of internal stress in the outer layers before demolding, so as to avoid distortion afterwards. Furthermore, increasing the temperature of the cavity walls as closely as possible to the melt temperature can improve the flowability of the injected plastic. It is thus possible to attain extreme flow-path/wall-thickness ratios [8.62, 8.63] as well as microstructured parts that have areas with micrometer dimensions [8.64]. Under certain circumstances, the heating time determines the cycle time in these applications.

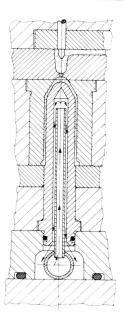

Figure 8.58   Cooling system sealed with O-ring [8.56]

Approaches to such dynamic mold cooling in which the mold is actively heated and cooled have been in existence since the 1970s. These employ different heating systems, the most important of which are discussed below.

In so-called variothermal heating [8.65], two differently cooled liquid-cooling (oil) circuits are regulated by a valve. When oil serves as the cooling medium, its poor heat-transfer properties lead to long cycle times.

With electric heating too, e.g., heating cartridges, heating is based on the principle of thermal conduction. The heating system is more efficient because of selective local heating of those areas in the mold that briefly need higher temperatures; this contrasts with Delpy's variothermal heating [8.65, 8.66], which provides for global heating of the entire mold.

The temporal change in temperature in variothermal molds is shown in Figure 8.59 [8.67]. It can be seen with both solutions that the cycle time can essentially only be influenced during the heating phase, provided it may be assumed that the constant temperature in solution b is also generated by oil heating. While oil has poorer heat-transfer properties than water, it can serve as a heating medium at much higher temperatures than water. The use of water as heating medium is limited to temperatures of 140 °C or 160 °C, even when special equipment with pressurization is used.

The heating methods presented below are more efficient on account of their heat-transfer mechanism or the heat flux densities which they supply [8.62]: induction or radiant heaters (infrared (IR) radiators, flame).

Induction heating can transfer particularly high heat fluxes (30,000 W/cm²) since the energy is introduced into the material for heating directly by turbulent flow. The volume to be heated up must furthermore be electrically conducting [8.68]. As described in Tewald [8.62], inductive heating occurs with the mold open. To this end, an inductor shaped to the mold contour is traversed into the mold halves. After the inductor has been

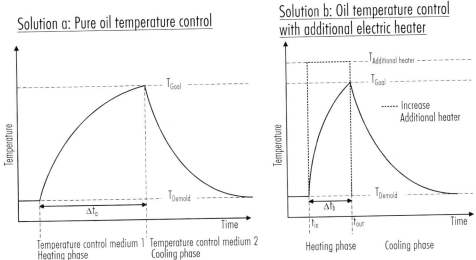

Figure 8.59   Temperature changes in variotherm molds as a function of time [8.67]

Table 8.3   Transferable heating efficiency of different types of heating [8.68]

Type of heating	Example	Possible transferable heating efficiency [W/cm²]
Convection	Hot air device	0.5
Radiation	Infrared heater	8
Thermal conduction	Burner	1,000
Induction	Inductor	30,000

removed, the mold closes and the plastic is injected into the cavity, whose surface is hot [8.62, 8.63].

Radiant heating using IR (ceramic or vitreous quartz radiators) or halogen radiators have so far predominantly been used as heating systems for thermoforming, but also have a high potential for providing additional heating for a dynamic system.

A look at the theoretically transferable heating efficiency (Figure 8.60) reveals that shorter heating times may be expected with these other mechanisms of heat transfer than with thermal conduction.

The advantage, therefore, of induction and radiant heating lies particularly in the fact that the mold surface can be selectively heated up for a brief period. The heat does not penetrate deep into the mold platens and so the cooling time is not essentially prolonged.

8.7.5   Empirical Compensation of Corner Distortion in Thermoplastic Parts from Heat-Flux Differences

It is known from experience that distortion of box-shaped moldings can be avoided if the temperature of the core is lower than that of the cavity. This method tries to compensate

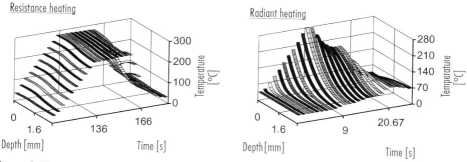

Figure 8.60 Heating processes

distortion, e.g. in samples. It is, howeve,r not recommended, because distortion occurs later during operation.

8.7.5.1 Cold Core and Warm Cavity

A low core temperature cools the part on this side so rapidly that ultimately, the remaining melt is located in the center of the corner section. This (apparently!) prevents distortion (Figure 8.61) [8.84].

Such an unavoidable eccentric cooling, however, may result in distortion of the straight faces of the part. In fact, this can be noticed with long side walls. Even with corners free from distortion, a slight warpage of very long walls becomes noticeable, as it occurs with asymmetrically cooled plates. There is another restriction to this method. If by design a core contains inside as well as outside corners, this method must inevitably fall because it can only deal with inside corners. In general, this method should be rejected because high residual stresses are generated in the molding even if distortion is prevented. The consequences may be brittleness, the risk of stress cracking, and distortion during use.

8.7.5.2 Modification of Corner Configuration

If the heat content of the internal corner is reduced and/or the heat-exchange surface enlarged, any other adjustment of heat fluxes becomes unnecessary. A "dam effect"

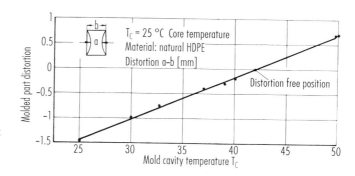

Figure 8.61 Distortion at constant core and variable cavity temperature [8.1]

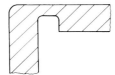

Figure 8.62   Avoiding distortion by changing the corner geometry [8.1]

(Figure 8.62) reduces the tendency to distortion further. Even with unfavorable gate position the filling process can be positively affected. The effect of orientation on distortion is eliminated in unreinforced materials.

A disadvantage is the weakening of the corners and an increase in mold costs.

If the function of the part and required cosmetic appearance permit it, the radii of the corners can be enlarged to approach the desired cooling conditions.

Another method of suppressing distortion of corners is reinforcing the side walls with ribs or doming them. This does not eliminate stresses in the corner areas, though, causing brittleness and sensitivity to stress cracking.

8.7.5.3   Local Adjustment of Heat Fluxes

The laws of heat conduction and transfer offer the following options for adjusting heat fluxes:

Improvement of the heat conductivity in the area between corner and cooling-channel wall. This can be realized in steel molds by inserting suitable materials with a higher thermal conductivity (e.g., copper inserts, Figure 8.63).

Making the distance between corner and channel wall as short as possible or lowering the coolant temperature. This means an additional cooling circuit in the corner area.

These relationships can also explain the occurrence of sink marks (e.g., in connection with ribs) and indicate methods of eliminating them. The delayed solidification at the base of a rib usually does not cause distortion for reasons of symmetry, but results in a more or less noticeable sink mark on the opposite side because of the volume deficit.

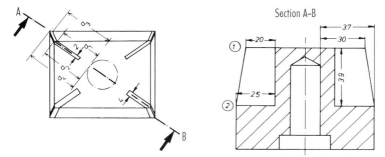

Figure 8.63   Mold with copper inserts [8.1]

8.8   Calculation for Heated Molds for Reactive Materials

These molds are only designed in accordance with the desired heating time. For this, empirical information is available, e.g., 20 to 30 W/kg mold weight is reported in the literature [8.69]. There is also a formula:

$$P = \frac{m \cdot c \cdot \Delta T}{t \cdot s}$$

(8.56)

Where

P   Wattage to be installed,
m   Mass of heated mold or mold section,
c   Specific heat $c_{steel} = 0.48$ (kJ/kg · K),
ΔT Temperature interval of heating,
t   Time of heating,
s   Efficiency ~ 0.6.

It is possible, of course, to perform a more detailed calculation if the design should be more precise. This can be done with a numerical solution by dividing the mold into finite elements. Examples can be found in the literature [8.70–8.73]. Apparently the finite boundary method is even better suited.

8.9   Heat Exchange in Molds for Reactive Materials

8.9.1   Heat Balance

The most important basis for calculating the heating system of a mold is the knowledge of its heat balance because the mold temperatures for elastomers and thermosets are 100–150 °C higher than for thermoplastics. Aside from exceptions one can expect the losses to the environment to be instrumental here. With reference to [8.13, 8.14] where a heat balance is established for molds for thermoplastics at higher temperatures, an energy balance will be set up by considering the heat fluxes for the quasi-steady state of operation (Figure 8.64).

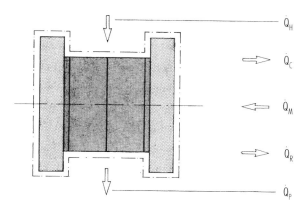

$\dot{Q}_H$

$\dot{Q}_C$

$\dot{Q}_M$

$\dot{Q}_R$

$\dot{Q}_P$

Figure 8.64 Heat flow assessment [8.16]

The equation for a mold is:

$$\dot{Q}_H - \dot{Q}_C + \dot{Q}_M - \dot{Q}_R - \dot{Q}_P = 0. \tag{8.57}$$

If the terms $\dot{Q}_C$ and $\dot{Q}_R$ are combined to a common power loss $\dot{Q}_L$ and $\dot{Q}_M$ with $\dot{Q}_P$ to $\dot{Q}_{Mo}$ then the equation is divided into three important areas:

– heat exchange with the environment ($\dot{Q}_L$),
– heat exchange with the molding ($\dot{Q}_{Mo}$),
– heat exchange with the heater ($\dot{Q}_H$).

$$\dot{Q}_L + \dot{Q}_{Mo} - \dot{Q}_H = 0. \tag{8.58}$$

To determine the losses, one should fall back on segmentation as proposed by [8.15] (Figure 8.65). The following assumptions are used.

The surroundings of the cavity should have a constant temperature (shaded areas in the picture). Now the heat flux is wanted, which is generated with a specified geometry. It is assumed that the segments can emit heat only through the outside faces. A heat exchange among segments is excluded. However, the segments can be composed of several layers so that an external insulation may be considered. Since the flow of heat loss and the pertinent temperature development are interdependent, the heat flux has to be calculated by iteration. A computer is best suited to solve this problem [8.28]. Thermal conductance is used in the calculation, which is determined for the respective pyramidal segment. The procedure is pictured with Figure 8.66.

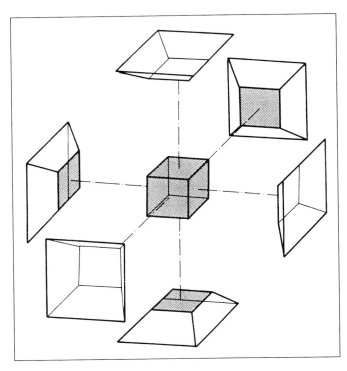

Figure 8.65 Breakup into segments [8.16]

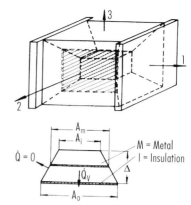

Figure 8.66 Evaluation of heat losses [8.16]

$$\text{ThC} = \frac{1}{A_m} \cdot \left(\frac{\Delta}{k}\right)_M + \frac{1}{A_o} \cdot \left(\frac{\Delta}{k}\right)_I \tag{8.59}$$

Δ Thickness of respective layer.

$$\text{TRR} = \frac{1}{A_o} \cdot \frac{1}{\varepsilon \cdot C_R} \tag{8.60}$$

$$\text{TTR} = \frac{1}{\alpha \cdot A_o} \tag{8.61}$$

$$\dot{Q}_C = (T_{max} - T_o)/\text{ThC} \tag{8.62}$$

$$\dot{Q}_R = \left[\left(\frac{T_{abs.o}}{100}\right)^4 - \left(\frac{T_{abs.\,E}}{100}\right)^4\right] \Big/ \text{TRR} \tag{8.63}$$

$$\dot{Q}_{Co} = (T_o - T_E)/\text{TTR} \tag{8.64}$$

$$\dot{Q}_L = \dot{Q}_C + \dot{Q}_{Co} + \dot{Q}_R \tag{8.65}$$

With ThC thermal conductance, TRR thermal radiation resistance and TTR thermal transfer resistance.

The losses calculated for each segment are combined to a total loss. The area A_m is introduced as a median value with which a constant median thermal conductance results for the whole segment. One can also use a variable thermal conductance as a function of areas. This leads to a solution by integration over the height of the segment [8.28]. The simplification creates slightly diverging results, which are on the safe side, though. Calculating with the median is preferable because it is much less complicated.

Attention should be paid to the heat-transfer coefficient, which can be determined for free convection with the surface temperature and the height of the mold (Figure 8.67).

The employed laws of heat transfer are only partly deduced from the equations of conversation. The major part was determined empirically [8.38, 8.74, 8.75, 8.76].

In the range of 0.4 to 0.6 m a transition of the convection from laminar to turbulent is noticeable. If a heat-transfer coefficient of 8 W/(m² K) is used, the calculated losses are higher than the real ones. With bigger molds, in contrast, one comes up with too small values. This may result in a heater which is too weak.

The energy which can be exchanged with the molding is the result of a simple calculation if one assumes that the mass of the molding m is brought up from the original temperature of the material to the mold temperature within the cycle time t_c. The specific thermal capacity is considered an average. The heat set free by the reaction is neglected in this consideration. This simplification is permissible for elastomers. For thermosets, the quantity of heat released may lead to a temperature increase of several degrees Celsius, however, as may be seen from Equation (8.66).

$$\dot{Q}_{Mo} = m \cdot c_p \cdot \frac{T_W - T_M}{t_C} \qquad\qquad (8.66)$$

With the heat losses according to Equation (8.64) and the energy exchanged with the molding Equation (8.65), the heat which has to be supplied by the heating system is established now. It is a steady figure, with which the mold remains in a "thermal balance". These estimates do not allow a statement about the temperature distribution or the behavior of the mold when heated up.

8.9.2   Temperature Distribution

If the temperature uniformity is considered, one has, in the first place, to confirm the assumptions with which the losses were calculated. With this confirmation the heat losses can be taken as an assured design criterion. This three-dimensional temperature field in the mold interior is not directly accessible. Therefore, one looks at cross sections of this temperature field and has, thus, transformed the three-dimensional physical problem into two-dimensional "patterns". They can be treated with electrical analogue models or a pattern of resistance paper. Nowadays, differential methods [8.77] and FEM programs [8.71] are used. Examples of the use of FEM programs are given in Chapter 14. Figure 8.68 is based on such a differential method. An instant was intentionally selected when a complete equilibrium of temperatures has not yet been established. One can very well recognize that temperature differences in the molding area are already partially leveled and that the isotherms are perpendicular to the segment boundaries. This confirms the assumption that no heat flux crosses the segment boundaries. Controlling the uniformity in the cavity region can be done by scanning a sectional plane of the mold, which makes processing by a difference procedure possible. The result is a set of isotherms, which are interpreted as temperature differences in the cavity wall. Adverse positioning of cooling lines can, thus, be avoided from the beginning. The significance of an effective insulation can also be demonstrated with such temperature profiles. This method of computation was originally developed for molds for thermoplastics and is now assigned to molds with a heat exchange by liquids.

The initial effort of input is reduced with a CAD application by so-called grid generators, which automatically produce the grid work for the difference method.

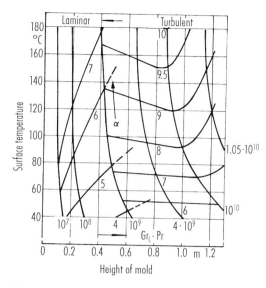

Height of mold

Figure 8.67   Coefficient of heat transfer for vertical flat surfaces [8.16]

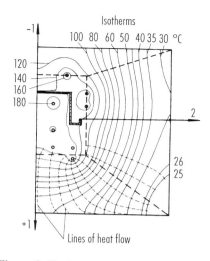

Figure 8.68   Temperature development in a mold [8.16]
○ Heating channel,
– – – – Segments

These considerations are less important for electrically heated molds because more significance is due to transient temperature variations. For this and other practical reasons heater cartridges are placed relatively close to the outside. This calls for sufficient insulation because otherwise the operating temperature can only be maintained with extremely high losses and will still be superimposed by fluctuations.

In this case a check on the actual mold temperature is highly recommended. Just regarding the temperature, set with the controller, as mold temperature, as it is still frequently done, is certainly insufficient because real temperatures may deviate from the set value by 20 °C [8.78, 8.79].

The effects of too low a temperature are best discussed by means of Figure 8.69. Lowering the temperature by 10 °C causes a severe reduction in the degree of curing. In this case the degree of curing drops to only 50% in the center of the molding while more than 85% was achieved with the required temperature. For this reason, an exact supervision of the mold temperature is indispensable and a good control highly recommended.

8.10 Practical Design of the Electric Heating for Thermoset Molds

According to [8.69] the installed wattage should be 20 to 30 W/kg to achieve an acceptable heating-up time and a stable temperature control. The heating elements have to be distributed uniformly throughout the mold. For electric-resistance heating the distribution should be checked by computer simulation. Large molds demand 8 to 16 heating circuits. Heating rods or tapered heater cartridges, as mentioned in Section 6.10.1.6 for

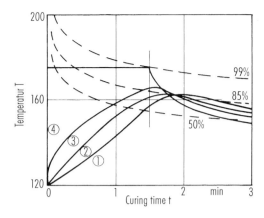

Figure 8.69   Diagram presenting temperature, time and rate of cross-linking for a Phenolic resin [8.16]
Part thickness 10 mm (c_p = f(T))
Distance from cavity wall:
① 5 mm,
② 4 mm,
③ 2.5 mm,
④ 0 mm

hot-runner manifolds, are suitable. They are installed as specified there. To achieve a stable temperature control each heating circuit needs a minimum of one thermocouple at least 12 to 15 mm away from the nearest heating element and at a distance from the cavity surface accordingly so that cyclic heat variations are attenuated and so recorded.

Large molds are often heated by steam. One can find an appropriate computation in [8.16]. In all cases an insulation around the entire mold is necessary especially against the clamping platens of the machine.

Electric heating systems can be selected according to their wattage whereas the dimensions (e.g., of the heater cartridge) are variable within a certain range. If the wattage needed for a fast heat-up results in a constant temperature in the quasi-steady range, then the wattage has to be reduced. This is mostly achieved with the switching rate but works at the cost of the service life. Thyristor-controlled concepts can be recommended because they always switch at the zero point of the AC wave. They are almost free of wear.

It is still important to obtain a good adjustment between the controlled system mold and controller when selecting a controller. The mold can be considered a controlled system of the first order with a time lag. A good approximation for the time constant is the heating-up time in accordance with the "adiabatic" heating formula. The parameters for the controller can be determined with the heating function according to [8.80, 8.81].

Although it is simple to find the dimensions of an electric heating system, the temperature has to be supervised, nevertheless. Any deviation results in relatively large temperature variations. Liquid-heating systems work the other way around: the temperature of the feed line remains within narrow limits, provided the necessary capacity can be transmitted. In contrast to a direct electric heating, the geometry of the heating system is particularly instrumental here. In addition one has to ensure a small temperature difference between feed line and return. Because of the small temperature differences at the heating channels, they can be placed closer to the cavities but the heat-up time is longer since the temperature differential remains small, especially if the heat exchanger only controls the feed temperature.

To achieve rapid, uniform heating, special heating platens may be used [8.82] (Figure 8.70). Here, standard tubular heaters introduce heat into thermal conduction tubes that distribute the heat rapidly and uniformly over a wide surface area. This can greatly reduce temperature differences across the platen surface, relative to conventional heating.

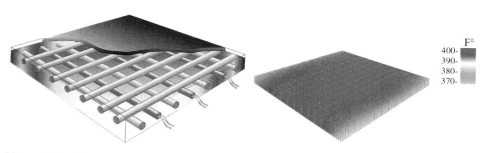

Figure 8.70   Temperature control by means of thermally conductive tubes (Acrolab) [8.82]

References

[8.1]   Schürmann, E.: Abschätzmethoden für die Auslegung von Spritzgießwerkzeugen. Dissertation, Tech. University, Aachen, 1979.

[8.2]   Kretzschmar, O.: Rechnerunterstützte Auslegung von Spritzgießwerkzeugen mit segment-bezogenen Berechnungsverfahren. Dissertation, Tech. University, Aachen, 1985.

[8.3]   Grigull, U.: Temperaturausgleich in einfachen Körpern. Springer, Berlin, Göttingen, Heidelberg, 1964.

[8.4]   Linke, W.: Grundlagen der Wärmeübertragung. Reprint of lecture, Tech. University, Aachen, 1974.

[8.5]   Wübken, G.: Einfluß der Verarbeitungsbedingungen auf die innere Struktur thermo-plastischer Spritzgußteile unter besonderer Berücksichtigung der Abkühlverhältnisse. Dissertation, Tech. University, Aachen, 1974.

[8.6]   Beese, U.: Experimentelle und rechnerische Bestimmung von Abkühlvorgängen beim Spritzgießen. Unpublished report, IKV, Aachen, 1973.

[8.7]   Döring, E.: Ermittlung der effektiven Temperaturleitfähigkeiten beim Spritzgießen von Thermoplasten. Unpublished report, IKV, Aachen, 1977.

[8.8]   Derek, H.: Zur Technologie der Verarbeitung von Harzmatten. Dissertation, Tech. University, Aachen, 1982.

[8.9]   Sönmez, M.: Verfahren zur Bestimmung des Druckverlustes in Temperiersystemen. Unpublished report, IKV, Aachen, 1977.

[8.10]   Gröber, H.; Erk, S.; Grigull, U.: Die Grundgesetze der Wärmeübertragung. Springer, Berlin, Göttingen, Heidelberg, 1963.

[8.11]   Carlslaw, H.; Jaeger, J. C.: Conduction of Heat in Solids. Oxford University Press, Oxford, 1948.

[8.12]   Menges, G.; Hoven-Nievelstein, W. B.; Schmidt, W. Th.: Handbuch zur Berechnung von Spritzgießwerkzeugen. Kunststoff-Information, Bad Homburg, 1985.

[8.13]   Catić, I.: Wärmeaustausch in Spritzgießwerkzeugen für die Plastomerverarbeitung. Disser-tation, Tech. University, Aachen, 1972.

[8.14]   Wübken, G.: Thermisches Verhalten und thermische Auslegung von Spritzgießwerk-zeugen. Technical report, IKV, Aachen, 1976.

[8.15]   Kretzschmar, O.: Auslegung der Temperierung von Spritzgießwerkzeugen für erweiterte Randbedingungen. Unpublished report, IKV, Aachen, 1981.

[8.16]   Paar, M.: Auslegung von Spritzgießwerkzeugen für vernetzende Formmassen. Disser-tation, Tech. University, Aachen, 1973.

[8.17]   Prömper, E.: DSC-Untersuchungen der Härtungsreaktion bei Phenolharzen. Unpublished report, IKV, Aachen, 1983.

[8.18]   Buschhaus, F.: Automatisierung beim Spritzgießen von Duroplasten und Elastomeren. Dissertation, Tech. University, Aachen, 1982.

[8.19]   Kamal, M. R.; Sourour, S.: Kinetics and Thermal Characterization of Thermoset Cure. Polymer Engineering Science, 13 (1973), 1, pp. 59–64.

[8.20]   Langhorst, H.: Temperaturfeldberechnung. Unpublished report, IKV, Aachen, 1980.

[8.21] Murray, P.; White, J.: Kinetics of the Thermal Decomposition of Clay. Trans. Brit. Ceram. Soc., 48, pp. 187–206.

[8.22] Nicolay, A.: Untersuchung zur Blasenbildung in Kunststoffen unter besonderer Berücksichtigung der Rißbildung, Dissertation, Tech. University, Aachen, 1976.

[8.23] Heide, K.: Dynamische thermische Analysemethoden. Deutscher Verlag für die Grundstoffindustrie, Leipzig, 1979.

[8.24] Differential Scanning Calorimetry (DSC). Publication, DuPont, Bad Homburg, 1988.

[8.25] Borchert and Daniels Kinetics. Publication, DuPont, Bad Homburg, 1982.

[8.26] Standard Test Method for Arrhenius Kinetic Constants for Thermally Unstable Materials. ASTM E 698–79.

[8.27] Piloyan, Y. O.; Ryabchikow, J. B.; Novikova, O. S.: Determination of Activation Energies of Chemical Reactions bei Differential Thermal Analysis. Nature, 212 (1966), p. 1229.

[8.28] Feichtenbeiner, H.: Berechnungsgrundlagen zur thermischen Auslegung von Duroplast- und Elastomerwerkzeugen. Unpublished report, IKV, Aachen, 1982.

[8.29] Kamal, M. R.; Ryan, M. E.: The Behaviour of Thermosetting Cornpounds in Injection Moulding Cavitics. Polymer Engineering and Science, 20 (1980), 13, pp. 859–867.

[8.30] Feichtenbeiner, H.: Auslegung eines Spritzgießwerkzeuges mit 4fach Kaltkanalverteiler für Elastomerforrnteile. Unpublished report, IKV, Aachen, 1983.

[8.31] Lee, J.: Curing of Compression Moulded Sheet Moulding Compound. Polymer Engineering and Science, 1981, 8, pp. 483–492.

[8.32] Schneider, Ch.: Das Verarbeitungsverhalten von Elastomeren im Spritzgießprozeß. Dissertation, Tech. University, Aachen, 1986.

[8.33] Baldt, V.; Kramer, H.; Koopmann, R.: Temperaturleitzahl von Kautschukmischungen – Bedeutung, Meßmethoden und Ergebnisse. Bayer Information for the Rubber Industry, 50 (1978), pp. 50–57.

[8.34] Kenig, S.; Kamal, M. R.: Cooling Molded Parts – a rigorous analysis. SPE-Journal, 26 (1970), 7, pp. 50–57.

[8.35] Sors, L.: Kühlen von Spritzgießwerkzeugen. Kunststoffe, 64 (1974), 2, pp. 117–122.

[8.36] Bird, R.; Stewart, W. E.: Transport Phenomena. John Wiley and Sons, New York, 1962.

[8.37] Eck, B.: Strömungslehre. In: Dubbels Taschenbuch für Maschinenbau. Vol. 1. Springer, Berlin, Heidelberg, New York, 1970.

[8.38] Renz, U.: Grundlagen der Wärmeübertragung. Lecture, Tech. University, Aachen, 1984.

[8.39] Hausen, H.: Neue Gleichungen für die Wärmeübertragung bei freier und erzwungener Strömung. Allg. Wärmetechnik, 9 (1959).

[8.40] Weinand, D.: Berechnung der Temperaturverteilung in Spritzgießwerkzeugen. Unpublished report, IKV, Aachen, 1982.

[8.41] Kretzschmar, O.: Thermische Auslegung von Spritzgießwerkzeugen. VDI-IKV Seminar "Computer Assisted Design of Injection Molds", Münster, 1984.

[8.42] Ott, S.: Aufbau von Programmen zur segmentierten Temperierungsauslegung. Unpublished report, IKV, Aachen, 1985.

[8.43] Thermische Werkzeugauslegung. Anwendungstechnische Information ATI, 892, Bayer AG.

[8.44] Wübken, G.: Thermisches Verhalten und thermische Auslegung von Spritzgießwerkzeugen. Dissertation, Tech. University, Aachen, 1976.

[8.45] Hengesbach, H. A.: Verbesserung der Prozeßführung beim Spritzgießen durch Prozeßüberwachung. Dissertation, Tech. University, Aachen, 1976.

[8.46] Mohren, P.; Schürmann, E.: Die thermische und mechanische Auslegung von Spritzgießwerkzeugen. Seminar material, IKV, Aachen, 1976.

[8.47] Temperiersysteme als Teil der Werkzeugkonstruktion. Publication, Arburg heute, 10 (1979), pp. 28–34.

[8.48] Rotary-Kupplung. Prospectus, Gebr. Heyne GmbH, Offenbach/M.

[8.49] Das Wärmerohr. Prospectus, Méchanique de l'Ile de France.

[8.50] Temesvary, L.: Mold Cooling: Key to Fast Molding. Modern Plastics, 44 (1966), 12, pp. 125–128 and pp. 196–198.

[8.51] Stöckert, K.: Formenbau für die Kunststoffverarbeitung. Carl Hanser Verlag, Munich, 1969.

[8.52] Spritzgießen von Thermoplasten. Publication, Farbwerke Hoechst AG, Frankfurt, 1971.

[8.53] Publication. AGA Gas GmbH, Hamburg, 1996.

[8.54] Westhoff, R.: Innovative Techniken der Werkzeugtemperierung – Hilfe bei der Senkung von Stückkosten. In: Kostenreduktion durch innovatives Spritzgießen. VDI-Verlag, Düsseldorf, 1997.

[8.55] System CONTURA – Publication. Innova Engineering GmbH, Menden, 1997.

[8.56] Mörwald, K.: Einblick in die Konstruktion von Spritzgießwerkzeugen. Brunke Garrels, Hamburg, 1965.

[8.57] Joisten, S.: Ein Formwerkzeug für Zahnräder. Die Maschine, 10 (1969).

[8.58] Spritzguß Hostalen PP. Handbook, Farbwerke Hoechst AG, Frankfurt/M., 1965.

[8.59] Temperieren von Spritzgießwerkzeugen. Information, Netstal-Maschinen AG, Näfels/ Switzerland, No. 12, June 1979, pp. 1–11.

[8.60] Friel, P.; Hartmann, W.: Beitrag zum Temperieren von Spritzwerkzeugen. Plastverarbeiter, 26 (1975), 9, pp. 491–498.

[8.61] Michaeli, W.; Kudlik, N.; Vaculik, R.: Qualitätssicherung bei optischen Bauteilen. Kunststoffe, 86 (1996), 4, pp. 478–480.

[8.62] Tewald, A.; Jung, A.: Dynamische Werkzeugtemperierung beim Spritzgießen. F & M, 102 (1994), 9, pp. 395–400.

[8.63] Tewald, A.; Thissen, U.: Angußloses Spritzgießen dünnwandiger Mikrohülsen. Reprint of lecture. Congrès Européen Chronométrie Biel/Bienne, October 1996, pp. 109–114.

[8.64] Michaeli, W.; Rogalla, A.: Kunststoffe für die Mikrosystemtechnik. Ingenieur-Werkstoffe, 6 (1997), 1, pp. 50–53.

[8.65] Delpy, U.: Einfluß variabler Formtemperaturen auf die Eigenschaften von Spritzlingen aus amorphen Thermoplasten. Dissertation, Univ. Stuttgart, 1971.

[8.66] Delpy, U.; Wintergerst, S.: Spritzgießen mit veränderlicher Werkzeugtemperatur. Lecture, 2nd Plastics Conference at Stuttgart, September 30, 1971.

[8.67] Rogalla, A.: Analyse des Spritzgießens mikrostrukturierter Bauteile aus Thermoplasten. Dissertation, RWTH, Aachen, 1998.

[8.68] Benkowsky, G.: Induktionserwärmung. Verlag Technik, Berlin, 1990.

[8.69] Keller, W.: Spezielle Anforderungen an Werkzeuge für die Duroplastverarbeitung. Kunststoffe, 78 (1988), 10, pp. 978–983.

[8.70] Weyer, G.: Automatische Herstellung von Elastomerartikeln im Spritzgießverfahren. Dissertation, Tech. University, Aachen, 1987.

[8.71] Ehrig, E.: Möglichkeiten der thermischen Auslegung von Duroplastwerkzeugen am Beispiel eines Kaltkanalverteilers. 8th International Symposium on Thermoset, Würzburg, 1996.

[8.72] Maaß, R.: Die Anwendung der statistischen Versuchsmethodik zur Auslegung von Spritzgießwerkzeugen mit Kaltkanalsystem. Dissertation, RWTH, Aachen, 1995.

[8.73] Diemert, J.: Einsatz physikalisch gestützter Prozeßmodelle zur Verbesserung der Qualitätsüberwachung beim Elastomerspritzgießen. Unpublished report, IKV, Aachen.

[8.74] Gröber, H.; Erk, S.; Grigull, U.: Die Grundgesetze der Wärmeübertragung. 3rd Ed., Sauerländer, Aaran, Frankfurt/M., 1980.

[8.75] Holman, J. P.: Heat Transfer. McGraw Hill Koga Kusha, Ltd. 4th Ed., 1976.

[8.76] VDI-Wärmeatlas. VDI, Fachgruppe Verfahrenstechnik (ed.), VDI-Verlag, Düsseldorf, 1957.

[8.77] Lichius, U.: Rechnerunterstützte Konstruktion von Werkzeugen zum Spritzgießen von thermoplastischen Kunststoffen. Dissertation, Tech. University, Aachen, 1983.

[8.78] Rauscher, W.: Praktische Überprüfung von Rechenmodellen zur thermischen Auslegung von Elastomerwerkzeugen. Unpublished report, IKV, Aachen, 1984.

[8.79] Rellmann, J.: Inbetriebnahme eines Duroplastspritzgießwerkzeugs. Unpublished report, IKV, Aachen, 1984.

[8.80] Recker, H.: Regler und Regelstrecken. In: Messen und Regeln beim Extrudieren. VDI-Gesellschaft Kunststofftechnik (ed.), VDI-Verlag, Düsseldorf, 1982.

[8.81] Wiegand, G.: Messen, Steuern, Regeln in der Kunststoffverarbeitung. Lecture material, Tech. University, Aachen, 1970.

[8.82] Publication. ACROLAB Engineering, Windsor, Ontario, Kanada, 1996.

[8.83] For QMC New Cooling Software. Plastics Technology, 30 (1984), 9, pp. 19–20.

[8.84] Simsir, E.: Vermeidung von Formteilverzug. Unpublished report, IKV, Aachen, 1977.

[8.85] Leibfried, D.: Untersuchungen zum Werkzeugfüllvorgang beim Spritzgießen von thermoplastischen Kunststoffen. Dissertation, Tech. University, Aachen, 1970.

9 Shrinkage

9.1 Introduction

If plastics are processed by injection molding, deviations of the dimensions of the molding from the dimensions of the cavity cannot be avoided. These deviations from the nominal size are summarized under the term shrinkage.

9.2 Definition of Shrinkage

In the injection-molding technique, shrinkage is the difference between an arbitrary dimension in the cavity and the corresponding dimension in the molding with reference to the cavity dimension.

$$S = \frac{l_C - l_M}{l_C} \cdot 100\% \tag{9.1}$$

Of course, this definition is not unambiguous (Figure 9.1) [9.1].

On one side, the dimensions of the cavity change from thermal expansion ($0 \rightarrow 1$) and mechanical loading during operation ($1 \rightarrow 2$), on the other side, the effect of time on the dimensions of the molding has to be taken into consideration ($2 \rightarrow 5$).

One distinguishes the demolding shrinkage (point 3), which is measured immediately after the molding has been ejected, and the processing shrinkage (point 4). The processing shrinkage is measured after storing the molding in a standard climate for 16 hours [9.2]. In this context the cavity dimension has to be determined at an ambient temperature of $23\,^\circ C \pm 2\,^\circ C$.

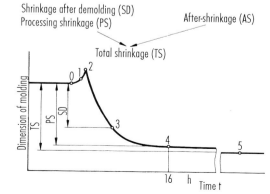

Figure 9.1 Dimensional changes as a function of time [9.1]
0 Dimension in cold mold,
1 Dimension in hot mold,
2 Dimension in mold under clamping force and holding pressure,
3 Dimension of molding after demolding,
4 Measurement of processing shrinkage (DIN 16901),
5 Dimension after storage

After extended storage another dimensional change may occur from the effect of temperature changes and especially from post-conditioning. It is called post-shrinkage ($4 \rightarrow 5$). This change is caused by relaxation of residual stresses, re-orientation and post-crystallization in crystalline materials. Except in crystalline materials, it is negligibly small, though. The sum of processing shrinkage and post-shrinkage is called total shrinkage. If additional dimensional deviations from moisture absorption or higher temperatures of use have to be taken into account at the time of acceptance, post-treatment and conditions of measurement have to be negotiated between molder and customer. In addition, one can distinguish shrinkage in dependence on the direction of flow (Figure 9.2). Radial processing shrinkage is shrinkage in the direction of flow, tangential shrinkage is that perpendicular to the direction of flow.

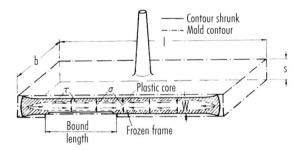

Figure 9.2   Frozen model [9.7]

The difference in processing shrinkage is the difference between radial and tangential shrinkage and is a measurement of the anisotropy of the shrinkage. The shrinkage in thickness is measured as section thickness, but it is usually not of interest in practice. For measuring, any kind of mechanical or optical instrument can be used, but a possible error from the measuring force should be taken into account for soft materials.

If the dimensions in Equation (9.1) are replaced by the volumes of cavity and molding, one talks about volume shrinkage [9.3].

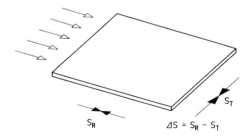

Figure 9.3   Magnitude of shrinkage depending on direction of flow
S_R Radial shrinkage,
S_T Tangential shrinkage,
S   Shrinkage difference

$$S_V = \frac{\upsilon_C - \upsilon_M}{\upsilon_C} = \frac{v(100 \text{ kPa, T}) - v(100 \text{ kPa, T}_R)}{v(100 \text{ kPa, T})}$$ (9.2)

υ = Specific volume of the material (Figure 9.7)

Longitudinal and volume shrinkage are related to one another but because of anisotropy (dependency of shrinkage on direction), linear shrinkage cannot be calculated from volume shrinkage. Another problem is the impossibility to measure volume shrinkage.

It is possible to make an assumption of shrinkage from volume shrinkage of thermoplastics in the following methode:

The shrinkage in direction of thickness H of an injection molded part
$$S_H \cong (0.9 \div 0.95)\, S_V$$
The shrinkage in the direction of length L
$$S_L \cong (0.05 \div 0.1)\, S_V$$

For crosslinking polymers exists a special standard for shrinkage (see source [9.4]).

9.3   Tolerances

The question of attainable tolerances is a cause of many complaints and, in extreme cases, has to be the subject of negotiations between molder and customer. Tolerances should never be closer than required for the perfect functioning of the part in use.

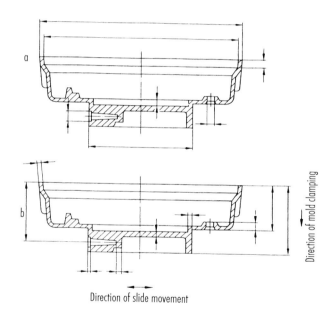

Figure 9.4   Mold-related (top) and not-mold-related (bottom) dimensions [9.2]

Direction of mold clamping

Direction of slide movement

Tolerances are closely related to shrinkage but also to the nature of the particular plastic. Close tolerances can only be expected with precise machine and mold control. Therefore they are beyond any action the mold maker can take. He has to try, however, to meet the required dimensions assuming normal processing conditions. If they can only be met under extreme conditions, the results are not in the center of the tolerance range. Then this range can easily be exceeded.

In addition, one has to differentiate between dimensions connected to the mold and those not connected to the mold. Dimensions connected to the mold are those which are determined by duplicating one mold part (Figure 9.4 top). Dimensions not connected to the mold are generated by the interaction of parts movable towards each other (stationary

STANDARDS AND PRACTICES OF PLASTICS CUSTOM MOLDERS	Engineering and Technical Standards ABS

NOTE: The Commercial values shown below represent common production tolerances at the most economical level. The Fine values represent closer tolerances that can be held but at a greater cost.

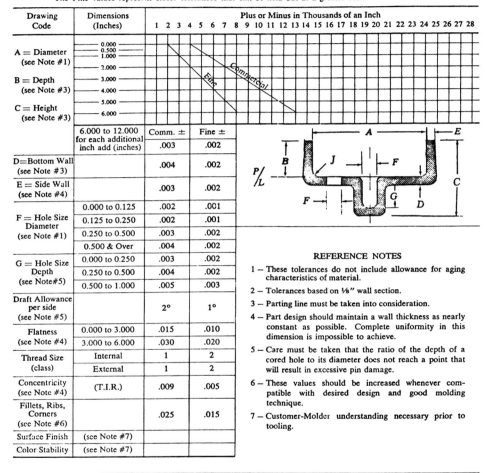

Drawing Code	Dimensions (Inches)			Plus or Minus in Thousands of an Inch 1 2 3 4 5 6 7 8 9 10 11 12 13 14 15 16 17 18 19 20 21 22 23 24 25 26 27 28
A = Diameter (see Note #1) B = Depth (see Note #3) C = Height (see Note #3)	0.000 0.500 1.000 2.000 3.000 4.000 5.000 6.000			
	6.000 to 12.000 for each additional inch add (inches)	Comm. ± .003	Fine ± .002	
D=Bottom Wall (see Note #3)		.004	.002	
E = Side Wall (see Note #4)		.003	.002	
F = Hole Size Diameter (see Note #1)	0.000 to 0.125	.002	.001	
	0.125 to 0.250	.002	.001	
	0.250 to 0.500	.003	.002	
	0.500 & Over	.004	.002	
G = Hole Size Depth (see Note#5)	0.000 to 0.250	.003	.002	
	0.250 to 0.500	.004	.002	
	0.500 to 1.000	.005	.003	
Draft Allowance per side (see Note #5)		2°	1°	
Flatness (see Note #4)	0.000 to 3.000	.015	.010	
	3.000 to 6.000	.030	.020	
Thread Size (class)	Internal	1	2	
	External	1	2	
Concentricity (see Note #4)	(T.I.R.)	.009	.005	
Fillets, Ribs, Corners (see Note #6)		.025	.015	
Surface Finish	(see Note #7)			
Color Stability	(see Note #7)			

REFERENCE NOTES

1 – These tolerances do not include allowance for aging characteristics of material.

2 – Tolerances based on 1/8″ wall section.

3 – Parting line must be taken into consideration.

4 – Part design should maintain a wall thickness as nearly constant as possible. Complete uniformity in this dimension is impossible to achieve.

5 – Care must be taken that the ratio of the depth of a cored hole to its diameter does not reach a point that will result in excessive pin damage.

6 – These values should be increased whenever compatible with desired design and good molding technique.

7 – Customer-Molder understanding necessary prior to tooling.

Figure 9.5 Practical tolerances on dimensions of articles molded from ABS (Courtesy of the Society of the Plastics Industry)

and movable mold half, slides) (Figure 9.4 bottom). This differentiation takes into account the lower accuracy, which results from movable mold components; they do not have exactly reproducible end positions.

Standards for tolerances are given in the form of tables in the "Plastics Engineering Handbook" [9.11]. These tables (an example is shown in Figure 9.5) were developed by the Society of the Plastics Industry, Inc. and are based on data obtained from representative material suppliers and molders.

Table 9.1 Coordination of tolerance groups with molding materials [9.2]

Moldings made of:	Tolerance groups		
	For common tolerances	For dimensions with allowances entered in the drawing	
		Grade 1	Grade 2
Acetal (polyoxyrnethylene)[1] (unfilled), part length: <150 mm	140	130	120
Acetal (polyoxymethylene)[1] (unfilled), part length: ≥150 mm	150	140	130
Acetal (polyoxymethylene)[1] (filled)	130	120	110
Acrylic	130	120	110
Diallyl phthalate compounds (with inorganic filler)	130	120	110
Polyethersulfone (unfilled)	130	120	110
Polyethylene[1] (unfilled)	150	140	130
Polyethylene terephthalate (amorphous)	130	120	110
Polyethylene terephthalate (crystalline)	140	130	120
Polyethylene terephthalate (filled)	130	120	110
Polyphenylene oxide	130	120	110
Polyphenylene oxide-styrene mixture (filled, unfilled)	130	120	110
Polyphenylene sulfide (filled)	130	120	110
Polypropylene[1] (unfilled)	150	140	130
Polypropylene[1] (filled with glass ribers or talc)	140	130	120
Polypropylene impact copolymer (unfilled)	140	130	120
Polystyrene	130	120	110
Polysulfone (filled, unfilled)	130	120	110
Polyvinyl chloride (without plasticizer)	130	120	110
Polyvinyl chloride (with plasticizer)	No information at present		
Styrene-acrylonitrile (filled, unfilled)	130	120	110
Styrene-butadiene copolymers	130	120	110
Fluorinated ethylene propylene	150	140	130
Thermoplastic polyurethanes (hardness 70–90 Shore A)	150	140	130
Thermoplastic polyurethanes (hardness > 50 Shore D)	140	130	120

[1] Take next higher tolerance group for unfilled, crystalline thermoplastics and wall thicknesses more than 4 mm.

Pertinent information concerning tolerances may also be received from the British Standard BS 4042. The following data are provided with reference to the German Standard DIN 16 901.

Depending on the molding material one can determine several tolerance groups (Table 9.1). Within each group a distinction is made between two grades of accuracy for entered allowances above and below nominal sizes. For each tolerance group allowances depending on nominal sizes can be taken from a second table (Table 9.2). In this table a

Table 9.2 Coordination of tolerances with tolerance groups [9.2]

Tolerance group from Table 1	Code letter[1]	Range of nominal dimensions (more than/until)								
		0 1	1 3	3 6	6 10	10 15	15 22	22 30	30 40	40 53
Common tolerances										
160	A	±0.28	±0.30	±0.33	±0.37	±0.42	±0.49	±0.57	±0.66	±0.78
	B	±0.18	±0.20	±0.23	±0.27	±0.32	±0.39	±0.47	±0.56	±0.68
150	A	±0.23	±0.25	±0.27	±0.30	±0.34	±0.38	±0.43	±0.49	±0.57
	B	±0.13	±0.15	±0.17	±0.20	±0.24	±0.28	±0.33	±0.39	±0.47
140	A	±0.20	±0.21	±0.22	±0.24	±0.27	±0.30	±0.34	±0.38	±0.43
	B	±0.10	±0.11	±0.12	±0.14	±0.17	±0.20	±0.24	±0.28	±0.33
130	A	±0.18	±0.19	±0.20	±0.21	±0.23	±0.25	±0.27	±0.30	±0.34
	B	±0.08	±0.09	±0.10	±0.11	±0.13	±0.15	±0.17	±0.20	±0.24
Tolerances for dimensions with entered allowances										
160	A	0.56	0.60	0.66	0.74	0.84	0.98	1.14	1.32	1.56
	B	0.36	0.40	0.46	0.54	0.64	0.78	0.94	1.12	1.36
150	A	0.46	0.50	0.54	0.60	0.68	0.76	0.86	0.98	1.14
	B	0.26	0.30	0.34	0.40	0.48	0.56	0.66	0.78	0.94
140	A	0.40	0.42	0.44	0.48	0.54	0.60	0.68	0.76	0.86
	B	0.20	0.22	0.24	0.28	0.34	0.40	0.48	0.56	0.66
130	A	0.36	0.38	0.40	0.42	0.46	0.50	0.54	0.60	0.68
	B	0.16	0.18	0.20	0.22	0.26	0.30	0.34	0.40	0.48
120	A	0.32	0.34	0.36	0.38	0.40	0.42	0.46	0.50	0.54
	B	0.12	0.14	0.16	0.18	0.20	0.22	0.26	0.30	0.34
110	A	0.18	0.20	0.22	0.24	0.26	0.28	0.30	0.32	0.36
	B	0.08	0.10	0.12	0.14	0.16	0.18	0.20	0.22	0.26
Precision molding	A	0.10	0.12	0.14	0.16	0.20	0.22	0.24	0.26	0.28
	B	0.05	0.06	0.07	0.08	0.10	0.12	0.14	0.16	0.18

[1] A for dimensions not connected to the mold, B for dimensions connected to the mold

Range of nominal dimensions (more than/until)

53 70	70 90	90 120	120 160	160 200	200 250	250 315	315 400	400 500	500 630	630 800	800 1000

Common tolerances

53 70	70 90	90 120	120 160	160 200	200 250	250 315	315 400	400 500	500 630	630 800	800 1000
±0.94	±1.15	±1.40	±1.80	±2.20	±2.70	±3.30	±4.10	±5.10	±6.30	±7.90	±10.00
±0.84	±1.05	±1.30	±1.70	±2.10	±2.60	±3.20	±4.00	±5.00	±6.20	±7.80	±9.90
±0.68	±0.81	±0.97	±1.20	±1.50	±1.80	±2.20	±2.80	±3.40	±4.30	±5.30	±6.60
±0.58	±0.71	±0.87	±1.10	±1.40	±1.70	±2.10	±2.70	±3.30	±4.20	±5.20	±6.50
±0.50	±0.60	±0.70	±0.85	±1.05	±1.25	±1.55	±1.90	±2.30	±2.90	±3.60	±4.50
±0.40	±0.50	±0.60	±0.75	±0.95	±1.15	±1.45	±1.80	±2.20	±2.80	±3.50	±4.40
±0.38	±0.44	±0.51	±0.60	±0.70	±0.90	±1.10	±1.30	±1.60	±2.00	±2.50	±3.00
±0.28	±0.34	±0.41	±0.50	±0.60	±0.80	±1.00	±1.20	±1.50	±1.90	±2.40	±2.90

Tolerances for dimensions with entered allowances

53 70	70 90	90 120	120 160	160 200	200 250	250 315	315 400	400 500	500 630	630 800	800 1000
1.88	2.30	2.80	3.60	4.40	5.40	6.60	8.20	10.20	12.50	15.80	20.00
1.68	2.10	2.60	3.40	4.20	5.20	6.40	8.00	10.00	12.30	15.60	19.80
1.36	1.62	1.94	2.40	3.00	3.60	4.40	5.60	6.80	8.60	10.60	13.20
1.16	1.42	1.74	2.20	2.80	3.40	4.20	5.40	6.60	8.40	10.40	13.00
1.00	1.20	1.40	1.70	2.10	2.50	3.10	3.80	4.60	5.80	7.20	9.00
0.80	1.00	1.20	1.50	1.90	2.30	2.90	3.60	4.40	5.60	7.00	8.80
0.76	0.88	1.02	1.20	1.50	1.80	2.20	2.60	3.20	3.90	4.90	6.00
0.56	0.68	0.82	1.00	1.30	1.60	2.00	2.40	3.00	3.70	4.70	5.80
0.60	0.68	0.78	0.90	1.06	1.24	1.50	1.80	2.20	2.60	3.20	4.00
0.40	0.48	0.58	0.70	0.86	1.04	1.30	1.60	2.00	2.40	3.00	3.80
0.40	0.44	0.50	0.58	0.68	0.80	0.96	1.16	1.40	1.70	2.10	2.60
0.30	0.34	0.40	0.48	0.58	0.70	0.86	1.06	1.30	1.60	2.00	2.50
0.31	0.35	0.40	0.50								
0.21	0.25	0.30	0.40								

difference is also made between dimensions connected to the mold and those not connected to the mold.

9.4   Causes of Shrinkage

The intrinsic cause for shrinkage of injection-molded parts is the thermodynamic behavior of the material (Figure 9.6). It is also called p-v-T (pressure-volume-temperature) behavior and characterizes the compressibility and thermal expansion of plastics [9.5].

There is a basically different p-v-T behavior between two classes of materials (amorphous and crystalline). As a melt, both classes show a linear dependency of the specific volume on the temperature. For the solid phase, however, there are considerable differences. On the basis of crystallization the specific volume decreases exponentially with decreasing temperature while amorphous materials also have a linear dependency in the solid phase. This difference is the reason for the greater shrinkage of crystalline thermoplastics.

To assess the process with respect to shrinkage, the change in state in a p-v-T diagram is very helpful. Pressure and temperature during the process are recorded isochronously in a p-v-T diagram (Figure 9.7).

Following the volume filling of the cavity ($0 \rightarrow 1$) the material is compacted in the compression phase without substantial change in temperature ($1 \rightarrow 2$). The magnitude of the locally attainable pressure in the molding depends on the magnitude of the holding pressure exerted by the machine and on the resistance to flow in the cavity.

Subsequently, the molding steadily cools down ($2 \rightarrow 3$). Related to this is a volume contraction, which can be partly compensated by the holding pressure, which supplies additional melt to the cavity through the liquid core of the solidifying molding. If no more melt can be fed into the cavity, e.g., by a solidified gate, the change in state is isochoric ($3 \rightarrow 4$).

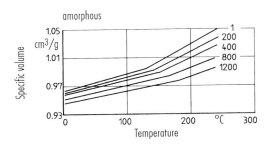

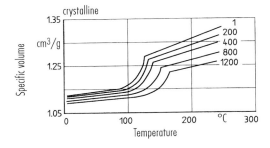

Figure 9.6   p-v-T diagram of an amorphous (top) and a crystalline (bottom) thermoplastic material

Figure 9.7 Change of state in the p-v-T
diagram [9.3]
0 → 1   Volumetric filling,
1 → 2   Compression,
2 → 3   Effect of holding pressure,
3 → 4   Isochoric pressure drop down to
temperature $T_{100\,kPa}$,
4 → 5   Cooling to demolding temperature T_E,
5 → 6   Cooling to room temperature T_R,
4 → 6   Volume shrinkage.

$$S_V = \frac{\overline{V}_{100\,kPa} - \overline{V}_E}{\overline{V}_{100\,kPa}} \cdot 100\%$$

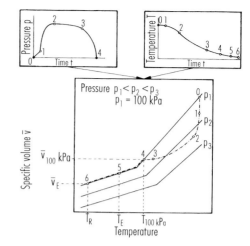

The point where the 100-kPa line is met (point 4) establishes the local volume shrinkage.
A higher volume shrinkage occurs if this point is in the range of larger volumes. Since
the volume shrinkage is equivalent to the shrinkage potential, a larger volume shrinkage
also results in a higher longitudinal shrinkage.

After the 100-kPa line has been reached, any further change of state is isobaric. At the
time the molding is ejected (point 5) and constraints from the surrounding cavity cease.

9.5   Causes of Anisotropic Shrinkage

Dimensional changes of a molding in the mold are restricted or prevented by a force-
locking clamping of the mold halves [9.6] and shrinkage is non-uniform (anisotropic). A
distinction must be drawn between internal and external constraints of contraction.

External restriction of shrinkage is a mechanical restriction against a change of shape
by the surrounding mold. The restriction of shrinkage and the related stress relaxation
result in a lower level of shrinkage. The shrinkage of a restricted part is less than that
of a restricted one and there is, moreover, less dependence on the process parameters
(Figure 9.8).

The mechanical restriction, of course, is effective only as long as the molding is still
in the mold. After ejection, restricted dimensions can also shrink freely. Therefore the
temperature of demolding is a characteristic for the change in mechanical boundary
conditions and for the shrinkage and distortion behavior.

Internal restriction of shrinkage is due to both cooling-related internal stresses and to
orientation.

Molecule orientations affect shrinkage in two ways. On the one hand, the coefficients
of linear expansion dependent on orientation cause a difference in shrinkage. On the
other, re-orientation (contraction) in the direction of orientation contributes to an
increase in shrinkage. The molecular orientation is determined by the process
parameters, and primarily by the type and location of gating, but exerts much less of an
influence than does, e.g., fiber orientation in fiber-filled molding compounds.

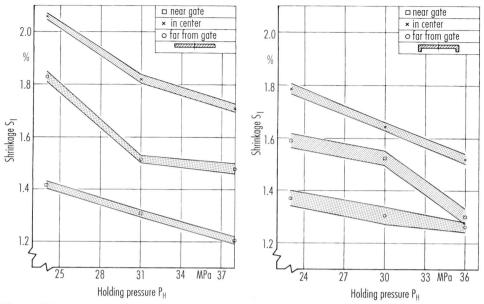

Figure 9.8   Shrinkage of a free (left) and a confined (right) circular plate

Oriented by the flow processes, the fibers hinder shrinkage primarily in the direction of orientation because they have a lower thermal expansion and greater stiffness relative to the matrix material (Figure 9.9) [9.8]. Through the use of fibers, shrinkage may be reduced by up to 80%. However, no further reductions in shrinkage behavior are observed at additions of more than 20% fiberglass.

Incorporation of fillers such as glass beads and mineral powder leads to isotropic shrinkage. The reduction in overall shrinkage that occurs is due to the lower compressibility of the material as a whole.

Aside from fiber orientation, molecular orientation in the direction of flow leads to anisotropic shrinkage.

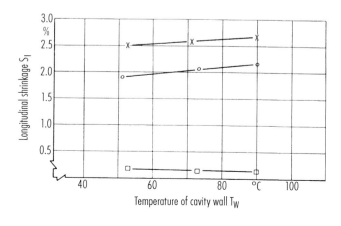

Figure 9.9   Effect of glass fibers and spheres on shrinkage
Part: bushing,
Material: PBPT unreinforced (x),
30% glass spheres ($\bigcirc$),
30% glass fibers ($\square$),
$T_M = 251\ °C$,
$P_W = 33$ MPa.

The low thermal conductivity of plastic results in the temperature profile shown in Figure 9.10. Different cooling conditions exist for the different layers and so volume contraction also varies. Due to mechanical coupling between the layers, thermal contraction in the longitudinal and transverse directions is restricted. This restriction does not exist in the direction of thickness, with the result that most of the volume shrinkage takes the form of shrinkage of the cross-section. Restrictions of shrinkage of the same kind in longitudinal and transverse direction result in the same shrinkage, provided no warpage orientation of molecules or fibers occurs.

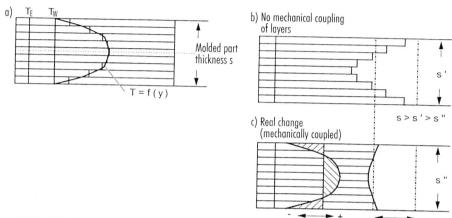

Figure 9.10   Model of stress buildup [9.12]

9.6   Causes of Distortion

Distortion is one result of anisotropic shrinkage. Frequently, it is caused by asymmetric cooling relative to the part thickness. A higher wall temperature on the top side as shown in Figure 9.10 leads, for example, to higher temperatures in the upper layers and, via greater volume contraction, to deflection towards the warmer side. This differential cooling may also be caused by inserts, such as decorative material in in-mold decoration.

Corner distortion (Figure 9.11; see also Figure 8.46) is due to poorer heat dissipation towards the inside of a corner. This has the effect of reducing the corner angles. Similarly, differences in the thickness of ribs will displace the temperature profile from its symmetrical center position and result in distortion of the moldings (Figure 9.12).

This cooling-induced distortion can be prevented by altering the mold temperature, where necessary, by relocating the cooling channels or using mold inserts of different material. Inner corners and thick ribs need to be cooled better than other part sections (see also Section 8.6).

To avoid distortion caused by orientation, the gate should be repositioned or the flow path modified by changing the wall thickness. Distortion in flat parts can be counteracted by applying thin bracing ribs.

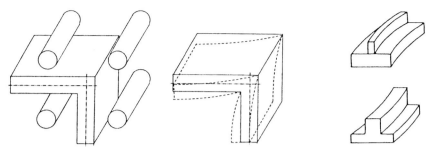

Figure 9.11   Corner distortion through poorer heat dissipation in internal corners

Figure 9.12   Distortion caused by differences in wall thickness

9.7   Effect of Processing on Shrinkage

Other than by modifications to the mold and a change of material, the molder can only influence shrinkage and distortion by making changes to the process. From the p-v-T diagram, it can be seen that pressure and temperature are the main factors affecting shrinkage. Design changes will affect these parameters and thus also the shrinkage. A survey of the influences exerted by various parameters is shown in Figure 9.13 [9.11].

With amorphous as well as with crystalline thermoplastics, the holding pressure exerts the greatest effect on shrinkage (Figure 9.8).

Under holding pressure, the material in the cavity is compressed and the volume contraction from cooling is compensated by additional melt supply. The influence of the holding pressure is shown in the p-v-T diagram charting the progress of the process (see Figure 9.7). If the holding pressure is increased, the process is shifted to lower specific volumes, reaching at lower specific volumes the 100-kPa line at which the part undergoes lower shrinkage. The influence of the holding pressure is degressive, however; in other words, the reduction in shrinkage decreases with increase in holding pressure.

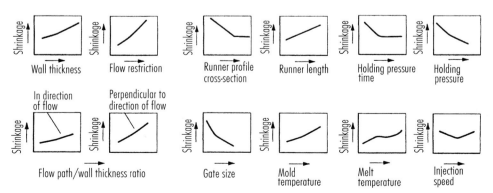

Figure 9.13   Relationships between shrinkage and characteristic parameters [9.11]

With an increase in holding pressure, a reduction in shrinkage of up to 0.5% can be obtained in crystalline materials. With amorphous materials, a reduction of just 0.2% max. is feasible because of their overall lower level of shrinkage.

The second major influence on shrinkage is the temperature of the material.

Theoretically, a higher injection temperature has two opposing effects on shrinkage: on the one hand, a higher temperature results in a higher thermal contraction potential of the material (see also Figure 9.7 and [9.7]) and, on the other, the decrease in melt viscosity causes a better transfer of pressure and with this a reduction in shrinkage. Given sufficiently long holding-pressure stages, the effect of improved cavity pressure predominates in the case of crystalline materials (Figure 9.14).

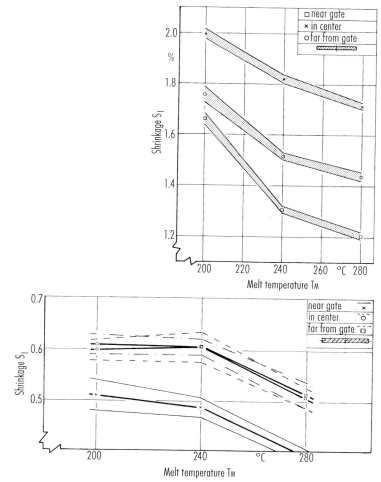

Figure 9.14   Effect of melt temperature on shrinkage (top: Crystalline, bottom: Amorphous material)

With crystalline materials, a reduction in shrinkage can be obtained of up to 0.5%; with amorphous plastics, the figure is up to 0.15%. All other parameters determine the shrinkage behavior via the pressure and temperature. While a greater wall thickness leads to better pressure transfer, the poor thermal conductivity of plastics makes volume contraction more noticeable at high temperatures and increases the shrinkage. Restrictions to flow impair pressure transfer and so increase shrinkage. In contrast, larger runner profile cross-sections and thicker gates make for better pressure transfer. A large runner length leads, just as does a large flow path to wall thickness ratio, to a drop in pressure and thus to greater shrinkage. Hot runners, however, reduce shrinkage.

The influences of the holding pressure time can be used to again illustrate the most important criterion concerning shrinkage. As the holding pressure time increases, the forcing of additional material into the cavity reduces shrinkage. This can only happen, however, as long as the melt, particularly the gate and sprue, has not frozen. Prolonging the holding pressure time beyond that has no further effect. For this reason, a part for homogeneous molding materials should always be gated at the thickest point and the wall thicknesses should be such that holding pressure can take effect even in those areas furthest away from the sprue.

With glass-reinforced materials, there are some particularities (Figure 9.15). In the direction of fiber orientation, it is not possible to affect shrinkage by modifying processing parameters, as the rigidity of the fibers exerts an extremely strong influence. The effect perpendicular to the direction of fibers is approximately the same as it is with the matrix material only.

9.8   Supplementary Means for Predicting Shrinkage

The simplest way to estimate shrinkage for dimensioning a mold is to consult tables (Table 9.3). They are provided by the raw-material suppliers in the data sheets for their respective materials.

However, the partly wide range of listed data is problematic because it does not allow a sufficiently accurate prediction of shrinkage; nor are pertinent process parameters known or configurations of moldings from which the shrinkage was obtained. Transfer to other configurations is, therefore, difficult.

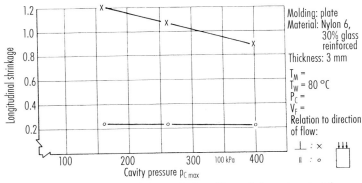

Figure 9.15   Effect on shrinkage of glass fiber reinforced materials

Table 9.3   Shrinkage of some thermoplastics [9.10]

Material	Shrinkage %	Material	Shrinkage %
Nylon 6	1–1.5	Polycarbonate	0.8
Nylon 6-GR	0.5	Polyoxymethylene (Acetal)	2
Nylon 6/6	1–2	Polyvinyl chloride, rigid	0.5–0.7
Nylon 6/6-GR	0.5	Polyvinyl chloride, soft	1–3
Low-density polyethylene	1.5–3	Acrylonitrile-butadiene-styrene	0.4–0.6
High-density polyethylene	2–3	Polypropylene	1.2–2
Polystyrene	0.5–0.7	Cellulose acetate	0.5
Styrene-acrylonitrile	0.4–0.6	Cellulose acetate butyrate	0.5
Polymethyl methacrylate (Acrylic)	0.3–0.6	Cellulose propionate	0.5

A more accurate prediction can be made on the basis of a collection of data gained through experience. This is the most reliable method for predicting linear shrinkage so far. In such a collection the shrinkage data of all parts are listed which have been produced in the past as well as their processing conditions. Because of different restrictions to shrinkage and geometry elements, families of dimensions are formed.

If the mold for a similar part has to be designed, these data can be used for dimensioning.

Another and increasingly more accurate estimate of shrinkage behavior is provided by FEA simulation [9.12] (see also Chapter 14). Exact dimensioning of parts with the aid of this method is not yet possible. However, the process and the part can be optimized in respect to shrinkage and distortion behavior. The ability to predict part dimensions and volume shrinkage, but more so the temperature and pressure behavior, are important tools for accomplishing this. Since several physical parameters such as crystallization behavior cannot as yet be determined, some material data cannot be determined with sufficient accuracy batch fluctuations cannot be allowed for, and it is still not possible to make an exact predictive simulation of part dimensions.

References

[9.1]   Hoven-Nievelstein, W. B.: Die Verarbeitungsschwindung thermoplastischer Formmassen. Dissertation, Tech. University, Aachen, 1984.
[9.2]   German Standard: DIN 16901: Kunststoff-Formteile Toleranzen und Abnahmebedingungen für Längenmaße.
[9.3]   Schmidt, Th. W.: Zur Abschätzung der Schwindung. Dissertation, Tech. University, Aachen, 1986.
[9.4]   German Standard: DIN 53464: Prüfung von Kunststoffen. Bestimmung der Schwindungseigenschaften von Preßstoffen aus warmhärtbaren Preßmassen.
[9.5]   Geisbüsch, P.: Ansätze zur Schwindungsberechnung ungefüllter und mineralisch gefüllter Thermoplaste. Dissertation, Tech. University, Aachen, 1980.
[9.6]   Zipp, Th.: Erfahrungsanalyse zur Ermittlung des notwendigen Werkzeugübermaßes beim Spritzgießen. Unpublished report, IKV, Aachen, 1985.
[9.7]   Stitz, S.: Analyse der Formteilbildung beim Spritzgießen von Plastomeren als Grundlage für die Prozeßsteuerung. Dissertation, Tech. University, Aachen, 1973.
[9.8]   Menges, G.; Hoven-Nievelstein, W. B.; Zipp, Th.: Erfahrungskatalog zur Verarbeitungsschwindung thermoplastischer Formmassen beim Spritzgießen. Unpublished report, IKV, Aachen, 1984/85.

[9.9]   Baur, E.; Schleede, K.; Lessenich, V.; Ort, St.; Filz, P.; Pötsch, G.; Groth, S.; Greif, H.:
         Formteil- und Werkzeugkonstruktion aus einer Hand – Die modernen Hilfsmittel für den
         Konstrukteur. Contribution to 14th Technical Conference on Plastics, Aachen, 1988.
[9.10]   Strack Normalien für Formwerkzeuge. Handbook, Strack-Norma GmbH, Wuppertal.
[9.11]   Frados, J.: Plastics Engineering Handbook. Van Nostrand Reinhold, New York, 1976.
[9.12]   Pötsch, M. G.: Prozessimulation zur Abschätzung von Schwindung und Verzug thermo-
         plastischer Spritzgussteile. Dissertation, RWTM, Aachen, 1991.

10 Mechanical Design of Injection Molds [10.1]

10.1 Mold Deformation

Injection molds are exposed to a very high mechanical loading but they are only allowed elastic deformation. Since these molds are expected to produce parts that meet the demands for high precision, it is evident, therefore, that any deformation of the mold affects the final dimensions of a part as well as the shrinkage of the plastic material during the cooling stage. Besides this, undue deformation of a mold can result in undesirable interference with the molding process or actuation of the mold.

Effects on the quality of the molding:

– Mold deformation results in dimensional deviations and possible flashing.

Effects on the function of the mold:

– Deformation of the mold, especially transverse to the direction of demolding and larger than the corresponding shrinkage of the molding, results in problems in mold-opening or ejection from jamming.
– Thus, the rigidity of a mold determines the quality of the moldings as well as reliable operation of the mold.
– Common molds are assembled from a number of components, which as a whole provide rigidity to the mold by their interaction. The components of a mold are compact bodies and both bending and shear strains have to be considered in design. They are still sufficiently slender, though, that, with some exceptions, permissible stresses need not be taken into account because of their small permissible deformation.

10.2 Analysis and Evaluation of Loads and Deformations

a) As a general principle, molds have to be designed with their permissible deformation in mind.
b) Dynamic deformations do not occur, which is why a large number of equations are used for working out the static load and deformation behavior.
c) Complex configurations make injection molds statically indeterminate systems. For calculating the expected deformation one can either use a finite-element method for a closed approximation or – much simpler and sufficiently exact – divide the molds into separate elements [10.1, 10.2]. Since only elastic deformations are admitted, the individual elements can be considered springs and the whole system computed as a set of springs. The individual "springs" are then added up to yield the overall deformation (cf. Sections 10.4 to 10.6).

10.2.1 Evaluation of the Acting Forces

The acting forces are:

a) Closing and clamping forces exerted by the machine.
b) The maximum cavity pressure. It acts via the molding compound in the mold cavities and the runner system on the mold, which may deform mainly by bending.

Two problems may arise:

1. The part is jammed in the mold when cavity pressure forces, acting perpendicularly to the mold axis, bend the walls further than the molding compound at these points shrinks in thickness after cooling to demolding temperature.
2. Under the effect of the cavity pressure forces in the direction of the mold axis, unpermissibly large gaps in the mold parting line could occur into which melt could penetrate and flash could form.

   The maximum cavity pressure in thermoplastics and normal operating conditions is always the maximum injection pressure.

   With elastomers and thermosets, the maximum cavity pressure usually only occurs after filling because now the molding compound is heated further both by the hotter mold wall and by the liberated heat of reaction, and it expands more. Only when cross-linking progresses further does a more or less large reduction in volume occur. This volume expansion due to heating is also the reason that these molding compounds almost always leave behind flash on the parts. While there are suitable appropriate controls for preventing this, they are rarely used. Many molders still afford themselves the luxury of expensive, very often manual finishing of every part produced.

   To calculate the formation of such gaps more accurately, it is not sufficient to use the simple calculations employing spring stiffness for the molds alone because the deformations of the press (clamping unit) contribute considerably to overall deformation.

   A great deal of work has been done in recent years that allows the formation of gaps in mold parting lines to be calculated very accurately [10.3, 10.4]. They always consist in adding up the deformations of the molds in the axial direction and those of the entire clamping unit. These calculations are admittedly much more complicated than the simple calculation that will be described below. An accurate calculation moreover requires that the molder determines the deformation of the clamping unit, including all its elements. To date, this deformation is usually not quoted by the machine manufacturers. The method of measuring and calculating the deformation of the clamping unit is described in [10.4]. There is, however, a relatively simple method that will be explained later. This method yields results that lie on the safe side.

c) Mold opening and ejection forces:
   These forces are usually much smaller. They only need to be considered when designing the ejection system. If present, however, the ejector pins must also be allowed for, as there are two frequently ignored dangers here:

   – the pins can buckle outwards and be destroyed,
   – the pins can punch through the molded part.

Calculations for these are provided in Chapter 12.

10.3 Basis for Describing the Deformation

The mold forms a link in the closed system of the clamping unit.

The following distinction has to be made to obtain characteristic deformations in dependence of the forces from injection pressure and clamping:

1. Which elements are relieved by the effects of the cavity pressure?
2. Which elements are loaded further by effects from the cavity pressure?

If deformations and forces parallel to the direction of the clamping force are considered, the following equivalent diagram (Figure 10.1) of the clamping unit including the mold is obtained.

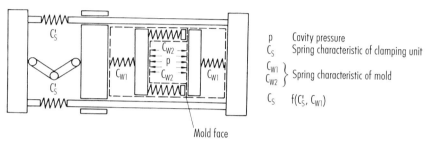

p   Cavity pressure
C_S' Spring characteristic of clamping unit
C_{W1}
C_{W2} } Spring characteristic of mold

C_S $f(C_S', C_{W1})$

Mold face

Figure 10.1   Equivalent diagram of clamping unit and mold [10.5]

The elements with the spring rate (load per unit deflection) C_{W1} and C_S are first stressed by the clamping force and then, in addition, by the reactive forces from cavity pressure.

Therefore, the machine platens exhibit the same deformation response as the tie bars of the clamping unit, taking the parting line of the mold as reference line.

That part of the mold with the spring rate C_{W2} (cavity area) is first stressed by the clamping force but then more or less relieved by the reactive forces from cavity pressure.

With the simple calculation, it is assumed that the mold faces are just in contact with one another when the additional elongation of the clamping unit ΔL_S and the decrease in compression of the cavity ΔL_W are equal.

10.3.1 Simple Calculation for Estimating Gap Formation

$$\Delta L_S = \Delta L_W \tag{10.1}$$

$$\Delta F = C \cdot \Delta L \tag{10.2}$$

$$\frac{\Delta F_S}{C_S} = \frac{\Delta F_W}{C_{W2}} \tag{10.3}$$

$$p \cdot A_M = \Delta F_S + \Delta F_W \tag{10.4}$$

$$p \cdot A_M = C_S \cdot \Delta L_S + C_W \, \Delta L_W \tag{10.5}$$

$$p \cdot A_M = \Delta L \cdot (C_S + C_W) \tag{10.6}$$

p   = Cavity pressure,
A_M = Projected part area.
ΔL_S = Deformation of clamping unit
ΔL_W = Deformation of mold
ΔF = Change in spring force
C   = Spring characteristic
S   = Index for clamping unit
W   = Index for mold

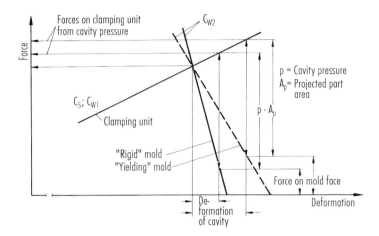

Figure 10.2
Characteristic deformation of mold and clamping unit in direction of clamping [10.1, 10.5]

The resulting characteristic deformations are depicted in Figure 10.2. The cavity deformation in the direction of clamping has a considerable effect on the quality of the molding. It does not only depend on the rigidity of the mold but also on that of the clamping unit. Under the reactive forces from cavity pressure a notable rigidity of the clamping unit results in

1. a small deformation of the cavity in the direction of clamping,
2. higher stresses in the clamping unit,
3. higher forces in the clamping surfaces.
Cases 2. and 3. occur only if there is no overload protection (e.g., with a fully hydraulic clamping unit).

High rigidity of the mold results in

1. small cavity deformation in clamping direction,
2. lower stresses in the clamping unit.

For these reasons it makes sense to design the mold with a high resilience.

10.3.2 More Accurate Calculation for Estimating Gap Formation and Preventing Flash

Flash in the mold faces can have two major causes:

1. Deflection of mold platens that are unsupported, primarily above the free space for the ejector plate. Particularly at risk in this respect are the molds for large-area parts, and multicavity molds.
2. So-called mold breathing, i.e. the uniform opening of the mold faces due to a combination of inadequate pressure exerted by the clamping unit (press) and deformation of the mold.

Whereas mold deformation may be determined with sufficient accuracy by the simple method of spring stiffness of the mold elements (see Sections 10.4 to 10.6), no information is available about deformation of the clamping units. This is usually not provided by the machine manufacturers and is extremely laborious to determine.

The clamping units, including toggle presses and locked hydraulic presses, are actually all very much softer than indicated in Figure 10.2. In effect, it is not, as suggested by Figure 10.2, just the tie bars that are placed under load but also a great many other elements, such as joints and link pins in the toggle presses and the locking elements and plates, the compressibility of the oil etc in the hydraulic presses. These all have to be included because they are all much softer. For this reason, the spring diagram for a clamping unit is more like that shown in Figure 10.3. One can now use a program developed by Krause et al. [10.3] to make very accurate calculations or, to avoid performing the tedious compilation work, use a simple, dependable method derived from the work of Krause et al.

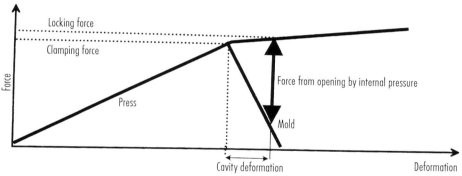

Figure 10.3   Realistic mold spring diagram

Flash is caused by deformation of the edge cavity in the mold. For the gap formation, its deformation must be considered with that of the mold in a combined spring diagram. One such diagram is shown in Figure 10.4. Since the rigidity of the mold edge is very high relative to that of the press, deformation diagrams like the example shown in Figure 10.4 are obtained. There would be hardly any error involved in ignoring the slight slope of the spring characteristic of the press above its drop after the clamping force is exceeded and employing instead a constant holding force of the same magnitude as the clamping force (i.e., as for a hydraulic machine without locking in which a constant pressure of the same magnitude as the clamping force is maintained by the pumps). This means that the **mold-opening forces due to the mold cavity pressure must be smaller than the clamping force.**

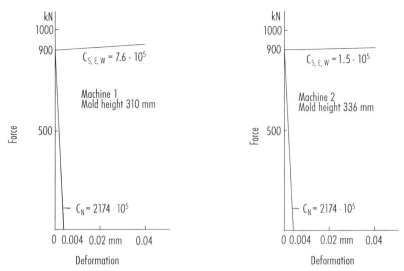

Figure 10.4   Realistic spring diagram for two normal mechanical clamping units with a typical mold [10.3]

10.4   The Superimposition Procedure

A complete mold base is generally composed of different components, which are exposed to different loads. It is useful, therefore, to dissect the mold into characteristic elements and consider their elastic behavior. This results in a simple method for determining the total deformation (Figure 10.5).

10.4.1   Coupled Springs as Equivalent Elements

The superimposition procedure represents a superimposition of individual deformations. All components of a mold base (plates, spacers, supports) are considered springs with a certain rigidity (Figure 10.6).

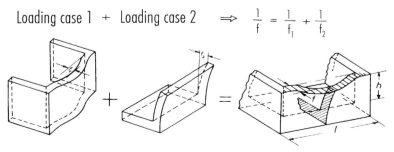

Figure 10.5   Dissection of a mold element [10.1, 10.5]

As already mentioned, bending and shear have to be taken into account in this context. If we look at the mathematical relation which describes the spring behavior

$$F = C \cdot f \rightarrow f = \frac{F}{C}$$

(10.7)

and the reaction of a plate to bending and shear, then the total deformation is

$$f = \underbrace{\frac{p_D \cdot L^4 \cdot 12}{384 \cdot E \cdot s^3}}_{\text{Bending}} + \underbrace{\frac{p_D \cdot L^2 \cdot 2.66}{8 \cdot E \cdot s}}_{\text{Shear}}$$

(10.8)

$$f = p_D \left(\frac{L^4 \cdot 12}{384 \cdot E \cdot s^3} + \frac{L^2 \cdot 2.66}{8 \cdot E \cdot s} \right)$$

(10.9)

In the case of the deflection of a plate, all quantities which depend on the geometry of the plate remain constant. The term of Equation (10.9) in brackets corresponds to the constant C of a spring.

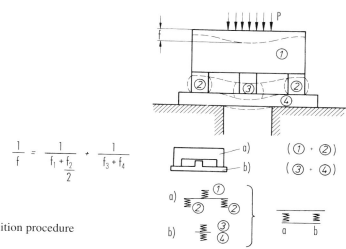

$$\frac{1}{f} = \frac{1}{\dfrac{f_1 + f_2}{2}} + \frac{1}{f_3 + f_4}$$

Figure 10.6   Superimposition procedure [10.1, 10.5]

All cases of loading in any section of a mold have similar correlations. This allows mold elements to be considered springs.

10.4.1.1 Parallel Coupling of Elements

With parallel, coupling, all components exhibit the same deformation under different loads. The total load is allocated to individual loads (Figure 10.7):

$$\frac{1}{f} = \frac{1}{f_1} + \frac{1}{f_2}$$

(10.10)

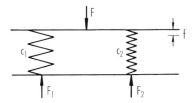

Figure 10.7   Parallel system of elements [10.5]

10.4.1.2 Elements Coupled in Series

All components are deformed by the same loads (Figure 10.8).

$$F = F_1 = F_2 \tag{10.11}$$

Hence, all springs are loaded by the full magnitude of the acting force and not proportionately. The resulting spring travel is

$$f = f_1 + f_2 \tag{10.12}$$

The total deformation is the sum of the individual deformations.

Thus, the possible number of loading cases can be reduced to three basic cases:

1. single load,
2. parallel coupling (Figure 10.7),
3. coupling in series (Figure 10.8).

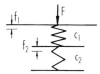

Figure 10.8   System of elements in series [10.5]

Figures 10.5 and 10.6 demonstrate how the total deformation can be determined by combining the basic cases.

10.5   Computation of the Wall Thickness of Cavities and Their Deformation

The configuration of all parts can be reduced to simple shapes. If all possible cavity and core configurations are analyzed with this assumption in mind, we can select the following typical geometries with the goal of obtaining a method for estimating dimensions:

1. circular cavities and cores,
2. cavities and cores with plane faces as boundaries.

If the existing loading cases are analyzed, the causes of deformations can be reduced to a few cases. The basis for this simplified calculation is the dissection of the mold component to be dimensioned into two characteristic equivalent beams as is done with a

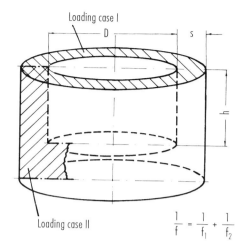

Figure 10.9   Dissection of a cylindrical
mold component [10.1, 10.5]

$$\frac{1}{f} = \frac{1}{f_1} + \frac{1}{f_2}$$

plate with three edges built-in (Figure 10.5), or with a cylindrical cavity with integrated bottom (Figure 10.9).

Diagrams are supplied for various cases of loading based on equations from the theory of elasticity (Figures 10.11 and 10.14 to 10.16). The required wall thickness – if steel is the material of choice – for cavities, cores and plates can immediately be obtained from them if the permissible deformation is taken as a parameter. To play it safe, the deformations from both characteristic cases of loading have to be computed. That wall thickness has to be chosen that results in the smallest deformation.

10.5.1 Presentation of Individual Cases of Loading and the Resulting Deformations

In Figure 10.10 the loading cases are presented schematically. Suitable combinations can be used to calculate the deformation of all occurring configurations and wall thicknesses. Formulae which result from the theory of elasticity were taken as basis for computing the deformation:

$$\sigma = E \cdot \varepsilon \tag{10.13}$$

and

$$\tau = \frac{Q}{A} = G \cdot \gamma \tag{10.14}$$

Because the components of an injection mold are exclusively compact bodies with thick walls, shear deformation besides deflection has to be considered by all means.

For the preparations of the diagrams, steel with a modulus of elasticity of 210 GPa and an internal cavity pressure of 60 MPa were assumed. If a different pressure should be considered, the deformation can easily be recalculated because deformation is linearly related to cavity pressure.

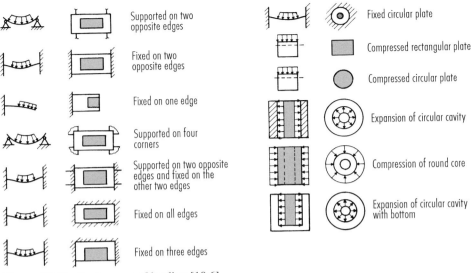

Figure 10.10   Basic cases of loading [10.6]

10.5.2 Computing the Dimensions of Cylindrical Cavities

Krause [10.8] offers a very accurate method for calculating the necessary thickness of the walls of mold cavities (transverse deformation) that is also available as software.

For the usual design in the absence of such software, the following dimensioning suggestions are made. These should be readily understandable and sufficiently accurate. The elastic expansion of a circular cavity can be taken from Figure 10.11 for the loading case I (Figure 10.9), which represents the following equation:

$$\Delta r_N = \frac{p_D \cdot r_{Ni}}{E} = \left[\frac{1 + \dfrac{r_{Ni}^2}{r_{No}^2}}{1 - \dfrac{r_{Ni}^2}{r_{No}^2}} + \frac{1}{m} \right] \qquad (10.15)$$

Δr_N = Expansion of cavity,
r_{Ni} = Inside radius,
r_{No} = Outside radius,
p_D = Injection pressure,
E   = Modulus of elasticity,
m   = Reciprocal of Poisson's ratio.

The elastic expansion of a cavity according to the loading case II (Figure 10.9) is computed from a relation presented in Figure 10.14 [10.9 to 10.11].

$$f = \frac{12 \cdot p_D \cdot h^4}{8 \cdot E \cdot s^3} + \frac{p_D \cdot h^2 \cdot 2.6 \cdot 1.2}{2 \cdot E \cdot s} \qquad (10.16)$$

f = Deflection,
h = Depth of cavity,
s = Wall thickness of cavity.

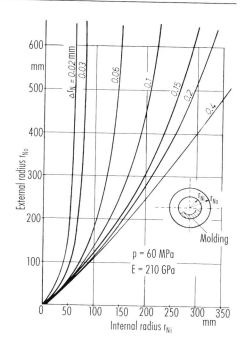

Figure 10.11   Expansion of cylindrical cavities [10.1, 10.5]

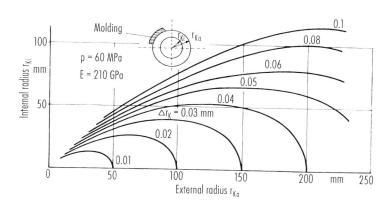

Figure 10.12
Compression of
cylindrical cores
[10.1, 10.5]

The radial compression of a core can be determined with Figure 10.12

$$\Delta r_K = \frac{p_D r_{Co}}{E} = \left[\frac{1 + \dfrac{r_{Ci}^2}{r_{Co}^2}}{1 - \dfrac{r_{Ci}^2}{r_{Co}^2}} + \frac{1}{m} \right] \tag{10.17}$$

Δr_C = Compression of core,
r_{Co} = Outside radius of core,
r_{Ci} = Inside radius of core.

10.5.3 Computing the Dimensions of Non-Circular Cavity Contours

If the boundaries of cavities are plane faces, the following loading cases can be present (Figure 10.13):

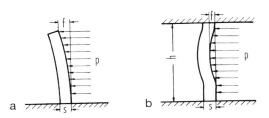

a   b   **Figure 10.13**   Two loading cases [10.11]

$$\text{a)}\quad f = \frac{12 \cdot p \cdot h^4}{8 \cdot E \cdot s^3} + \frac{p \cdot h^2 \cdot 2.66}{2 \cdot E \cdot s} \cdot 1.2 \qquad\qquad \text{(Figure 10.14)}\quad (10.18)$$

$$\text{b)}\quad f = \frac{12 \cdot p \cdot h^4}{384 \cdot E \cdot s^3} + \frac{p \cdot h^2 \cdot 2.66}{8 \cdot E \cdot s} \cdot 1.2 \qquad\qquad \text{(Figure 10.15)}\quad (10.19)$$

The loading case of common injection molds is between case a and case b. Friction acts on the mating surfaces of the clamped mold. This restricts the expansion of the cavity but does not necessarily result in a rigid, inflexible condition.

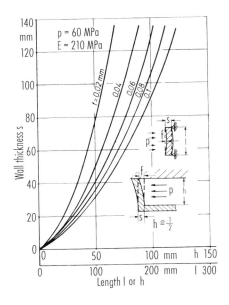

Figure 10.14   Deformation from deflection and shear [10.1, 10.5]

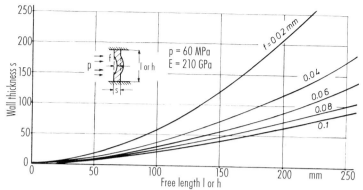

Figure 10.15   Deformation of rectangular plates fixed on two edges from deflection and shear [10.1, 10.5, 10.7]

10.5.4 Computing the Dimensions of Mold Plates

The deflection of the stationary clamping plate is determined by assuming the case of a plate with all four edges built in (Figure 10.16):

$$f = \frac{p_D \cdot D^4 \cdot 12}{1138 \cdot E \cdot s^3} + \frac{p_D \cdot D^2 \cdot 2.66 \cdot 1.2}{16 \cdot E \cdot s} \qquad (10.20)$$

D = Diameter of opening in top clamping plate,
s = Thickness of plate.

The dimensions of a core-retainer plate result from rectangular plates built-in at two opposite edges from Figure 10.16:

$$f = \frac{12 \cdot p_D \cdot h^4}{384 \cdot E \cdot s^3} + \frac{p_D \cdot h^2 \cdot 2.66 \cdot 1.2}{8 \cdot E \cdot s} \qquad (10.21)$$

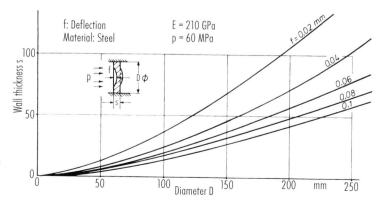

Figure 10.16
Deflection of built-in circular plates [10.5, 10.7]

h = Free length,
s = Thickness of core-retainer plate.

10.6   Procedure for Computing Dimensions of Cavity Walls under Internal Pressure

The following steps are necessary:
1. Computing of expected shrinkage
   Before the wall thickness of the cavity can be established, the expected shrinkage in the cavity has to be determined because it is the basis for the maximum permissible cavity expansion. According to Chapter 9, the theoretical shrinkage can be computed from the p-v-T diagram (Figure 9.4). A very good reference is the information provided by the material supplier. It has to be noted, however, that those data are for longitudinal shrinkage. The shrinkage in thickness may be calculated with the aid of the following equation:

$$S_W \geq 9 \cdot S_L \tag{10.22}$$

Where
S_W = Shrinkage of the wall thickness,
S_L = Shrinkage in length (data provided by manufacturer).

2. Dimensioning the thickness of cavity walls
   The elastic deformation of the cavity has to be smaller than the expected shrinkage. They are based on the maximum permissible deformation of a wall. This requirement is met by procuring the necessary wall thickness for a given deformation from Figures 10.11, 10.12, and 10.14–10.16.

10.7   Deformation of Splits and Slides under Cavity Pressure

(See also design examples in Section 12.9)
   Design formulae have been compiled not only by employees of the Institute for Plastics Processing (IKV) [10.6] but also by Krause et al., along with a software program that allows easy accurate checking or simulation [10.12].

10.7.1 Split Molds

This special design of of injection mold permits large-area undercuts to be demolded and is very common (see also Section 12.9). Figure 10.17 shows that the split mold has two principal parting lines perpendicular to each other. The one primary parting plane A is, as usual, aligned perpendicularly to the direction of closing. The second is located between the splits and is aligned parallel to the direction of closing. With split molds, the clamping unit must take on much greater forces than in a mold with just one parting plane because, as Figure 10.22 shows, in addition to the force from the cavity pressure acting on the clamping parting line there is the vertical component of the forces from the cavity pressure acting on the splits.

The first step in the calculation is to determine the minimum clamping force and then to choose the machine.

It is already necessary at this stage to decide on the type of splits because functionality and flash-free parts require that the splits be closed so tightly and pre-stressed that no gaps can form under the internal pressure. This is only possible if enough air is left during the vertical closing movement – Krause [10.2] calls this a "functional gap" (Figure 10.17, labeled a) – that the splits do not sit proud in the frame even after elastic deformation under the locking force. To carry out these calculations, a spring model of the split mold is made, as shown in Figure 10.21. (To perform a highly accurate calculation, the program created by Krause et al. [10.8] may be used. This user friendly program employs a step-by-step approach and contains the strength calculations for any bolted joints and the consequences of elastic deformation of these bolts.)

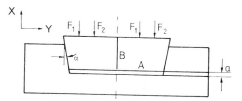

Figure 10.17   Section from a split mold (principle) [10.6]
a = Functional gap

CADMOULD [10.13] which contains a program for designing split molds is somewhat simpler to use but has proved to be of sufficient accuracy. As shown below, these calculations can be understood by users, without their having to use the program itself*.

The goal is to compute the load on a split mold (Figure 10.18).

The distribution of forces flows entirely through the split (element (1)), which is thereby compressed. The counterforce is applied by the two locking mechanisms (2) and (3), which are both subjected to the same deformation. Thus, elements (2) and (3) are connected in parallel. The split is connected in series with these two elements. This is

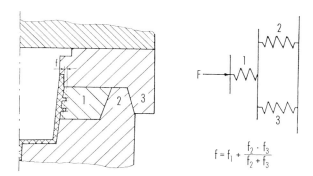

Figure 10.18   Spring diagram for deformation of a split mold [10.6]

$$f = f_1 + \frac{f_2 \cdot f_3}{f_2 + f_3}$$

---

* Extract from "Handbook for Computing Injection Molds", published by KI-Verlag, 1985 [10.6]

shown in the equivalent circuit diagram on the right. Deformations f_1 to f_3 are deformations under the total load, where f_1 is pure compression, and f_2 and f_3 are deflections
of beams or plates clamped on one side.

Aside from having to deal with the general dimensioning problems, the split mold
designer has to cope with problems involving the clamping forces because overpacking
has to be avoided in several parting lines. This will now be explained (Figure 10.17). In
this case, both the dimensions of the gap a and the angle α are important, the latter very
often being pre-set at 18°.

Section 10.2.1 discussed the basic deliberations for this special case; here, instead of
deflection, the rigidities of the various elements are connected in series. These
deliberations help to estimate the complex conditions obtaining in split molds.

To make the following discussion clearer, the total clamping force will be divided into
the components F_1 and F_2.

The clamping force in parting line A (Figure 10.19) is given by:

$$F_{cx} = A_x \, p \qquad\qquad (10.23)$$

where
$F_{cx} = 2 \cdot F_2$
p = Cavity pressure,
F_{cx} = Clamping force in the x-direction,
A_x = Area of the part in the x-direction.

The clamping force in parting line B (Figure 10.20) is given by:

$$F_{cy} = A_y \, p$$
$$F_{cy} = F_1/\tan \alpha \qquad\qquad (10.24)$$

where
F_{cy} = Clamping pressure in the y-direction,
A_y = Projected area of part in the y-direction.

The clamping force F_c computes to

$$F_c = 2 \, (F_1 + F_2) \qquad\qquad (10.25)$$

For the rest of the calculation, it is important to look at the interaction of the elements.
To clarify this, the equivalent circuit diagram for the system is shown in Figure 10.21.

The circled numbers represent the equivalent springs that are loaded in the clamping
direction x. The numbers in the squares represent equivalent springs that are loaded
perpendicularly to the clamping direction y.

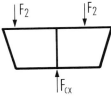

Figure 10.19   Loading on jaws in parting
line A (in clamping direction) [10.6]

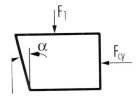

Figure 10.20   Loading on jaws in parting line
B (right angles to clamping direction) [10.6]

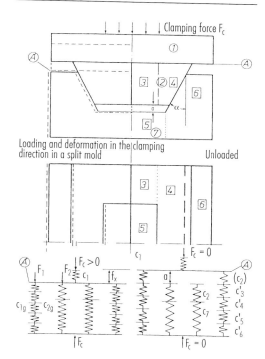

Figure 10.21 Loading and deformation in clamping direction of split mold [10.6]

The stiffness values marked with ' are parameters that have been converted from the transverse direction into the clamping direction.

The following relationships may be derived from Figure 10.21.

$$F_1 = C_{1g} \cdot f_x \tag{10.26}$$

Substituting 10.24, the horizontal component is:

$$F_{cy} = \frac{C_{1g} \cdot f_x}{\tan \alpha} \tag{10.27}$$

$$F_2 = C_{2g} \cdot (f_x - a) \tag{10.28}$$

Substituting 10.23, we get

$$F_{cx} = 2 \cdot C_{2g} \cdot (f_x - a) \tag{10.29}$$

$$\frac{F_{cx}}{F_{cy}} = 2 \tan \alpha \frac{C_{2g} \cdot (f_x - a)}{C_{1g} \cdot f_x} \tag{10.30}$$

From Equations (10.28) and (10.29), it follows:
Substituting Equations (10.25), (10.26), and (10.28), the unknown parameter f_x is given by:

$$\frac{F_c}{2} = F_1 + F_2 = C_{1g} \cdot f_x + C_{2g} \cdot f_x - C_{2g} \cdot a$$

$$\frac{F_c}{2} + C_{2g} \cdot a = f_x \cdot \left(C_{1g} + C_{2g} \right)$$

$$f_x = \frac{F_c + 2 C_{2g} \cdot a}{2 \cdot \left(C_{1g} + C_{2g} \right)} \tag{10.31}$$

$$\frac{F_{cx}}{F_{cy}} = 2 \cdot \tan \alpha \cdot \frac{C_{2g}}{C_{1g}} \cdot \left[1 - \frac{1 + \dfrac{C_{1g}}{C_{2g}}}{1 + \dfrac{F_s}{\left(2a \cdot C_{2g} \right)}} \right] \tag{10.32}$$

We now need to determine the individual stiffness values C_{1g} and C_{2g}.
    Figure 10.22 shows the deformation of the system in the y-direction (f_y) on closing of the mold over the distance f_x.

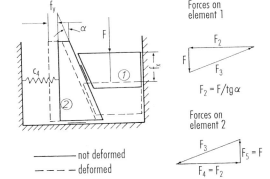

Figure 10.22   Deformation at right angles to clamping direction [10.6]

The spring constant that occurs in the y-direction is C'. This can be normalized by the constant factor $\tan^2 \alpha$ to the x-direction.
    Thus:

$$F = C \cdot \Delta l$$
$$F = C' f x$$
$$f_x = f_y / \tan \alpha$$

$$f_y = F/C = \frac{F}{\tan \alpha \cdot C_4} \tag{10.33}$$

$$f_x = \frac{F}{\tan^2 \alpha \cdot C_4} \Rightarrow F = C_4 \tan^2 \alpha \cdot f_x$$

$$C' = C_4 \tan^2 \alpha$$

This normalized spring constant C' can be used to describe the deformation in a transverse direction.

I) The equivalent springs 3, 4, 5 and 7 are connected in series (Figure 10.21 bottom):

$$\frac{1}{C_{1g}} = \frac{1}{C_3'} + \frac{1}{C_4'} + \frac{1}{C_5'} + \frac{1}{C_6'}$$

(10.34)

For the purposes of simplicity, the following geometry will be assumed (Figure 10.23):

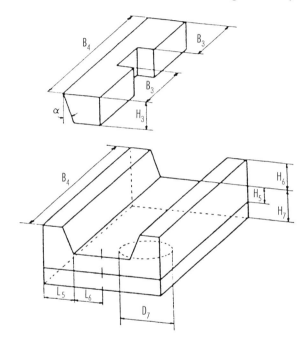

Figure 10.23   Geometry [10.6]

II) The equivalent springs 2 and 7 (Figure 10.21) are also connected in series:

$$\frac{1}{C_{2g}} = \frac{1}{C_2} + \frac{1}{C_7}$$

(10.35)

Once the individual stiffness values have been determined, Equation (10.36) may be used to calculate the gap width a.

$$a = \frac{F_{cy}}{C_{1g}} \cdot \tan\alpha - \frac{F_{cx}}{C_{2g}}$$

(10.36)

Here, the forces F_x and F_y must be calculated using the cavity pressure (pressure x projected area).

10.8   Preparing for the Deformation Calculations

Examples

When the simple computational premises such as those described above are used for dimensioning injection molds, it must be borne in mind from the outset that certain simplifications have to be made with respect to description of geometry and load. Ultimately, the designer must use his engineering knowledge to decide himself on such simplifications. Some examples will serve to illustrate this.

If we consider the case of two bolted mold plates (Figure 10.24), we have to decide whether these two plates are to be regarded as one plate with the sum of the individual thicknesses or if this loading instance has to be defined as a parallel circuit of the individual plates.

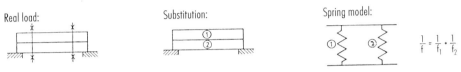

Figure 10.24   Simplified view of loads using bolted plates; connected in parallel [10.6]

This depends most definitely on the number of bolts or shear pins or guide pins, i.e., the question is whether the plates can still be displaced to a certain extent. The final decision can therefore only be reached in practice.

Calculations of the two loading cases can prove of great assistance to the designer when deciding on the design. The results of one example are shown in Figure 10.25.

Case A clearly shows that the deformation in a parallel circuit (two plates loosely on top of each other) is twice as high, so the designer should consider the use of supports.

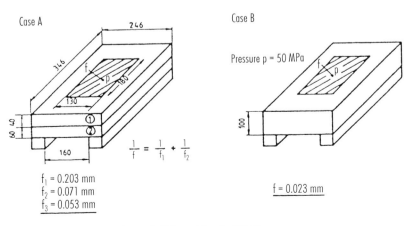

Figure 10.25   Comparison of different loads [10.6]

Table 10.1 Individual rigidity within the system [10.6]

Load	Deformation	Rigidity
	Compression $$\frac{F}{2 \cdot B_3 \cdot H_3} = \frac{\Delta L_3}{L_3} \cdot E$$	$$C_3 = \frac{2 \; B_3 \cdot H_3}{L_3} \cdot E$$ $$C_3' = C_3 \cdot \tan^2\alpha$$
	Compression $$\frac{F}{B_4 \cdot H_3} = \frac{\Delta L_4}{L_4} \cdot E$$	$$C_4 = \frac{B_4 \cdot H_3}{L_4} \cdot E$$ $$C_4' = C_4 \cdot \tan^2\alpha$$
	Extension $$\frac{F}{H_5 B_4} = \frac{\Delta L_5}{L_5} \cdot E$$	$$C_5 = \frac{B_4 \cdot H_5}{L_5} \cdot E$$ $$C_5' = C_5 \cdot \tan^2\alpha$$
	Flexure and thrust $$\Delta L_{6\,tot} = \frac{F}{EB_6}\left(\frac{17}{16} \cdot \left(\frac{H_6}{L_6}\right)^3 + 1.17 \cdot H_6/L_6\right)$$	$$C_6 = \frac{B_6 \cdot L_6}{H_6\left(1.06\left(\frac{H_6}{L_6}\right)^2 + 1.17\right)} \cdot E$$ $$C_6' = C_6 \cdot \tan^2\alpha$$
	Compression $$\frac{F}{2 \cdot L_3 \cdot B_3 + L_4 \cdot B_4} = \frac{\Delta H_3}{H_3} E$$	$$C_2 = \frac{2L_3 \cdot B_3 + L_4 \cdot B_4}{H_3} \cdot E$$
	Flexure and thrust $$\Delta H_7 = \frac{4 \cdot F \cdot 12 \cdot D_7^2}{1138 \cdot \pi \cdot E \cdot H_7^3} + \frac{4 \cdot F \cdot 2.6 \cdot 1.2}{16 \cdot \pi \cdot E \cdot H_7}$$	$$C_7 = \frac{4.028 \cdot H_7}{1 + 0.054 \cdot \left(\frac{D_7}{H_7}\right)^2} \cdot E$$

To summarize the approach, a geometric simplification of the real loading case must be made and then the equivalent spring system for this chosen system has to be decided on.

In many instances, an equivalent loading that is covered by computational possibilities first has to be found for the actual loading.

This can be a problem even in very simple cases, as shown in Figure 10.26 for a plate with a sprue gate.

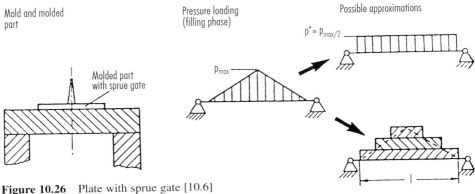

Figure 10.26   Plate with sprue gate [10.6]

The comparison of the two possibilities quickly shows up the limits that deformation values impose.

Since the choice obviously requires a certain amount of experience, a few tips will now be given concerning the simplified approach. Figure 10.27 shows schematic simplifications for frequently recurrent loading cases. The table is by no means complete.

If a definitive determination of the load proves to be difficult, the system should be calculated under minimal and maximal load to establish precise limits.

10.8.1 Geometrical Simplifications [10.15]

Often, mold cavities have really complicated contours. One such example is provided in Figure 10.28.

In this case, too, comparative computation of the deformations using two different equivalent wall thicknesses yields sufficient information to help the designer reach a decision.

The same approach is used in the example shown in Figure 10.29. In this case, the simplification is a plate clamped on one side with equivalent thickness.

A further problem is posed by the weakening of the mold plates brought about by drilled cooling channels; only FEA can take this into account. Again, mold elements with different equivalent wall thicknesses may be used (Figure 10.30).

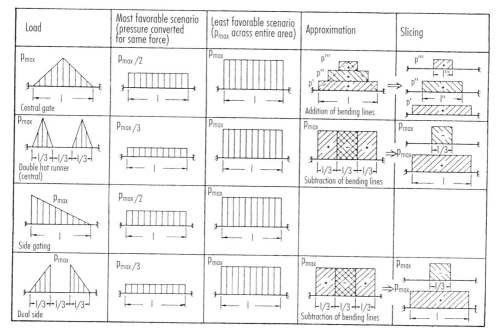

Load	Most favorable scenario (pressure converted for same force)	Least favorable scenario (p_{max} across entire area)	Approximation	Slicing
p_{max} / Central gate	$p_{max}/2$	p_{max}	Addition of bending lines	
p_{max} / $l/3 + l/3 + l/3$ / Double hot runner (central)	$p_{max}/3$	p_{max}	p_{max} / $l/3 + l/3 + l/3$ / Subtraction of bending lines	p_{max} / $l/3$
p_{max} / Side gating	$p_{max}/2$	p_{max}		
p_{max} / $l/3 + l/3 + l/3$ / Dual side	$p_{max}/3$	p_{max}	p_{max} / $l/3 + l/3 + l/3$ / Subtraction of bending lines	p_{max} / $l/3$

Figure 10.27 Compilation of simplified load scenarios [10.6]

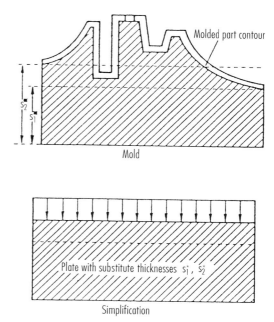

Molded part contour

S_2 S_1

Mold

Plate with substitute thicknesses s_1^*, s_2^*

Simplification

Figure 10.28 Simplification of complicated mold contours [10.6]

Real mold–section

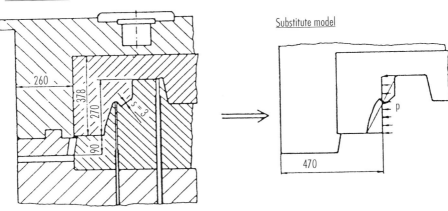

Figure 10.29   Simplified contour of a fender [10.15]

Frame with floor, with and without cooling channels

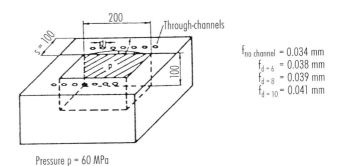

$f_{no\ channel}$ = 0.034 mm
$f_{d\,=\,6}$ = 0.038 mm
$f_{d\,=\,8}$ = 0.039 mm
$f_{d\,=\,10}$ = 0.041 mm

Figure 10.30   Estimate of deformations when cooling channels present [10.15]

Pressure p = 60 MPa

10.8.2 Tips on Choosing Boundary Conditions

Various boundary conditions often have a major effect on the results. The designer obtains valuable information by varying these conditions and performing the necessary computations. Primarily the following conditions can be modified:

a)   clamping conditions,
b)   material characteristics,
c)   pressure,
d)   loading area,
e)   split followers or solid material.

If pressure (load) and geometry have been established for a mold element, a comparison of different clamping conditions reveals that deformations can change by orders of magnitude (Table 10.2).

Table 10.2 Varying the clamping conditions [10.6]

Flexure Bilaterally mounted plate		Flexure Bilaterally clamped plate		Flexure Plate clamped on all sides	
Length	596 mm	Length	596 mm	Length	596 mm
Breadth	446 mm	Breadth	446 mm	Breadth	446 mm
Pressurized length	596 mm	Pressurized length	596 mm	Pressurized length	596 mm
Pressurized breadth	446 mm	Pressurized breadth	446 mm	Pressurized breadth	446 mm
Pressure	25 MPa	Pressure	25 MPa	Pressure	25 MPa
Thickness	116 mm	Thickness	116 mm	Thickness	116 mm
X-coordinate	298 mm	X-coordinate	298 mm	X-coordinate	298 mm
Deflection	1.51051 mm	Deflection	.415842 mm	Deflection	.11835 mm

Practitioners very often over-estimate the influence of hardened tool steel versus unhardened tool steel with regard to deflection and compression behavior. While normal steel has a modulus of elasticity of 210,000 N/mm^2, the modulus of elasticity of hardened steel is 215,000 N/mm^2 or roughly 7% higher. Accordingly, the deformation behavior changes only slightly.

The situation is totally different, however, when, e.g., beryllium-copper is used. The modulus of elasticity decreases to 130,000 N/mm^2. This is a change of approx. 40% and will naturally have a corresponding influence on the deformation behavior.

As the calculations have just shown, the pressure is directly proportional to the deflection, i.e. a doubling of pressure leads to a doubling of deflection.

The influence of the loaded area is not as strong. Table 10.3 shows the different relations for a given geometry.

Table 10.3 Varying the loaded area [10.6]

Flexure Bilaterally mounted plate		Flexure Bilaterally clamped plate		Flexure Plate clamped on all sites	
Length	596 mm	Length	596 mm	Length	596 mm
Breadth	446 mm	Breadth	446 mm	Breadth	446 mm
Pressurized length	596 mm	Pressurized length	446 mm	Pressurized length	298 mm
Pressurized breadth	446 mm	Pressurized breadth	298 mm	Pressurized breadth	223 mm
Pressure	50 MPa	Pressure	50 MPa	Pressure	50 MPa
Thickness	116 mm	Thickness	116 mm	Thickness	116 mm
X-coordinate	298 mm	X-coordinate	298 mm	X-coordinate	298 mm
Y-coordinate	223 mm	Y-coordinate	223 mm	Y-coordinate	223 mm
Deflection		Deflection		Deflection	
	.2367 mm		.159065 mm		9.28907E-02 mm

If the loading area is reduced by a factor of 2, the deflection is roughly 2:1.34; for an area ratio of 4:1, the deformation is reduced by a factor of roughly 4:1.56.

Often the designer is faced with the problem of either not being able to work out the contour of the molded part from the solid material or, for economic or technological reasons, not wanting to do so, and instead he resorts to a split follower. In such cases, he weakens the mold and the geometry of the split follower has to be optimized. Figure 10.31 provides an example of one way of performing the calculation in such a case.

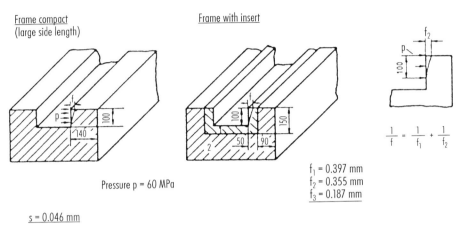

$$f_1 = 0.397 \text{ mm}$$
$$f_2 = 0.355 \text{ mm}$$
$$f_3 = 0.187 \text{ mm}$$

Pressure p = 60 MPa

s = 0.046 mm

Figure 10.31   Mold with and without insert [10.6]

10.9   Sample Calculations

While sample calculations were provided in the previous section, for the sake of clarity, the process will be repeated here for several components.

The following numerical figures can be reproduced with the aid of the equations quoted[*].

Figure 10.18 showed a spring equivalent model of a split mold. We will now calculate the opening for the simplified geometry of the split locking mechanism sketched in Figure 10.32.

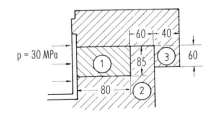

p = 30 MPa

Figure 10.32   Simplified geometry of a split locking mechanism [10.6]

---

[*] The expressions used stem from the use of the program PLASTISOFT-POLI-M, but another program (e.g., CADMOULD) could equally be used.

It will be assumed that the split is 60 mm broad and the mold is 200 mm long.

Compression of element (1)
The pressure of the split face can be used to calculate a compressive loading force.
$F = p \cdot A$,
$F = 30$ MPa $\cdot$ 85 mm $\cdot$ 70 mm,
Input/output (Table 10.4, left).
(2) For element (2), the load is a plate clamped on one side. For simplicity, it will be assumed that the pressure is uniform.
Input/output (Table 10.4, center).
(3) For element (3), the conditions are the same as those for (2).
Input/output (Table 10.4, right).

Table 10.4 Input/output data for calculating the expansion of a split locking mechanism [10.6]

Flexural modulus of elasticity
210000 N/mm²
Poisson's ratio $\mu = 0.3$

Compression		Flexure		Flexure	
Force	178500 N	Unilaterally clamped plate		Unilaterally clamped plate	
Side length A	70 mm	Length	85 mm	Length	60 mm
Side length B	85 mm	Breadth	200 mm	Breadth	200 mm
Height	80 mm	Pressurized length	85 mm	Pressurized length	60 mm
Compression		Pressurized breadth	70 mm	Pressurized breadth	70 mm
	1.14286E-02 mm	Pressure	30 MPa	Pressure	30 MPa
		Thickness	60 mm	Thickness	40 mm
		X-coordinate	85 mm	X-coordinate	60 mm
		Deflection		Deflection	
			2.58864E-02 mm		2.08406E-02 mm

With this information, the total deformation

$$f = f_1 + \frac{f_2 \cdot f_3}{f_2 + f_3}$$

can be determined quantitatively:

$$f = 1.14 \cdot 10^{-2} + \frac{2.5 \cdot 2.08 \cdot 10^{-4}}{(2.5 + 2.08) \cdot 10^{-2}}$$

$f = 0.023$ mm

As a second example, consider the mold half shown in Figure 10.33.
The given geometry of the mold chassis is as follows:
Plate (1)   246 · 190 · 36,
Support (2) = (3) 190 · 38 · 56,
Plate (4)   254 · 190 · 22.

Let the diameter of the aperture beneath plate (4) be: d = 125 mm.

Let the cavity pressure be: p = 50 MPa
and let it act along the entire length between the supports; in the other direction, the loaded length is 100 mm.

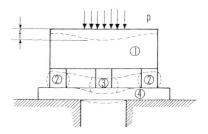

Figure 10.33   Principle structure of mold [10.6]

a) Determination of individual deformations
(1) Plate clamped on two sides.

This type of loading is assumed because, in the closed condition, the plate is clamped under the clamping force (effectively, the load is intermediate between resting on top and clamped).
Input/output (Table 10.5, left).

(2) Compression of the supports.
The total force acting on the mold is
F = p · A,
F = 60 MPa · 170 · 100 mm^2,

Table 10.5   Input/output data for calculating the deformation of the example in Figure 10.35 [10.6]

Flexural modulus of elasticity
210000 N/mm^2
Poisson's ratio μ = 0.3

Flexure		Compression		Flexure	
Bilaterally clamped plate				Circular plate	
Length	170 mm	Force	1.02E+06	Diameter	125 mm
Breadth	190 mm	Side length A	190 mm	Load diameter	125 mm
Pressurized length	170 mm	Side length B	38 mm	Pressure	50 MPa
Pressurized breadth	100 mm	Height	56 mm	Thickness	22 mm
Pressure	50 MPa	Compression		Radius	0 mm
Thickness	36 mm		3.76731E-02 mm	Deflection	.105776 mm
coordinate	85 mm				
Deflection	.103027 mm				

F = 1020 kN,
Input/output (Table 10.5, center).

(3) Compression of the central support.
$f_3 = f_2$,

(4) Deflection of the plate above the aperture,
Input/output (Table 10.5, right).

b) Total deformation
With the aid of the individual deformations, the total deformation can now be determined
(see Superimposition of Mold Elements).

$$\frac{1}{f} = \frac{1}{f_1 + \dfrac{f_2}{2}} + \frac{1}{f_3 + f_4}$$

The value of the deformation is:
f = 0.0659 mm.
This calculation shows that the third support (3) can reduce the deflection of plate (1) by
0.077 mm.

Calculation of the plate thickness for the same deflection leads to a value of approx.
45 mm. The designer can now decide which solution he prefers.

As another example, let us determine the gap width of a split mold.

The objective is to design a split mold for the molded part shown in Figure 10.34. The
pressure on the mold is 60 MPa.

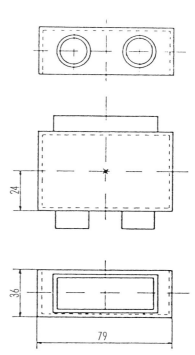

Figure 10.34   Example: molded part box [10.6]

Due to the part geometry, the following geometric parameters may be assumed for a normal split mold (Figure 10.35).
Half width of part: L_3 = 24 mm,
Length of part: B_F = 79 mm,
Height of part: H_3 = 36 mm.

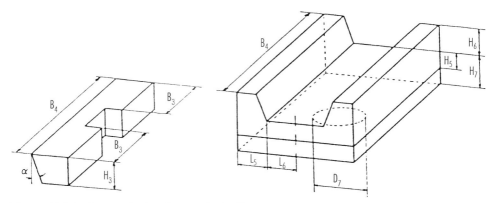

Figure 10.35   Geometric data for standard split molds [10.6]

A search among appropriate standard mold units (e.g. KB 246246/36/2764) yields a mold with the following design dimensions:
L_B = 70 mm,
α = 18°,
B_4 = 245.7 mm,
B_F = $B_4 - 2B_3$ = 79 → B_3 = 83.35 mm,
L_5 = 58.95 mm,
H_5 = 31.3 mm,
Chosen platen or backing plate 27 mm,
L_6 = 64.05 mm,
H_6 = 34.7 mm,
H_7 = H_5 + 27 mm = 58.3 mm,
The aperture for the ejector (D_7) will be ignored in this example.

All geometric parameters have been established and the calculation can now be performed.
The first step is to determine the requisite gap width (Figure 10.36). The input/output data are also shown in Figure 10.36.
The projected areas and the cavity pressure are the basis for the two clamping forces F_X and F_Y, which in turn yield the overall clamping force and thus a gap width of 0.04 mm.
The overall clamping force quoted is the force that the machine has to apply to prevent the mold from opening in two parting lines.
Let us now perform a variant of this calculation to investigate the influence of an oversize L_B. For this, the angle will be held constant.
L_B = 70.01 mm, 70.1 mm, 71 mm.

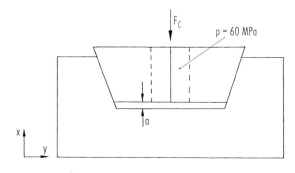

Normal split mold
Geometric mold:

L_B = 70 mm	α = 18 °
B_3 = 83.35 mm	L_3 = 24 mm
H_3 = 36 mm	B_4 = 245.7 mm
L_5 = 58.95 mm	H_5 = 31.3 mm
L_6 = 64.05 mm	H_6 = 34.7 mm
H_7 = 58.3 mm	D_7 = 0 mm

Gap width
Pressure: 60 MPa
Clamping force in x-direction F_X
227.52 kN
Clamping force in y-direction F_Y
170.64 kN
Total clamping force: F_C 338.409 kN
Gap width a: 4.31992E-02 mm

Figure 10.36 Calculation of the gap width of a split mold [10.6]

Input/output (Table 10.6).

Table 10.6 Influence of oversize L_B on the gap width a at constant taper [10.6]

Input/output Modified data:	Modified data:	Modified data:
L_B = 70.01 mm	L_B = 70.1 mm	L_B = 71 mm
α = 18°	α = 18°	α = 18°
Gap width	Gap width	Gap width
Pressure: 60 MPa	Pressure: 60 MPa	Pressure: 60 MPa
Clamping force in x-direction: F_X	Clamping force in x-direction: F_X	Clamping force in x-direction: F_X
227.52 kN	227.52 kN	227.52 kN
Clamping force in y-direction: F_Y	Clamping force in y-direction: F_Y	Clamping force in y-direction: F_Y
170.64 kN	170.64 kN	170.64 kN
Total clamping force: F_C	Total clamping force: F_C	Total clamping force: F_C
338.409 kN	338.409 kN	338.409 kN
Gap width a:	Gap width a:	Gap width a:
4.32025E-02 mm	4.32325E-02 mm	4.35309E-02 mm

The calculation clearly illustrates the following:
– The overall clamping force remains constant since α remained constant.
– The rigidity of the split becomes smaller and so a larger gap has to be observed to maintain the sealing force in the parting line x. The changes are negligibly small, however.

Varying the bevel angle yields the following results.

Input/output (Table 10.7).

Table 10.7   Influence of oversize L_B on the gap width a at varying taper [10.6]

Input/output		
Modified data:	Modified data:	Modified data:
L_B = 70.01 mm	L_B = 70.1 mm	L_B = 71 mm
α = 18°	α = 18°	α = 18°
Gap width	Gap width	Gap width
Pressure: 60 MPa	Pressure: 60 MPa	Pressure: 60 MPa
Clamping force in	Clamping force in	Clamping force in
x-direction: F_X	x-direction: F_X	x-direction: F_X
227.52 kN	227.52 kN	227.52 kN
Clamping force in	Clamping force in	Clamping force in
y-direction: F_Y	y-direction: F_Y	y-direction: F_Y
170.64 kN	170.64 kN	170.64 kN
Total clamping force: F_C	Total clamping force: F_C	Total clamping force: F_C
318.966 kN	338.409 kN	358.525 kN
Gap width a:	Gap width a:	Gap width a:
5.32458E-02 mm	4.31992E-02 mm	3.59815E-02 mm

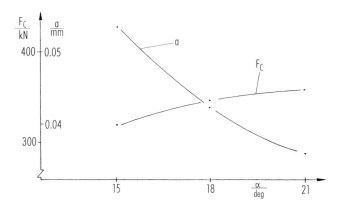

Figure 10.37   Total clamping force and gap width as a function of taper [10.6]

A plot of the results (Figure 10.37) reveals, as expected, that as the angle increases the more it becomes necessary to increase the overall clamping force in order to avoid opening in the y-clamping face. On the other hand, the gap width decreases.

Since inaccuracies in production may yield other gap widths, the real conditions change. This is simulated below by assuming that the real gap width is 0.02 mm.

The mold is to be used on a 4000 kN injection molding machine.

The question now arises as to whether the mold opens during the pressure loading in these conditions.

Input/output (Table 10.8).

Table 10.8 Clamping force calculation for 0.02 mm gap width and clamping force of 400 kN [10.6]

Normal split mold	Clamping forces
Geometric data:	F_C: 400 kN
L_B = 70 mm	Gap width: .02 mm
α = 18°	Clamping force in
B_3 = 83.35 mm	x-direction: F_X
L_3 = 24 mm	344.311 kN
H_3 = 36 mm	Clamping force in
B_4 = 245.7 mm	y-direction: F_Y
L_5 = 58.95 mm	85.696 kN
H_5 = 31.3 mm	
L_6 = 64.05 mm	
H_6 = 34.7 mm	
H_7 = 58.3 mm	
D_7 = 0 mm	

The clamping force components in the x- and y-directions may be converted into a maximum pressure via the projected areas.

$$P_{Xmax} = \frac{F_x}{A_x}$$

$$P_{Ymax} = \frac{F_y}{A_y}$$

$$A_X = B_F \cdot 2L_3 = 3792 \text{ mm}^2$$
$$A_Y = B_F \cdot H_3 = 2844 \text{ mm}^2$$

$$P_{Xmax} = \frac{344.311 \cdot 10^2}{3792 \cdot 10^{-6}} \text{ MPa} = 90.799 \text{ MPa} \quad P_{Ymax} = \frac{85.696 \cdot 10^2}{2844 \cdot 10^{-6}} \text{ MPa} = 30.132 \text{ MPa}$$

$$P_{Xmax} > 60 \text{ MPa} \qquad\qquad P_{Ymax} < 60 \text{ MPa}$$

The result shows that there is insufficient clamping force in the clamping face Y and that overpacking will occur there. The machine setter will therefore choose another machine with a higher clamping force (75 t).

Input/output (Table 10.9):

The results may be used to calculate the maximum permissible pressures again.
P_{Xmax} = 180.189 MPa,
P_{Ymax} = 36.103 MPa.

It turns out that, even when the overall clamping force is almost doubled, transfer of the clamping force in the y-direction is not appreciably improved when the gap is too narrow. The only way to avoid the problem is to increase the gap.

Clamping forces	**Table 10.9** Clamping
F$_C$: 750 kN	force calculation for
Gap width:: .02 mm	0.02 mm gap width and
	clamping force of
Clamping force in	750 kN [10.6]
x-direction: F$_X$	
683.277 kN	
Clamping force in	
y-direction: F$_Y$	
102.676 kN	

Postmachining of the mold yields a gap width of 0.095 mm. The calculation should now reveal what the conditions for the 40 t machine are.

Input/output (Table 10.10).

Again, checking the maximum applicable sustainable pressure makes the situation clearer.

P_{Xmax} = 48.199 MPa,
P_{Ymax} = 117.540 MPa.

In this case, in which the gap is too large, the clamping force in the x-parting line is not high enough to keep the mold closed.
The remedy is to switch to a larger machine (75 t).
Input/output (Table 10.11).

P_{Xmax} = 137.588 MPa,
P_{Ymax} = 123.512 MPa.

By way of summary, a larger machine does not always help in practice to avoid flash. In any event, the flash must be located and then a decision taken as to whether the gap width needs to be changed or whether increasing the machine clamping force will prove helpful.

Clamping forces	**Table 10.10** Clamping force calculation for
F$_C$: 400 kN	0.095 mm gap width and clamping force of
Gap width: .095 mm	400 kN [10.6]
Clamping force in	
x-direction: F$_X$	
182.769 kN	
Clamping force in	
y-direction: F$_Y$	
334.284 kN	

Table 10.11   Clamping force calculation for 0.095 mm gap width and clamping force of 750 kN [10.6]

Clamping forces
F_C: 750 kN
Gap width: .095 mm
Clamping force x-direction: F_X 521.734 kN
Clamping force y-direction: F_Y 351.265 kN

10.10   Other Loads

All deliberations so far, in addition to dimensional changes of the plastics material (shrinkage), have taken into account deformation of the mold resulting from loads caused by injection and holding pressure. Production-related as well as thermal deformation and the effects of molding-machine operation have been ignored. These will now be discussed briefly below.

10.10.1   Estimating Additional Loading

1. Effects Arising from Mold Making:
   These effects result from inevitable machining tolerances and their impact on assembly of individual components for the completion of a mold. If several plates cannot be ground at the same time on a coordinate grinder, then important dimensions can deviate by up to 0.02 mm per 100 mm. Deviations from parallelism are of the same magnitude. With an increase in the number of times a workpiece has to be clamped and adjusted the accuracy naturally decreases.

   Besides this, not all distortions (like those resulting from hardening, removing large volumes of material or shrink fitting) can be corrected by additional machining. Restrictions on movements, shifting of cores, and others are the consequences. Examination is suggested with conventional shop methods such as making a contact image of fits as well as taking the measurements of a cavity with a cast using a non-shrinking Bi-Sn alloy [10.16]. Measuring deformations calls for a consideration of the machining operation during mold making. Only then can the data be associated with the loads.
2. Thermal Effects During Operation of the Mold
   These result from temperature differences between mating components and cause stresses from the difference in thermal expansions. Here:

$$\delta = \frac{\Delta L}{L_0} = (\alpha_1 - \alpha_2) \cdot T \tag{10.37}$$

or if there are differences in temperature,

$$\delta = \frac{\Delta L}{L_0} = \alpha_1 \cdot T_1 - \alpha_2 \cdot T_2 \tag{10.38}$$

If this deformation is restricted, stresses are generated, which can be calculated from the equilibrium of forces.

A typical example is the manifold of a hot-runner mold, for which the dislocation has to be computed at least for the position of the gates.

3. Effects Arising from Molding Machines:
   Since injection molds are usually compact bodies, machine effects can usually be omitted from the computation. It is difficult to measure them anyway because they vary from machine to machine.

References

[10.l]   Schürmann, E.: Abschätzmethoden für die Auslegung von Spritzgießwerkzeugen. Dissertation, Tech. Univ., Aachen, 1979.

[10.2]   Kretzschmar, O.: Rechnergestützte Auslegung von Spritzgießwerkzeugen mit Segment-bezogenen Berechnungsverfahren. Dissertation, RWTH, Aachen, 1985.

[10.3]   Krause, H.; Starke, B.: Gratbildung durch Plattenverformung rechnerisch ermitteln. Plastverarbeiter, 43, 10, pp. 133–141 and 11, pp. 90–94.

[10.4]   Krause, H.: Modellbetrachtungen zur Werkzeugatmung beim Spritzgießen. Plastverarbeiter, 44, 8, pp. 58–64 and 9, pp. 82–90.

[10.5]   Döring, E.; Schürmann, E.: Thermische und mechanische Auslegung von Spritzgießwerkzeugen. Internal report, IKV, Aachen, 1979.

[10.6]   Menges, G.; Hoven-Nievelstein, W. B.; Kretzschmar, O.; Schmidt, Th. W.: Handbuch zur Berechnung von Spritzgießwerkzeugen. Verlag Kunststoff-Information, Bad Homburg, 1985.

[10.7]   Zawistowski, M.; Frenkler, D.: Konstrukcja form wtryskowych do tworzyw termoplastycznych (Design of injection molds for thermoplastics). Wydawnictwo Naukowo-Techniczne, Warszawa, 1984.

[10.8]   Krause, H.: Programm zur Berechnung der Querverformung. Plastverarbeiter 42 (1991), 3, pp. 126–132 and 4, pp. 103–107.

[10.9]   Barp; Freimann: Kreisförmige Platten. Escher Wyss AG, Zürich.

[10.10]   Timoschenko, S.: Strength of Materials, 1 and 11. Van Nostrand, London, 1955/56.

[10.11]   Bangert, H.; Mohren, P.; Schürmann, E.; Wübken, G.: Konstruktionshilfen für den Werkzeugbau. Lecture at the 8. Kunststofftechnisches Kolloquium, IKV Aachen; Industrie Anzeiger, 98 (1976), pp. 678–681 and pp. 706–710.

[10.12]   Arndt, St.; Krause, H.: Spritzgieß-Backenwerkzeuge. Berechnungsprogramin zum Konstruieren und Betreiben. Plastverarbeiter, 41 (1990), 11, p. 144ff and 12, p. 52f.

[10.13]   Rechenprogramm, Kunststofftechnische Software GmbH., Kaiserstr. 100, 52134 Herzogenrath, Phone: 02407 5088.

[10.14]   Henkel, W.: Untersuchung der Steifigkeit verschiedener Backenverriegelungen. IKV, 1982.

[10.15]   Bangert, H.: Mechanische Auslegung für komplizierte Anwendungsfälle. VDI-IKV-Seminar "Rechnerunterstütztes Konstruieren von Spritzgießwerkzeugen", Münster, 1984.

[10.16]   Dick, H.: Übersicht über einfache Verfahren zur Herstellung von Prototypenwerkzeugen für die Kunststoffverarbeitung. Unpublished report, IKV, Aachen, 1976.

11 Shifting of Cores

11.1 Estimating the Maximum Shifting of a Core

A problem in the mechanical design of molds is the determination of core shifting in parts which are like cups or sieves or more complex parts containing such configurations.

An eccentric mounting of a core due to inaccuracy during production or an asymmetric gating causes a lateral loading of the core. The resulting deformation of the center axis leads to a shift at the tip of the core. Dimensional accuracy of the molding and the demolding process are adversely affected.

Mounting of cores (on one or two sides) and kind and location of the gate have, among others, a significant influence on the mold concept. This should make it clear that knowledge of a core shift may assume great importance already in the development stage when the mold concept is being determined. The design engineer has options available to estimate this core shift. This will be discussed in the following.

The computer module "core shift" can be called during the phase of quantitative mold design to clarify the question of the size of the core mount or the permissible machining tolerance. Examples of computer programs for this are to be found in [11.1–11.4 and 11.14]. For simple geometries, an analytical approach involving the use of formulas from the fields of statics and the strengths of materials is also possible [11.3].

The shifting of a core in practice is composed of

– a core shift at the mount,
– a relative deformation of the center axis of the core proper.

It is also assumed that both mold halves are properly aligned. Both deformations can be superimposed as shown with Figure 11.1.

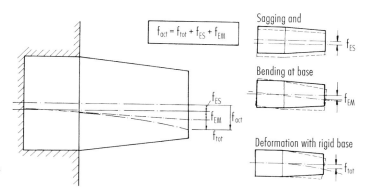

Figure 11.1 Super-position of deformations of a core [11.5]

Part of the shifting of a core generated during mold filling can recover after the filling process during the cooling stage. The magnitude of this recovery depends on processing parameters, material characteristics and geometrical conditions. One can only expect a minor recovery with thin sections, slender cores, low injection speeds (quick freezing) and materials with a high modulus of elasticity. Therefore the maximum shift of cores during injection is considered in the following.

For a variety of core configurations (circular and rectangular, with or without cooling line) and cases of loading (kind and location of gates) computing modules have been generated [11.5–11.7], which are employed in Software such as CADMOULD. Some of these cases will be presented.

Theoretically computed data for core shifting or limits for the height-to-diameter ratio of cores are in good agreement with results from practical experience. Another verification of theoretical data could be indirectly made by comparing the description of the flow pattern and the pressure consumption with actual measurements [11.8, 11.9].

By way of example, core shifting of circular cores with lateral pinpoint gate at the base (rigid mount) will be calculated below.

11.2   Shifting of Circular Cores with Lateral Pinpoint Gate at the Base (Rigid Mount)

The computation is based on a number of assumptions [11.6]:
1. The pressure profile declines linearly along the length H_C of the core (Figure 11.3).
2. A slightly tapered shape of the core is substituted by a cylindrical one.
3. A core cooling is substituted by a through hole. The stabilizing effect of the section without hole is neglected.
4. In cores with an inserted cooling line, the insert is not considered load-supporting.
5. The weight of the core is not taken into account.
6. Mounting of the core is assumed rigid for the time being.
7. A building-up effect remains neglected at present. The effect of the core shift on the filling pattern during injection is not considered. Comments on assumptions 1 and 7 are given in [11.6, 11.7].

The basic procedure for determining the core shift is illustrated with Figure 11.2.

– The core shift $f_{v,10}$ at the tip of the core is computed for a reference pressure $p^* = 10$ MPa at the gate and a linear pressure profile along the core.
– To consider the supporting effect on the core by the melt from the base up, the shape factor K_1 is introduced. This coefficient includes the various height-to-diameter ratios.
– Establishing the pressure factor K_2 considers the real pressure for filling the mold.
– Computing the whole deflection with

$$f_v = K_1 \cdot K_2 \cdot f_{v,10} \qquad (11.1)$$

Based on this procedure, computer programs were assembled, which provided the core-shift data presented in the following pictures [11.6].

Figures 11.3 and 11.4 demonstrate the core shift for polystyrene dependent on the core diameter and various core lengths. If the diameter remains below certain values, the shifting increases rather rapidly (refer to design limits in Section 11.4).

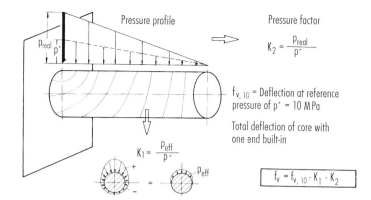

Figure 11.2
Computation of
core shifting with
rigid base [11.5]

With increasing core length and the same core diameter shifting of the core increases as
well. In any case, the core shift decreases with increasing wall thickness as is shown with
the crosshatched area between the curves for 1 and 2 mm wall thickness.

Comparing Figure 11.3 with Figure 11.4 demonstrates the effect of a hollow core of
cooling line ($D_{Ci} = 0.6\ D^*_C$). Similar diagrams for other materials and processing
parameters can be obtained with such a program.

In [11.7] a first theoretical computation of the effect of the build-up phenomenon on
the core shift was carried out. According to the results and in contrast to all expectations,

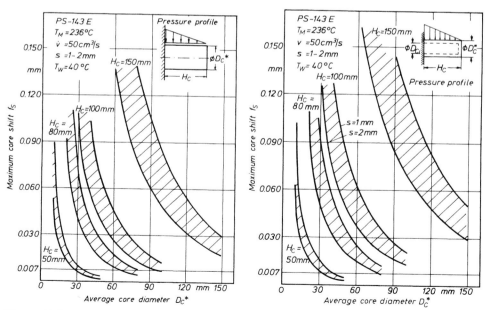

Figure 11.3   Maximum shifting of a core
with full circular cross-section [11.5]

Figure 11.4   Maximum shifting of a
core with full annular cross-section
($D_{Ci} = 0.6\ D^*_C$) [11.5]

it does not add to the shifting of cores. The core-shifting data are even somewhat smaller than without regarding this effect. A first explanation is given by two kinds of opposing effects from the core shift during injection and its consequences on the rheological conditions. First the melt flows into the nonuniform cavity. From this, a higher load on one side of the core can be expected, but then the melt can flow more easily into the enlarged gap. This means less total pressure demand and relief for the core. Both effects almost cancel one another out. The effect of a build-up cannot be measured by itself because it is present at all measurements.

The theoretical computation of a core shift with a build-up taken into account for the case of a laterally gated core was carried out – similar to Section 11.3.1 (Figure 11.6) – by a stepwise computation of the filling process.

For estimating the maximum core shift in design the presented results (without build-up effect) are sufficient, especially since these results are even somewhat higher.

In [11.5], a calculation of core shifting for further core geometries, mounting and loading types is presented. These are:

– Core shifting of circular cores with lateral pinpoint gate at the core tip (rigid mount).
– Core shifting of rectangular with lateral pinpoint gate at the base (rigid mount).
– Shifting of cores related to mounting (exemplified by a circular core laterally gated at the base).

11.3 Shifting of Circular Cores with Disk Gates (Rigid Mount)

In [11.7] the shifting of circular cores without cooling line but with build-up effect is computed for

– cores with disk gate,
– cores with lateral pinpoint gate at the base,
– cores with lateral pinpoint gate at the tip.

Now the case "core with disk gate" will be discussed (Figure 11.5). Compared with previously treated cases, estimating the core shift calls for a different procedure here. The reason for this is an eccentric core mount linked to the machining process. This causes the initial core shift. It is assumed, of course, that temperature, pressure, and velocity (viscosity) are the same in the whole gate region.

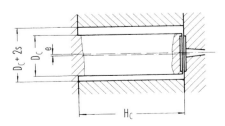

Figure 11.5 Core with disk gate [11.5]

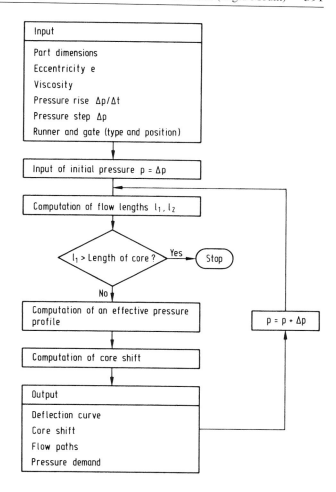

Figure 11.6   Flow chart of an algorithm for computing a core shift (with building-up effect) [11.5]

11.3.1 Basic Examination of the Problem

In all cases previously discussed, the maximum core shift shortly before the end of the filling process was discussed, at the time when the advancing flow front had reached the full length of the core. Besides this, the core shift occurring during mold filling will also be computed. This can be done by dividing the filling process into small steps and computing the occurring deformation after each step. Thus, the whole problem to be resolved consists of

– rheological problem (establishing a flow pattern) and
– mechanical problem (computation of deformation).

The way of computing the core shift including the build-up effect is presented with the flow chart in Figure 11.6. After the data of part and material, the pressure rise $\Delta p/\Delta t$ at the gate and the pressure intervals Δp are entered. By introducing pressure intervals the whole filling process is divided into single steps. As a first step the flow pattern is computed for the pressure p = Δp by determining two characteristic lengths l_1 and l_2 (Figure 11.7). After this the effective pressure profile (Figures 11.8 and 11.9) and the core shift are calculated. Subsequently, the pressure is raised by one step and the computation continued. By now, the increased wall thickness (from the core shift) has to be taken into account. If the flow front has reached the full length of the core, the computation is terminated [11.7].

11.3.2 Results of the Calculations

Based on the presented procedure for a resolution of the problem, sub-programs have been developed, which can be employed within the CADMOULD system to compute the core shift.

The length of the core has a considerable effect on the core shift (Figure 11.10). With increasing length the shifting is enhanced, first slowly but then, with even longer cores, more rapidly. The rapid increase in core shift commences at a length-to-diameter ratio of the core of about

$$\frac{H_C}{D_C} \approx 5 \tag{11.2}$$

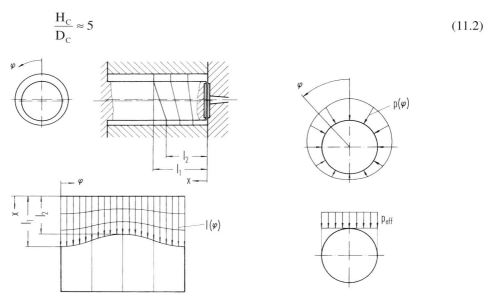

Figure 11.7   Molding with instantaneous flow pattern (projected onto a plane) [11.5]

Figure 11.8   True pressure (top) and effective pressure (bottom) in cross-section [11.5]

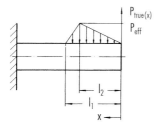

Figure 11.9   Effective instantaneous pressure profile [11.5]

This ratio is largely independent from any eccentricity because the latter affects core shifting linearly. The limit ratio also retains its validity with different wall thicknesses. Even with adhering to this ratio, the resulting core shift may already be too much. Similar limits are also published in [11.10].

Thin sections in a part promote an asymmetrical flow into the cavity and produce larger deformations (Figure 11.11).

If, however, the mentioned length-to-diameter ratio is observed, wall thickness has little effect.

The influence of eccentricity can be recognized in Figure 11.12. In a normal case this eccentricity is the deviation of the center axis of the core from the center axis of the cavity caused by inaccurate machining.

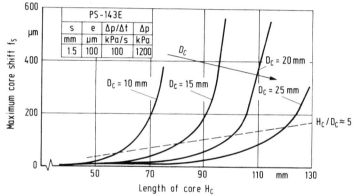

Figure 11.10 Maximum shifting of cores with varying diameters depending on core length (disk gate) [11.5]

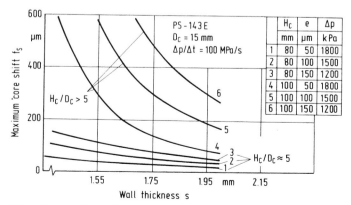

Figure 11.11 Maximum shifting of cores depending on wall thickness (disk gate) [11.5]

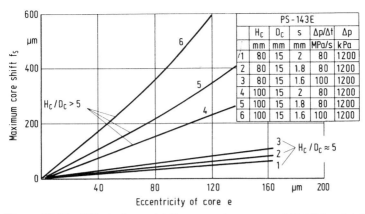

Figure 11.12   Effect of eccentricity on maximum core shift (disk gate) [11.5]

Shifting of the core increases about linearly with increasing eccentricity. Therefore, known deformation and eccentricity in one case permits the deduction of a deformation from the eccentricity in another case:

$$f(e_1) = f(e_2) \cdot \frac{e_1}{e_2}$$ (11.3)

If the maximal permissible eccentricity is known, the tolerance limit for mounting the – in an ideal case centric – core is also established.

11.4 Shifting of Cores with Various Types of Gating (Rigid Mount)

In [11.7] design limitations are presented for cores gated in different ways. If they are exceeded, the maximum core shift increases disproportionately.

– Core with disk gate $H_C/D_C \approx 5$ (11.4)

– Pinpoint gate lateral at the core tip $H_C/D_C \approx 1.5$ (11.5)

– Pinpoint gate lateral at the core base $H_C/D_C \approx 2.5$ (11.6)

As expected, the disk gate is the most favorable solution with respect to core shifting. If the eccentricity is kept very small, extremely small absolute values of core shifting can be obtained. Then one can possibly accept greater length-to-diameter ratios of cores. In [11.11] a ratio of 5 to 15 is mentioned for precision injection molding and of 1 to 5 for technical moldings of average difficulty.

In Figure 11.13 the deformations are shown for three kinds of gating as they have been computed [11.7] under otherwise almost identical conditions. The smallest results are obtained with a disk gate even with an assumed eccentricity of e = 100 µm. The largest deformation is brought on by a pinpoint gate at the core tip. The result of the pinpoint gate at the core base can be placed between the other two types of gating and is frequently the best solution for multi-cavity molds, too. Besides the already mentioned

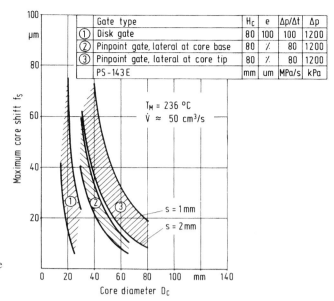

Figure 11.13 Maximum core shift with different gate types [11.5]

methods of gating there are some more actions which can be taken to reduce excessive core shifting such as

– two-sided instead of one-sided core support,
– deeper core mounting,
– reduction of eccentricity with disk gate,
– modification of injection parameters,
– reduction or sacrifice of cooling line.

The order of the measures roughly matches their efficacy.

11.5 Shifting of Inserts

The approach presented in the previous chapters (coupling of rheological and mechanical mold design) for calculating core shifting can also be used to calculate the deformations of insert parts due to the pressure distribution that exists in the mold. This is of major importance, for example, in molds for rubber-metal or rubber-plastic parts.

11.5.1 Analytical Calculation of Deformation of Metal Inserts Using a Cylindrical Roll Shell as an Example [11.3]

These parts, which are used as transport rolls, are made by injection molding in which rubber is molded around a cylindrical insert core. The part geometries in question have a variable length/diameter ratio and are gated via film gates of various widths (Figure 11.14) (Gate position shown in Figure 11.15).

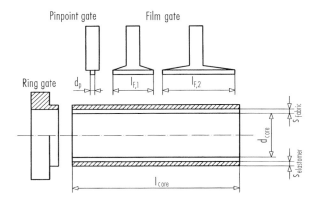

Figure 11.14   Dimensions and gating arrangement for molded parts [11.14]

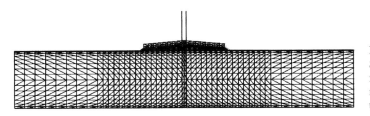

Figure 11.15
Structured geometry of molded part 1c with network refinement in the gate area [11.3]

As already explained in the previous section, a rheological mold design is performed first, to determine the filling pressure distribution in the cavity. The boundary conditions of the simulation calculation arise essentially through the given material-specific coefficients of the material models and the processing parameters of the injection molding process.

As an example, Figure 11.15 shows the finite element net used for a part geometry and Figure 11.16 shows the corresponding computed filling pressure distribution from a simulation. The pressure distribution of a number of part geometries was calculated at different filling levels.

Then, the deformation calculation was performed analytically for this part with the aid of the formulae provided by statics and the strengths of materials.

11.5.1.1 Evaluation of the Deflection Line for Different Part Geometries

Figure 11.17 shows the maximum mold insert deflection against filling level for the sample calculation. It can be seen that the deflection of the mold core rises until the filling level is roughly 50%. At around 50%, the flow fronts converge in the area opposite the gate. As the mold continues to fill up, the pressure difference between the core upper surface or gate side and the core lower surface decreases, as a result of which the deflection of the mold core decreases towards the end of filling. It can generally be said that the narrower the film gate, the higher is the deflection on account of the higher pressure losses.

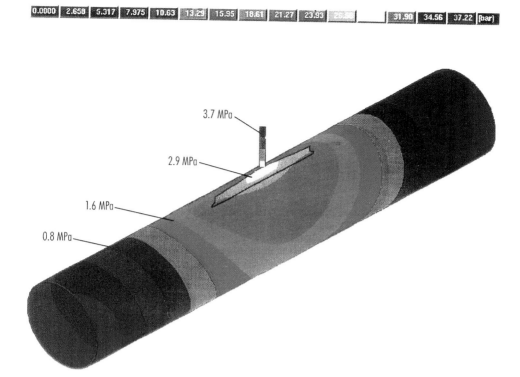

| 0.0000 | 2.658 | 5.317 | 7.975 | 10.63 | 13.29 | 15.95 | 18.61 | 21.27 | 23.93 | 26.58 | | 31.90 | 34.56 | 37.22 | [bar] |

3.7 MPa

2.9 MPa

1.6 MPa

0.8 MPa

Figure 11.16   Representation of filling pressure requirement at end of filling [11.3]

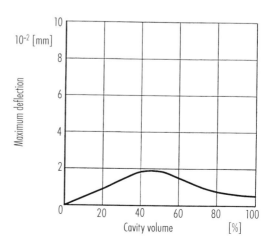

Figure 11.17   Maximum deflection of the mold core as a function of filling volume [11.3]

The qualitative change in deflection of the mold geometries at different diameter/length ratios is shown in Figure 11.18. From the function curves, it can be seen that the deflection of the mold cores increases until that filling level is reached which corresponds with the converging of the flow fronts in the area opposite the gate. After filling of the cavity just after convergence of the flow fronts is complete (diameter/length ratio of approx. 0.5), the maximum deflection is attained at the end of filling.

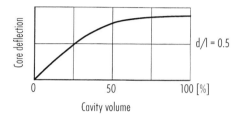 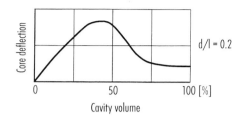

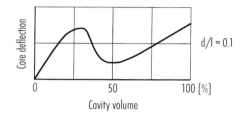

Figure 11.18   Change in deflection for different geometries [11.3]

In the case of a diameter/length ratio d/l of approx. 0.2, once the flow fronts have converged, further filling of the cavity is accompanied by a reduction in the pressure difference between part upper face or gate side and part lower face, as a result of which deflection of the mold core decreases again towards the end of filling.

Molded part geometries with a d/l ratio less than 0.1 pass through a deflection minimum and then undergo a renewed rise in deflection that may be attributed to other components of the pressure difference between the sprue side and the mold clamping half. The temporary maximum deflection on convergence of the flow fronts can be exceeded at the end of filling.

These studies show the deflection that may be expected for a given pressure distribution and the influence that the part geometry has on the maximum deflection and the course of deflection.

However, due to multiple integration of the pressure function, this analytical approach can only be applied to parts whose geometry comprises simple basic bodies (cylinders, plates) and whose pressure load can be described by simple mathematics.

The use of an FE analysis program is advisable for more complex geometries or loading.

11.6 Design Examples for Core Mounting and Alignment of Deep Cavities

The significance of proper core mounting is evident from the previous sections. It is obvious that the best conditions for a core to be positively secured against the cavity is achieved by making the core and retainer plate in one piece. This can only be done by machining an appropriately large piece of stock, and material losses are accordingly high. For large molds it cannot be done in any other way. For smaller molds, cores and retainer plates are usually made separately. Then the core has to be carefully and immovably mounted to the retainer plate. Simple circular cores are machined with a collar, which provides alignment. A flange over this collar secures the core in the retainer plate (Figure 11.19). Figure 11.20 depicts a mold with a divided core to form ribs and facilitate machining. This core is attached to the retainer plate by wedges. Among all possible options for mounting cores, this method approaches best the desired unity of core and retainer plate.

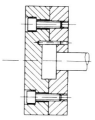

Figure 11.19   Core mounting with compressed flange in retainer plate

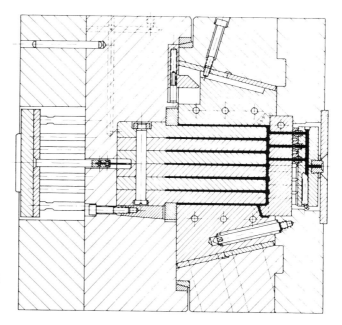

Figure 11.20   Multisectional core mounted with wedges [11.12]

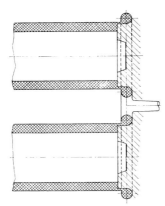

Figure 11.21   Ring gate permits support of core at both ends [11.13]

Even with the best possible core mounting, long and slender cores still require support at both ends (Figure 11.21).

References

[11.1]   Guth, W.; Schenk, R.; Schroiff, V.: Deformation von Einlegeteilen beim Umspritzen simulieren. Kunststoffe, 84 (1994), 3, pp. 244–248.

[11.2]   Knaup, J.: Werkzeugverformung beim Füllvorgang vorausberechnen. CAD – CAM – CIM (1994), 5, pp. 64–68.

[11.3]   Michaeli, W.; Ehrig, E.: Berechnung der Verformung von Metalleinlegeteilen in Spritzgießwerkzeugen aufgrund der in der Kavität vorliegenden Fülldruckverteilung Innovative Technologien. In: Elastomerverarbeitung, VDI-Z, Integrierte Produktion, VDI-Verlag, Düsseldorf, 1996.

[11.4]   Pesavento, M.: Analyse des Herstellungsprozesses eines Thermoplast/Elastomer-Verbundbauteils zur Verbesserung der Formteilqualität. Unpublished report, IKV, Aachen, 1995.

[11.5]   Bangert, H.: Systematische Konstruktion von Spritzgießwerkzeugen und Rechnereinsatz. Dissertation, RWTH, Aachen, 1981.

[11.6]   Schreuder, S.: Rechnerunterstützte Konstruktion von Spritzgießwerkzeugen (Kernversatz). Unpublished report, IKV, Aachen, 1987.

[11.7]   Behrenbeck, U. P.: Erstellung eines Rechenprogramms zur Ermittlung des Kernversatzes während des Einspritzvorganges unter Berücksichtigung des Aufschaukeleffektes. Unpublished report, IKV, Aachen, 1980.

[11.8]   Schmidt, L.: Auslegung von Spritzgießwerkzeugen unter fließtechnischen Gesichtspunkten. Dissertation, RWTH, Aachen, 1981.

[11.9]   Menges, G.; Lichius, U.; Bangert, H.: Eine einfache Methode zur Vorausbestimmung des Fließfrontverlaufes beim Spritzgießen von Thermoplasten. Plastverarbeiter, 31 (1980), 11, pp. 671–676.

[11.10]   Menges, G.; Mohren, P.: Anleitung zum Bau von Spritzgießwerkzeugen. Carl Hanser Verlag, Munich, 1974.

[11.11]   Schlüter, H.: Verfahren zur Abschätzung der Werkzeugkosten bei der Konstruktion von Spritzgießwerkzeugen. Dissertation, RWTH, Aachen, 1981.

[11.12]   Lindner, E.: Spritzgießwerkzeuge für große Teile. Mitteilung aus dem Anwendungstechnischen Laboratorium für Kunststoffe der BASF, Ludwigshafen/Rh.

[11.13]   Kegelanguß, Schirmanguß, Ringanguß, Bandanguß. Technische Information, 4.2.1, BASF, Ludwigshafen/Rh.

[11.14]   Wüst, A.; von Diest, K.; Voit, T. H.: Und er bewegt sich doch. Kuststoffe, 87 (1997), 12, pp. 1762–1764.

12 Ejection

12.1 Summary of Ejection Systems

After the molding has solidified and cooled down, it has to be removed from the mold. It would be ideal if gravity could separate the part from cavity or core after mold opening. The molding is kept in place, however, by undercuts, adhesion and, internal stresses and, therefore, has to be separated and removed from the mold by special means.

Ejection equipment is usually actuated mechanically by the opening stroke of the molding machine. If this simple arrangement effected by the movement of the clamping unit is not sufficient, ejection can be performed pneumatically or hydraulically [12.1 to 12.3]. Manually actuated ejection can only be found in very small or prototype molds and for small series if little force suffices for actuating ejection and an exact cycle is of no concern.

The ejector system is normally housed in the movable mold half. Mold opening causes the mechanically actuated ejector system to be moved towards the parting line and to eject the molding. Precondition for this procedure is, of course, that the molding stays on or in the movable mold half. This can be achieved by undercuts or by letting the molding shrink onto a core. Taper and surface treatment should prevent too much adhesion, however.

Retaining the molding on the movable half becomes a problem if the core is on the stationary side. This should be avoided or more elaborate demolding systems are needed. Figure 12.1B is an example.

Figure 12.1 summarizes the usual ejector systems, as they are used for smaller moldings:

A Standard system for small parts.
B Direction of ejection towards movable side. Stripping is used but usually for circular parts only.
C Demolding at two parting lines for automatic operation including separation of gate.
D Demolding of parts with local undercuts (slide mold).
E Demolding of large, full-side undercut (split-cavity mold).
F Unscrewing molds for threads.
G Air ejectors usually provide support. Breaking is done mechanically.

Finally, one has to take into account that larger moldings are demolded by pushing them out, but they must not be ejected. They are removed manually or by robot after being loosened.

Presentation of mold	Ejection method	Components of operation	Application
A 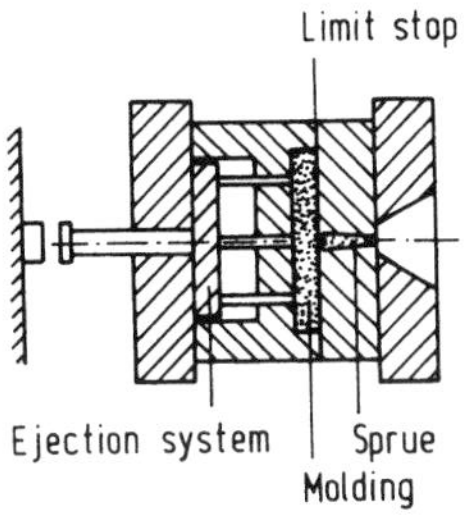	During opening stroke thrust in direction of demolding. Ejection with pins, sleeves or stripper plate.	Mechanical, hydraulic, pneumatic, manual, machine stop, lifting cylinder, cam, pivot, inclined plane, thrust plate. Also two-stage or mixed ejection.	Moldings of all kinds without undercut.
B 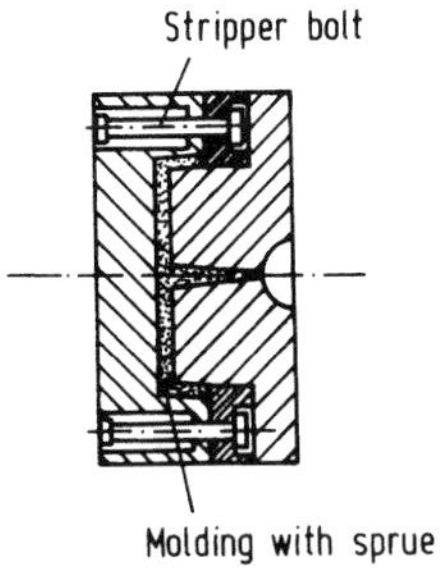	During opening stroke pull in direction of demolding. Ejection with stripper plate.	Mechanical, hydraulic, pneumatic. Stripper bolt, lifting cylinder, pin-link chain.	Cup-like moldings with internal gating.

Figure 12.1 Summary of ejection methods

C	During opening stroke thrust in direction of demolding. Ejection with pins, sleeves or stripper plate.	Mechanical, stripper bolt.	Moldings with automatic gate separation.
D	During opening stroke thrust in direction of demolding. Ejection with pins, sleeves or stripper plate after release of undercut.	Mechanical, cam pins, lifter, slide mechanism. Likewise hydraulic.	Flat parts with external undercuts e.g. threads.
E	During opening stroke thrust in direction of demolding. Ejection with pins.	Mechanical: toggle, latches, links, pins, springs, cams. Hydraulics as separate actuators.	Parts with external undercuts (ribs) or openings in side walls, e.g. crate for bottles.

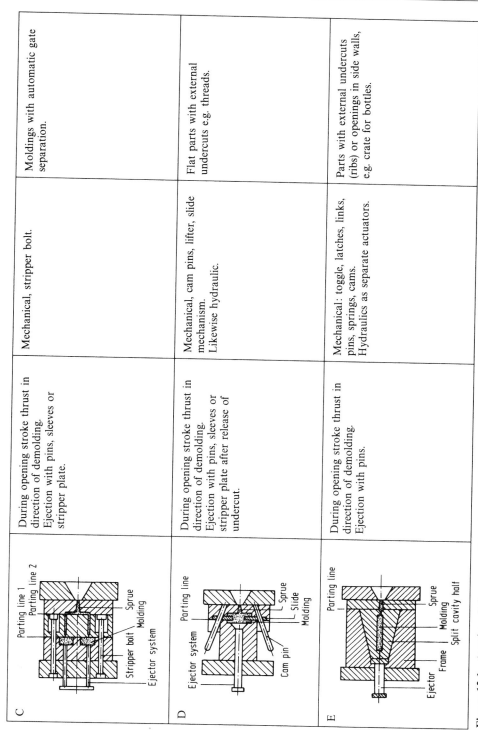

Figure 12.1 (continued) Summary of ejection methods

404   12   Ejection

Presentation of mold	Ejection method	Components of operation	Application
F Parting line, Gear, Molding, Core, Lead screw, Ejector system	Forming mold components are screwed off the molding in a closed or opened mold. Then ejection with pins or sleeves depending on shape of molding.	Mechanical: gear train with belt or chain drive, racks, coarse lead screws with nuts. Separate electric or hydraulic drive. Rarely manual e.g. with changing cores.	Parts with internal or external threads.
G Stripper ring, Molding, Ejector stroke, Air inlet	Thrust in direction of demolding causes a first release followed by ejection with compressed air.	Mechanical-pneumatic in stages.	Cup-like, deep parts.

Figure 12.1   (continued) Summary of ejection methods

12.2 Design of the Ejection System – Ejection and Opening Forces [12.4]

12.2.1 General Discussion

After the geometry and the mass of a part have been established, the release forces can be determined. The position of the part within the mold must also be known. For a detailed design of the ejection system (number, location and type of ejecting elements) it is important to know the release forces. The magnitude of the release force may also suggest the necessity for changing the position of the part in the mold and, therefore, the whole ejection system. Besides this, knowing the release force and the parameters affecting it, provides the possibility of reducing this force by making minor changes in the part configuration.

Basically, two kinds of forces can be expected:

– Opening forces: they are generated if the mold is jammed by too little shrinkage or too much deformation.
– Release forces which are subdivided into:

  a) Loosening forces: they are present for all parts with cores and are generated by the shrinking of the molding onto the core. They can also be noticed with thin slender ribs with little taper. Here they may cause a fracture of the lamellae which form the ribs.
  b) Pushing forces: they can arise from too little taper of a core and the resulting friction between part and core.

Thus, opening forces are less responsible for production difficulties than release forces, competent dimensioning provided.

The parameters affecting the release forces are presented in Figure 12.2. It is evident that changes in the release forces can be expected from four groups of effects. In experiments with sleeve-like parts (Figures 12.3, 12.8 and 12.9) direction and magnitude of the effect on the release force could be demonstrated [12.6, 12.7]. Figure 12.4 summarizes the results of these experiments. The various effects are discussed with respect to their efficacy in reducing the release forces and rated from 0 to 3, with 0 having no to 3 having a very strong effect. The arrows indicate whether the respective variable parameters have to be set higher or lower to reduce the release forces. The physical explanations for these effects cannot be discussed here because they would exceed the scope of this book. Reference is made to the original papers [12.6, 12.7].

The design engineer is mostly left to his own experience. This leads to the selection of cores with the greatest permissible taper especially if materials with high shrinkage are to be processed.

Figure 12.5 presents the suitable taper depending on the magnitude of linear shrinkage. This taper has to be increased for undercuts in the form of surface roughness. The expert relies on a necessary increase in taper of 0.5 to 2% per 2/100 mm roughness depending on viscosity and shrinkage of the melt. The higher values have to be applied to crystalline materials.

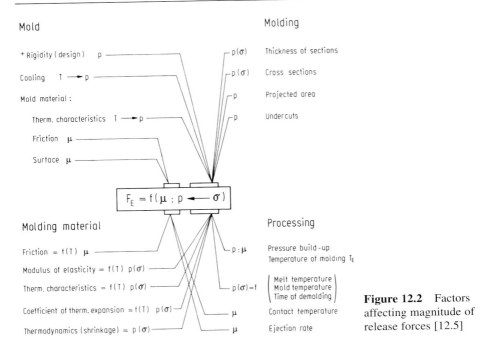

Figure 12.2 Factors affecting magnitude of release forces [12.5]

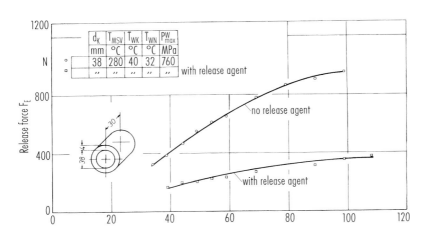

Figure 12.3   Effect of mold release agent on magnitude of release force (PP) [12.6]

Parameter	Magnitude of effect	Direction of change	Remarks
Cooling time t_E	3	↓	—
Average temperature at ejection $\bar{T}_E$	3	↑	—
Core temperature $\bar{T}_{WK}$	3	↓	$\bar{T}_E$ = const.
	3	↑	t_K = const.
Cavity wall temperature $\bar{T}_{WN}$	3	↓	$\bar{T}_E$ = const.
	3	↑	t_K = const.
Melt temperature T_M	0 – 1	↑ ↓	—
Injection pressure p_E	1	↑ ↓	—
Injection rate v_E	1 – 2	↑	t_K = const.
Holding pressure P_N	1 – 2	↑	t_K = const. or $\bar{T}_E$ = const.
Holding pressure time t_N	0 – 1	↑	t_K = const.
Ejection rate v_{out}	1 – 2	↑	t_K = const. or $\bar{T}_E$ = const.
Use of release agent	3		Effect increases with increasing cooling time

(vertical axis label: Magnitude of effect from 0 to max. 3)

Figure 12.4 Options for reducing release forces of sleeves [12.6, 12.7]

12.2.2 Methods for Computing the Release Forces

12.2.2.1 Coefficients of Static Friction for Determining Demolding and Opening Forces

For sleeves or box-shaped parts which shrink upon cores the release force can generally be determined with the normal stress present at the time of ejection and a coefficient of friction.

$$F_R = f \cdot p_A \cdot A_C \tag{12.1}$$

f = Coefficient of friction,
p_A = Contact pressure between molding and core,
A_C = Core-surface area.

The magnitude of the coefficient of friction f depends, in essence, on the pairing plastic – steel but also on some processing parameters. This coefficient is affected by the contact between the solidified surface layer and the mold surface at the time of demolding. Only measurements in the mold itself, under real processing conditions and without preceding separation between molding and mold surface, can be used in Equation (12.1) if realistic values are to be obtained. For mold inserts, which were made by EDM and polished, the coefficient of friction was determined in dependence on the surface roughness and are presented in Table 12.1.

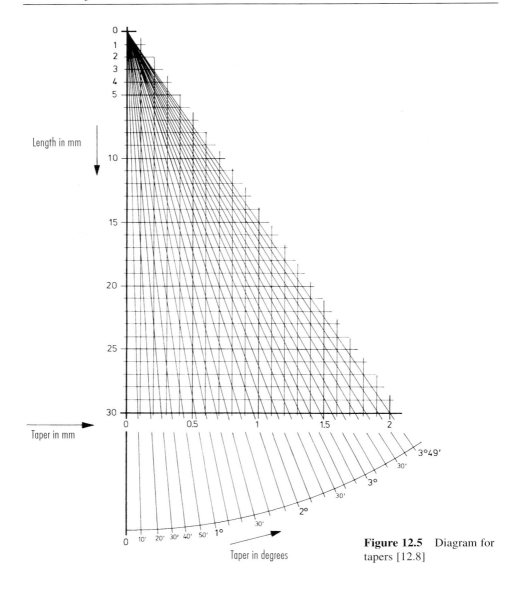

Figure 12.5   Diagram for tapers [12.8]

The static friction coefficients in Table 12.1 were determined from the breaking forces that occurred immediately at the start of the demolding process for the part in the injection mold [12.9, 12.10].

The values in Table 12.1 are the maximum values for the friction coefficient determined using various process parameters. Since the process parameters are not known exactly at the start of the mold design and cannot always be reliably observed during operation when known and since, moreover, the f values show scatter, the respective value in the Table should be multiplied by a factor of 1.5 to 2.

Table 12.1   Coefficient of static friction dependent on roughness height [12.9, 12.10]

Material	Coefficient of friction for roughness height		
	1 μm	6 μm	20 μm
PE	0.38	0.52	0.70
PP	0.47	0.50	0.84
PS	0.37	0.52	1.82
ABS	0.35	0.46	1.33
PC	0.47	0.68	1.60

The scope for influencing the f value is shown in Figure 12.6, taking PE as an example. The extent of the influence is divided into 4 steps, 0–3, where

0: No influence,
1: Slight influence,
2: Moderate influence,
3: Strong influence.

The direction of the arrow in the column "Direction of change" denotes whether the pertinent influential parameter has to be increased (↑) or decreased (↓) in order to obtain lower f values. The dependency shown here for PE apply in the main to PP, PS, ABS and PC as well [12.9, 12.10].

Aside from the coefficient of friction, the surface pressure between core and molded part must be determined. It may be calculated theoretically [12.11–12.13] or with the aid of a very simple method based on experience (shrinkage values).

Factor	Magnitude of effect	Direction of change		Remarks
Contact pressure	0–1 (2)	↑	↓	More notable effect (2) is the exception with $R_a = 35$ μm
Ejection rate	0–1 (2)	↑	↑	Considerable effect with very rough surfaces
Cooling time	(0) 1–2		↑	(0) with release agent
Average cavity-wall temperature	2–3		↓	
Melt temperature	0–1 (2)	↑	↓	No definite statement possible
Holding pressure	0–1 (2, 3)		↓	Considerable effect with rough surface and with increasing holding pressure raising pw_{max}
Release agent	1–3		↑	
Surface finish	1–3		↓	Deviations possible in rare cases

Figure 12.6   Options for reducing the coefficient of static friction of polyethylene (PE)

12.2.2.2 The Estimation Method for Cylindrical Sleeves

For practical purposes another method has been developed to quickly estimate the release forces. They can be determined with sufficient accuracy, e.g., for sleeves [12.6, 12.12], which demand high release forces by their nature.

The assumption is made that the design engineer has to be able to establish an appropriate core diameter, which corresponds to the final internal diameter of the part. From the resulting diameter differential an absolute upper limit for the release force can already be estimated by means of an equilibrium of forces. This will be explained with a thin-walled sleeve as an example [12.6] (see flow chart in Figure 12.10).

The shrinkage of the molding is restricted by the core. This causes a build-up of stresses in the cross-sections of the part, which results in forces normal to the surfaces restrained from shrinking. The stored energy-elastic forces can recover spontaneously with demolding. The resulting contraction of circumference or diameter causes a measurable decrease in the inside diameter of the sleeve. The relative decrease of the part's diameter or circumference is:

$$\Delta d_r = \Delta C_r = \frac{d_C - d_i(t_E)}{d_C} \tag{12.2}$$

Wherein

ΔC_r Relative change in circumference,
d_C Core diameter,
$d_i(t_E)$ ID of sleeve immediately after demolding.

The circumferential reduction, measured immediately after demolding is directly associated with the tensile stress in the cross-section of the part as long as the molding was still on the core. Its computation is simple; the thin-walled sleeve does not require Poisson's ratio to be applied (Figure 12.7):

$$\sigma = E \cdot \varepsilon \qquad \text{(Hooke's law)} \tag{12.3}$$

$$\varepsilon = \frac{\Delta L}{L} \tag{12.4}$$

or in this case (Figure 12.7):

$$\varepsilon = \frac{\Delta d}{d_C} = \Delta d_r = \Delta C_r \tag{12.5}$$

$$\sigma_\varphi = \Delta C_r \cdot E\,(T) \tag{12.6}$$

Hence

$$p_A = \sigma_\varphi \frac{d_o - d_i}{d_C} \tag{12.7}$$

$$p_F = \sigma_\varphi \frac{s_M}{r_C} = \frac{E(T) \cdot \Delta C_r \cdot s_M}{r_C} \tag{12.8}$$

with

s_M Wall thickness of part,
r_C Radius of core.

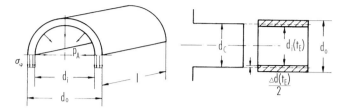

Figure 12.7 Computation
of release forces for sleeves
[12.6]

With the coefficient of friction f, the necessary release force is obtained from

$$F_E = f \cdot p_A \cdot A_C \tag{12.9}$$

$$F_E = f \cdot E(\overline{T}_E) \cdot \frac{\Delta d\ (t_E)}{d_C} \frac{s_M}{r_C} \cdot d_C \cdot \pi \cdot L \tag{12.10}$$

F_E is always the breaking force which is necessary to initiate the ejection movement and is related to static friction. This release force takes only the force for demolding the part into account regardless of the friction of the ejector system.

Depending on mold size and the kind of ejector system, the forces from friction in the system ΣF_S, which vary in magnitude, have to be added to the release forces if the total power demand for demolding has to be computed (selection of a machine with adequate ejection force).

The demand on ejection force for n cavities results from

$$F_{mach/eject} = n \cdot F_E + \Sigma F_S \tag{12.11}$$

This method was used for a sleeve made of ABS (d_C =38 mm, d_o = 46 mm, L = 30 mm) and compared with measurements of the release force. The results were satisfactory as demonstrated in Figure 12.8 [12.6].

Because the design engineer usually does not know in advance which change in diameter he can expect, he can calculate this change with the information of linear shrinkage (raw-material supplier) and use it as a starting point for his calculation (Figure 12.9) as follows:

Point A shows the mold open with the molding still on the core producing contact pressure from cooling. B represents the state immediately after ejection. A decrease in diameter results from the spontaneous energy-elastic deformation:

$$\Delta d_r = \frac{d_C - d_i(t_E)}{d_C} \tag{12.12}$$

Stages A and B correspond to one and the same point for T_E on the 100 kPa-line in a p-v-T diagram because they do not differ thermodynamically from one another. Cooling down to room temperature results in the stage C, for which a corresponding volume

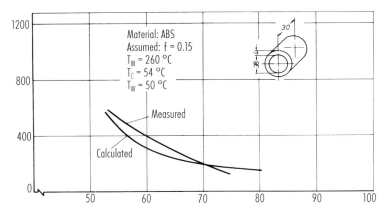

Figure 12.8   Release forces on sleeves (measured and estimated) [12.6]

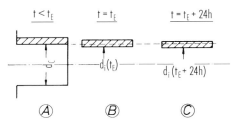

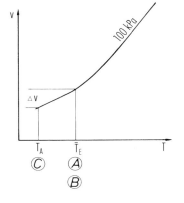

Figure 12.9   Change in internal diameter of a
sleeve [12.6]
A   Immediately after freezing
B   After pushing off
C   24 h after demolding

contraction ΔV can be taken from the diagram. State B can be derived from the state C
using the following procedure:

For sleeves, which were demolded at a known average temperature of the molding,
the additional relative volume contraction was computed by means of the p-v-T diagram:

$$\Delta s_V = \frac{V(\overline{T}_E) - V(T_A)}{V(\overline{T}_E)}$$

(12.13)

Experiments with various materials and sleeves with 4 mm section thickness have
demonstrated that the ratio between circumferential and volume contraction, the factor
K, was constant for all kinds of thermoplastics [12.6]:

$$\frac{\Delta d_r \left|_{t_E}^{t_E + 24\,h}\right.}{\Delta s_V} = K \tag{12.14}$$

For ABS $K \approx 0.43$,
For PS $K \approx 0.7$ (with a wall thickness $s_M = 4$ mm),
For PP $K \approx 0.6$.

From this results the procedure for estimating release forces as presented with Figure 12.10 [12.6].

12.2.2.3 Rectangular Sleeves

Although the previous computations were compounded for cylindrical sleeves, they can also be used for approximating the release forces for rectangular ones. According to the method of estimation (Section 12.2.2.2) the relationship between cylindrical sleeves and rectangular ones with the same wall thickness can be expressed by the ratio

$$\frac{F_{E\,\bigcirc}}{F_{E\,\square}} \approx 0.785 \tag{12.15}$$

There is a correlation between the diagonal (and the width-over-height ratio) of a rectangular sleeve and the diameter of a cylindrical one. Thus, for rectangular sleeves one can proceed in the same manner as for cylindrical sleeves by taking half the diagonal instead of the radius [12.11].

12.2.2.4 Tapered Sleeves

For slightly tapered sleeves the contact pressure p_A can be computed in the same way as for cylindrical sleeves by using the average diameter (as shown with Figure 12.10). Figure 2.11 demonstrates how much the release force is reduced by the angle α. The ratio between the release forces for a tapered and a cylindrical sleeve is plotted against α. One recognizes that a decrease in f reduces the release force disproportionately if the taper is kept constant.

Tapering is important for reducing the release force but it also ensures that adhesion between part and core only exists for a short distance (less hazard of damage because of smaller energy demand) [12.4].

12.2.2.5 Summary of Some Basic Cases

Figure 12.12 lists the correlations for estimating the release forces based on the considerations in Sections 12.2.2.2 and 12.2.3. They allow reliable estimates to be made for virtually all cases.

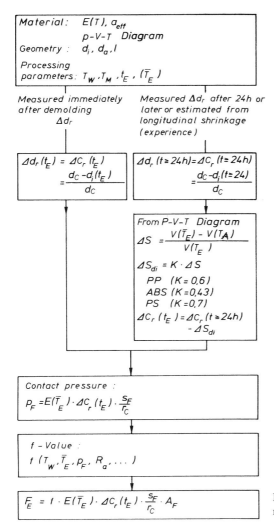

Figure 12.10   Flow chart for computing release forces on sleeves [12.6]

12.2.3 The Release Forces for Complex Parts Exemplified with a Fan

How the previously presented method of estimating release forces of sleeves can be applied to more complex parts will be demonstrated with a five-blade fan as an example. Other parts of different shape and geometry can be treated similarly [12.12].

In contrast to simple sleeves, here the release forces are also affected by ribs, top and inside hub. Figure 12.13 shows schematically the fan and its breakdown into basic geometries for estimating the release force.

The external sleeve is most significant for estimating the release force. If the diameter difference of part/core is known from experience – and it has always been necessary for

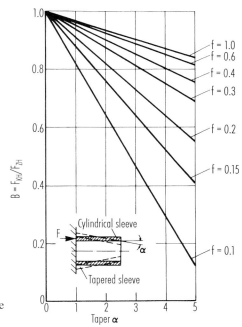

Figure 12.11   Estimated release
force of tapered sleeves [12.7]

other reasons for the designer to know this – the release force can be estimated according
to Section 12.2.2.2.

At first the configuration of the fan under consideration (5 blades) is broken down into
basic geometries (Figure 12.13).

The external sleeve, the top, and the 5 ribs have to be taken into account for
computing the release force. The small internal sleeve (hub) can be neglected but not the
top. The following consideration should illustrate the computation. As long as the
molding is still on the core (stage 1) the total force is transmitted from the core to the
sleeve; there is an equilibrium of forces. At the moment of demolding this force
vanishes. The part diameter decreases immediately (stage 2). This also reduces the length
of the ribs and the diameter of the top. Altogether there is again an equilibrium of forces.

If one now computes the forces necessary for the elongation of sleeve (including the
top) and ribs, the sum of these forces is equal to the force transmitted by the core to the
part in stage 1. It is exactly this force together with the coefficient of friction that
accounts for the release force (Figure 12.13).

Taking the longitudinal shrinkage S_1 in percent (here 1.5% for polypropylene) the
diameter differential sleeve/core can be approximated:

$$\Delta d \ (t \geq 24 \ h) = d_C \cdot \frac{S_1}{100} \tag{12.16}$$

d_C Diameter of core,
S_1 Longitudinal shrinkage in %.

With $d_C = 74.2$ mm and $S_1 = 1.5\%$ the result is $\Delta d(t \geqq 24 \ h) = 1.113$ mm.

Geometry	Estimate of release forces or moments
Open cylindrical sleeve 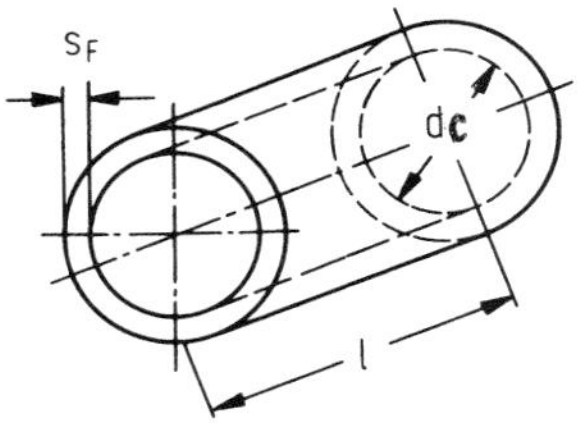 	From Figure 12.10: $$F_R \approx f \; E\,(\overline{T}_E) \cdot \left[\frac{s_l\,(\%)}{100} - K \cdot \Delta s_v \right] \cdot 2\,s_F\,\pi\,l$$ $s_l = \Delta d_r \;(t \geqq 24\,h)$ s_l : Shrinkage of ID ($t \geqq 24h$) PP : $K \approx 0.6$ ABS: $K \approx 0.43$ PS : $K \approx 0.7$ $$\Delta s_v = \frac{v\,(\overline{T}_E) - v\,(T_v)}{v\,(\overline{T}_E)}$$ $v\,(\overline{T}_E)$, $v\,(T_v)$: Specific volume from P-V-T diagram If $K\Delta s_v$ is not known, eliminate term. This results in estimate of absolute max. value for F_R.
Closed cylindrical sleeve 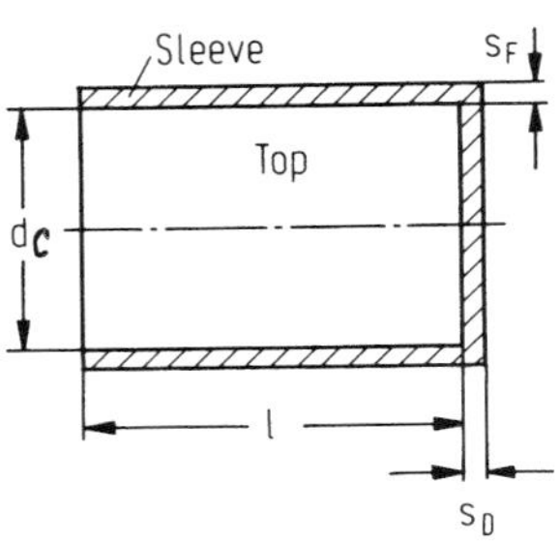 	a) With core venting : $$F_R \approx f\,E\,(\overline{T}_E) \cdot \left[\frac{s_l\,(\%)}{100} - K\Delta s_v \right] \cdot \left[2\pi\,s_F\,l + \frac{d_K\,\pi\,s_D}{1 - \nu} \right]$$ b) Without core venting : $$F_R \approx f\,E\,(\overline{T}_E) \cdot \left[\frac{s_l\,(\%)}{100} - K\Delta s \right] \cdot \left[2\pi\,s_F\,l + \frac{d_K\,\pi\,s_D}{1 - \nu} \right] + \frac{d_K^2\,\pi}{4}\,p_u$$ p_u : Negative pressure ($p_{u\,max} = 100\,kPa$)
Open rectangular sleeve 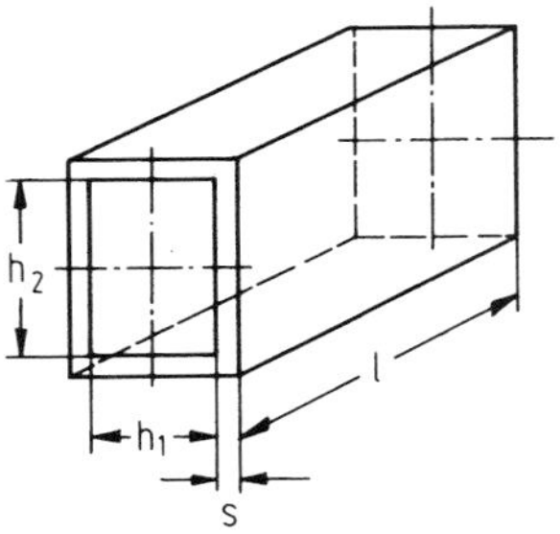 	$$F_R \approx f\,E\,(\overline{T}_R) \left[\frac{s_l\,(\%)}{100} - K\Delta s_v \right] \cdot 8\,s\,l$$

Figure 12.12   Summary of various basic cases [12.4] (continued on next page)

Geometry	Estimate of release forces or moments

Closed rectangular sleeve

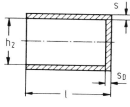

h_1 see previous figure

a) With core venting :

$$F_R \approx f\, E(\overline{T}_\varepsilon) \cdot \left[\frac{s_l(\%)}{100} - K\Delta s_v \right] \cdot \left[8sl - \frac{2s_D(h_1+h_2)}{1-f} \right]$$

b) Without core venting :

$$F_R \approx f\, E(\overline{T}_\varepsilon) \cdot \left[\frac{s_l(\%)}{100} - K\Delta s_v \right] \cdot \left[8sl - \frac{2s_D(h_1+h_2)}{1-f} \right] + h_1 h_2 p_u$$

p_u = Negative pressure ($p_{u\,max}$ = 100 kPa)

Threaded sleeve
Saw-toothed thread
Direction of demolding ⟶

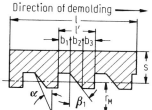

Two different cases of shrinkage are considered. The highest torque is taken for design purposes

Case 1 :
  Root rests upon crest

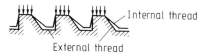

Static torque :

$$M_{EH} = f_0\; 2r_M^2\, \pi l p_F \qquad p_F = E(\overline{T}_\varepsilon) \cdot \left[\frac{s_l(\%)}{100} - \Delta K s_v \right] \cdot \frac{s_F}{r_M}$$

Sliding torque :

$$M_{EG} = \frac{2\pi l r_M^2\, p_F}{\dfrac{\cos\alpha}{f} - \dfrac{\sin\alpha}{\cos\beta_1}(1-\sin\beta_1)}$$

f_0 Coeff. of static friction; f Coeff. of sliding friction

Case 2 :
Internal thread rests uniformly upon external thread.

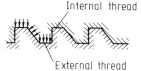

Static torque :

$$M_{EH} \approx \frac{2\pi l r_M^2\, p_F \left[b_1 + b_2 + b_3 \left(\dfrac{1+\cos\beta_1}{\sin\beta_1} \right) \right]}{l'\left(\dfrac{\cos\alpha}{f_0} - \dfrac{\sin\alpha}{\cos\beta_1} \right)}$$

Sliding torque :
Same as above with f_0 replacing f for sliding

Figure 12.12   (continued) Summary of various basic cases [12.4]

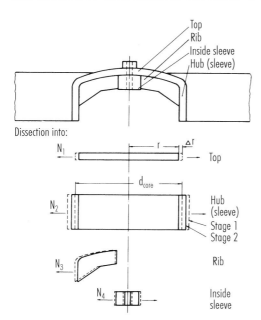

Figure 12.13   Analysis of a fan for computing release forces [12.12]

For comparison, the diameter differential of several fans was measured after 24 hours. The average was 1.1 mm. With this starting value the release force could be computed.

Besides the sleeve, the five ribs and the top also have to be taken into consideration. The release force is calculated from

$$F_E = f \cdot N,$$ (12.17)

and the normal force N is

$$N = N_{sleeve} + N_{rib} + N_{top}$$ (12.18)

As an approximation

$$N = p_A d_C \cdot \pi \cdot L + 5 \; E(\overline{T}_E) \frac{d_C - d_i(t_E)}{2 \, L_R} \, b_R s_R \cdot N_{top}$$ (12.19)

Core diameter (sleeve): d_C = 74.2 mm,
Wall thickness (sleeve):   s   = 2.3 mm,
Length of sleeve: L = 29.8 mm,
Wall thickness (top): s_T = 2.0 mm,
Length of rib: L_R = 31.0 mm,
Width of rib: b_R = 12.0 mm,
Wall thickness (rib): s_R = 1.5 mm,

$$N_{sleeve} = p_A \cdot d_C \cdot \pi \cdot L \text{ and } d_i(t_E)$$ (12.20)

can be determined with the flow chart Figure 12.10. The release forces estimated by employing Equations (12.17) and (12.19) are presented with Figure 12.14. Because of venting in the core, no additional force from vacuum formation was present. This may have to be reconsidered with a different design.

Test results for this method are likewise shown in Figure 12.14. Even for long cooling times the release force is adequately computed by this method.

How much sleeve, rib and top contribute to the release force can also be seen in Figure 12.14.

The contact pressure on the fan under release forces is given by

$$p = \frac{F}{A}$$

(12.21)

For the ejector pins

$$p_A = \frac{S \cdot F_{E\,max} \cdot 4}{n \cdot D_E^2 \cdot \pi}$$

(12.22)

with

p_A Contact pressure,
S   Safety factor (1.5),
n   Number of ejector pins (4),
D_E Diameter of ejector pins (8 mm).

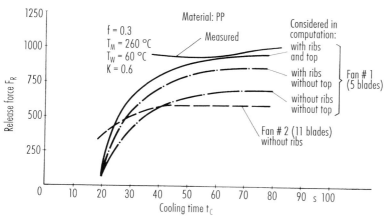

Figure 12.14 Estimated and measured releases forces for fan [12.12]

Hence, for $F_{E\,max}$ = 1000 N (Figure 12.14)
p_A = 7.46 MPa.

This contact pressure has to be compared with the permissible pressure, which depends on the plastic material, the wall thickness of the part, the release temperature, and the overall rigidity of the molding. Exact limits have not yet been determined. Therefore, the relative depth of penetration under the release force will be calculated here:

From

$$\sigma = E \cdot \varepsilon$$

and

$$\varepsilon = \frac{\Delta L}{L_0} = \frac{\Delta s}{s_W} \tag{12.23}$$

the approximate result is

$$\Delta s = \frac{p_A \cdot s_W}{E(\overline{T}_E)} \tag{12.24}$$

Δs Depth of penetration,
s_W Wall thickness (2 mm).

Hence

$$\Delta s = 0.021 \text{ mm}$$

or

$$\varepsilon = \frac{\Delta s}{s_W} \ 100\% = 1.05\%$$

The small values of Δs and ε indicate that no problems need be expected.

Finally, there is the buckling of the ejector pins under cavity pressure to be considered.

According to Figure 12.27, and assuming

$p_{W \, max}$ = 100 MPa,
D_E = 8 mm,

the critical length is 280 mm. This dimension must not be exceeded.

12.2.4 Numerical Computation of Demolding Processes (for Elastomer Parts)

For the mechanical design and dimensioning of injection molded parts, the finite elements method (FEA) has proved to be a powerful tool for the designer [12.14]. However, FEA is rarely used for the numerical computation of demolding processes of thermoplastic injection-molded parts. The major reason for this is that the frictional relationships between mold wall and part surface, as described in Section 12.2.2.1, are influenced by a large number of factors. For this reason, the friction conditions during part design can often only be estimated, with the result that FEA can only yield rough computations. The outlay involved in FEA calculations is not always offset by the benefit gained.

With elastomer parts, however, FEA can be a useful tool for assessing demolding processes. Unlike thermoplastics, elastomers are noted for their high elasticity. This special material property is not only exploited in the technical applications for which the parts are used. When rubber parts are made with undercuts by the master mold method (e.g., injection molding, transfer molding), often demolding occurs without the undercut of the mold being eliminated, i.e., without the undercut being exposed by slides or splits. This type of demolding is possible due to the high deformation capability of elastomers. The resulting cost-efficient production of parts with undercuts opens up greater design scope than is actually used in practice. A large number of rubber parts with undercuts find technical application (e.g. rotary shaft seals, spring elements of car engine bearings, expansion bellows).

Nevertheless, for elastomers, the demoldability of undercuts is limited by the maximum sustainable material load. The local stresses and strains that actually occur in the part during the demolding process are dependent on the following parameters:

– undercut geometry,
– elastomer material,
– frictional conditions between part and mold wall,
– processing parameters,
– nature and mode of action of the force exerted by the demolding device on the part (e.g., by ejector system or handling device).

If the elastomer material is over-stressed during demolding, this may cause the part to fail immediately (e.g., the part may tear in the undercut area) or it may suffer damage that makes the part no longer adequate in terms of service life or maximum sustainable load. The design of rubber parts can therefore not just proceed on the basis of loads in practical use alone. Demoldability of undercuts must also be assessed at the design stage. At this stage, the loads are generally completely different from those which affect the part in practical use.

At the moment, dimensioning of the undercuts is usually carried out on the basis of the designer's experience or that of trial and error. Often, this fails to fully exploit the potential of rubber as a material, or damage is sometimes introduced unnoticed into the part during demolding. With the aid of FEA calculations, however, the designer is put in the position of being able to estimate the maximum possible undercut as a function of the geometry, the material, and the demolding kinematics. A motivation for applying FEA to this problem is that FEA can already be applied to complex geometries during the design stage and that local peak loads in the part can be simulated under major deformation, frictional contact and non-linear material behavior.

The potential and advantages of using FEA to simulate demolding will be illustrated in the following example in which the dimensions of a rotary shaft seal have to be determined, whose geometry is shown in Figure 12.15. For a certain type of rotary shaft ring demolding, local crack formation was discovered in a large number of parts (up to 30%). They are labeled in the part geometry shown in Figure 12.15.

The rotary shaft seal is made in an automated injection process. The mold employed is shown in Figure 12.16 (starting situation). It consists of four parts: the upper part (item 1), the middle part (item 3, the lower part (item 2b), and the mandrel (item 2a). Demolding is initiated by raising the upper part. Then the lower part with mandrel is lowered. The traverse speed of the mold parts is roughly 0.02 m/s. The rotary shaft seal remains during both stages in position on the middle part. This is now swiveled into the manipulating area of the machine, where the seal is pushed out of the mold by a slide bar.

This demolding process was simulated with the aid of FEA [12.15]. Figure 12.17 shows, by way of example, the results of the deformation simulation for traversing the

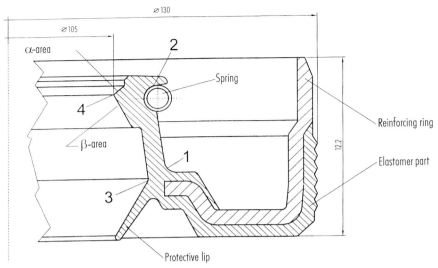

Figure 12.15   Cross-section of a radial shaft seal showing damage that occurred during demolding: 1–4 (in order of damage frequency)

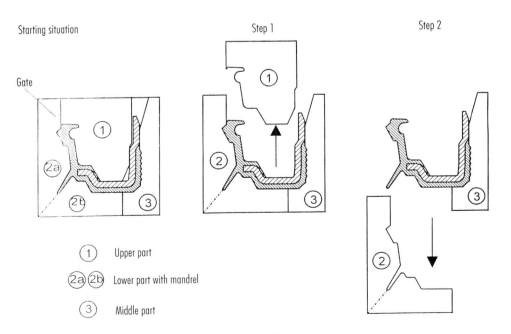

Figure 12.16   Demolding process for the radial shaft seal

Demolding step 2

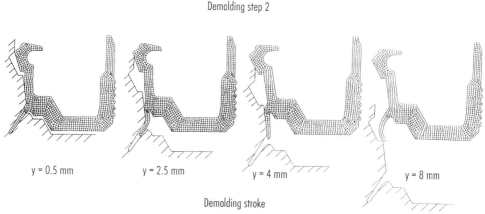

y = 0.5 mm   y = 2.5 mm   y = 4 mm   y = 8 mm

Demolding stroke

Figure 12.17 Calculated deformation of the part cross-section during demolding of the upper mold part (see mold part in Figure 12.13)

lower mold and the mandrel with protective lip as undercut. Other results of the FEA analysis are the stress and strain distributions in the part. The locations of the calculated peak loads are found to coincide with crack formation zones observed in practice (inner side of the annular spring seat and protective lip connection; Figure 12.15). The calculated maximum stresses and maximum strains match or exceed the strength limits for the elastomer material that were determined in the rapid tearing test. FEA analysis can therefore help to predict failure of the part during demolding.

The demolding forces which the part exerts on the mold wall can also be calculated, provided that the value of the coefficient of sliding friction is known with sufficient

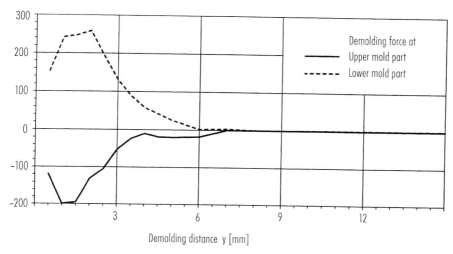

Demolding distance y [mm]

Figure 12.18 Calculated demolding forces at the upper and lower mold parts for part shown in Figure 12.13

accuracy. Figure 12.18 shows the change in reaction forces for the simulated demolding process. The demolding stage at which the reaction force is a maximum correlates with the point in time at which there is maximum internal stress in the part. It is thus possible to identify the most critical phase of the demolding process.

To avoid part failure during demolding, process and geometry variants are calculated with the aid of FEA. Reversing the sequences of the mold (demolding of the lower part followed by demolding of the upper part) does not lead to any improvement (this was confirmed in practice). In contrast, slight changes to the undercut geometry bring about major reductions in peak loads. If the undercut of the protective lip is reduced by 20%, without the lip being shortened, the stress peaks are below the value of the yield stress (Figure 12.19).

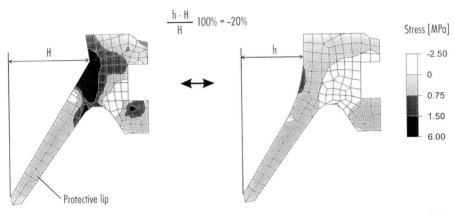

Figure 12.19   Reduction in material stress during demolding as a result of design optimization (see mold part in Figure 12.13)

12.2.5 Estimating the Opening Forces

Only a completely accurate design ensures adequately small opening forces, which do not cause interruptions of the production or damage to molds and moldings. Because considerable opening forces should not occur, methods of their computation as proposed in the literature are dispensed with. If required, they can be taken from this literature [12.4].

Should there, nevertheless, be a problem of this kind at the start-up of a new mold, trying the following measures is recommended:

1. Reducing the injection pressure to the level required to just fill the cavity. If the part can now be demolded perfectly, then increase the injection pressure to the level just beneath where the problems occur again.
2. If method 1 fails, the mold is too weak and must be strengthened or modified. To this end, the necessary dimensions for the critical areas are calculated with the aid of the formulae quoted in Chapter 10.

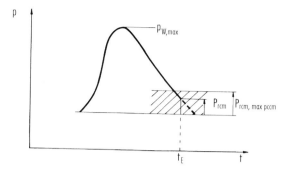

Figure 12.20   Pressure on cavity side wall under remaining pressure during mold opening

12.2.5.1 Changes of State in a p-v-T Diagram for Molds with Different Rigidities

Figure 12.21 presents the qualitative change in state for molds with different rigidities (transverse to the direction of clamping). The presentation is somewhat simplified because the point of gate freezing (sealing point) also changes with different rigidities in spite of identical machine settings. The response from rigid and resilient molds to the change of state after the sealing point has been reached and its effect on the mold-opening forces should become clear, though. The more rigid the mold is, the smaller are the mold-opening forces which can be expected [12.16].

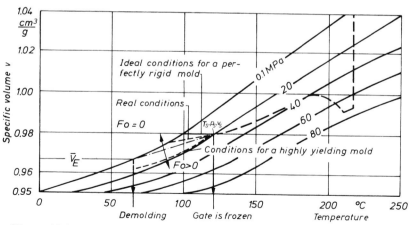

Figure 12.21   Effects from molds with varying rigidity [12.16]

12.2.5.2 Indirect Opening Forces

The release force in slide molds, acting perpendicular to the direction of mold opening, is generated by the opening force through the cam pins (Figure 12.83).

Estimating the release force Q (e.g., for a sleeve positioned accordingly) can be done with the Equations in the Chapter "Slide Molds". Using the relationships shown in Figure 12.83 the necessary opening force can then be computed.

12.2.5.3 Total Opening Force

The total opening force is composed of

- Forces from friction in the mold F_{of},
- Forces from acceleration F_{oa},
- Direct opening forces F_{od},
- Indirect opening forces F_{oi}

$$F_{o, \text{machine/mold}} = F_{of} + F_{oa} + F_{od} + F_{oi} \qquad (12.25)$$

12.3   Types of Ejectors

12.3.1   Design and Dimensions of Ejector Pins

To demold a molding, ejector pins are the most frequently employed elements. They are on the market as standards in many variations and dimensions [12.17 to 12.21]. Ejector pins are mostly made of hotwork die steel (AISI H-13 type) and hardened or gas nitrided to achieve a high surface hardness of about 70 Rc. Adequate hardening and good surface quality prevent seizing in the mold and ensure long service life. Molybdenum disulfide should be applied to the pin surface during maintenance work to improve function under adverse conditions.

Nitrided ejector pins are primarily used in molds for thermosets and for lengths of more than 200 mm.

For shorter lengths (under 200 mm) and low operating temperatures, ejector pins of annealed tool steel are also in use. Their hardness is 60 to 62 Rc at the shaft and 45 Rc at the head.

The heads of ejector pins are hot-forged. This produces a uniform grain flow and avoids sharp corners, which would weaken the pin by a notch effect.

There are two basic types of ejector pins according to their intended use (Figure 12.22):

a) Straight cylindrical pins are the most common for all ejection forces. The cylindrical head reduces the hazard of being pressed into the ejector plate. They are usually available in diameters from 1.5 to 25 mm or 3/64 to 1 in. and in lengths up to 635 mm or 25 in.

b) Shoulder-type or stepped ejector pins are employed if only a small area of the molding is available for ejection and little force is needed. The stepped shaft raises the buckling strength. Common diameters are from 1.5 to 3.0 mm or 3/64 to 7/64 in.

in standard lengths up to 355 mm or 14 in. with standard shoulders of 1/2 or 2 in. length.

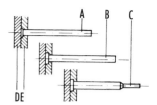

Figure 12.22   Schematic presentation of ejector pins [12.2]
A Ejector pin with conical head and cylindrical shaft,
B Ejector pin with cylindrical head and cylindrical shaft,
C Shoulder type ejector pin,
D Ejector plate,
E Ejector retainer plate

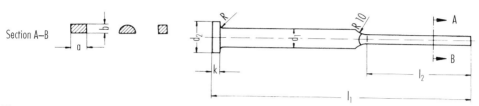

Figure 12.23   Ejector pins with noncylindrical shafts [12.17]

Special ejector pins are available if the tip of the pin has to be adapted to the contour of the molding (Figure 12.23). These pins have to be secured against twisting and guided by special elements if a certain length is exceeded (Figure 12.24).

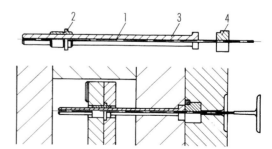

Figure 12.24   Blade ejector with leader elements [12.20]
1 Blade ejector, 2 Bushing, 3 Sleeve, 4 Leader block

Sleeve ejectors are presented with Figure 12.25. (They are also available as standards). All ejector pins are precision honed for close tolerances to ensure smooth sliding in the mold. Their fit in the mold depends on the plastic to be molded and the mold temperature (refer also to Chapter 7 "Venting of Molds").

In heated molds, attention should be paid to the fact that ejector pins should not be actuated before the proper mold temperature has been attained. Figure 12.26 presents an example for an ejector pin assembly. When dimensioning ejector pins, eventual

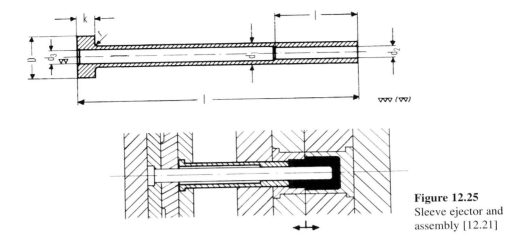

Figure 12.25
Sleeve ejector and
assembly [12.21]

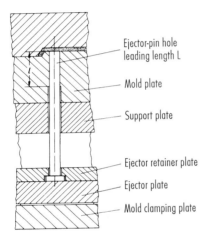

Ejector-pin hole
leading length L

— Mold plate

— Support plate

— Ejector retainer plate

— Ejector plate

— Mold clamping plate

Figure 12.26   Ejector pin assembly [12.1]

instability problems should be considered due to the slenderness of the pins. The
diameter is therefore computed from [12.12, 12.22]:

$$d \geqq 0.000836 \cdot L \cdot \sqrt{p} \tag{12.26}$$

For steel and p = 100 MPa pressure

$$d \geqq 0.028 \cdot L \tag{12.27}$$

In this equation, L is the unguided length of the pin. For safety reasons and because the
guided length is usually short, the total length should be taken as L especially for thin
ejector pins. The diameter of an ejector pin dependent on critical length of buckling and
injection pressure can be taken directly from Figure 12.27, which is derived from
Equation (12.27).

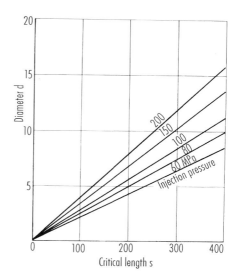

Figure 12.27 Suggested diameter of ejector pins depending on critical length of buckling and injection pressure [12.12, 12.22]

12.3.2 Points of Action of Ejector Pins and Other Elements of Demolding

Figure 12.28 considers the compressive strength of the molding and the ejection force. One can conclude from this diagram whether or not the part can withstand the release force without being damaged. Pins have to distribute the ejection forces uniformly to the molding so that they can take the forces without being distorted or punched. The points

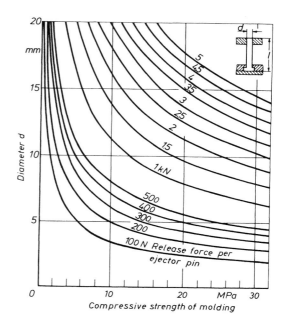

Figure 12.28 Suggested diameter of ejector pins depending on part strength and release force [12.12, 12.22]

of action have to be sufficiently close together and at places of high rigidity to avoid distortion of the molding. This could eventually result in stretching. Best suited are intersections of ribs. Figure 12.29 presents a number of examples.

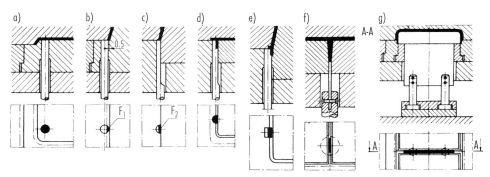

Figure 12.29   Locations for ejector-pin action [12.23]

Number, design (acting surface of ejector pins on the molding as large as possible), and placement of ejectors depend on the configuration of the molding as well as on the processed molding material. Rigidity and toughness are decisive factors [12.3].

Every ejector leaves a visible mark on the molding. This has to be taken into consideration in the design of the ejector system. Even more so since flashing may occasionally be noticed at the point of action because of inferior workmanship. Although flash can be removed, the operation leaves traces which interfere with appearance. There-fore, special attention has to be paid to a close fit between ejector pin and hole.

This fit is less critical with thermoplastics than with thermosets because of the low mold temperatures for thermoplastic materials. Small moldings, especially those with a central cylindrical core, do not offer much useful surface for the action of ejector pins. They are ejected with ring ejectors or ejector sleeves for better use of the ejection force. Sleeves may act upon the whole circumference of the molding (Figure 12.30). They are more expensive than pins, though. For large moldings, close tolerances are mandatory;

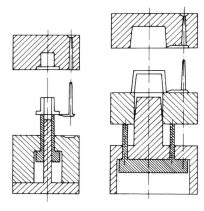

Figure 12.30   (left side) Ejection with stripper ring or sleeve [12.24]

Figure 12.31   (right side) Ejection with stripper plate [12.24]

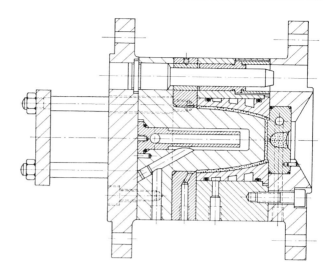

Figure 12.32 Tumbler mold
with stripper ring [12.3]

otherwise flashing occurs between sleeve and pin, which impedes demolding and calls
for expensive postoperation to remove flash.

Another option for letting the ejection force act along the whole circumference is
provided by a stripper plate (Figure 12.31). It is useful for stripping off circular as well
as differently shaped moldings. Because of the expensive fitting work required, ring
ejectors, ejector sleeves and stripper plates are primarily used for cylindrical parts.

The mating surfaces between core and stripping device are usually tapered in
accordance with the configuration of the part to achieve good sealing in the closed mold
and reduce wear of these surfaces (Figure 12.32). Tapered faces also facilitate the return
movement. A small step (Figure 12.33) prevents damage to the polished core surface
during ejection or return [12.3].

Figure 12.34 demonstrates an ejector system that acts upon the molding in several
planes simultaneously. This system is particularly useful for ejecting deep moldings of

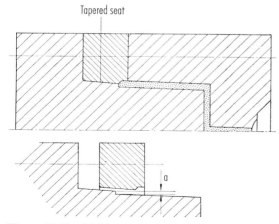

Figure 12.33 Mold core and stripper ring [12.3]

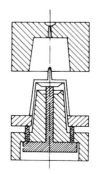

Figure 12.34 Simultaneous
ejection in more than one plane
[12.24]

less rigid materials. The "disk" ejector at the bottom also provides for venting and air access during ejection (against generating a vacuum). For complete removal, the molding in this example has to be taken off by hand or ejected by compressed air exiting sideways from a center hole.

"Disk" ejectors (Figure 12.35) should be employed whenever the disk diameter has to be larger than 20 mm. They are also very useful for ejecting deep moldings, which have to be lifted from the inside at the bottom. The tapered seat provides a good seal against the plastic melt. A taper angle of 15° to 45° has been used successfully. Disk ejectors can be very well cooled with bubblers (Figure 12.35).

To support the demolding process, venting pins can be employed. Figure 12.36 shows a venting pin, which is actuated by compressed air. The air pushes the pin back and opens the exit into the cavity space. The entering air prevents vacuum formation and facilitates the demolding process at the same time. This design is not suited, however, for soft and sticky materials [12.1].

To avoid the formation of vacuum, which acts against the demolding process, it has been suggested to convert the ejector plate into a piston, which would compress the air and makes it flow past the ejector pins under the molding [12.25].

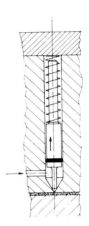

Figure 12.36   Venting pin acting as air ejector [12.1]

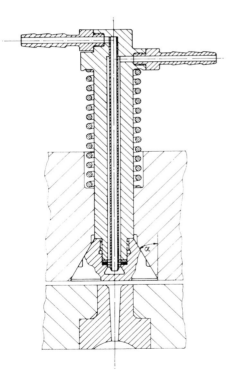

Figure 12.35   Disk or "mushroom" ejector with bubbler [12.1]

12.3.3 Ejector Assembly

Ejector pins together with ejector plates, retainer plates, several stoppers, and a return mechanism form the ejector assembly. If several ejector pins act on the molding, they all have to be actuated at the same time. They are, therefore, assembled in ejector plates and move simultaneously with them. Unevenly actuated pins would cause the molding to be cocked. It would jam in the mold.

Ejector pins are retained in the ejector-retainer plate, which is bolted to the ejector plate. This plate is actuated by a bolt connected to the ejection system of the machine. Stop pins limit the travel of the ejector assembly during demolding.

Ejector pins have to have sufficient lateral play in the retainer plate that they can adjust themselves to the holes in the mold plates (Figure 12.37). This is essential because all plates have different temperatures during operation. The ejector plates are hardened so that the pins cannot dent them. They should not be carried by the ejector pins because this may cause the plates to tilt and the pins to jam. Therefore, the ejector plates have to be guided [12.19]. Special leader pins or bolts can be used for this purpose. At the onset of demolding the whole ejection force is concentrated at the center of the ejector plates. For a simultaneous action of all ejector pins the force must be transmitted uniformly by the ejector plates. Therefore, they have to be sufficiently rigid to make sure that they do not deflect and are dimensioned accordingly. When the mold is closed, the ejector system has to return to its original position without damage to the pins from the opposite mold half. This is usually accomplished by return pins, spring loading, or a toggle mechanism (refer to Section 12.6).

Figure 12.37 illustrates an ejector assembly, which uses an ejector bolt as leader pin.

Normally, especially in single-cavity molds, the ejector assembly is located in the center of the mold. It moves in a "hollow" space within the mold. This may create the

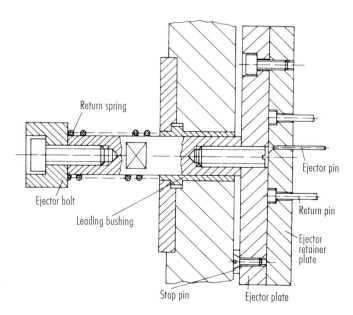

Figure 12.37 Ejector system actuated and guided by ejector bolt

hazard of excessive deflection of the cavity plate which can be avoided by an eccentric arrangement of two or more ejector bolts. As additional advantage, especially in multi-cavity molds, a central support pillar for the cavity plate can be used (Figure 12.38).

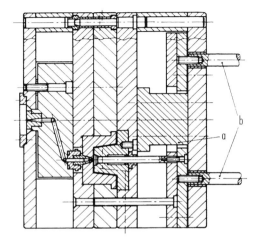

Figure 12.38   Mold with support pillar a for mold plate and ejector systems b positioned at each side [12.26]

12.4   Actuation of the Ejector Assembly

12.4.1 Means of Actuation and Selection of Places of Action

As shown in Figures 12.1A and B there are two directions of ejector action. The preferred one is in the direction towards the nozzle. In a minority of cases, ejection is accomplished in the direction towards the movable side. The reason for this is mostly the solution of a problem with marks on the viewing side.

There is also the other option to eject the molding by an air ejector, either by itself or in support of a mechanical system. Figure 12.1G presents such an ejector. It is preferred with cup-like moldings to supply air under the bottom and prevent formation of a vacuum, which would counteract the ejection movement. A stronger air blow can even completely remove the cup from the mold.

There is, finally, the possibility for split-cavity molds to break the molding and lift it by the movements of the cavity halves. One can find this occasionally for large moldings such as body parts for the automotive industry. The molding is then carried off by a manipulator or a robot. This has the additional advantage of avoiding the hollow space in the mold and an eventual reduction in rigidity.

12.4.2 Means of Actuation

In the majority of all cases the ejector assembly is actuated mechanically by the opening stroke of the molding machine. The molding is broken loose by a thrust when the mold hits the ejector bolt in the machine (Figure 12.39). The ejector pins push the molding towards the parting line until it drops out of the mold by gravity. This kind of actuating the ejector assembly causes little difficulty in design and is the least costly solution. It is

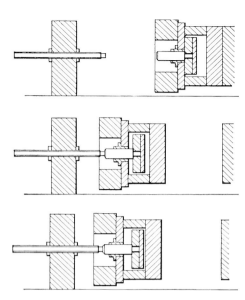

Figure 12.39 Schematic presentation of the demolding process initiated by the opening stroke of the machine [12.27]

useless for fragile parts, though. The initial thrust may damage the moldings. Besides this, the ejection procedure is very noisy.

The design depicted in Figures 12.40 and 12.41 releases the moldings far more carefully. Toggle and links are mounted outside the mold and no openings in the plates

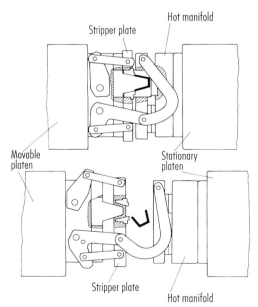

Figure 12.40 Stripper plate is actuated by toggles mounted at the mold [12.28]

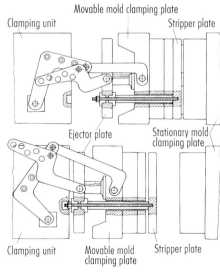

Figure 12.41 Ejector system is actuated by toggles mounted at mold and machine [12.28]

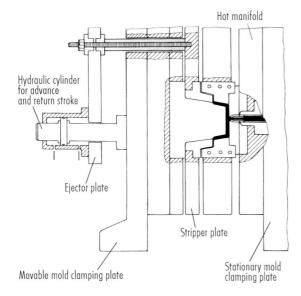

Hot manifold

Hydraulic cylinder
for advance
and return stroke

Ejector plate

Movable mold clamping plate

Stripper plate

Stationary mold
clamping plate

Figure 12.42   Stripper plate,
hydraulically actuated [12.28]

affect the rigidity of the mold. The release procedure is gently initiated immediately with
the onset of the mold opening stroke. This system, too, is rather simple in its design.
Because the whole assembly is located outside the mold, it is easily accessible and
maintained. Special return equipment, is not required. The application of this design is
limited to stripper plates, though.

Besides mechanical actuation, the ejector system can be operated pneumatically or
hydraulically (Figure 12.42). These systems are more expensive because they often need
special, additional equipment, but they operate smoothly and can be actuated at will.
Release force and velocity are adjustable as the conditions require. Special devices or

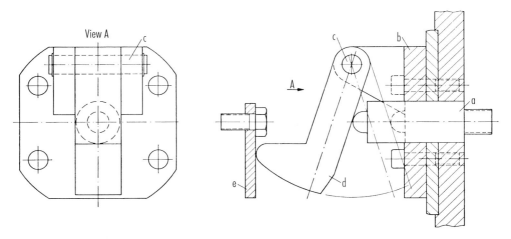

View A

Figure 12.43   Ejector step-up for 35 mm stroke [12.25]
a Ejector bolt, b Bracket, c Pivot pin, d Lever, e Flap

safeguards for returning the ejector assembly with mold closing is not needed if double-acting cylinders are employed.

For proper functioning of all systems, an ample stroke of the ejector plates is required. The plates have to advance the ejector pins (or other means of ejection) sufficiently far towards the parting line that gravity can act on the molding. Only then is a fully automatic operation possible.

In very deep molds (buckets) the ejector stroke may not be sufficient to completely release the molding. Then a combined release method is often employed. The part is first partially released by mechanical operation of the ejector assembly and then blown off the core by compressed air. If no compressed air is available, the part has to be removed manually after breaking.

A combined, stepwise release method is also used if especially high breaking forces are needed. The step-up ejector in Figure 12.43 increases the ejection force two to three times [12.25]. After loosening the molding, it is advanced in a second step or taken off by hand.

12.5 Special Release Systems

12.5.1 Double-Stage Ejection

Large but thin-walled parts often have to be demolded in several stages. This is especially the case if ejector pins cannot act at places where the moldings cannot withstand the forces without damage. An example is presented with Figure 12.44. At first the molding is broken loose by the stripper ring. To prevent formation of a vacuum under the bottom, the ejectors are moved likewise and support the bottom. The element that is used for double ejection is introduced in the literature as ball notch [12.1]. During, demolding the ejector bolt a moves against a fixed stop and so actuates the ejector system f. At the same time the ejector system g is taken along by means of the engaged balls e. Thus, stripper plate and ejector pins simultaneously remove the part from the core. By now both ejector plates have advanced so far towards the parting line that the fixed bolt c has become too short to keep the balls apart. They drop out of the recess and only the ejector plate f is actuated further. Its ejector pins finally release the part.

Because of high wear, the balls (ball bearings), the bushing, and the bolt c have to be hardened. To ensure proper function of the mold, attention has to be paid to the dimensions and the arrangement of the individual elements so that the balls are forced into a rolling motion. The diameter of the balls has to be larger than the diameter of the bolt [12.1].

Figure 12.45 presents a typical two-stage ejector for separating tunnel gates from the molding.

12.5.2 Combined Ejection

Another version of double-stage ejection is the possibility shown with Figure 12.46. During mold opening the part is first stripped off the core mechanically. Final ejection is done with compressed air. This system has the advantage of lower mold costs compared

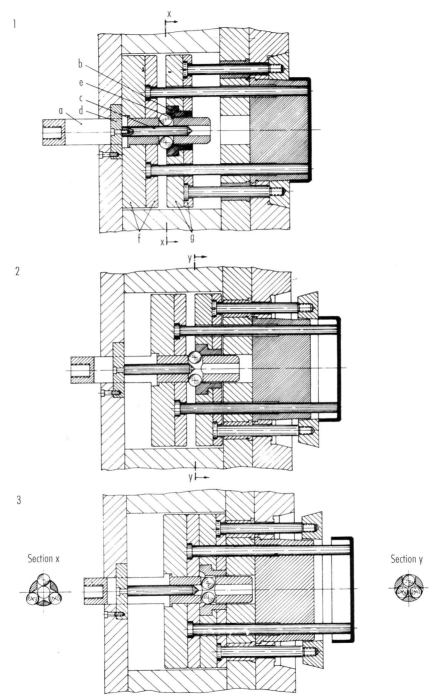

Figure 12.44   Two-stage ejection actuated by ball catch [12.1] a Ejector bolt, b Bushing, c Fixed bolt, d Mounting plate, e Balls, f and g Ejector and ejector retainer plates

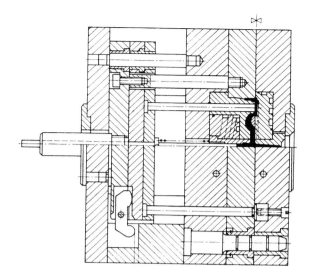

Figure 12.45 Demolding of tunnel gates [12.23]

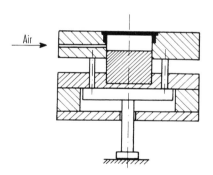

Figure 12.46 Combined ejection [12.24]

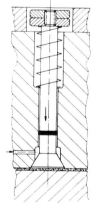

Figure 12.47 Disk or "mushroom" ejector pneumatically actuated (air ejector) [12.1]

with a fully mechanical system and gentler ejection because the release pressure (air) acts upon the entire surface area. It is primarily utilized where the length of the ejection stroke is insufficient for complete demolding (deep parts). The position of the air inlet is arbitrary.

A similar design is presented with a pneumatically operating "mushroom" or "disk" ejector. The compressed air first lifts the disk and the part breaks loose; then it flows past the disk and ejects the molding completely (Figure 12.47).

12.5.3 Three-Plate Molds

If multi-cavity molds or molds for multiple gating of one part are employed, the runner system has to be separated from the molding inside the mold during mold opening and ejected to achieve a fully automatic operation. Therefore, the mold has to have several parting lines at which the mold is opened successively. The ejection movement can be actuated in different ways. Most common is the stripper bolt or the latch bar.

12.5.3.1 Ejector Movement by Stripper Bolt

Figure 12.48 shows a three-plate mold in open (left) and closed position (right) [12.29]. The mold is first opened in the plane of parting line 1. One has to ensure that the part remains still on the core. Thus it is separated from the gate or gates. After a certain opening stroke the floating plate is taken along by the bolt B_1 and the mold opens in the plane of parting line 2. The runner system is still kept by undercuts until it is ejected by an ejector bar which is actuated by bolt B_2.

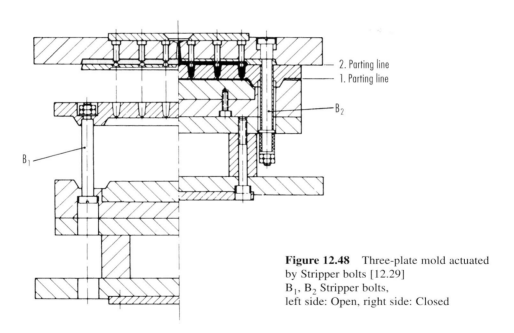

2. Parting line
1. Parting line

B_2

B_1

Figure 12.48   Three-plate mold actuated by Stripper bolts [12.29]
B_1, B_2 Stripper bolts,
left side: Open, right side: Closed

12.5.3.2 Ejector Movement by Latch

This device first locks the floating plate with a latch. After the opening stroke has advanced a certain distance, the release bar unlocks and the mold opens at the parting line 2. Figure 12.49 shows the opening procedure of a mold with a latch. Figure A is the closed position. The latch a locks the floating plate g. The latch can pivot around the bolt

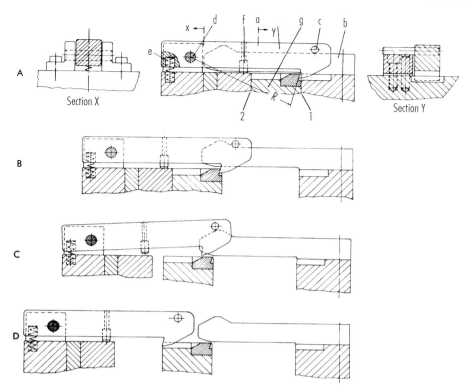

Figure 12.49 Latch assembly [12.1], Explanation of A to D in text.
1, 2 Parting lines, a Latch bar, b Release bar, c Guide pin, d Pivot pin, e Spring, f Stop pin,
g Floating plate

d. It is kept in a horizontal position by the spring e and the stopper f as long as the mold is closed. During mold opening the release bar b lifts the bolt c (Figure B) and releases the latch a (Figure C). With the continuing movement the mold can, therefore, open at the parting line 2. Thus, molding and runner are ejected separately. Because of the occurring high wear, latch and release bar, as well as the stop at the floating plate, have to be made of hardened steel. Such molds can be employed for part weights up to 1 kg. For larger sizes, pneumatic locking and hydraulic opening are preferable [12.1].

In all molds which open in several planes the floating plates have to be precisely guided and aligned so that the cavity surfaces are properly engaged and not damaged. The latch assembly has to be mounted in such a way that it does not interfere with the molding dropping out of the mold by gravity after demolding.

12.5.3.3 Reversed Ejection from the Stationary Side

Some molds are designed in such a way that the molding remains in the stationary mold half. These molds have to have a different demolding action. Demolding takes place by stripping the part off the core. The stripper plate can be actuated by a stripper bolt (Figure 12.50), which is attached to the movable mold half by a pin-link chain, or by

Figure 12.50   Demolding from
stationary half with stripper bolts

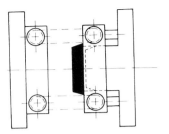

Figure 12.51   Demolding from
stationary half with pin-link chain

hydraulic or pneumatic action. Thus the ejection occurs by traction in the direction of
demolding (Figure 12.51).

There are disadvantages, though. The accessibility of the mold is poor. Two other
options are shown with Figure 12.52. The ejector is actuated by a lever or a crank.

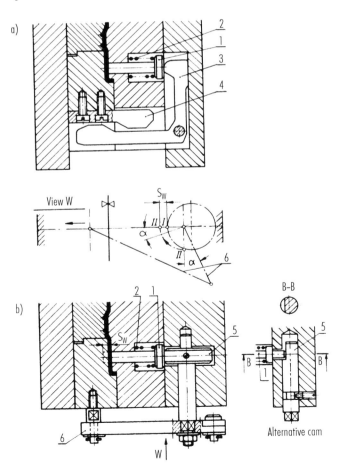

Figure 12.52   Ejector
actuation by lever or crank
for flat moldings which stick
to the stationary side
a) Lever: 1 Ejector, 2 Return
spring, 3 Lever, 4 Cam plate,
b) Crank: 1 Ejector, 2 Return
spring, 5 Cam disk, 6 Crank
[12.23]

12.6 Ejector Return

When the mold is being closed, the advanced ejector pins, stripper plates, etc. have to be returned on time into their position for a closed mold. Otherwise the ejection assembly or the opposite mold half may be damaged. The return can be achieved by various means, either by return pins, by springs, or special return devices.

The most reliable solution for returning the ejector assembly is provided by return pins. Ejector pins with cylindrical head and shaft can be used as return pins. They are either nitrided or annealed, and are kept in the ejector plates like ejector pins. During mold closing they are pushed back by the opposite mold half (Figure 12.53) or by pins mounted in that mold half (Figure 12.54) and return the entire ejector assembly.

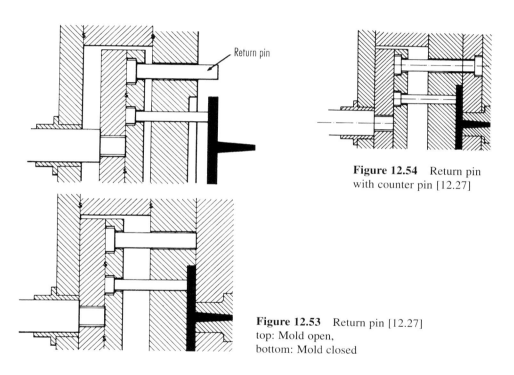

Figure 12.54   Return pin with counter pin [12.27]

Figure 12.53   Return pin [12.27]
top: Mold open,
bottom: Mold closed

Such counter pins are recommended because of the ease of their replacement as parts subject to wear.

In other molds the ejector assembly is returned during closing by springs (Figure 12.37). The springs have to be sufficiently strong to reliably overcome the sometimes considerable friction on ejector and guide pins. If the spring force is insufficient, the mold is damaged during closing. The service life of a spring is limited and depends on the kind of loading and the stress, and also on the number of loading cycles. With such molds it is advisable to provide for a return safeguard. Therefore, a combination of return spring and return pin is frequently used. Since return pins are often

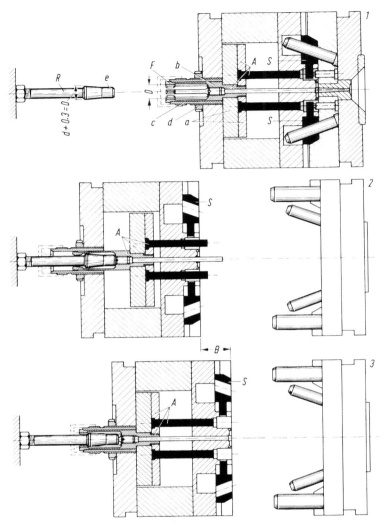

Figure 12.55　Pawl pin [12.30]

1) Mold is closed. Slides are returned. Ejector system A is in returned position.

2) Mold is open. Slide travel is complete. Return bolt R has been inserted in catch F, has actuated ejector system A and ejected moldings. Fingers of catch have locked behind bolt head.

3) Closing mold has positively returned ejector system after having moved the distance B (ejector travel + 5 mm). Slides can be moved now without restriction.

obstructive, electric limit switches are also employed as safeguards. They shut the machine down if the ejector assembly is not completely returned. Besides this, mechanically operating return devices have been developed, which are presented in Figures 12.55 and 12.56.

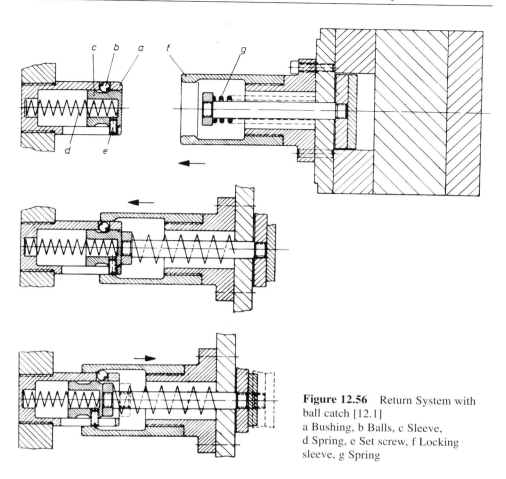

Figure 12.56   Return System with ball catch [12.1]
a Bushing, b Balls, c Sleeve, d Spring, e Set screw, f Locking sleeve, g Spring

In the slide mold in Figure 12.55 a pawl pin b is screwed into the ejector plates a instead of an ejector bolt [12.30]. This pin is surrounded by a clamping sleeve c, which is connected to the ejector housing by a fine thread and secured by a slotted nut d. When the mold opens, the pawl pin hits the profiled tip of the knockout bolt e in the machine, which spreads the catches of the pawl pin. As soon as the machine bolt has dipped into the hollow pawl pin, the catches snap back behind its collar. This creates a positive connection between pin and machine bolt. The machine bolt can now return the ejector assembly while the mold is being closed. At the end of the return stroke the catches free the machine bolt. The length of the stroke can be accurately determined by adjusting the bushing. This system works very reliably.

The return system in Figure 12.56 operates with a ball notch [12.1]. The machine does not carry a knockout bolt but the bushing a, which accommodates the balls b. The small bolt c with a profiled surface can slide in this bushing. A spring d keeps the bolt under tension and the set screw e stops it at the foremost position. In this position the balls catch a recess in the bolt. During mold opening the bushing a enters the sleeve f, which is attached to the movable plate. The bolt c hits an ejector bolt. The engaged balls provide for a solid connection and the ejector plates are moved towards the parting line. An additional stroke frees the balls and the ejector assembly returns under the effect of the spring g provided that spring g is stronger than spring d. During closing of the mold the bushing is retracted and the spring d pushes the bolt into its initial position, The disadvantage of this system is its limited ejection stroke; its advantage the possibility of returning the ejector assembly in a mold still open.

12.7   Ejection of Parts with Undercuts

The question of how to demold parts with undercuts depends above all on the shape and the depth of the undercut. They determine whether the undercut can be directly demolded or special arrangements have to be made to free the undercut with slides, a split cavity, or by screwing it off. Parts that cannot be demolded directly call, therefore, for expensive tooling and possibly additional equipment to the molding machine. Consequently, one should first investigate whether or not undercuts can be avoided by a minor design change of the part such as clever use of a taper or an opening in a side wall. Examples are presented in Figure 12.57.

In the following, such moldings with undercuts, which still can be demolded directly, are discussed first. Snap fits and threads belong to these relatively rare cases.

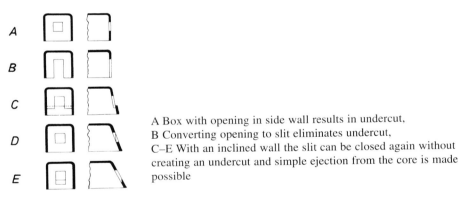

A Box with opening in side wall results in undercut,
B Converting opening to slit eliminates undercut,
C–E With an inclined wall the slit can be closed again without creating an undercut and simple ejection from the core is made possible

Figure 12.57   A change in part design results in a less expensive mold [12.31]

12.7.1 Demolding of Parts with Undercuts by Pushing Them off

Demolding of parts with undercuts by pushing them off without eliminating the undercut of the mold (Figures 12.58 to 12.60) is only possible by deforming the part sufficiently to overcome the undercut. This must not cause a plastic deformation. Top view of Figures 12.58 to 12.60 refer to Figure 12.63.

Table 12.1 lists some data of permissible elongations, which can be equated to the maximal permissible undercuts in thin-walled parts. Other references [12.32, 12.34] state larger permissible elongations but then demolding is not reliable under all possible conditions.

Figure 12.58 Highest strain in a molding during ejection can be expected opposite the tip of the ejector pins [12.24]

Figure 12.59 Highest strain in a molding during ejection from an undercut [12.24]

Figure 12.60 Disk or "mushroom" ejector for ejecting part with negative undercut [12.24]

12.7.2 Permissible Depth of Undercuts for Snap Fits

In practice, one can encounter plenty of moldings with undercuts for snap fits. Three basic shapes predominate: hooked, cylindrical, and spherical moldings. Independent of the kind of undercut there is a linear correlation between the depth of the undercut H and the elongation ε. Consequently the permissible depth of the undercut H_{perm} is limited by the maximum permissible strain ε_{perm} at the elastic limit of the respective plastic material (Table 12.2). The correlations summarized in Figure 12.61 allow the determination of the maximum permissible undercut for all three basic configurations.

It is also important that sections containing undercuts can be elongated or compressed without restrictions. Furthermore, the angle of a snap-joint element has to point in the direction of demolding because otherwise with an angle of $\alpha_2 = 90°$ for a nondetachable joint, demolding is not feasible anymore (Figure 12.62); the undercut would be shorn off. According to publications, suitable joint angles are $\alpha_1 = 10$ to $45°$ [12.32].

Table 12.2 Permissible short-term elongation of thermoplastics

Material	Permissible elongation or maximum undercut %
Polystyrene	< 0.5
Styrene – acrylonitrile – copolymer	< 1.0
Acrylonitrile – butadiene – styrene – copolymer	< 1.5
Polycarbonate	< 1
Nylon	< 2
Polyacetal	< 2
Low-density polyethylene	< 5
Medium-density polyethylene	< 3
High-density polyethylene	< 3
Polyvinyl chloride, rigid	< 1
Polyvinyl chloride, soft	< 10
Polypropylene	< 2

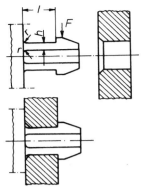

Hook-like parts

$$\blacksquare \quad Rectangular\ cross\text{-}section \quad H_{perm} = \frac{2}{3} \cdot \frac{l^2}{h} \frac{\varepsilon_{perm}}{100}$$

$$\blacktriangle \quad Semicircular\ cross\text{-}section \quad H_{perm} = 0.578 \frac{l^2}{r} \varepsilon_{perm}$$

$$\text{Cross-section:} \atop \text{third of a circle} \quad H_{perm} = 0.580 \frac{l^2}{r} \varepsilon_{perm}$$

$$\text{Cross-section:} \atop \text{quarter of a circle} \quad H_{perm} = 0.555 \frac{l^2}{r} \varepsilon_{perm}$$

These formulae also approximate annular cross-sections. Compared with the exact calculation the error is smaller than 10%.
l = Length of hook, h = Height of hook,
r = Radius, ε_{perm} = Elongation (%)

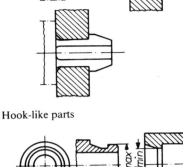

Cylindrical parts

$$H_{perm} = D_{max} - D_{min}$$

Refer to cylindrical parts

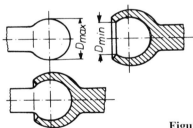

Figure 12.61   Calculating dimensions of snap fits [12.35]

It should also be mentioned that an undercut can be significantly enlarged if rigid, solid cross-sections such, as a cylinder, are divided into flexible elements by slitting (Figure 12.63). For the resulting hook-like parts the release forces are considerably lower. The strain of the outer fiber from bending should be smaller than specified in Table 12.2.

Special attention should be given to the location where the release force acts on. It is best to let the forces act immediately at the undercut on a large area of contact. Therefore, stripper plates are particularly recommended.

Joint angle α_1 and holding angle α_2
Detachable connection

Non-detachable connection for $\alpha_2 = 90°$

Direction of
demolding

Figure 12.62   Angles for snap fits
[12.35]

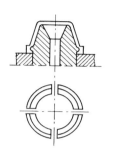
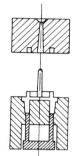

Figure 12.63   Mold with undercut (left) and cavity
on movable side (right) [12.24]

12.8 Demolding of Threads

External threads for modest demands can often be formed by slides. Thus, one deals with
the undercut less expensively than by unscrewing. This option does not exist for internal
threads, though. Occasionally, they can be stripped off. This will be discussed first.

12.8.1 Demolding of Parts with Internal Threads

12.8.1.1 Stripper Molds

The possibility of molding parts with internal threads in a stripper mold is very limited.
It depends on the molding material and the design of the thread. Parts made of materials
with a low modulus of elasticity, such as PA-6, PP, and especially soft PE, can be
demolded by stripping them off the core if they have a suitable thread. The rules of
Section 12.7.1 are the criteria for this. Generally a thread depth of 0.3 mm can still be
demolded by stripping, especially if it is a knuckle thread.

12.8.1.2 Collapsible Cores

Occasionally a collapsible core can be used for small parts. The thread is contained in a
split sleeve, which can be stressed or relieved by a tapered pin or sleeve (Figure 12.64).
An additional stripper plate is needed here. Besides its high cost, this mold component

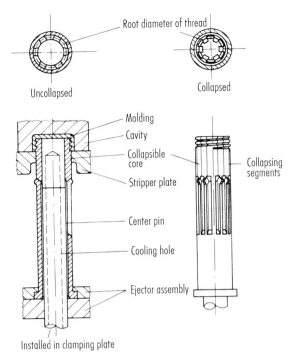

Uncollapsed

Collapsed

Root diameter of thread

Molding
Cavity
Collapsible core
Stripper plate
Center pin
Cooling hole
Ejector assembly

Collapsing segments

Installed in clamping plate

Figure 12.64   Collapsible core [12.36]

has more disadvantages. Marks from the segments of the split core can hardly be avoided and reduce the quality of sensitive parts.

Figure 12.65 shows another possibility of producing and demolding internal threads with a split core. The core is a metal pin with a cap of silicone rubber. The thread is worked into the silicone rubber cap. If the mold is closed, the rubber cap is stressed by the pin. Both form an accurate core for a thread. When the mold opens, the metal pin is retracted and the rubber cap collapses; the molding is released from the core. This mold is considerably cheaper than the previous one with split core and other unscrewing devices but its service life is not particularly long. This disadvantage has little importance because the caps are inexpensive and can easily be replaced. Cooling is a problem, though, and one has to put up with longer cooling times.

The rubber caps can be deformed more than permissible under pressure from the entering plastic melt. This may limit the dimensional accuracy.

12.8.1.3 Molds with Interchangeable Cores

Interchangeable cores are frequently used especially for parts with internal threads, if only short production runs are required. These cores have a tapered shaft, usually with 15° taper, with which they are inserted into an appropriate receptacle by the machine operator [12.1, 12.38]. Such molds are relatively inexpensive. They are primarily employed in cases where high dimensional accuracy is called for. At the end of a cycle, the cores pull the molding out of the cavity. Then molding and core together are taken from the mold. Now the part can be screwed off the core either manually or with the help

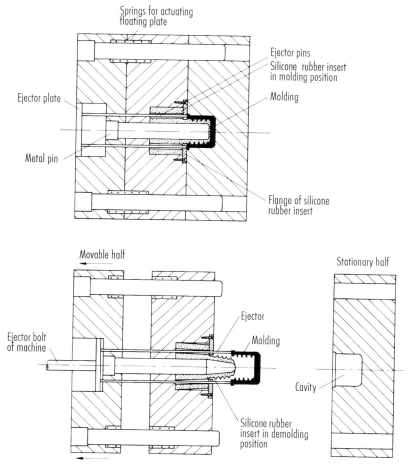

Figure 12.65   Collapsible core made of a metal pin with silicone-rubber cap [12.37]

of a suitable device such as a crank handle or auxiliary motor. The number of cores needed has to be adapted to the molding cycle including the time for cooling the cores and warming them up. One should postpone the demolding from the core until the part has attained room temperature in order to keep its shrinkage as small as possible [12.3, 12.39].

12.8.2 Molds with Unscrewing Equipment

High quality threads can only be molded economically and in large quantities by using an unscrewing device. The mold components which form the threads, generally cores for internal and sleeves for external threads, can be rotated in the mold. During demolding they are either screwed off or out of the part while the mold is either open or closed. The part has to be designed in such a way that it can be protected against rotation.

In all such molds attention has to be paid to an exact mounting and alignment of cores and drive units. Insufficiently supported cores, particularly slender ones, can be shifted from their center position by the entering plastic material more easily than rigidly mounted cores. This would impede or even prevent the unscrewing process because the driving torque becomes insufficient for breaking off the core from the deformed molding. There is also a deviation from the desired geometry of the part. One distinguishes between semi- and fully automatic molds, which will now be discussed in greater detail.

12.8.2.1 Semiautomatic Molds

In semiautomatic molds, the part is demolded by manually operated unscrewing devices. The rotation is transmitted to the thread forming core by a gear drive, a V-belt, or a chain. The mold can be a single or multi-cavity one. Care should be taken, though, that the force at the crank shaft does not exceed 150 N [12.1]. Among the multitude of possible design features, a mold with chain drive is depicted in Figure 12.66. At present, this system is only met here and there.

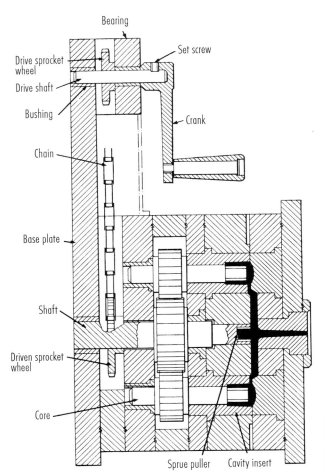

Figure 12.66   Mold with unscrewing device manually operated by crank and chain [12.27]

12.8.2.2 Fully Automatic Molds

The drive force for these molds is transmitted from the opening movement by a lead screw with coarse threads or a rack. Separate drives such as electric, pneumatic, or hydraulic drives are common, too. The latter are often actuated by separate controls.

Unscrewing Molds with Racks
The number of threads that can be molded in these molds is limited by the diameter of the part, the force of the machine or the rack drive, and the stroke of the rack. The stroke is actuated by the opening movement of the machine (Figure 12.67) or by a separate hydraulic or pneumatic actuator.

The functioning of such molds is only flawless if the rack has an accurate pitch, and rack and driven pinion have precise bearings and guides. Otherwise there is a threat that teeth are skipped. This threat is especially great if the required rotation is not transmitted directly to the core but at an angle (Figure 12.67) with bevel gears or another rack.

In the mold in Figure 12.68 the racks are actuated by an external drive (hydraulic or pneumatic). This design has the particular advantage of being able to demold the threads in the closed mold.

Molds with Coarse Lead Screws
Coarse lead screws which are driven by nuts mounted in the machine are considered the simplest, and at the same time, most reliable driving elements for thread-forming cores. They do not cause extended setup time and need no special control of the stroke. During the demolding process they transform the opening movement of the molding machine into a rotary action and drive the cores. Opening stroke and force of the machine determine kind, diameter and number of threads, and the number of cores, which can be demolded. As a general rule 15 threads can be readily demolded by this method. With diameters up to 10 mm, even some more threads are acceptable because a smaller pitch is then used [12.41].

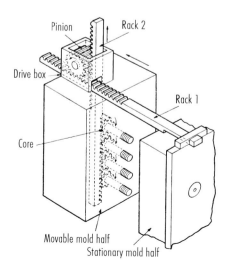

Figure 12.67   Mold with unscrewing device actuated by racks and pinion [12.27]

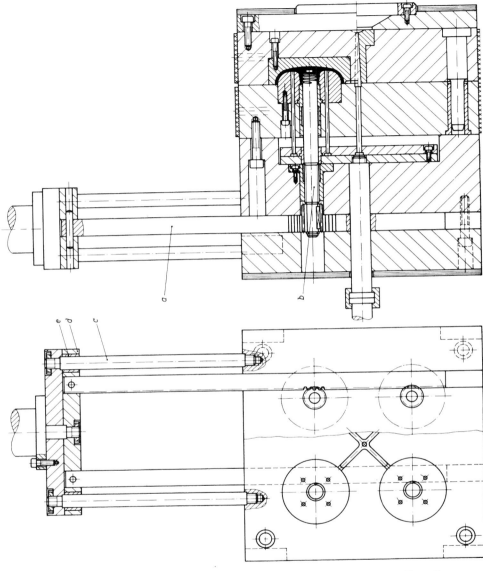

Figure 12.68 Four-cavity mold with unscrewing device actuated by racks [12.40]
a Rack, b Threaded core with lead screw, c Spacer, d Mounting plate, e Leader bushing

Coarse lead screws are available on the market in a number of diameters (20 to 38 mm) with 5 to 12 left or right hand stub threads and a pitch of 80 to 200 mm [12.41, 12.43]. An increasing lead angle improves the efficiency, reduces the contact pressure, and increases the service life of screw and nut. It is advisable, therefore, to work with a spur-gear drive built into the mold. The pitch needed for demolding a thread can be computed with

$$p = \frac{s_1 - s_2}{n}$$

(12.28)

s_1 Opening stroke of machine,
s_2 Length of ejector stroke,
n Number of threads to be demolded,
p Pitch.

With a value of $p < 60$ mm and for multi-cavity molds, a built-in drive should be used. The transmission ratio of the drive is calculated with

$$i = \frac{n \cdot p}{s_1 - s_2}$$

12.29

Some design features are presented in the following.

In the molds in Figures 12.69 and 12.70 the coarse lead screws are mounted in the molds with tapered-roller bearings and the nut is firmly attached to a cross-tie bar of the machine. The design of Figure 12.69 represents a special case because the coarse lead screw is at the same time the thread core of the mold. In contrast to the Figures 12.70 to 12.72, the core does not move axially during demolding. The turning lead screw moves the molding axially and pushes it from the core. It has to be kept in the cavity so that it cannot follow the rotation.

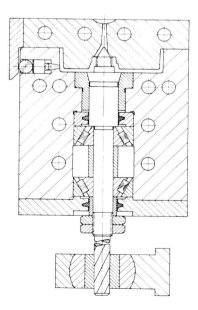

Figure 12.69   Coarse lead screw rotates in mold and also functions as thread-forming core [12.1]

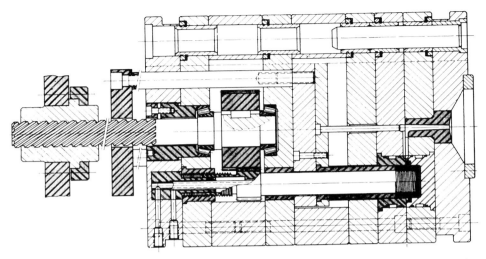

Figure 12.70   Mold with unscrewing device – rotating coarse lead screw is mounted in mold [12.41]

With the design of Figures 12.70 to 12.72, the threaded cores are moved axially. To improve the bearing the cores have another thread with the same pitch at the opposite end, with which they screw themselves into a so-called leader bushing during mold opening. While in Figure 12.70 the nut is mounted in the machine outside the mold, Figures 12.71 and 12.72 present two very elegant and reliable design features. In both

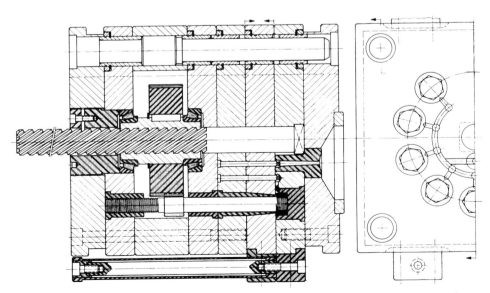

Figure 12.71   Mold with unscrewing device – coarse lead screw is mounted stationary and nut rotates in bearing [12.41]

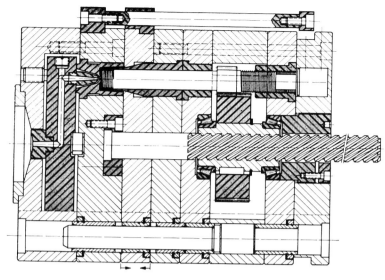

Figure 12.72   Hot runner mold with coarse lead screw and nut mounted in mold. Nut rotates in bearing [12.41]

cases the screw is positively attached to the mold and secured against rotation. With mold opening the nut, which forms the hub in the pinion, rolls off the stationary screw. Because a centrally located screw in Figure 12.71 interferes with the sprue bushing, it was placed off center. This problem was overcome with a hot-runner mold (Figure 12.72).

Molds with Separate Unscrewing Drive

In the automatically operating molds presented so far, the number of threads to be demolded was limited by the stroke of the clamping unit. Demolding with the mold closed was the exception. Frequently demolding threads is practical only in a closed mold. This is the case if the threads are located in different planes or at angles to one another (Figure 12.73). For such complex parts, and if the thread is rather deep, a separate drive independent of the stroke of the molding machine is indispensable. For this reason special electrically and hydraulically operated drive units have been developed. The accuracy of switching on or off such devices has to meet particularly high requirements. If the motor is turned off before the core has reached its end position, considerable flashing would occur. If it is turned off late, the whole torque would be transmitted to the core and jam it. The mold could be damaged [12.42].

If the cores have to be plane with the bottom of the cavity, one often uses a hardened collar as stop. Demolding requires a considerably higher torque, though [12.44]. Damage cannot be excluded in the long run either. Less disastrous are the consequences of a wrongly positioned cores in blind holes such as those for molding threaded caps. This only causes a deviation from dimensions. Occasionally threads are also laterally drilled through. To mold such parts, a lateral slide has to dip into the core. Then the end position of the core has to be mechanically secured. As a simple solution, it is suggested that a spur gear of the drive is provided with an axial hole (between teeth and hub). Closing the

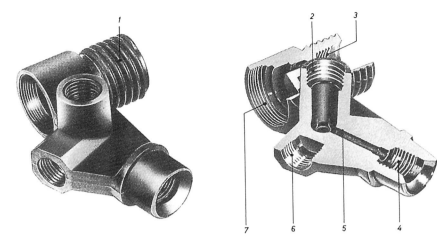

Figure 12.73   Cock-valve body [12.43]
1 External thread, 2 Internal thread and flow channel, 3 Internal thread and flow channel,
4 Internal thread, 5 Flow channel, 6 Internal thread, 7 Internal thread with sealing seat

mold causes a tapered pin to be inserted into this hole. If there is only a small deviation in the position of the gear, it will be corrected. If the pin cannot be inserted, a safety switch is actuated and the mold opens again [12.45].

Figure 12.74 depicts a mold with an electrically operated unscrewing unit that is flange-mounted. This equipment consists of a worm drive with an electric motor with mechanical brake, an automatic switch, and a special clutch. The clutch takes care of a smooth positioning of the core independent of the accuracy of switching. During demolding the moving toothed ring remains in a positively engaged position and transmits the full motor power. The whole torque is available for unscrewing the core. When the core is returned (closing of mold), the gear ring is taken along by a friction cone which can be adjusted to a torque just sufficient to take the core out of the threaded bushing. This design permits the core to strike the collar in the mold even if the drive shaft should move by an additional angle of 30° until the final stop of the motor. This would be a multiple of the listed switch accuracy of 3.5° [12.44]. The automatic switch takes this special demolding technique into account. It is a complete control unit, operates without limit switches (usually a dangerous switching element in injection molding), and can accept additional duties such as control of platen movement [12.42]. The automatic control can switch electric motors in various sizes up to 4 kW.

A special feature of the mold of Figure 12.74 is the simultaneous demolding of the sprue by the components b, c, d, and e during the unscrewing movement. Usually the opening movement of the clamping unit would be used.

Figure 12.75 shows a hydraulically operated unit. It was developed for molding machines with core-pull control. This makes an additional hydraulic system for unscrewing unnecessary.

The depicted unit consists, in essence, of a continuously variable hydraulic motor. With a line pressure of 10 MPa a torque of 170 N · m is available. With constant oil temperature the accuracy of the device is ± 2° [12.45].

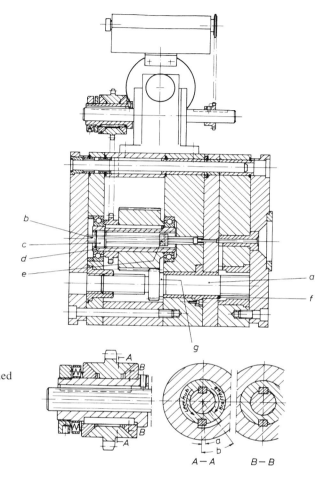

Figure 12.74 Mold with attached unscrewing device, electrically operated [12.44]
a Threaded core, b Keyed shaft, c Movable threaded piece, d Bearing sleeve, e Set collar, f Turn lock, g Auxiliary switch

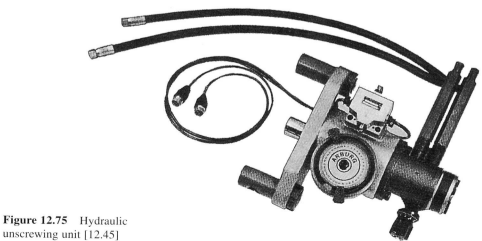

Figure 12.75 Hydraulic unscrewing unit [12.45]

12.8.3 Demolding of Parts with External Threads

External threads can basically be produced in unscrewing molds in the same manner as internal threads. Examples are provided with Figures 12.76 and 12.77. With the design of Figure 12.76 two threads with different pitch, and with that of Figure 12.77 an internal and an external thread are both demolded simultaneously. In each case the sleeve forming the external thread is rotated by a lead screw. It frees the molding, which is kept in the mold, with an axial motion.

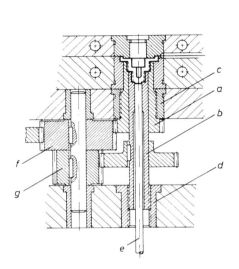

Figure 12.76   Molding with two external threads with different pitch [12.1]
a, b Threaded sleeves, c, d Nuts, e Ejector, f, g Gears

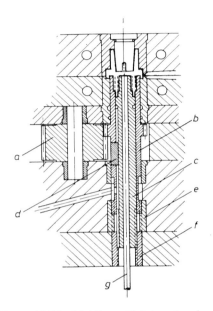

Figure 12.77   Molding with internal and external thread [12.1]
a Pinion, b Sleeve, c Threaded core, d Key, e, f Nuts for axial motion, g Ejector

Unscrewing molds are structures of highest precision. Therefore they are expensive. They should only be employed for molding external threads in high quality parts, for which marks from a parting line cannot be tolerated and large quantities justify the expenses.

In many cases external threads can also be formed by slides especially if the unavoidable marks from a parting line are acceptable.

12.9   Undercuts in Noncylindrical Parts

12.9.1 Internal Undercuts

Figure 12.78 shows such a part as an example, a cover with undercuts in two opposite walls. In simple cases it can be demolded with a collapsible core. The core is composed of several oblique segments, which are stressed or relieved by a wedge. There are no undercuts in the range of the wedge. The employment of such a design assumes a certain minimum size of the mold.

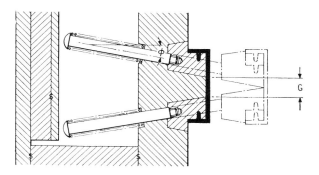

Figure 12.78   Mold with split core [12.27]

12.9.2 External Undercuts

Ribs, cams, flanges, openings, blind holes, as well as threads can form external undercuts. Parts with such undercuts are produced in molds in which that part of the forming contour, which creates the undercut, is laterally moved for demolding. This frees the undercut. Such so-called slide or split-cavity molds are discussed in the following section. Slides actuate a core that forms a locally limited undercut (e.g., a blind hole). Split cavities form whole sides of parts with undercuts (e.g., ribs). Both design features have one thing in common; they have to be built very rigid and leaders and interlocks have to be fitted with special care. Yielding molds expand under the cavity pressure during injection and melt can enter into the parting line. The same happens with inadequately fitted slides. Undesirable flash at the part is the least serious consequence, although it calls for a postmolding operation. It is also possible that such high bending and shear stresses are generated in these components or the actuating elements and, of course, in the mold base, that they are severely damaged and made useless. Slides can be guided with T-grooves, dovetail grooves, or leader pins. Particular significance should be attached to their operating properties in emergencies. Lubricating movable components with molybdenum disulfide can result in staining of moldings or discoloring of the melt. Thus, its use is limited.

   By properly pairing suitable materials, easy sliding has to be ensured and wear inhibited. Aluminum bronze has been successfully used in molds of medium size. In large molds, slide properties have been improved by build-up welding of bronze on the sliding surfaces [12.46]. Hardened and adjustable rails can counteract wear of wedges for proper clamping.

During use, the molds are subjected to the effects of heat from the hot melt, the hot manifold in hot-runner molds, and the heat exchange system. Thermal expansion can cause jamming of sliding components if not taken into account during design. Either the whole mold has to be kept at the same temperature by connecting the movable parts to the heat exchange system or such fits have to be chosen that no inadmissible gaps are generated at parting lines or guide surfaces during operation. Special design features of such molds, are presented in the following sections.

12.9.2.1 Slide Molds

During demolding, slides are positively actuated either by leader pins or lifters, or less frequently by direct hydraulic action. In addition, there are special design features which are discussed below.

Figures 12.79 and 12.80 present the design and installation of leader pins and lifters and characteristic dimensional data for the assembly. The function of these two mold types is demonstrated in Figures 12.81 and 12.82. While a leader pin moves the slide simultaneously with mold opening, a lifter allows a delayed onset of the lateral movement.

Commonly available leader pins can be used. Their dimensions are determined by their loading from release forces (Section 12.2), the weight of the slides and the resistance from friction. Figure 12.83 depicts a leader pin and the forces acting upon it during an upward movement of the slide. The force that acts upon the leader pin can be computed from the motion of a body on an oblique plane (Figure 12.84).

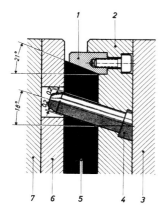

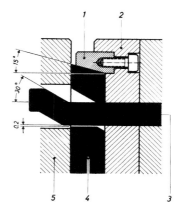

Figure 12.79   Schematic presentation of a cam pin assembly [12.2]
1 Heel block, 2 Mold plate, 3 Clamping plate, 4 Cam pin, 5 Slide, 6 Mold plate, 7 Support plate

Figure 12.80   Design with lifter [12.2]
1 Heel block, 2 Mold plate, 3 Lifter, 4 Slide, 5 Mold plate

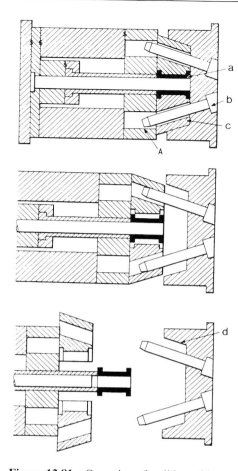

Figure 12.81   Operation of a slide mold with cam pins [12.27]
a Molding, b Cam pin, c Slide, d Clamping surface

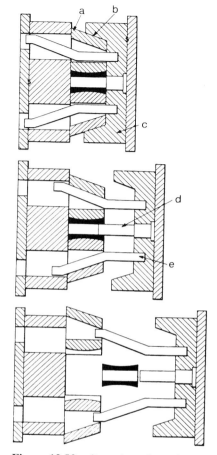

Figure 12.82   Operation of a mold with lifters [12.27]
a Slide, b Clamping surface, c Stationary mold half, d Core, e Lifter

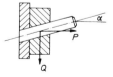

Figure 12.83   Forces acting on a cam pin

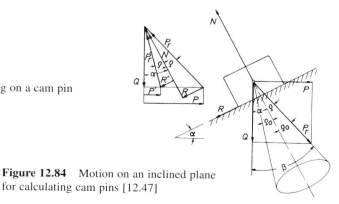

Figure 12.84   Motion on an inclined plane for calculating cam pins [12.47]

The resultant from the individual forces P_R follows from

$$P_R = \sqrt{Q^2 + P^2} \qquad\qquad (12.30)$$

Substituting for P

$$P = Q \cdot \tan(\alpha + \rho), \qquad\qquad (12.31)$$

results in

$$P_R = \sqrt{Q^2 + Q^2 \cdot \tan^2(\alpha + \rho)}, \qquad\qquad (12.32)$$

or

$$P_R = Q\sqrt{1 + \tan^2(\alpha + \rho)}. \qquad\qquad (12.33)$$

Where

$$\tan \beta = \tan(\alpha + \rho); \qquad\qquad (12.34)$$

$\tan \beta$ should generally not exceed the value of 0.5 [12.49].
From the diagram of forces with $R = f \cdot N$

$$\tan \rho = \frac{R}{N} = f \cdot \frac{N}{N} = f \qquad\qquad (12.35)$$

The coefficient of friction f is 0.1 for steel moving against steel. This allows the angle for the inclination of the pins to be computed.

The resulting force perpendicular to the pin determines the cross section of the pin and is computed with

$$A = \frac{Q}{\tau} \qquad\qquad (12.36)$$

or in this case

$$A = \frac{Q\sqrt{1 + \tan^2\beta}}{\tau}\cos\alpha \qquad\qquad (12.37)$$

In Figure 12.85 the opening force acting upon the pin is plotted against time. The full force acts upon the pin only at the moment of breaking the part loose [12.48].

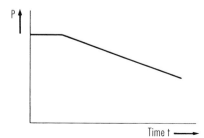

Time t ⟶

Figure 12.85   Load upon a cam pin plotted against time [12.48]

In order to evade overloading the cam pins, one has found out that the most favorable range for the inclination of cam pins is between 15 and 25 °C. Larger angles facilitate mold opening while smaller ones provide higher clamping forces. One has to make a compromise, which is determined by the mold size and the clamping and opening forces.

The method of securing the slide with an inserted heel block against shifting under injection pressure is sufficient in most cases (Figure 12.79). The advantage of this design is its simplicity compared with machining a solid plate (Figure 12.81) and faster interchangeability after it is worn out. The angle of the face should be kept 2 to 3° steeper than the corresponding angle of the cam pin to compensate for a possible play between pin and hole [12.1, 12.2, 12.46] and to cause clamping in the closed state and with this a firm seat of the slide.

During mold closing the slides have to be returned to their original position. This is accomplished either by the cam pin (Figure 12.86) or with the face of the mold. The last method is used if only very short cam pins are acceptable for design reasons. This design calls for means to keep the slide in place (with a ball notch in Figure 12.87) so that the pin accurately enters the hole in the slide when the mold is closed.

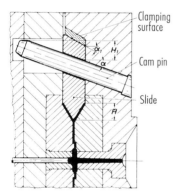

Figure 12.86   Return of slides with cam pins [12.1]
α and α₁ Inclination of cam pin and clamping surface, H Undercut, H₁ Opening stroke of slide

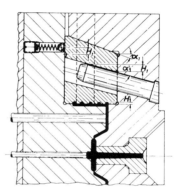

Figure 12.87   Return of slides with clamping surface [12.1]
α and α₁ Inclination of cam pin and clamping surface, H Undercut, H₁ Opening stroke of slide

The length of the cam pin depends on the required opening stroke. Figure 12.87 shows a short pin. A long stroke requires an appropriately long pin. Then, as Figure 12.86 demonstrates, the pin has to enter deep into the movable mold half and an opening has to be provided accordingly.

To obtain longer slide travel with a shorter lead, the angle of inclination has to be increased. Since an angle of 25° should not be exceeded (increased wear, poor force transmission), another design has to be found. Figure 12.88 presents a slide mold in which rollers c are attached to the slide a. They are guided in oblique grooves b. It is the advantage of this design that the angle of the oblique grooves can be increased to more than 45°. This results in the desired longer slide travel with shorter guide ways. Because of the rolling motion, there is considerably less friction and wear than with sliding cam pins.

Another interesting variation is shown in Figure 12.89. The mechanism for actuating the slides is mounted outside the mold. This provides more space for the cavity. The device operates with two racks with helical gearing perpendicular to one another. They are engaged within a leader block. This equipment and its variations are commercially available as standards.

In contrast to pins, lifters can delay the onset of the slide motion. The mold can open for a certain distance, which depends on the contour of the lifter, before the slides are laterally moved and the undercuts are freed for demolding. This makes it possible to achieve a partial ejection of the molding from the core while the slides are still closed; e.g. for deep sleeves or tumblers. After the slides have released the part, it can be ejected pneumatically.

The angle of the slope of the lifters should be between 25 and 30°. The angle of the clamping faces may be more acute. This increases the clamping force. Experience

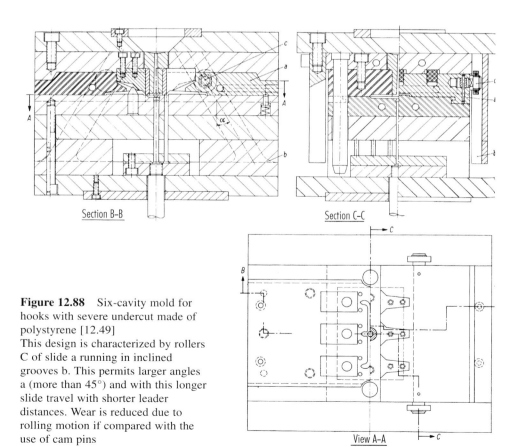

Figure 12.88 Six-cavity mold for hooks with severe undercut made of polystyrene [12.49]
This design is characterized by rollers C of slide a running in inclined grooves b. This permits larger angles a (more than 45°) and with this longer slide travel with shorter leader distances. Wear is reduced due to rolling motion if compared with the use of cam pins

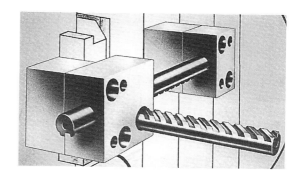

Figure 12.89 Slide mold operated by two beveled racks and guide block [12.50]

indicates an angle of about 15°. Figures 12.90 and 12.91 show two examples of applications and the appropriate lifters [12.1, 12.2].

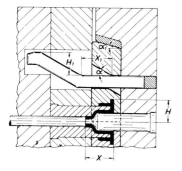

Figure 12.90 Mold with lifter [12.1]
H Undercut, H_1 Opening stroke of slide,
α_1 Inclination of clamping surface, α Inclination
of lifter, x Height of molding, x_1 Opening stroke;
after this stroke the lifter initiates the motion of
the slide

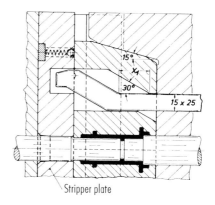

Stripper plate

Figure 12.91   Mold with lifter [12.1]
x_1 Initial opening stroke
(see explanation to Figure 12.90)

12.9.2.2 Split-Cavity Molds

If entire side faces have to be moved away to allow a stripping of the part from the main core, a split cavity is used. Typical examples are boxes with outside ribbing and openings in the side wall. Even in cases where the clamping force of the machine is insufficient, the base of a split cavity mold can provide part of the needed clamping force.

Figure 12.92 demonstrates the mode of operation of a split-cavity mold. The wedge shaped sliding components are guided in a frame, which is adequately dimensioned so that it cannot expand under the cavity pressure during injection. Otherwise flashing is inevitable. The slides are guided in a dove-tail or T groove. The possibilities of compensating for wear or keeping it small have already been discussed in connection with the guides for cam pins. The slides should have a taper of 10 to 15°. This range has been successfully used in practice. Smaller angles may lead to jamming in the frame under the effect of the clamping force, while larger angles react against the clamping force. These molds are locked by the opposite mold half, which also can accommodate protruding slides in a tapered recess (Figure 12.93). The same effect can be obtained with a design according to Figure 12.94. Here the expansion of the frame during injection is prevented with a tapered fit between movable and stationary mold half [12.1].

The opening motion of a split-cavity mold can be positively initiated by the opening movement of the molding machine or by a separate actuator. The machine movement acts upon the slides with toggles or latches (Figure 12.92), ejector pins (Figures 12.93

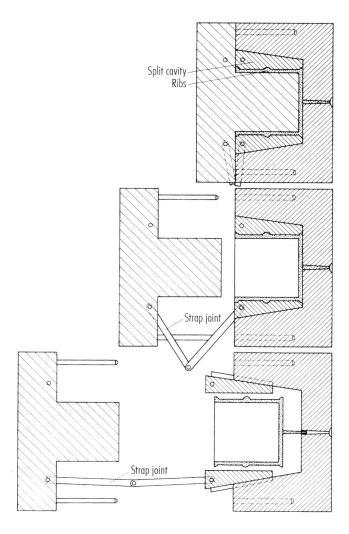

Figure 12.92   Operation of a split-cavity mold actuated with strap joints [12.51]

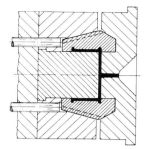

Figure 12.93 Countertaper to improve mold sealing [12.1]

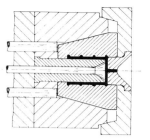

Figure 12.94 Tapered seat in the stationary half prevents excessive breathing of mold [12.1]

and 12.94), links (Figure 12.95), springs (Figure 12.96) or lifters (Figure 12.97). Hydraulic cylinders can be employed as separate actuators (Figures 12.98 to 12.100).

If ejector pins, links or springs are used as actuators, an additional demolding mechanism is usually needed. With toggles, latches or cams the lateral opening motion can be delayed until the molding is stripped off the core. Thus, additional ejection is generally not required.

It should be mentioned, finally, that spring operated split-cavity molds allow only a short opening stroke, which is limited by stops (Figure 12.96).

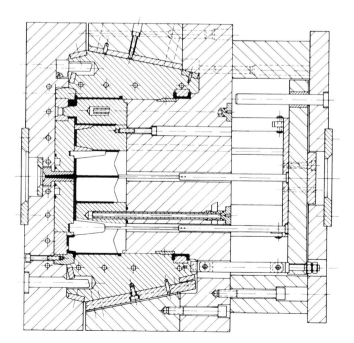

Figure 12.95 Split cavity actuated by links [12.51]

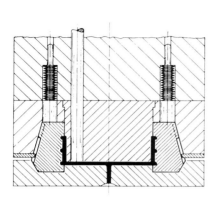

Figure 12.96   Split cavity actuated by
springs, stroke limited by stop pins [12.1]

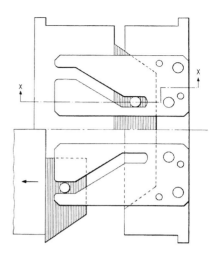

Figure 12.97   Split cavity operated by
cam [12.1]
top: Mold closed,
bottom: Mold opened

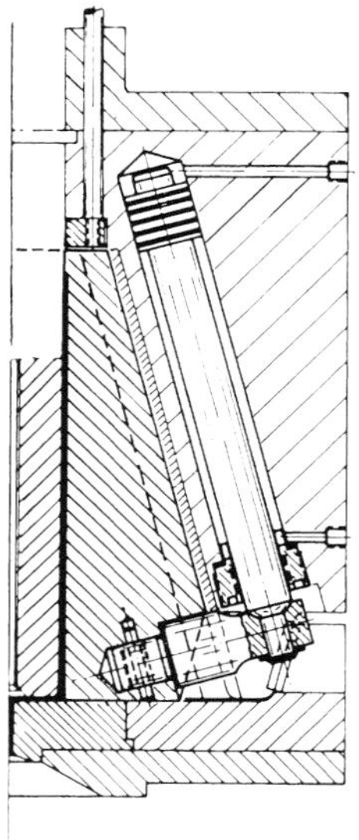

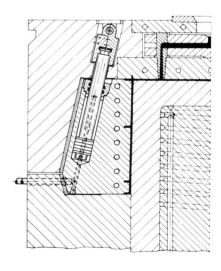

Figure 12.98   Split cavity operated by
hydraulic cylinder [12.1]

Figure 12.99   Split cavity operated by
hydraulic cylinder [12.1]

Figures 12.98, 12.99, and 12.100 illustrate split-cavity molds, which are actuated by
hydraulic cylinders. They should be controlled in the framework of a fully automatic or
a programmed control of the machine to increase reliability of operation. The hydraulic
cylinder has to be mounted in such a way that no lateral forces act upon the piston. They

would impede its function. For this reason the piston cannot accept any guiding functions while moving the slides. Adequate cooling of the slides has to keep the temperature of the hydraulic system low and provides for fast cooling of the injected material [12.1]. Each slide has to be sufficiently cooled and should have a separate cooling circuit with temperature control.

12.9.3 Molds with Core-Pulling Devices

When producing large pipe fittings the cores have to be pulled before the part is demolded. Figure 12.101 pictures a sectional view of such a mold.

At first the core 3 is pulled followed by core 4. Because pulling of the core 4 poses the hazard of tearing the socket off the pipe, the forming component 6 is pushed from the core 3 by the piston 7 until the core 4 is loosened from the fitting. The component 6 is taken along by the screw 8 after a short free travel.

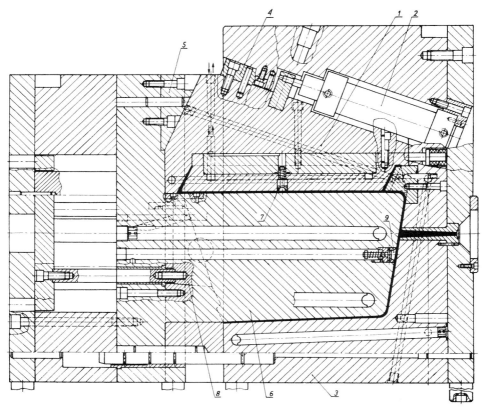

Figure 12.100   Operation of split cavity by hydraulic cylinder [12.23]
1 Split cavity half, 2 Hydraulic cylinder, 3 Frame, 4 Guide bar, 5 Clamp bar, 6 Core, 7, 9 Air valves, 8 Stripper ledge

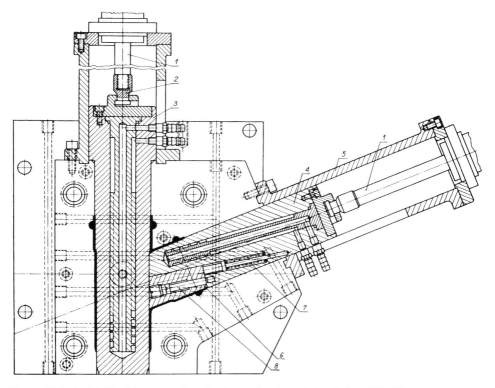

Figure 12.101   Mold with core pullers for the production of pipe fittings [12.23]
1 Pull rod hydraulically operated,
2 Tie-in of pull rod to core,
3 Core for main pipe,
4 Core for socket,
5 Extended core guide,
6 Socket release,
7 Piston for socket release,
8 Holding pin for socket release

References

[12.l]   Möhrwald, K.: Einblick in die Konstruktion von Spritzgießwerkzeugen. Garrels, Hamburg, 1965.
[12.2]   Mink, W.: Grundzüge der Spritzgießtechnik. Kunststoffbücherei. Vol. 2. Zechner & Hüthig, Speyer, Wien, Zürich, 1966.
[12.3]   Entformungseinrichtungen. Technical Information, 4.4, BASF, Ludwigshafen/Rh., 1969.
[12.4]   Bangert, H.: Systematische Konstruktion von Spritzgießwerkzeugen und Rechnereinsatz. Dissertation, Techn. Univ., Aachen, 1981.
[12.5]   Kaminski, A.: Messungen und Berechnungen von Entformungskräften an geometrisch einfachen Formteilen. In: Berechenbarkeit von Spritzgießwerkzeugen. VDI-Verlag, Düsseldorf, 1974.
[12.6]   Karakücük, B.: Ermittlung von Entformungskräften bei hülsenförmigen Formteilen. Unpublished report, IKV, Aachen, 1979.

[12.7]   Schlattmann, M.: Messung von Entformungskräften. Unpublished report, IKV, Aachen, 1978.
[12.8]   Spritzgießtechnik. Publication, Chemische Werke Hüls AG, Marl, 1979.
[12.9]   Yorgancioglu, Y. Z.: Ermittlung von Entformungsbeiwerten beim Spritzgießen von Thermoplasten (PS, ABS, PC). Unpublished report, IKV, Aachen, 1979.
[12.10]   Ribbert, E. J.: Ermittlung von Entformungsbeiwerten beim Spritzgießen von Thermoplasten (PP, PE). Unpublished report, IKV, Aachen, 1979.
[12.11]   Schrender, S.: Ermittlung von Entformungskräften beim Spritzgießen von Thermoplasten. Unpublished report, IKV, Aachen, 1979.
[12.12]   Bangert, H.; Döring, E.; Lichius, U.; Kemper, W; Schürmann, E.: Bessere Wirtschaftlichkeit beim Spritzgießen durch optimale Werkzeugauslegung. Paper block VII at the 10th Tech. Conference on Plastics, IKV, Aachen, March 12–14, 1980.
[12.13]   Cordes, H.: Theoretische Ermittlung von Entformungskräften. Unpublished report, IKV, Aachen, 1975.
[12.14]   Aengenheyster, G.: Gestaltung und Dimensionierung von Verbundkonstruktionen mit Thermoplast- und Elastomerkomponente. Dissertation, RWTH, Aachen 1997.
[12.15]   Koos, W.: Finite – Elemente – Berechnung der inneren mechanischen Beanspruchungen in einem Radialwellendichtring bei der Entformung. Unpublished report, IKV, Aachen, 1993.
[12.16]   Benfer, W.: Algorithmus zur rechnerunterstützten mechanischen Auslegung eines Spritzgießwerkzeuges. Unpublished report, IKV, Aachen, 1980.
[12.17]   Präzisions-Schleifteile. Catalog, Drei-S-Werk, Schwabach.
[12.18]   Handbook of Standards, Sustau, Frankfurt.
[12.19]   Entformungseinrichtungen. Technical Information, 4.4, BASF, Ludwigshafen, 1969.
[12.20]   Catalog of Standards, Hasco, Lüdenscheid.
[12.21]   Handbook of Standards, Strack-Norma GmbH, Wuppertal.
[12.22]   Schürmann, E.: Abschätzmethoden für die Auslegung von Spritzgießwerkzeugen. Dissertation, Tech. Univ., Aachen, 1979.
[12.23]   Zawistowski, H.; Frenkler, D.: Konstrukcja form wtryskowych do tworzyw termoplastycznych (Design of injection molds for thermoplastics), Wydawnictwo Naukowo-Techniczne, Warszawa, 1984.
[12.24]   Morgue, M.: Moules d`injection pour Thermoplastiques. Officiel des Activités des Plastiques et du Caoutchoucs, 14 (1967), pp. 269–276 and pp. 620–628.
[12.25]   Gastrow, H.: Der Spritzgießwerkzeugbau in 100 Beispielen. 3rd Ed., Carl Hanser Verlag, Munich, 1966.
[12.26]   Lohmann, A.: Auswerfereinrichtungen an Spritzgießmaschinen. Kunststoffe, 59 (1969), 3, pp. 137-139.
[12.27]   Pye, R. G. E.: Injection Mould Design (for Thermoplastics). Iliffe Books, London, 1968.
[12.28]   Actuation Methods for Part Ejection. Prospectus, Husky GmbH, Hilchenbach/Dahlbruch, 1973.
[12.29]   Spritzguß-Hostalen PP. Handbook, Farbwerke Hoechst AG, Frankfurt.
[12.30]   Automatische Auswerfer-Rückzug-Einrichtung. Prospectus, Zimmermann, Lahr/Schwarzwald.
[12.31]   Kuroda, J.: Mold Designing and Construction for Automation and High Cycle Molding (1). Jpn. Plast. Age, 11 (1973), pp. 39–44.
[12.32]   Schnappverbindungen. Material sheet 3101.1, BASF, Ludwigshafen, 1973.
[12.33]   Halbzeugverarbeitung. Information for Tech. Application, Farbwerke Hoechst AG, Frankfurt, 1975.
[12.34]   Erhard, G.: Schnappverbindungen bei Kunststoffteilen. Kunststoffe, 58 (1968), 2, pp. 131–133.
[12.35]   Berechnen von Schnappverbindungen mit Kunststoffteilen. Information for Tech. Application, Farbwerke Hoechst AG, Frankfurt, 1978.
[12.36]   Collapsible Core. Prospectus DME Madison Heights (Detroit), USA, 1970.
[12.37]   New Collapsible – Core Tooling System. Br. Plast., 44 (1971), 9, pp. 195–196.
[12.38]   Spritzgießwerkzeuge. Information, H. Weidmann, Rapperswil AG, Switzerland, 1972.
[12.39]   Stoeckert, K.: Werkzeugbau für die Kunststoffverarbeitung. 3rd Ed., Carl Hanser Verlag, Munich, 1979.

[12.40]   Müller, M.: Vierfach – Abspindel-Werkzeug mit Zahnstangen. Kunststoffe, 66 (1976), 4, p. 201.

[12.41]   Steilgewindespindeln mit Muttern. Prospectus, Zimmermann, Mahlberg.

[12.42]   Entformung von Spritzteilen mit Gewinden. Plastverarbeiter, 30 (1979), 4, pp. 189–192.

[12.43]   Mink, W.: Grundzüge der Spritzgießtechnik. 5th Ed., Zechner & Hüthig, Speyer, 1979.

[12.44]   Schneckengetriebe mit Bremsmotor, Schaltautomat und Spezialkupplung. Prospectus, Zimmermann, Mahlberg.

[12.45]   Hydraulische Ausschraubeinheit zum Entformen von Gewindeteilen. Arburg heute, 5 (1974), 7, pp. 31–37.

[12.46]   Reimer, V. v.: Konstruktionselemente der Spritzgießformen. Ind. Anz., 93 (1971), 104, pp. 2657–2659.

[12.47]   Sass, F.; Bouché, Ch.: Dubbels Taschenbuch für den Maschinenbau. Vol. 1. Springer, Berlin, Göttingen, Heidelberg, 1958.

[12.48]   Catić, I.: Calcul dimensionnel rapide des broches inclinées. Plast. Mod. Elastomères, 17 (1965), pp. 99–105.

[12.49]   Trapp, M.: Bewegungselemente für Spritzgießwerkzeuge mit langen Schieberteilwegen. Kunststoffe, 63 (1973), 2, pp. 86–87.

[12.50]   Schiebermechanik. Prospectus Hasco, Lüdenscheid.

[12.51]   Kunststoffverarbeitung im Gespräch, 1: Spritzgießen. Publication, BASF, Ludwigshafen/Rh., 1979.

[12.52]   Lindner, E.: Spritzgießwerkzeuge für große Teile. Information from the Laboratory for Tech. Application of Plastics, BASF, Ludwigshafen/Rh.

13 Alignment and Changing of Molds

13.1 Function of Alignment

Injection molds are mounted onto the platens of the clamping unit of the injection-molding machine. The clamping unit opens and closes them during the course of the molding cycle. The molds have to be guided in such a way that all inserts are accurately aligned and the mold halves are tightly closed. Without proper insert alignment, molded parts would exhibit deviations in wall thickness; they would not have the required dimensions.

Because guiding molds with the clamping unit alone is generally not sufficient, injection molds also need a so-called internal alignment. It aligns both mold halves with the necessary precision and prevents their convolute joining.

13.2 Alignment with the Axis of the Plasticating Unit

Precise alignment is mandatory here. Otherwise the nozzle could not be sealed by the sprue bushing, and undercuts occurring at the sprue would interrupt operation. Therefore, alignments concentric with the sprue bushing are used almost exclusively. For this purpose the mold is equipped with a flange-like locating ring, which keeps the sprue bushing in the mold (Figure 13.1) and matches the corresponding opening in the machine platen, the diameter of which depends on the size of the machine platen [13.1].

Locating rings are readily available from producers of mold standards (see Chapter 17) and are machined from case-hardening steel or water-quenched unalloyed tool steel.

The locating ring has a tapered inner bore of appropriated dimensions to allow the nozzle tip to pass through.

A mold has usually only one locating ring. If both mold halves are equipped with locating rings, then a loose fit is needed on the movable side to better align both sides. This is only a help for setting up the mold.

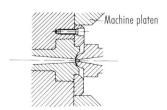

Figure 13.1   Locating ring

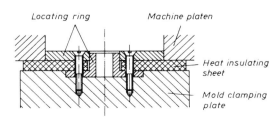

Figure 13.2   Split locating ring [13.5]

The locating ring is slightly press-fitted into the mold plate and fits the machine platen with a close sliding fit. Figure 13.2 shows a locating ring combined with a thermal insulation plate. This design is particularly useful for adding an insulating sheet to the mold. This is done for processing thermosets or thermoplastics if a high mold temperature is needed for molding precision parts.

13.3   Internal Alignment and Interlocking

The mold halves themselves have to be guided internally, by the tie bars of the machine, to obtain the needed accuracy. In small molds this is done with leader pins; pins which protrude from one mold half of the opened mold and slide into precisely fitting bushings in the other mold half during mold closing. This ensures a constant and accurate alignment of both surfaces of flat molds without shifting during injection and the production of moldings. In molds with deep cavities, especially those with long and slender cores, a shifting of the core can occur during injection in spite of exact alignment with leader pins. This has already been discussed in Chapter 11, "Shifting of Cores". Design examples for such molds were presented with Figure 11.21.

Figure 13.3 shows an example for positioning and mounting a leader pin and the appropriate bushing. Four leader "units" (pin and bushing) are usually required for proper alignment. To facilitate the assembly and to make sure that the mold is always correctly put together, one leader pin is offset or made in a different dimension [13.2–13.5]. The latter method may cause fewer difficulties, especially when standard mold parts are used. If two leader pins, one diagonally opposite the other, are made

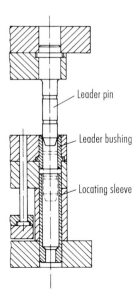

Leader pin

Leader bushing

Locating sleeve

Figure 13.3   Leader-pin assembly
[13.5]

longer, it is easier to slide the two halves together while placing the mold into the machine or during assembly. The leader pins are positioned as close to the edge as possible to gain space for the cavity and an adequate number of cooling lines.

Effective alignment is possible only if close tolerances are kept between leader pins and corresponding holes. This, however, causes considerable wear. Therefore it is not prudent to let the pins slide directly into the respective holes of the mold plates. As a matter of principle, leader bushings should be used to counteract wear and to enable worn-out parts to be exchanged easily. Leader bushings, like leader pins, are made of case-hardened steel with a hardness of 60 to 62 Rc. They are commercially available in various sizes and shapes. Wear can furthermore be reduced by lubricating the pins with molybdenum disulfide. For this purpose pins or bushings have oil grooves. Leader pins without lubrication (Figure 13.7) should only be used for rather small molds or special applications in slides or with ball bushings (Figure 13.9) Low-maintenance operation is achieved through the use of leader bushings with solid lubricant depots [13.2–13.4].

Leader pins and bushings are commercially available in various designs (Figure 13.5). Attention should be paid to the recommended fits (Figure 13.4). The length of leader pins depends on the depth of the cavity.

Leading has to begin before the mold halves are engaged. Therefore, a sufficient length must be selected. Shoulder leader pins (Figure 13.5) can, at the same time, pin mold plates together. Commercial availability is treated in Chapter 17.

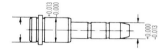

Figure 13.4   Tolerances for leader pins

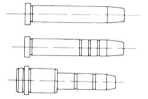

Figure 13.5   Common design of leader pins

The length of leader bushings depends on their diameter. It should be 1.5- to 3-times the inside diameter (Figure 13.6). The corresponding holes in the mold plate have to be drilled according to instructions. Figures 13.7 and 13.8 demonstrate the assembly of leader pins and bushings [13.4].

Figure 13.7 shows the system for the design of simple molds and for leading slides. A jig drill is needed for proper alignment of the holes in the various plates.

Figure 13.8 represents a system with shoulder leader pins. Bores for pin and bushing can be drilled in a single operation (equal diameters).

The system of Figure 13.9 is not used very often. It is carefree and ensures a precise and low-friction movement, but at additional expenses.

With leader sleeves, mold plates can be aligned and connected with one another which would otherwise not be engaged by leader pin or bushing (Figure 13.10). The diameter of the sleeve is kept the same as that of the bushing or the shoulder of a shoulder leader pin. Therefore, all holes can be machined in one pass. The internal diameter is adequately large to permit unrestricted entering of pins. The removal of the locating sleeves can be readily effected through the tapped hole at the end by means of an extractor, e.g., leader pin in Figure 13.11.

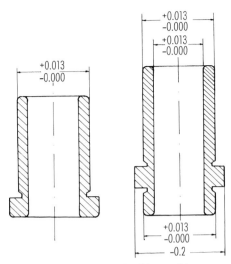

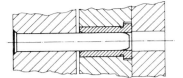

Figure 13.7   Leader pin without oil grooves [13.4]

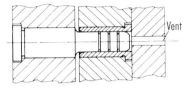

Figure 13.8   Shoulder leader pin [13.4]

Figure 13.6   Shoulder bushings with common tolerances

Among the multitude of leader-pin systems, those shown in Figures 13.11 and 13.12 should also be mentioned. Both are based on components already described: leader pin, bushing and sleeve. The system of Figure 13.11 has pins and bushings with threaded holes and tightening screws, which are propped on the opposite side by head supports. Plates not engaged by pin and bushing are lead by sleeves. This design is more expensive, however, than the one shown in Figure 13.3 but has some decisive advantages. The assembly of the individual plates is not done with screws, and additional drilling of holes is unnecessary. The plates are kept together with the tightening screws. At the same time, the mold plate can be utilized better for accommodating cavity and cooling lines [13.3, 13.5].

The system in Figure 13.12 shows a very different design of leader bushings and sleeves and their assembly in the mold. The bushings consist of three parts: the bushing proper, a retainer ring, and a ring nut. There are two locations for the retainer ring. The bushings can be mounted flush with the plate surface or protruding by 5 mm and

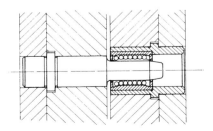

Figure 13.9   Leader pin assembly with ball bushing [13.4]

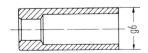

Figure 13.10   Locating sleeve [13.4]

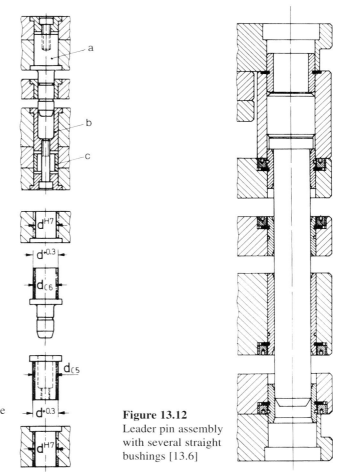

Figure 13.11   Leader pin assembly [13.5]
a Leader pin with tapped hole for retainer screw,
b Shoulder bushing with retainer screw,
c Tubular dowel

Figure 13.12
Leader pin assembly with several straight bushings [13.6]

tightened with the nut. In the protruding position any number of plates can be connected to one another. This kind of bushing can take over the job of bushings with or without retainer ring and can also be used as a leader sleeve. In this case, no ring nut is used. The nut connects the bushings reliably to the mold plates. This could save additional plates, which are needed in other systems for supporting the bushings. The height of the mold is lower. This may compensate the costs for the more elaborate design [13.32].

To ensure proper operation, no lateral forces should act upon the leader systems. If there is no lateral load, the required cross-section of the leader pins does not have to be computed. For oblique pins, especially those acting on slides, the necessary cross-section should be calculated. The same equations can be used as presented in Section 12.9.2.1 for slide molds.

13.4 Alignment of Large Molds

Occasionally, no leader pins are used in large and deep molds such as those for buckets and boxes. Guiding is left to the tie bars of the molding machine during opening and closing until short of complete engagement. Since this accuracy is insufficient for proper alignment, special arrangements become necessary. They are all characterized by the fact that the alignment does not begin sooner than shortly before the mold is closed.

Both mold halves brace one another when closed. Of special advantage is the "pot" design (Figure 13.13) and its variations (Figures 13.14 and 13.15) because it also reacts

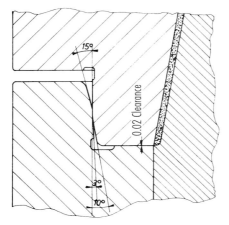

Figure 13.13   Interlock machined into solid material [13.4]

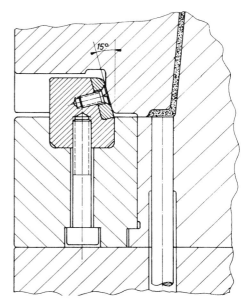

Figure 13.14   Interlock with attached gib [13.4]

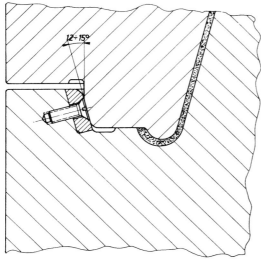

Figure 13.15   Interlock with inserted gib [13.4]

against forces from cavity expansion. The inserted ledges in the variations are easily replaceable after they are worn out. More variations are presented with Figure 13.16.

Frequently bolts are used as aligning interlocks fitted into both mold halves (Figure 13.17). Their center line is not in the plane of the parting line. Thus, both mold halves are braced against one another after clamping. Figure 13.18 shows such a mold. Instead of cylindrical alignment bolts, rectangular interlocks made of shock-resistant tool steel can be employed. Alignment of molds by such interlocks calls for high accuracy of machining, because later corrections are not practical. Frequently, tapered interlocks according to Figure 13.19 are finally used. The long design of the locating bolts and bushings, in contrast to the round design, allows different thermal expansion on the nozzle and clamping sides to be compensated.

If precise alignment and precision locating of the mold halves are necessary, the use of flat leaders with solid lubricant depots may be expedient (Figure 13.20). These permit

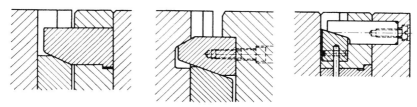

Figure 13.16   Modified interlocks [13.7]

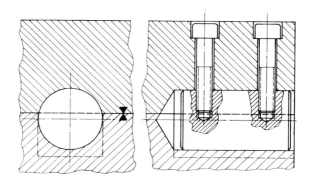

Figure 13.17   Cylindrical interlock [13.4]

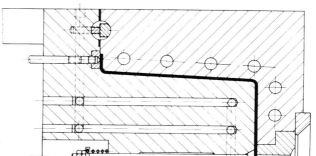

Figure 13.18   Alignment with cylindrical interlock [13.8]

precise centering even before the mold is completely closed, as well as compensation of differential thermal expansion of fixed and moving mold half. Flat leaders may also be combined with circular (conventional) leader elements (Figure 13.21).

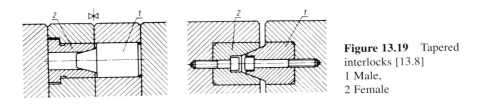

Figure 13.19   Tapered interlocks [13.8] 1 Male, 2 Female

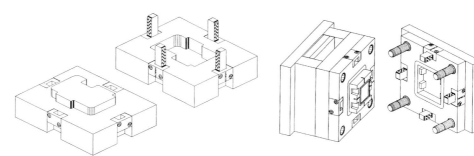

Figure 13.20   Flat leaders with solid lubricant depots [13.2]

Figure 13.21   Combinations of flat leaders with conventional leader elements [13.2]

13.5   Changing Molds

13.5.1 Systems for a Quick Change of Molds for Thermoplastics

Injection molds are usually mounted to the machine platens by mechanical clamping devices (conventional mold clamps with bolts) and connected to power- and water-supply lines. To do this the mold is either horizontally or vertically brought into the machine by a lifting device. Depending on size and weight of the mold and the number of connections, this leads to shutdown times, which may last from an hour to several days (Figure 13.22) [13.9]. Such secondary times affect the productivity considerably, especially in the case of small batch sizes and thus frequent mold changes. The development towards automation, the demand for more flexibility and better economics, lead by necessity to systems for a quick mold change [13.10–13.12]. In spite of this, such systems have not prevailed so far. There are two reasons for this. One is the lack of compatibility among the various systems on the market today [13.13, 13.14, 13.16, 13.17]. The second one is the need for a change of almost all molds used in a machine and the associated high costs.

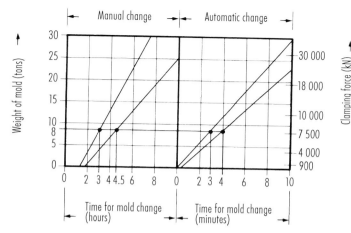

Figure 13.22 Cutback in down time with rapid-clamp systems for mold changes in injection molding machines [13.9]

A quick-change system consists of several components which allows changing injection molds either fully automatically or semi-automatically, controlled by an operator. Such components serve the function of

– detaching and fastening the mold at the machine platens,
– disconnecting and connecting the supply lines,
– bringing the mold into the clamping unit or taking it out.

From this follows the need for these means of a quick-change system:

– quick-clamping devices,
– quick-connection couplings,
– changing equipment.

Besides this, some more components are required for automation of mold changing, which have to be combined to one system (Figure 13.23). Only the combined action of all components permits flexible and automated injection molding [13.13].

Mold design is mostly affected by quick-connection couplings. Two solutions for quick-clamping devices have prevailed on the market. One can distinguish between adaptive and integrated clamping systems, which are usually actuated hydraulically. They can easily be inserted into a concept of flexible automation.

The adaptive clamping system has hydraulically actuated locking cylinders or ledges with integrated collets [13.9] mounted to the clamping platens of the machine, into which the precisely machined clamping plate of the mold is inserted. They are mostly chamfered or provided with a groove (Figures 13.24 and 13.25).

During clamping, the piston or ledge, which is also chamfered, is moved against a corresponding counter chamfer of the mold (Figures 13.24 and 13.25). The counter chamfer is about 5°. This angle causes self-locking (as long as no oil has dripped on it). For reasons of safety, clamping elements are therefore equipped with a proximity switch as a standard [13.9, 13.16–13.19, 13.21].

The integrated clamping system has a hydraulic locking device integrated into the clamping platens. It clamps the mold either directly via bolts mounted on base plates of the mold [13.22, 13.23] or via its own bolts, which press flat against the edges of the mold edges (platens) [13.23] (Figure 13.26).

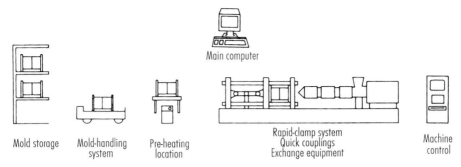

Mold storage   Mold-handling   Pre-heating
                system   location

Main computer

Rapid-clamp system
Quick couplings
Exchange equipment

Machine
control

Figure 13.23   Components of automatic mold change [13.13]

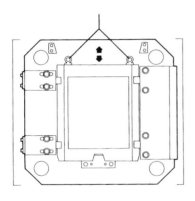

Figure 13.24   Adaptive rapid clamp system
Hydraulic jack (left),
Clamping ledge with integrated lugs, mold plate
chamfered (right) [13.9]

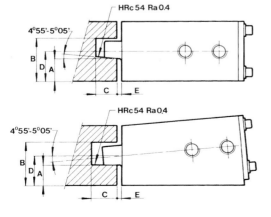

Figure 13.25   Adaptive rapid-clamp
system
A–E Variable dimensions of the individual
design,
top: Cylinder with sloped piston assuring
a self-locking clamp with correspondingly
sloped clamping face,
bottom: Tilted cylinder causes self-
locking clamp with level clamping face

The equipment to automate the mold change is so multifarious today that not everything can be presented here.

All systems presented so far operate basically with the same locking mechanisms. As soon as they are mounted to the clamping platen of the machine, they have to be looked at as a rigid system, which determines the size of the mold plate independent of the mold size. Thus, the base plates of small molds may become disproportionately large, and for large molds one eventually has to switch to a machine with greater clamping force.

The solutions presented so far assume a rectangular clamping plate of the mold. Figure 13.27 presents a design with hydraulic clamping slides acting on locating rings and integrated into the clamping platens of the machine. This system is suited for rectangular as well as for circular molds [13.24, 13.25]. Mechanical systems are possible for smaller molds [13.3, 13.19].

A manually operated quick-clamping device is on the market as a supplement composed of standards [13.2].

A totally different concept is the use of magnetic clamping plates. The surface of the mold is kept uniformly on the clamping plate by a magnetic holding force generated by a permanent magnet. No additional clamping devices are necessary. The system is designed such that the magnetic force is maintained even during a power failure, enabling the mold to be held securely. It only requires a short burst of current for switching the plate on and off.

The advantage of the system is that only the machine needs to be modified. No changes have to be made to the molds. A prerequisite, however, is the use of magnetizable materials. Furthermore, when the mold is being designed, it must be borne in mind that any requisite thermal insulation plates have to be integrated before the mold

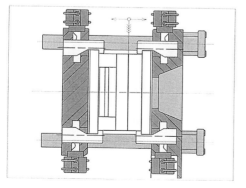
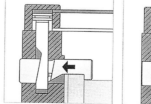
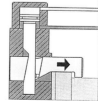

Figure 13.26   Integrated hydraulic rapid-clamping system [13.23]

clamping plate since the magnetic force only acts over a distance of roughly 12 mm in the plates most commonly employed in injection molding machines [13.26]. Proximity switches ensure that the mold is securely mounted on the machine. If, for instance, the mold moves by more than 0.2 mm from the magnetic plate during the opening process, the machine is stopped (Figure 13.28). According to the manufacturer, about 1,000 injection molding machines have been fitted out with magnetic plates.

Aside from the mold, there must be an automatic connection with the ejector system if spring-loaded ejectors are not used. For this purpose, hydraulic clamping systems and mechanical couplings are resorted to [13.2, 13.17, 13.27]

The use of quick-clamping systems for changing molds already results in a considerable shortening of the setup time, but it does not yet ensure a fully automatic change of molds. This is only made possible by employing quick-coupling systems for energy supply and for sensors. The following connections are required for

– heat-exchange medium (oil, water),
– hydraulic oil for slides, core pullers etc.,
– electric connections for heating (hot runners),
– connections for thermocouples and pressure sensors.

Design and installation on both the mold and the machine are set out in [13.27]. Coupling systems are designed as modular systems and consist, depending on size, of individual couplings for energy supply and sensor connections, guide pins, lock and docking cylinders (Figures 13.29 and 13.30). Thus, the systems can be assembled as demanded by the application. They are supplied as standards for manual and automatic operation.

Accurate assembly of these systems is mandatory for their trouble-free function. A small inexactness during assembly can lead to inadmissible displacement, which results in leakage and premature wear.

Therefore, coupling elements are floating in the carrier plates in order to eventually allow an adjustment. Thermal expansions can also be captured this way especially in large molds [13.9, 13.13, 13,18]. A mold-change equipment affects mold design, if at all, only insignificantly. They are accessories, which change molds and, depending on make-up, can accept the function of a mold-conveying system. The complete mold is always exchanged. Depending on the standard of the equipment, molds are immediately ready

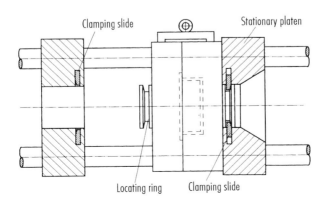

Figure 13.27 Rapid mold-clamping system.
Mold clamped to locating ring [13.24]

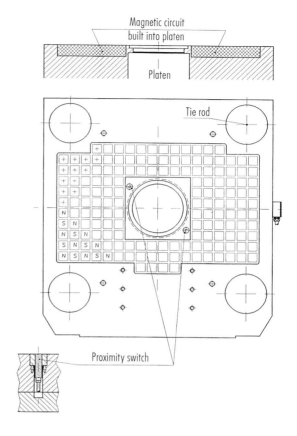

Figure 13.28 Magnetic mounting platen [13.26]

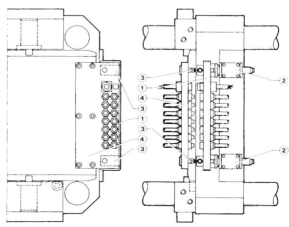

Figure 13.29 Mold with rapid-coupling device [13.9].
As soon as the mold is laterally placed into the machine and securely fastened the movable coupling half 1 is inserted into the coupling half 4 mounted to the mold by pressure cylinders 2. Guide pins align both halves. When they are fully connected, they are interlocked by locking cylinders 3.

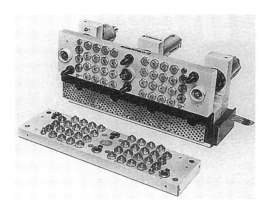

Figure 13.30   Multiple coupling with interlock and insertion equipment [13.9] The model presented here with 32 water, six hydraulic and two compressed-air connections is only one eighth of the complete equipment. The multiple-coupling system is part of the clamping ledge and is equipped with a central insertion cylinder and a locking cylinder on each side. The locking cylinders have hollow pistons which carry the guide pins of the system. As soon as the coupling halves are connected the pins become self-locking by means of a hydraulically actuated chuck.

for operation and are already brought up to working temperature during the change procedure.

The effectiveness of the quick-clamping systems is being continuously improved by the use of mold carriers. The mold, located on the carriers which are attached to the side of the injection molding machine, can be prepared for production before being inserted into the clamping unit [13.23] (Figure 13.31).

The system shown in Figure 13.31 has been developed for the production of small runs, prototypes, and test specimens. It consists of a changing frame and various slots that contain the mold cavities [13.28, 13.29]. The inserts can be locked by the quick-clamping device mentioned at the outset and attached with the quick-fit couplings to the power supply. A similar concept for multicavity molds is presented in [13.30].

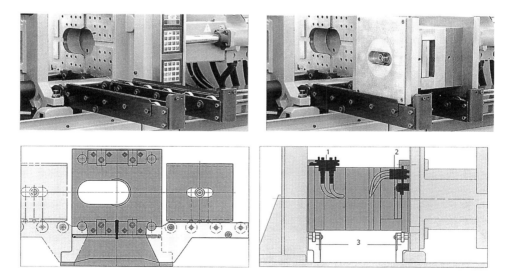

Figure 13.31   Mold carrier-enhances the effectiveness of rapid mounting systems [13.33]

This system cannot be transferred to all types of molds, but offers significant cost advantages for small lots and with simple molds.

13.5.2 Mold Exchanger for Elastomer Molds

A quick-change system for elastomers was designed (Figure 13.33) in accordance with the same concept as was illustrated in Figure 13.32 for thermoplastics. For a "mold

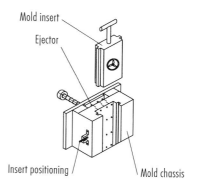

Figure 13.32   Mold-changing system for small production runs [13.29]

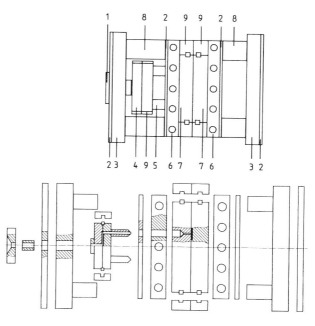

Figure 13.33   Modular mold design [13.31]
1 Locating ring, 2 Insulating sheets, 3 Clamping plates, 4 Cold manifold, 5 Nozzle, 6 Heated plate, 7 Cavity retainer plates, 8 Spacing blocks, 9 Guide ledges

change" only the forming mold plates are exchanged with this system. The changing plates are designed in such a way that they do not contain sensitive mold components, especially no heater elements or sensors. Thus, they can be carried to a cleaning bath without much effort immediately after the exchange. For the exchange of the mold plates an exchange system (Figure 13.34) was built, which operates fully automatically. It takes the plates from a preheating station and carries them to the mold base.

To ensure a fully automatic operation, a cold-runner manifold was used, which was largely separated from the remaining mold and kept at about the temperature of the injection unit. With this, the otherwise considerable material loss from the runner system could be circumvented. The manifold was designed so that it could be exchanged effortlessly.

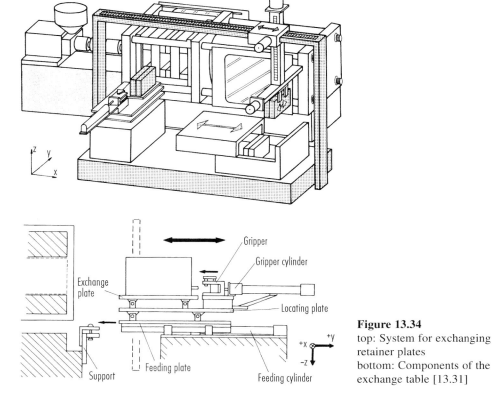

Figure 13.34
top: System for exchanging retainer plates
bottom: Components of the exchange table [13.31]

References

[13.1]   Euromap 2. Injection moulding machines. Mould fixing and connection dimensions. Europäisches Komitee der Hersteller von Kunststoff- und Gummimaschinen, VDMA, Frankfurt, 1995.
[13.2]   Standards, Hasco, 1997, Lüdenscheid.
[13.3]   Standards, EOC, Lüdenscheid, 1996.

[13.4]   Möhrwald, K.: Einblick in die Konstruktion von Spritzgießwerkzeugen. Garrels, Hamburg, 1965.
[13.5]   Handbook of Standards, Sustan, Frankfurt, 1966.
[13.6]   Standards, Prospectus, Zimmermann, Lahr/Schwarzwald.
[13.7]   Zawistowski, H.; Frenkler, D.: Konstrukcja form wtryskowych do tworzyw termoplastycznych (Design of injection molds for thermoplastics). Wydawnictwo Naukowo-Techniczne, Warszawa, 1984.
[13.8]   Lindner, E.: Spritzgießwerkzeuge für große Teile, Information from the Laboratory for Tech. Application of Plastics, BASF, Ludwigshafen/Rh.
[13.9]   Handbuch für den Einsatz von Werkzeugschnellwechselsystemen in Spritzgießbetrieben. Prospectus, Enerpac, Düsseldorf, 1987.
[13.10] CIM im Spritzgießbetrieb. Carl Hanser Verlag, München, 1992.
[13.11] Fernengel, R.: Konzept einer modernen Spritzgießfertigung. Kunststoffe, 84 (1994), 10, pp. 1361–1362 and pp. 1364–1367.
[13.12] Rationeller fertigen mit Werkzeugwechselsystemen Plastverarbeiter, 45 (1994), 2, pp. 87–89.
[13.13] Benfer, W.: Werkzeugwechselsysteme an Spritzgießmaschinen. Kunststoffe, 77 (1987), 2, pp. 139–149.
[13.14] Heuel, O.: Lösen Schnellwechsel-Systeme für Spritzgießformen alle Probleme? Kunststoffberater, 32 (1987), 11, pp. 22–25.
[13.15] Spannen der Werkzeuge vereinheitlicht. Plastverarbeiter, 38 (1987), 2, pp. 82–84.
[13.16] Bourdon, K.: Zuführungs- und Entnahmetechniken an der Spritzgießmaschine. SKZ-Conference, Würzburg, June 19–20, 1991, pp. 83–101.
[13.17] Schultheis, M.: Der Spritzgießbetrieb und seine Wettbewerbsfähigkeit. VDI-K, Düsseldorf (1992), pp. 113–131.
[13.18] Handbuch für den Einsatz von Werkzeugwechselsystemen in Spritzgießbetrieben. Prospectus, Enerpac, Veenendaal, NL.
[13.19] Prospectus. Stäubli AG, Seestraße 250, CH-8810, Horgen/Zürich.
[13.20] Thoma, H.: Rechnereinsatz und flexible Maschinenkonzepte. Kunststoffe, 75 (1985), 9, pp. 568–572.
[13.21] Schneller Werkzeugwechsel möglich. Plastverarbeiter, 37 (1986), 8, pp. 56–57.
[13.22] Prospectus. Engel, Schwertberg, Österreich, 1995.
[13.23] Prospectus. Arburg, Loßburg, 1996.
[13.24] DE PS 2938665 C2, Krauss-Maffei, 1984.
[13.25] Prospectus. Netstal AG, Näfels, Schweiz, 1996.
[13.26] Prospectus. Gühring. Spanntechnik, Komwestheim, 1997.
[13.27] Euromap 11. Automatic mould changing on injection moulding machines. Europäisches Komitee der Hersteller von Kunststoff- und Gummimaschinen, VDMA, Frankfurt, 1993.
[13.28] Backhoff, W.; Lemmen, E.: Stammwerkzeug mit wechselbaren Einsätzen. Information for Tech. Applications, 259/79, Bayer AG, Leverkusen, 1979.
[13.29] Sparmer, P. et al.: Flexibles Fertigungszentrum für das Spritzgießen kleiner Serien. Kunststoffe, 74 (1984), 9, pp. 489–490.
[13.30] Vollaustauschbares Werkzeug-Wechselsystem. Plastverarbeiter, 45 (1994), 2, pp. 42–43.
[13.31] Weyer, G.: Automatische Herstellung von Elastomerartikeln im Spritzgießverfahren. Dissertation, Tech. University, Aachen, 1987.
[13.32] Formenwechsel leichtgemacht. Kunstst. Plast., 32 (1985), 11, pp. 12–13.
[13.33] Halbwerkzeugverarbeitung. Technical Information. Farbwerke Hoechst AG, Frankfurt, 1975.

14 Computer-Aided Mold Design and the Use of CAD in Mold Construction

14.1 Introduction

Development work on the simulation of the injection molding process started in the mid-1970s when the first simple programs for programmable pocket calculators became available for calculating the pressure loss in specified flow channels. The geometry options then available were cylinders for the gate system and plates and circular segments for the molded part, depending on whether the melt flowed through a constant or a divergent channel (Figure 14.1).

Figure 14.1 Differently segmented geometries for calculating the pressure loss in the flow channel

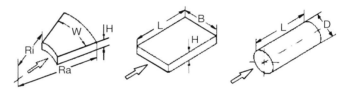

14.1.1 The Flow Pattern Method Pointed the Way Forward

Even 20 years ago, injection molders and mold builders were already confronted with the same problems as today, namely: where should the gates be located, how many gates should there be, and where can weld lines or even entrapped air occur. At that time, the so-called flow pattern method had been developed by the IKV Plastics Processing Institute of the Technical University of Aachen, which made it possible to simulate cavity filling with a compass and pencil on the basis of a developed view of the molded part. Once the flow pattern had been compiled, the developed view was cut out and glued to give the 3D molded part (Figure 14.2). A new flow pattern had to be compiled for each new gate position and this was naturally very time-consuming.

Working on from this, a joint research project was set up with industry under the name CADMOULD with the aim of developing a calculation model for use in the rheological, thermal, and mechanical layout of an injection mold. Those involved in the project were raw materials producers, machine producers, injection molders and producers of standard mold components. At the same time, MOLDFLOW in Australia also developed a system for rheological simulation. These initial programs simply produced tables showing the prevailing pressure losses, viscosities, shear rates and temperatures, by way of a result. This nonetheless marked the start of computer-aided simulation for injection molding.

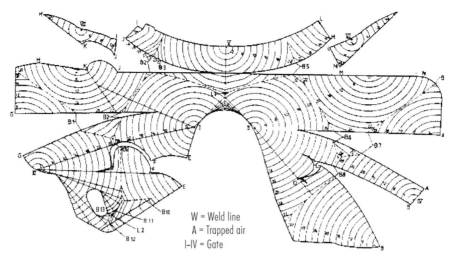

Figure 14.2   Flow pattern method – simulation of mold filling through a developed view of the molded part; diagram of wheel lining

Midway through the 1980s, computers were able to calculate flow patterns. Following this, the pace of development of simulation programs increased, and it was soon possible to calculate not only the filling phase, but also the holding pressure phase, as well as the fiber orientation, shrinkage, and warpage.

14.1.2 Geometry Processing Marks the Key to Success

Injection moldings are almost always shell-shaped, i.e., their thickness is very small in relation to their other dimensions. This makes it possible to perform the simulation in a so-called 3D shell model. In a 3D shell model, the molded part geometry is presented three-dimensionally, with the exception of the molding thickness. The thickness is simply allocated as a parameter. This model has proved its benefit for users in a large number of problem solutions ever since the first computed flow patterns became available, and is still used for calculations even now.

In the past (and, in some cases, today still), only 2D drawings were available for converting a molded part geometry into a 3D shell model. This meant that a pre-processor was required to convert the geometry in the appropriate manner and discretize linear, plane triangles – the so-called finite element network. To begin with, conversion of the geometry for a stacking crate took approximately the same amount of time as compiling a developed view on paper. Once the geometry had been compiled on the computer, however, it could be used several times over to calculate different gating variants, which meant that a considerable amount of time was saved on optimization.

With the development of CAD systems, interfaces gradually became available for the exchange of geometric data, such as IGES or VDA-FS, which further simplified the processing of the geometry.

14.1.3 Complex Algorithms Mastered

While the description of the geometry has essentially remained the same up to today, major advances have been made in the internal computation algorithms, which are not readily evident to the user. Compared with the situation at the outset, the computing time required for an individual molding has not really changed at all, as new and more accurate computation methods have been introduced.

Today, however, the calculations are carried out in layers over the thickness of the molded part, making allowance for intrinsic viscosity, temperature and compression – something which is not readily apparent to the inexperienced observer. The calculation results achieved by these methods tally very well with the situation in practice. If, however, the calculation bases from earlier times are used on present-day hardware, the computation results are achieved within a matter of seconds – including for complex geometries.

14.1.4 Simulation Techniques Still Used Too Infrequently

The simulation of the injection molding process is now regarded as a standard tool. The entire injection molding, process can be calculated, from the filling phase via the holding pressure, right through to the warpage of a molded part that has cooled to room temperature. Special processes can be simulated, such as two-component injection molding, injection compression molding, and the gas injection technique. The processing behavior of elastomers, thermosets and RIM materials can also be simulated today. Despite the extensive simulation options available at present, the processes and methods referred to above still hold considerable development potential.

Despite its invaluable advantages, process simulation is unfortunately used only by a small percentage of industrial companies. Surveys have shown that, on average, cycle time reductions of up to 15% can be achieved, and savings of up to 50% on the cost of mold alterations. Some 90% of the market is still not benefiting from the opportunities offered by simulation software, although an increasing number of companies are buying in simulation services in order to familiarize themselves with the advantages.

Over the past few years, a trend has emerged, with customers increasingly requiring their suppliers to conduct process simulations. This trend also continues further down the supplier line. In many cases, the simulation is used at the acquisition phase already. Increasing use is being made of high-grade yet, in some cases, difficult-to-process materials: requirements on the component have risen and design is imposing ever-greater demands. Prolonged experience and intuition will no longer suffice – there are too many questions that remain unanswered.

14.1.5 Simpler and Less Expensive

Low-cost software has always been available for those making the initial move into simulation. The fact that the majority of plastics injection molders and mold builders have not taken up this software is obviously not due to the investment involved, but rather to the elaborate processing required for the geometry prior to process simulation. The small and medium-sized companies of the sector are subject to such keen cost pressure that they do not have any suitably qualified personnel.

Starter packages from different software companies, such as CADMOULD RAPIDMESH (Figure 14.3) have been available for about a year now. These are not only inexpensive, but also considerably simplify geometry processing. Almost anyone can perform a simulation with a starter package.

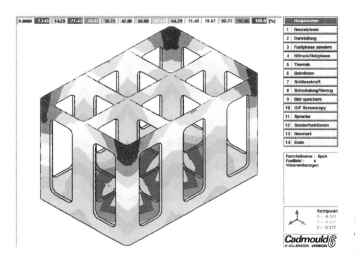

Figure 14.3 Filling pattern simulation for a bottle crate, designed with the program CADMOULD

This has been made possible through a different type of geometry description, taken from the field of rapid prototyping (STL file). A file of this type can be output from most 3D-CAD systems at the push of a button. With an STL file, these starter packages will automatically process the geometry and select the gate positions; they will calculate the flow pattern, the filling pressure and the residual cooling time and also establish the clamping force from the filling pressure. The simulation based on this model will only permit the filling phase to be calculated as yet, however.

14.1.6 The Next Steps already Carved out

The possibilities that exist for simulating the injection molding process in a 3D volume model have been described several times already (Figure 14.4). This technology is still right at its initial stage of development, at least as far as plastics are concerned. The advantage of this model is that no essential simplifications have to be made for the geometry model and hence the full range of physical effects can be described. Examples include the possibility of making allowance for gravitational and inertia effects (development of free jetting). Thick components and components with thick points can be correctly described with this process. A further advantage of the use of the CV-FDM (Control Volume – Finite Difference Method) is the problem-free adoption of CAD geometries and their fully automatic conversion into networks in a matter of minutes. Over and above this, the model always contains the entire shape, ensuring that full consideration is always given to the influence of the mold (e.g. cooling, corner warpage).

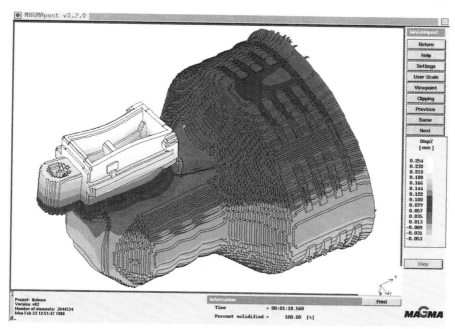

Figure 14.4   3D simulation (volume model); shrinkage and distortion of a lamp socket made of plastic.
Photo: Magma, Aachen

It can be assumed that injection molding simulation will be employed on an increasingly widespread basis in future. A prerequisite for this is a maximum of user friendliness, i.e., a geometry model that is compiled at the press of a button. An appropriate computation process must be available for each individual problem, giving a rapid overview or permitting a specific problem to be calculated as accurately as possible. At the same time, software of this type should offer support in the interpretation of the results, and also automatic optimization strategies.

A comprehensive simulation must cover the low-end ranges (e.g., Rapidmesh), the mid-range with simulations in a shell model, and the high-end, with volume-oriented software. It would be conceivable for the low-end installation to be installed at several points within the company (purchasing, development, marketing) and the other systems in the classical engineering departments.

The injection molding simulation software must additionally be optimally integrated in the company's environment. This means that fully automatic geometry and results interfaces need to be available to other development tools, such as structural and modal analysis systems, and also production, planning, and control, production data collecting system, quality assurance, and quality optimization systems.

14.2   CAD Use in Mold Design

14.2.1 Introduction

Through the consistent use of modern information systems, many companies have increased their competitiveness considerably in recent years. The success brought about by the introduction of a CAD system or the change-over to a more powerful one is frequently measured in terms of the savings made on time and costs during the design process. According to the trade literature, these amount to as much as 75% (e.g. [14.1–14.3]). Similarly, a marked, although less quantifiable improvement in product quality is reported.

While CAD systems were initially aimed at superseding the drawing board, the current trend is towards obtaining an exact copy of the product in the form of a three-dimensional model as early as possible during development and thus to generate a virtual prototype in order that this may be used, with the aid of computers, for further development stages. Necessary geometric models, e.g., for FEA simulation or rapid prototyping/rapid tooling, can be derived with little effort, in some cases, from the solid model. The borders between CAD and CAE in the conventional sense are therefore becoming very fluid.

A study commissioned in 1996 by the CAD CIRCLE [14.4] showed that only two out of every three companies use CAD systems. Mostly only 2D functions are employed and relatively little use is made of generated CAD data for other development stages, e.g., for technical documentation, quality assurance or NC processing. Thus, data that define a product are having to be generated repeatedly. This is expensive in terms of time and errors. Only a small fraction of the enormous potential inherent in CAD systems is currently being exploited.

14.2.2 Principles of CAD

14.2.2.1 2D/3D Model Representation

Compared with molded-part design and its sometimes complex description of freeform surfaces, mold design commands a much greater proportion of CAD activities in the field of classical drawing since the mold is largely made up of relatively simple geometric objects (rectangular, cylindrical, prismatic).

The CAD model of a design is a representation of the geometry in the computer. The type of internal representation leads to models with different information contents.

A basic distinction may be drawn between:

– 2D graphics systems,
– $2^1/_2$D graphics systems,
– 3D graphics systems.

The use of 2D systems is restricted to drafting at the screen level. The informational content is only slightly higher than that of a drawing. It is the task of the user to draw all necessary views and cross-sections one by one. The various views are independent of each other, with the result that they are not automatically self-consistent. The advantage of the 2D CAD drawing over a sketch is primarily that a major change does not entail having to do another complete drawing. Individual geometry elements can be changed selectively; similarly the representation of individual views can be revised.

$2^1/_2$D systems store, in addition, information about the component thickness. Work is also done initially in two-dimensions on the display screen. The third dimension is internally created by the computer by a displacement or rotation vector. Thus, consistency can be guaranteed between several views.

Only 3D systems describe the complete molded-part geometry. They can be divided up (Figure 14.5) according to different descriptive techniques:

− vector-oriented models (skeleton, line or wireframe models),
− surface-oriented models,
− volume-oriented models.

Wireframe models, unlike 2D models, do not have level restrictions. Apart from the elements of the 2D model, numerically calculable 3D-splines are available. Since only lines or curves are stored internally, there is no information about areas or volumes. For this reason, geometry processing functions such as cutting or visibility clarification are not possible.

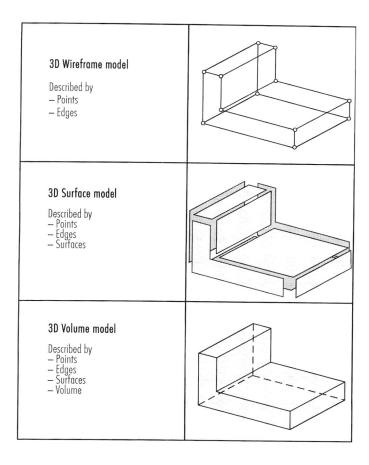

Figure 14.5   Types of geometric presentations in 3D CAD systems

With surface models, it is possible to describe arbitrary entities by means of the boundary areas. Interpolating, approximative, and analytical procedures are used for the area description. Apart from the surfaces that are analytically easy to capture, such as plane, cylinder, cone, sphere, pyramid, and toroid, the user frequently employs the following types:

– surface of rotation (rotation of a contour about a line),
– translation or profile surface (translation of a contour along a guideline),
– ruled surface (linking of two contours by curves),

The differences in the performance of 3D surface models come primarily to the fore when it is a matter of describing freeform surfaces (mathematically indeterminate areas; areas that have a different curvature in every point). In the past, it was usual to apply the methods of Coons and Bezier [14.6, 14.7] or B(ase) splines [14.8]. Newer CAD systems use more powerful algorithms for the surface description. In this connection, mention should be made of NURBS (Non-Uniform Rational B-Splines), a surface description method that allows both analytical and nonanalytical curves and surfaces to be described, with the result that all geometric operations may be performed with a uniform algorithm [14.9, 14.10].

The surfaces of arbitrary molded parts can thus be described with these functions. Information on which side of the area the volume (material) is located, however, is missing. Section operations can only yield intersection lines and do not generate the hatching of the sectioned volume automatically. Furthermore, only with the aid of this additional information would a visibility clarification be possible.

The solid model delivers the most complete parts description. Purely volume-oriented operations, like the determination of solid volumes, center of gravity, or moment of inertia, as well as the derivation of arbitrary section views, become possible. Depending on the type of geometry representation, the solid models can in turn be classified in various ways. The best known are the CSG, the B-REP, the FEA and hybrid models (Figure 14.6).

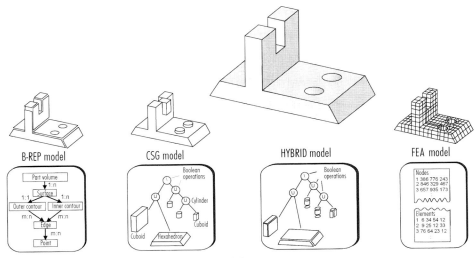

Figure 14.6   Describing parts by volume models

The CSG model (constructive solid geometry) is an entity-oriented solid model that is generated by the Boolean linkage of sub-entities [14.11]. Set-theory operations employed are union, difference, and intersection. Since only a tree structure with the generative and logical operation for the sub-volumes is stored, this model representation has a low memory space requirement. The history of the model structure remains comprehensible and additional modifications of individual elements, e.g., a cylinder diameter, can be carried out easily [14.12]. CSG models are basically highly suitable for the parameterized model construction and the linking of form-features (see Section 14.2.2.2). If partial design modifications have to be made to a part, however, there is no access to edges or points since there is no surface information in the data model. The shape of the surface is described only indirectly. Visible surfaces and shapes of entity edges are only determined for graphical output and not used further for calculations.

In B-REP models (boundary representation) a part is defined by its boundaries [14.11]. The bounded surface is defined by individual sub-areas that in turn are built up of points, lines and areas. Although the individual model elements can be accessed directly in order that modifications may be made to the surface, the B-REP model has no concern for its history. Surface models are used in applications requiring as accurate a description of the parts surface as possible, e.g., when a CAD model is to serve as the design geometry for the co-ordinate measuring technique.

FEA models approximate real parts through having finite elements. They are only mentioned here for the sake of completeness since these models are used exclusively for calculating the parts behavior of complex objects. A FEA net is not normally constructed, but rather is derived from one of the other models described here.

To exploit the advantages of the various models, nowadays CAD systems are being employed and developed that combine several representational forms in so-called hybrid models. A useful combination is that afforded by CSG and B-REP models as this allows complicated surfaces of sub-bodies to be described exactly while permitting the history to be understood since the sub-bodies are processed with a solid modeler [14.13].

14.2.2.2 Enhancing the Performance of CAD Models by Associativity, Parametrics, and Features

The various possibilities of computerized geometrical representation having been described in the previous section, let us now turn to the methods and properties of modern CAD systems that primarily contribute to rapid, consistent generation and modification of geometries. These are associativity, parametrics, and features.

Associativity

The term associativity stands for the relationship between two or more objects in which a change made to one is automatically performed in the linked (associated) objects. This includes linking of a three-dimensional model with the (two-dimensional) draft design derived from it. If an attribute such as the position of a drill hole is changed on a 3D model, this change immediately affects the various views of the draft. The model and draft always remain consistent as a result. Associativity can also be made to apply to several individual parts within a module. Provided the model of the module is constructed appropriately, a geometrical change made to a single part will affect other parts. Thus, changing the diameter of an ejector pin, for instance, can influence the pertinent drill holes of the mold platens. If the drill hole in the mold insert is moved, it also moves in the platens. Another example of associativity is that the contents of the module model always tallies with the piece list derived from it.

Parametrics
The use of parametric models helps to increase efficiency. The term parametric refers to the way in which elements in the CAD system can be generated and modified. In parametric models, it is possible to copy constraints in the computer model and also to vary every attribute of a geometrical element (position, dimension, color, material property, etc.) at any time during design. This approach allows the model to be easily adapted to altered boundary conditions, and supports the rapid generation of part variants and series (Fig. 14.7). Parametric relations can be generated not only within an individual part, but also between components of a module, which results in the associativity mentioned above [14.14].

Feature Processing
Features may be used for individual, repetitive geometrical, or functional elements (Figure 14.8). They are parameterized objects that are generated as application-related variants during the design process. They usually carry geometric and technological information (e.g., tolerances) as well as knowledge for the handling and processing of these variants [14.15]. In the case of form features, which serve initially to generate a certain component of the molded-part geometry, the designations employed frequently are the same as those of the design elements which they represent (e.g., drill hole or thread). Such features are system defaults, but can often be defined by the user ("user-defined features", UDFs) without the need for external programming.

14.2.2.3 Interfaces and Use of Integrated CAD

Interfaces are always used for transferring data from system A to system B. This applies equally to geometrical information (e.g., drawings, models), technological information (e.g., material information, NC programs), and organizational information (e.g., lists of

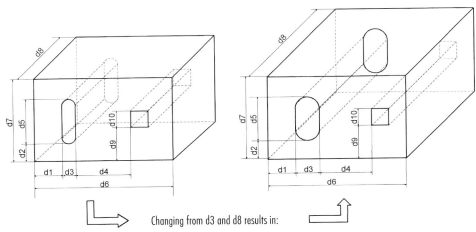

Changing from d3 and d8 results in:

Figure 14.7   Changes in the parametric geometry model

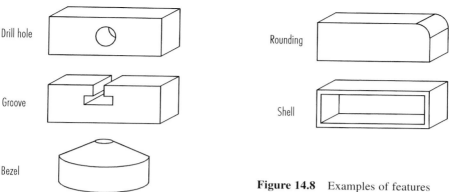

Drill hole

Groove

Bezel

Rounding

Shell

Figure 14.8   Examples of features

components) [14.16]. Systematic storage and transfer of this information may be aided by a product model. Ideally, a product model contains a computer-specific copy of a product throughout its life cycle. This supports interlinking of information from the areas of design, work planning, production planning and control, production, assembly, and quality assurance.

The exchange of product data is an important, if not the most important, aspect of integration of the development process. Since different computer systems generally employ different internal data models (see Figure 14.9), data to be swapped must be converted into the appropriate format. This is the purpose of interfaces.

Since every kind of data exchange between external systems is fundamentally prone to data loss and the post-processing of flawed or incomplete data records is extremely time consuming and expensive, the problem of interfaces in CAD/CAE/CAM enjoys an extremely high status.

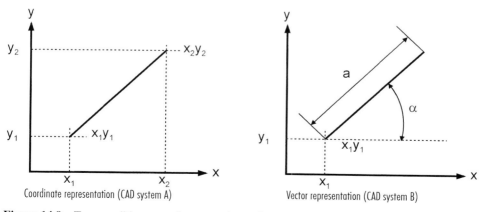

Coordinate representation (CAD system A)

Vector representation (CAD system B)

Figure 14.9   Two possible ways of representing a distance

Interfaces for data exchange may be system-specific ("native"), specific for several systems (e.g., DXF), standardized at national level (e.g., VDA-FS in Germany), or at international level (e.g., IGES, STEP). The data from system A are converted during data exchange into the interface format by a preprocessor. The postprocessor reads these data into system B. Figure 14.10 shows that native solutions require many more processors than is the case when standardized interfaces are used. For this reason, there have been many attempts in the past to develop neutral data exchange forms in the form of interface standards.

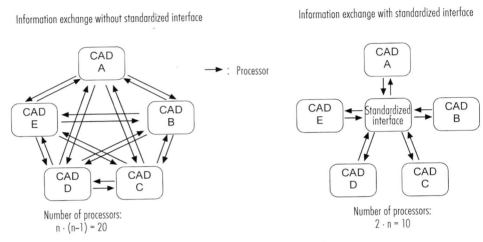

Figure 14.10   Reducing complexity through standardized interfaces

IGES

IGES (Initial Graphics Exchange Specification) was originally conceived for transferring drawings [14.17] and was later augmented with the description of spatial geometrical information (surfaces) [14.18]. Aside from geometrical data, text and dimensions can also be exchanged. The scope for copying freeform curves and surfaces is limited since only polynomials as far as the 3rd degree can be classified. Later versions of IGES permit, in principle, the exchange of solid models, texts and symbols, measurements, and drawing views. IGES is the most widespread standard in the world. The main criticism is its large size and the extent of interpretation needed for the interface specification which leads to a situation where hardly any processor offers total, generally compatible support. Frequent problems encountered with this interface are reproduced in e.g. [14.19]. The possibility of exchanging solids is rarely used in practice due to a lack of processors.

VDA-FS

The VDA surface interface was developed by the German automobile association (VDA) to bridge the weaknesses of the IGES interface. VDA-FS is used primarily for exchanging area-related data [14.16], is used on a large scale for data exchange by automotive manufacturers and their suppliers, and is primarily employed in Germany. Drawing information cannot be exchanged with VDA-FS.

VDA-IS

VDA-IS is a more precise subset of IGES (IS stands for IGES subset) defined by the VDA for the needs of the German automotive industry. The interface supports the exchange of geometrical and measurement elements as well as freeform surfaces. The implementation of accompanying convertors is thereby restricted to selected, essential functions, a fact which should increase the quality of the exchanged data.

VDA-PS

A programming interface developed by DIN and the VDA, VDA-PS in Germany is used for providing standardized and repetitive parts. VDA-PS contains the generative logic for the standardized parts.

SET

SET (Standard d'Exchange et de Transfert) was developed by the French as an improvement on IGES. More detailed data descriptive of the product can be copied, especially those required by the aerospace industry. SET is primarily used in France.

STEP

Several years ago under the auspices of ISO (ISO 10303) and drawing on the collective experiences of other interface concepts, development was begun on a universal data exchange format [14.20, 14.21]. STEP (STandard for the Exchange of Product model data) lays claim to being the sole all-embracing standard of the future and to superseding IGES and other formats. Aside from geometrical data, information from the entire product life cycle will be transferred in the long term by STEP. In particular, this will include organizational data. Through division into so-called partial models and through their application to individual application areas in the form of application protocols, STEP can be used to describe all kinds of product information. For example, Application Protocol AP 214 of the automotive industry will not only exchange geometrical data but also product structure data, kinematic data, NC data, material information, and surface properties [14.22]. Although not yet realized, it is conceivable for parametric models and form features to be exchanged on the basis of SET [14.23]. STEP has been used since 1996 for the productive exchange of solid models at international level. Furthermore, surface models, module structures, and organizational data can be transferred.

DXF

While DXF (draft exchange format) from the company AutoDesk is not an attempt by a standards committee to produce a standard, it has become the most important format for the 2D sector.

Native Formats

Despite all attempts to exchange geometric data between different systems via neutral interfaces, many sectors operate with "native" data so as to minimize possible sources of error. This means that the recipient of geometric data uses the same CAD system (and same version) as the transmitter. This situation necessitates high outlay on costs and personnel particularly for subcontractors working for different clients because of the large number of CAD systems employed.

Direct Interface

An alternative is the direct interface, which converts the model generated on CAD system A into that of system B. Since direct converters are developed especially for one system combination, the amount of information transferred is often high. The disadvantage is the usually high costs required for each individual interface (see Figure 14.10).

SAT

The ACIS solid modeler from the company Spatial Technologies is the core of a number of CAD systems. ACIS processes different geometric objects, such as wireframe, freeform, and solid models in a uniform data structure. The systems on top of this core permit direct data exchange via the internal SAT (Save ACIS Text) interface of the ACIS models [14.24]

Interfaces nowadays are indispensable to communications between the countless systems on the market and they are the most common means of data transfer [14.25]. Nevertheless, there has been a clear tendency developing in recent years to integrate CAE/CAM systems, that until now were stand-alone programs receiving their CAD data via interfaces, into comprehensive CAD packages. The purpose is to avoid those disadvantages associated with interfaces:

– It is not always possible to transfer all the required information (restrictions on the performance capability of the interfaces, and transfer losses).
– Data exist in various, redundant representations because they are duplicated. Considerable amounts of work are needed before all data have the same, up-to-date status.

In general, the aim should be to use as few different software systems and thus data formats as possible. Ideally, there would be one database which is accessed by all programs in the process chain. The CAD model thus consists no longer of purely geometric data, but instead is expanded by more detailed information. Integrated packages have the goal of supporting the entire product development cycle and offer the advantage of just one user interface with consistent handling.

14.2.2.4 Data Administration and Flow of Information

The areas of data administration and integration are very closely linked to each other and must always be considered in conjunction with the respective CAD system. This is particularly true of modern 3D CAD systems. As soon as the information for describing a mold no longer is restricted to conventional drawings, data administration becomes very important. It is not only the actual data administration but also the work processes, which so far have been directed at conventional drawings, that have to be rethought and adapted. Such deliberations yield a range of not immediately apparent consequences that are discussed below. The following points usually have to be considered:

– authoritativeness of information,
– data dependencies,
– archiving of information,
– job processes (e.g., approval procedures and modification documentation).

Whereas drawings used to be the authority for the definition of a product, nowadays they are usually derived from the mainly superordinate CAD model. The creation of such a

2D drawing simply requires information as to which 3D model is to be represented in which orientation and with which section view in the layer. The geometrical information (lines, hatching, etc.), which ultimately make up a view, no longer need to be created and stored explicitly, but instead are computed from the 3D geometry. Often, information is also taken directly from the CAD model without its being documented on a drawing, e.g., for NC programming. The CAD model is authoritative in this case. This does not mean, however, that drawings no longer play a role. If NC machines are not used for production, for example, classical drawings are required. To an extent depending on the company and the economics, working practice will be a mixture of direct production based on CAD data records and conventional production from drawings.

It must always be ensured that drawings and CAD models reflect the same stage of modification or development. This applies not just to CAD model and drawing, but also to all data (e.g., NC programs, computational models, etc.). All these dependent data must be updated when modifications are made. Since the CAD model contains authoritative information on the definition of the mold, there must be a process for approving the CAD models. Aside from approval, there is the question of documenting modifications to approved data to be considered.

In connection with modifications, the distribution of information plays an important role. Since interdependent data may in certain circumstance be used for different tasks at different places, not only must the corresponding data be available there, but also information concerning the development and approval stages. When modifications occur, those places with dependent data that are affected by the modifications, must be informed accordingly. This is vitally important when various development stages for shortening development times are performed in parallel.

The actual solution to data administration, archiving and modification processes is extremely dependent on company vagaries, the CAD system employed and the entire computer environment. The quality of the solution is crucial to the efficiency of the use of CAD and thus ultimately to product development. Most systems manufacturers offer data administration software tailored to the CAD system that takes account of generated data dependencies. Furthermore, there are powerful, adaptable engineering and product data management systems (EDM and PDM) available on the market that provide back-up for this problem area.

14.2.3 CAD Application in Mold-Making

CAD systems, especially the modern parametric solid modelers, offer numerous possibilities for an efficient, accelerated approach in the construction of mold-making and tool-making.

14.2.3.1 Modeling

As far as the use of CAD is concerned, there are generally three possible ways to construct a mold:

– 2D construction,
– hybrid construction,
– 3D construction.

In 2D construction, the entire mold construction is performed with the aid of a 2D CAD system. The result is drawings. All other stages that are necessary for attaining the finished mold essentially fall back on this source of information. Complex freeform surfaces can for instance be introduced into the mold by copy milling with the aid of physical models (copy aids).

In hybrid construction, the shape-giving mold parts are created with a 3D CAD system. Particularly for complex molded parts with a large number of freeform surfaces, this enables at least the NC programs for producing the mold insert or eroding electrodes to be made on the basis of the CAD data. The remaining mold buildup is performed with conventional tools (2D CAD, drawing board).

In 3D construction, the entire mold is created with the aid of a 3D CAD system. It is thus possible to attain the deepest process penetration with CAD data. In the ideal case, nearly all data that define the mold are stored in the computer model.

Although 3D geometries are much easier for the viewer to understand than complicated technical drawings, the generation of solid models frequently entails the use of 2D drafting. It is thus standard practice to create a cross-section of a profile as a sketch and then to convert it into a 3D object by translation or rotation. In this case, the 2D drafting stage is required for preparing for the solid modeling. On the other hand, once the 3D modeling is complete, work on the 2D draft may be necessary, for example, to represent sections and details in the form of workshop drawings for production of individual parts. This results in a constant switching between representational and modeling levels.

The design activities needed for an injection mold can be divided into two rough areas: the molded-part geometry is used to derive the shape-giving mold contour, and the two mold halves are built up around the mold inserts. These activities produce the typical demands on geometry generation shown in Table 14.1. Since the mold designer does not always have a geometry file of the molded part, he may also have to generate the "positive", which is why molded-part construction is also listed in the table. Modeling of the molded-part has numerous parallels with the design of eroding electrodes for mold production.

Useful special functions for model generation in mold design are explained below (Figure 14.11).

Shrinkage
The molded part is designed with nominal dimensions, whereas material shrinkage must be allowed for when the mold cavities are being designed. A suitable way of starting to design the cavities is to scale a copy of the molded-part geometry. Since scaling should compensate the material shrinkage, it is usually necessary in the case of highly anisotropic molded-part properties to allow for differential shrinkage in different spatial directions. In addition, a selective definition of shrinkage in which different areas of the model can be scaled with different factors is beneficial. It can often be desirable to exempt individual geometrical elements (e.g. perfectly circular cylinders) from anisotropic shrinkage as they would otherwise mutate into more complicated geometries whose production later would involve excessive work.

Mold-Parting Lines
The design of the mold-parting line is undoubtedly one of the most demanding tasks of mold design. The CAD system must, therefore, generally possess good surface functions. It is already possible nowadays in the case of simple molds to have the CAD system automatically compute all of the mold-parting line for a pre-determined mold-

Table 14.1   Demands on geometry generation

Molded part	Mold insert	Mold
Production of freeform surfaces	Importing of molded part model	Generation of prismatic and cylindrical bodies
Smoothing and delimiting of curves and surfaces	Generation and modification of tapers	Application of radii and bezels
Rounding of edges and corners	Inverting the molded part geometry into the shape-giving mold contours	Simple copying, mirroring, modification of geometries generated once
Intersection of several surfaces	Support for generating parting lines	Use of standard elements
Generation of tapers (drafts)	Modification of the CAD model through allowance for shrinkage	Use of libraries (standard parts and repetition parts)
Modelling of functional areas (ribs, snap-on connectors...)	Simple copying, mirroring, modifying of geometries generated once	
Simple modifying of wall thickness	Derivation of contours for field electrodes and full mold electrodes for eroding	
Simple copying, mirroring, modifying of geometries produced once	Redefining tolerances and surface information from function-orientation to production-orientation	

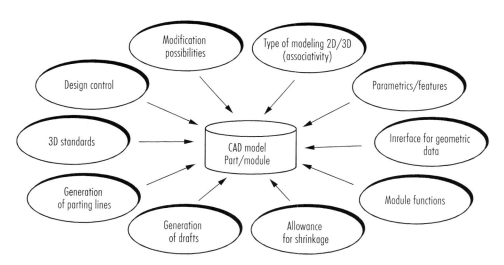

Figure 14.11   Support for generating the CAD model

opening movement. This function runs up against its limitations in the case of complex molded parts. The designer can, however, avail himself of help functions, such as the silhouette curve. This is a curve projected onto the molded part that represents the visible molded-part edge for a view parallel to mold parting. This potential parting curve of the molded-part can then be used to design the parting surface. An elegant way to generate the parting surface consists in slicing the "mold block" that surrounds the molded part with the defined parting surfaces and to generate the shape-giving mold inserts. Modifications to the molded part are automatically taken account of when the block is separated. The contour is therefore always up to date (Figure 14.12).

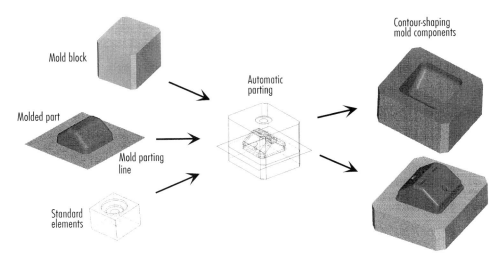

Figure 14.12   Principle of "Parting" with molded part and parting lines

Drafts
Several systems support the design of demolding drafts. Aside from simple conicity, functions such as non-constant drafts or drafts that tangentially run into existing surfaces are of great benefit to the designer, especially in the case of complex molded parts and mold-parting lines.

Gate System
The design of the gate system can be simplified with the aid of user-defined design elements known as UDFs (user-designed features; see Section 14.2.2.2). When such UDFs are stored in appropriate parametric form in a library it is possible to compose a complete gate system quickly and flexibly from individual components (runners, gates, sprues, etc.).

Modules
3D CAD systems that have extensive modular functions offer a distinct advantage. Individual parts of a mold are assembled into a complete mold with the aid of incorporation conditions. The entire module can then be modified, with the various parts

behaving associatively. For example, drill holes for ejector pins in the module can be "drilled" through several mold platens. By means of definable relationships, diameter and positions of the ejector pins, for instance, may be set in relation to the pertinent drill holes. This ensures that the drill holes are always aligned in all mold platens and that the size and position are adapted to the pins employed. Displacement of the ejector pins automatically leads to displacement of the ejector holes. If the CAD system is also capable of variant design and handling, exchanging the ejector pins automatically causes the drill holes to be adapted.

Standard Units

For many molds, it is possible to revert to standard structures contained in libraries of standard parts. Such a case would be a fully assembled mold to which only the gate, mold inserts, and ejector pins have to be added. But even molds that are not predefined as standards can be built up simply and used again as the basis for similar designs.

Design Control

Diverse design control functions make it possible to check if the design has been performed logically in terms of geometry on the one hand and plastics on the other. It is thus possible to test radii and demolding drafts as well as undercuts on the molded-part model and the mold insert.

Aside from the basic capability of generating and modifying geometric objects, CAD systems must also support manipulation by the user in a dependable manner and yield the expected results. For complex three-dimensional models, there are design aids available that support spatial movement and positioning and identification of salient points as reference points. Object snap with adjustable sensitivity is especially helpful in this regard.

Due to different demands imposed on design support in various development areas, molded-part geometries in practice are frequently generated on CAD systems other than those used for the pertinent molds. This raises the problem of data transfer with possible loss of data (see Section 14.2.2.3) that makes it necessary to repair the swapped models. Repairing of CAD models, also known as CAD finishing, necessitates the availability of diagnostic functions that can detect damaged or incomplete part surfaces as well as easy-to-use manipulation tools, such as dragging together of individual surfaces and insertion of surface sections to bridge gaps.

Data transfer over standard interfaces causes parametric information to be lost. If the geometric model has to subsequently be scalable or even to be used for making an efficient design variant, it is necessary to parameterize the imported data afterwards. Precisely in the case of complicated models, this may prove so difficult as to make a new design preferable. Furthermore, in many instances it can be better to de-parameterize the model so as to emphasize other geometric relations or constraints.

14.2.3.2 Integrated Functions for Mold-Making

When the fully described solid models of the molded part and corresponding mold are available, there are numerous ways of using the model information of the entire CAD/CAM process chain, as Figure 14.13 shows. Not only are classical areas such as the derivation of NC data or preprocessing for simulation computations supported, but different forms of representation can be chosen leading to marked increases in efficiency in the fields of preparatory work, quality control, technical documentation, right through

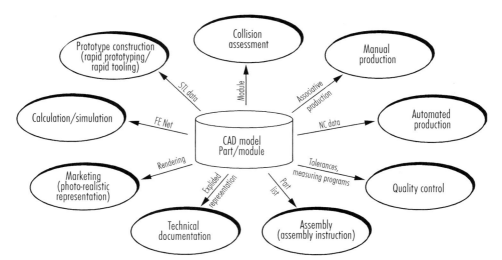

Figure 14.13   Using the CAD model in the process chain

to marketing. Direct generation of programs for the production of rapid tooling or stereolithographic parts is rapidly becoming widespread.

Collision View

Collision view affords a means of checking the assembly of the mold. It provides a simple means of detecting overlapping individual sub-entities. The virtual model can be used to check opening of the mold and movement of the slide bars and ejectors (see Figure 14.14). With complicated molds, it affords a timely way of checking the demolding process. It is also possible to plan part removal by handling equipment and to synchronize the opening movements of the mold. Furthermore, access of tools for assembly, installation and service activities can be verified.

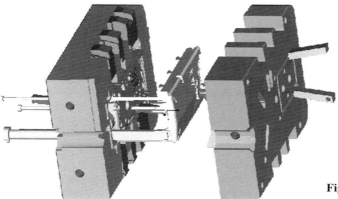

Figure 14.14   Collision assessment of a mold [14.26]

Draft Generation

Any number of views, straight or variable sections, and details can be made from the solid model. This remains indispensable for several production and assembly steps. Due to the associativity described in Section 14.2.2.2, up-to-date drawings can be made at any time from the master geometric model. If the system is capable of bi-directional associativity, a change in one dimension in the draft is immediately reproduced in the linked 3D model. However, wholly automatic derivation of drawings is still only possible in the case of simple objects.

NC Programming

If a 3D model of the mold inserts is available, appropriate NC programs for machining of cavities or for producing electrodes for erosion can be created. Thus it is possible on the CAD system to process tasks that are classified as work preparation, without the need for data transfer. Integrated CAD/CAM systems have add-on modules that allow common standard formats for NC codes to be generated without the intervening inter-facing step. Some systems also permit the production process to be simulated on screen. Before the NC codes are compiled, machine data such as dimensions, maximum displacement and the limits on the processing conditions (traverse, speed) must be entered.

Quality Check / Metrology

If the 3D model is supplemented with information about tolerances (degrees of fit, shape and position tolerances), this information would be suitable for performing the quality check later on. Just as with NC programming, appropriate measuring programs for coordinate-measuring machines can be compiled on the computer that allow the actual geometries of the finished parts to be evaluated for their dimensional accuracy relative to the computer model. Moreover, additional software can be used to perform a tolerance analysis on the module with a view to supporting the selection of meaningful tolerances for technically perfect and economical production.

Assembly Preparation / Technical Documentation

By positioning the individual parts in three-dimensional space, any number of representations and views of a mold can be generated, ranging from exploded views right through to the completely assembled module. In conjunction with the list of parts, which can be derived associatively from the module, it is thus possible with little effort to document the assembly process for each case. Often, the requisite tables or images are embedded into so-called office applications, such as word-processors. In particular, the persistent tendency to use CAD systems under MS Windows/Windows NT on personal computers speaks in favor of the increasing importance of coupling CAD applications and office applications [14.27, 14.28]. Other application possibilities consist, for example, in compiling maintenance instructions and service manuals.

Presentation / Marketing

The already mentioned incorporation of CAD model presentations into text documents benefits marketing, among other areas. For presentation purposes, the CAD model can be manipulated with so-called rendering software to produce an image resembling a photograph. This generates an early, realistic impression of the product (see Figure 14.15). Similarly, animated sequences of images (e.g., to show movements) can be created for advertising purposes.

FE Simulation

Several CAD systems offer the designer the possibility of preparing geometric models for further use in external simulation programs for thermal-rheological mold design. This occurs through the generation of a finite element net based on a CAD model. Since common FE programs for ambitious process simulations (e.g., CADMOULD, C-MOLD, MOLDFLOW) exclusively compute in 2D at the moment, there is no getting round a central layer model of the molded part or the cavity. The automatic generation of central layer models remains a problem to be adequately resolved [14.29]. Algorithms for automatically deriving the central layer, fail at the very latest in the case of complex geometries involving frequent abrupt changes in wall thickness and freeform surfaces. The experts still have to convert the model manually and perform the simplifications for the simulation. There are now CAD systems on the market that have integrated simulation programs. Some of these programs utilize STL data of the complete 3D geometry by way of geometry specification. This affords a means of finding the end positions of weld lines and occluded air in the mold. These modules are without exception designed as tools for rough assessments and they are directed more at designers to help them perform a preliminary estimate than at simulation experts, who are more interested in the most realistic prediction possible of the process behavior [14.30].

Prototyping

Now that rapid prototyping has largely become established in product development in recent years, rapid tooling is starting to grow in importance. Only with the advent of this process has it proved possible to produce close-to-series prototypes while making allowances for process influences and using the material that will later be employed. Most CAD models have the capability of converting a CAD model into an STL model. The STL format has now become the standard format in the field of rapid prototyping (rapid tooling).

14.2.3.3 Application-Specific Function Extension

The CAD systems currently on the market are generally universal types that can be used in a number of branches. To ensure that CAD is used efficiently for a specific application (e.g., the development of plastic parts of a certain product range), functional capabilities can be added to the CAD systems. This is achieved by integrating or expanding them with product-specific or company-specific application software. A prerequisite for the compilation and incorporation of such program modules is the presence of suitable data and program interfaces (e.g., FORTRAN or C) in the CAD system.

Adaptation of the performance capability of the CAD system to factory requirements includes the provision of macros (drafting and design macros), character sets, standard parts, the programming of standard procedures and of requisite variant parts [14.3].

Nowadays, plastic parts are often offered not just once on the market but as a whole range of parts that differ only in size and not in function. Successful parts are not just made once; it is standard practice to make generations of parts that appear at intervals with slight modifications. For such parts and the pertinent molds, plastics converters have a considerable amount of factory-specific know-how gained from computations, experiments, and practical experience of series production. This knowledge has to be rendered computer-readable and made available to the designer, e.g., in the form of menus of features for his CAD workstation.

Office chair
H. Miller Inc.

Mold
LS Mold Inc.

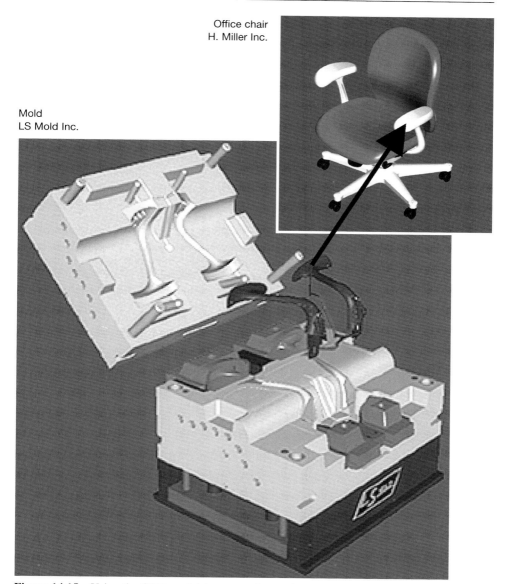

Figure 14.15   Using the CAD model to present the product

14.2.3.4 Possibilities Afforded to Concurrent Engineering through the Use of CAD

To an extent depending on the CAD system employed and the model representation used therein, there are considerable differences in the approaches taken in mold design. The following example depicts the approach adopted by Parametric Technology Corporation using the Pro/Engineer CAD system. It uses feature technology and has extensive integration capability along the entire process chain for the design and production of injection molds. The systematic approach employed has already been implemented by various plastics processing companies.

The "ideal" mold design presupposes a part design suitable for plastics. This includes above all, taking production needs into account, which can be achieved by intense cooperation between part and mold design departments. Effective cooperation between the two departments can be aided and carried out in parallel by suitable functions on the CAD system (Figure 14.16).

The basis for parallel development stages is the functional model. This CAD model is initially the result of function-finding, in which the essential functions of the molded part are defined and stored in the CAD model. Initially, details such as demolding drafts and general fillets are ignored. This functional model is already sufficient for providing a first assessment and for improving the mechanical, thermal, and rheological properties. Also, it contains enough information to permit the first steps in designing the mold to be taken. This model is continually improved and more details added by iterative methods.

The essential advantage of using an explicit CAD functional model is that the development stages can be carried out in parallel already at a very early point in time (Figure 14.17). This makes it possible to optimize the molded part early in the process in terms of mechanical, thermo-rheological, and production demands. Because initially the model has a simple geometric composition, the necessary steps for this are generally much easier and quicker to perform than would be the case for a completely detailed CAD model.

If FEA is used for mechanical, thermal, or rheological analyses, the absence of such details as demolding drafts and general fillets makes the necessary preparation of the model for the computation much simpler. The outlay on networking can be reduced. The results are generally good enough to provide enough information. Furthermore, FEA solid models allow simpler nets with fewer elements, a fact which makes the networking easier and drastically reduces the computation time.

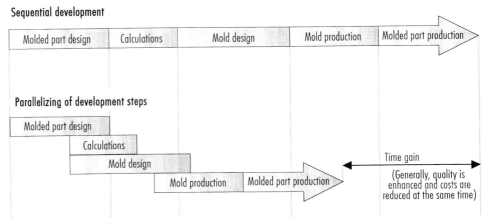

Figure 14.16   Parallelizing development steps

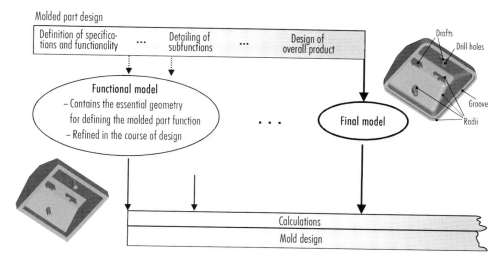

Figure 14.17   Using the functional model to integrate molded part and mold design

Concurrently, the functional model can be forwarded to the mold-maker. The CAD model can be used to make the first analyses from production aspects. By this stage at the latest, the principle mold-parting line is defined. With the aid of CAD functions, important information for a preliminary mold design can be determined very readily. This includes undercuts, molded-part volume, projected area, packaging dimensions, wall thickness, and an estimation of the flow path lengths.

This information provides a rough definition of the mold. But at this stage there is still the possibility of making design changes to the molded part for production reasons. Details of the molded part can thus be defined in parallel to the mold design that has already been started. Key to the capability to parallelize is the associativity of all data within the CAD system. Updating the model automatically causes all derived data to be changed.

The mold is assembled as a module from the various individual parts. The shape-determining components of the mold are derived directly from the molded-part model. All other components are taken as far as possible from libraries of injection molding standards. The overall mold is therefore generated from the definition of the functional areas, namely scaled molded part, parting lines (including slide bars and core inserts), mold platens, gate system, demolding system, temperature-control system, and diverse detailed elements such as guide pins, bolts, and springs (Figure 14.18).

14.2.4 Selection and Introduction of CAD Systems

The meteoric development of CAD systems has meant that many companies in the plastics industry wish to introduce a new CAD system or are looking to replace an existing system that no longer satisfies requirements. The sheer variety of systems on offer makes it exceedingly difficult for a company to select the system best suited to its needs. Studies show that market surveys and analyses do not pay sufficient attention to

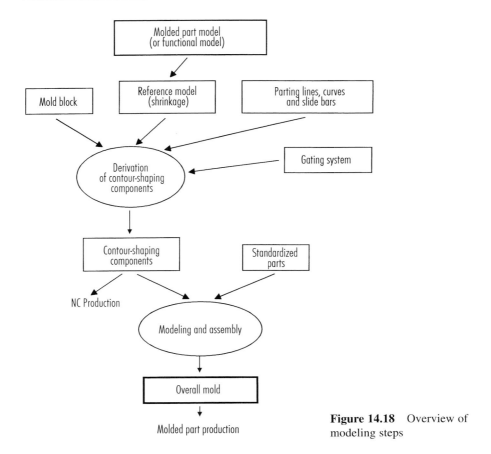

Figure 14.18   Overview of modeling steps

the vagaries of plastics processing [14.31, 14.32]. The selection and introduction of CAD systems poses a high risk due to the high investment sums involved and the pronounced effects on the entire company. Building on the considerations provided in previous sections, there now follows some advice on how to systematically select and introduce a system for a specific company.

14.2.4.1 Phases in System Selection

The introduction of CAD is a long-term project that initially costs more in terms of money and especially time than it brings in benefits to the users. There is no avoiding a certain systematic approach if the outlay and associated costs and loss of time are to be kept within limits. The choice and introduction of a CAD system usually follow the phases shown in Figure 14.19.

The first step consists in determining the needs of the company on the basis of a previously defined plan. This is followed by screening, using basic CAD functions weighted according to their importance. This type of information is to be found in market surveys, brochures, discussions at trade fairs, etc. (Figure 14.20). The final decision

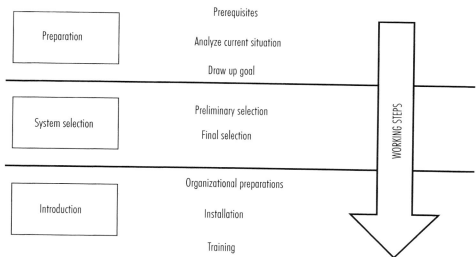

Figure 14.19   Phases in CAD selection

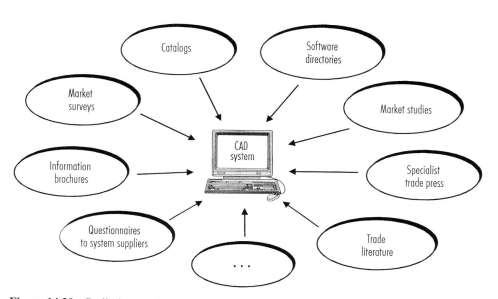

Figure 14.20   Preliminary selection of CAD systems

generally requires a more intensive benchmark test on the screened systems. The number of test candidates should not be too high (five at most) [14.28] as benchmarking is relatively time-consuming and can last up to several days for each candidate, depending on the extent of the testing. Once a CAD system has been chosen, there follows the introductory phase, the length of which is often underestimated. Experience shows that it can take up to several months, since exchanging the CAD program triggers a series of further changes, e.g., new hardware and network components, a new solution for data storage, linking of proven modules from the old system and, of course, employee training.

14.2.4.2 Formulating the CAD Concept

Planning the use of CAD entails systematically analyzing the factory and non-factory situation at the outset; a thorough stocktaking of the existing situation and a careful determination of the needs are crucial to selecting a system [14.33]. The analysis usually embraces the organizational structure, range of parts, and the way in which the processing of tasks is organized, including interfaces to other companies [14.34].

Even as the plan is being devised, as many persons as possible from those affected by the introduction of CAD should be included in the project. This ensures on the one hand that the needs of the various company divisions are taken into account. On the other, it brings about early contact with the new program and increases acceptance, defusing the change-over phase. The basic questions that the project team should ask of every individual system are [14.33]:

– How easy is it to change plans?
– What interfaces are needed to other systems?
– What is the development potential of the software?

In the past, the process of arriving at a concept for the intended use of CAD has been aided with checklists and lists of criteria (e.g., [14.3, 14.35–14.37]) whose purpose is to ensure that every major aspect of planning the introduction of CAD are covered. These aids are employed in discussions with all those affected to establish a specifications profile for the system to be chosen, which include these problem areas:

– computer plan, data storage,
– hardware, operating system,
– compatibility with existing system,
– interfaces to other applications and company divisions,
– interfaces to other companies (e.g., suppliers),
– system add-ons, additional functions,
– programming environment, expandability,
– design activities, compilation of drawings, model generation,
– geometry model in computer, accuracy,
– parametrics, variant design,
– features, macros,
– ergonomics, input and output devices,
– introductory phase, extent of training, user-friendliness,
– customer service, updates,
– long-term planned development processes and production technologies.

The formulated demands imposed on the future CAD system can then be assigned to various categories (essential, minimal, desirable requirements) of the specification list shown in Figure 14.21. An essential requirement would be, for instance, that the CAD system is fitted out with an IGES interface. A minimal (or maximal) requirement, by contrast, would be information on the permissible amount of time for the introductory phase, whereas the existence of an STL interface (for mold-making, an STL interface is

Specification	Type			Verifiability				Pass?	
	Essential	Minimal	Desirable	System supplier	Reference customer	Test scenario	Separate tests	Yes	No

Figure 14.21   Specification list for choosing a CAD system

an absolute necessity) would be considered desirable, if not absolutely necessary. In principle, it is also possible to rank the various requirements so as to use them later to determine how much alternative solutions satisfy them.

The requirements can further be differentiated by the way in which they are ultimately tested. There are four possibilities:

– the systems supplier provides the necessary information,
– the systems supplier names a reference customer who provides the desired information,
– the requirement is checked by modeling a part of practical use with the test system,
– the desired property is not covered by the practical test and is therefore checked in a separate test.

The system can now be chosen on the basis of this list. The result of the test conducted on each individual requirement is entered into the last column. See [14.3] for information on the processes involved in performing an economics analysis during the course of a benefit analysis.

14.2.4.3 Benchmarking

The final selection can only generally be made after benchmarking has been performed by evaluating a few pre-selected systems for their ability to help the company overcome a typical design problem. This may be a representative part or a module. In many instances, however, it is more efficient to develop a test part that matches the entries on the specification list and is, therefore, better suited for testing the performance and handling of the pre-selected CAD systems.

In the benchmark test, the software supplier designs the test part in the presence of his potential customer, who then evaluats the design process and result. Checking the interfaces is also part of the procedure: experience shows that the existence of an interface is not a guarantee of trouble-free data transfer.

A few major characteristics for mold design are listed below, whose implementation in a study revealed marked differences between the systems tested [14.31, 14.32].

– Fillet problems: The generation and representation of merging radii causes difficulties for some systems.
– Duplication of an element: it should be possible to duplicate a geometry element that has been generated (e.g. drill hole). In some systems, it is not possible to break the link between the original and the duplicated elements. This may be necessary for compensating e.g., differential shrinkage.
– Identification of changes: when changes are made to a model or drawing, it is often desirable to automatically emphasize the modified areas so that the differences from the previous version may be seen more easily.
– Working in two-dimensional views and sections: many systems do not allow section work. The sections can be generated at will, but modifications made therein do not affect the 3D geometry, i.e., there is no bidirectional associativity.
– Transposing the tolerance field from function-oriented to production-oriented: for NC production, it is often necessary to dimension according to tolerance field centers.
– Representation of standard parts: care should be taken to ensure that standard parts such as threads are depicted in compliance with the standard when the drawing is compiled.
– Demolding drafts: the attaching of demolding drafts is absolutely indispensable for mold-making. Many systems offer very good tools here. Often, the behavior of the bordering areas and radii poses a problem during automatic generation.
– Changing the radius of a bezel: the production of an injection mold often requires a bezel to be used instead of a radius, or vice versa. This presents difficulties for many systems.
– Scaling: to compensate the expected processing shrinkage, it is often necessary to scale with different factors in the x-, y- and z-axes. This is not supported by many systems. It must also be remembered, however, that with such scaling it is possible that radii turn into more complex freeform surfaces, which is not always desirable due to higher production costs.
– Hardware: when CAD benchmarking is performed at a systems supplier's, usually the best computer systems are used. The actual configuration should be borne in mind in order that comparisons may be made.

14.2.4.4 CAD Introduction

The work involved in introducing a new CAD system is often underestimated. Modern powerful CAD systems not only bring about a change in the designer's work practices, but also introduce incisive changes into factory procedures and organizational forms.

The success of introducing such a system depends heavily on a positive attitude on the part of the employees at all company levels, i.e., from users through to management [14.33]. At least in the initial phase, it is a good idea to form individual project teams to get the employees involved. Active collaboration during the change-over promotes motivation and increases the level of acceptance towards the new system.

The following stages are part of the introductory phase:

Planning of the Introduction
Without detailed planning, it is not possible to switch over to the new CAD system efficiently. Assessable milestones need to be incorporated into planning as a way of measuring the success of the introductory phase. This may help to identify problems at an early stage and perhaps allow countermeasures to be implemented.

Process Re-Engineering

The introduction of a new CAD system provides an opportunity to take a fresh look at the entire development process and to render it more efficient. Only the planned incorporation of the manifold possibilities offered by modern CAD systems can fully unlock the vast potential that they offer for shortening development time, improving quality and reducing development and downstream costs.

Training

All employees must be familiarized with the new system before they can use it safely and productively on a daily basis. Although in-house training can be performed, courses held outside the familiar environment prove to be more effective. There, the employees are not distracted by their daily working environment and are able to fully concentrate on the new system.

Coaching

In order to be able to work as efficiently as possible, it is often helpful to integrate socalled coaching phases into the employees' training plan. During pilot applications, these phases can be used very effectively to demonstrate new approaches in working and developing with the CAD system. There is also the opportunity to examine and improve the re-engineering.

Data Administration

The modified data administration has to be implemented concurrently with system introduction. Interfaces to other company areas must be installed and checked.

Allowance for Company Specifics

The further use or transfer of existing design data must be initiated. It is not uncommon for this to necessitate the development of a special interface for reading in existing data. Furthermore, company-specific solutions must be incorporated into the new design environment by, for instance, taking add-in programs from the old system that were developed in-house and adapting them to the new CAD system.

The company's expertise in handling the new CAD system must be built up step by step. It is, therefore, beneficial to bring in experts during the introductory phase. Such services are offered by most CAD systems producers. Consultation provided by the systems manufacturer during the introductory phase has proven to speed up the introduction. In particular, process re-engineering with the aid of specialists supplied by the systems manufacturer is more reliable and more comprehensive as they are better able to assess the possibilities which the system offers. Additionally, they have experience of previous projects with other customers that will help to reduce the risk of errors or judgment.

References

[14.1]   Sendler, U.: Varianten aus dem 3D-Baukasten. CAD/CAM, 1997, No. 2, pp. 94–96.
[14.2]   Weule, H.; Krause, F.-L.; Kind, C.; Ulbig, S.: Nutzeffekte rechnerunterstützter Werkzeuge in der Produktentwicklung. ZWF, 92 (1997), 3, pp. 81–85.
[14.3]   Einführungsstrategien und Wirtschaftlichkeit von CAD-Systemen. VDI-Richtlinie, 2216, VDI-Verlag, Düsseldorf, 1994.

[14.4]   Stand der C-Technik-Anwendung in Deutschland. Studie des Instituts für Management-Praxis im Auftrag des CAD-CIRCLE, Winterthur, 1996.
[14.5]   Rooney, J.; Steadman, P.: CAD-Grundlagen von Computer Aided Design. R. Ouldenbourg-Verlag, Munich, Vienna, 1990.
[14.6]   Grieger, L.: Graphische Datenverarbeitung – mathematische Methoden. Springer, Berlin, Heidelberg, New York, 1987.
[14.7]   Piegl, L.: Hermite- and Coons-like Interpolation Using. Bezier approximation form infinite control points. Computer-aided Design. Vol. 20. No. 1, 1988, pp. 2–10.
[14.8]   Piegl, L.; Tiller W.: Curve and surface constructions using rational B-Splines. Computer-aided Design. Vol. 19. No. 9, 1987, pp. 485–498.
[14.9]   Walter, U.: Was sind NURBS ? Eine kleine Einführung CAD/CAM, 3 (1989), pp. 96–98.
[14.10]   Farin, G.: From Conics to NURBS: A tutorial and survey. IEEE Graphies & Applications, September 1992, pp. 78–86.
[14.11]   Pahl, G.: Konstruieren mit 31-CAD-Systemen – Grundlagen, Arbeitstechnik, Anwendungen. Springer-Verlag, Berlin, Heidelberg, New York, 1990.
[14.12]   Mortenson, M. E.: Geometrie Modeling. Wiley, New York, Chichester, Brisbane, Toronto, 1985.
[14.13]   Casale, M. S.: Free-form Solid Modeling with Trimmed Surface Patches. IEEE Computer Graphics and Applications. Vol. 7. No. 1, 1987, pp. 33–43.
[14.14]   Parametrik. CAD/CAM, 1996, 6, pp. 73–76.
[14.15]   Features verbessern die Produktentwicklung: Integration von Prozeßketten. (VDI-Berichte, 1322). VDI-Verlag, Düsseldorf, 1997.
[14.16]   Scholz-Reiter, B.; von Issendorf C.: CAD-Schnittstellen in der Praxis. CIM Management, 10 (1992), 2, pp. 23–30.
[14.17]   Jäger, K.-W: Schnittstellen bei CAD/CAE-Systemen – Grundlagen, Anwendungsbeispiele, Problematik, Lösungsansätze, VDI-Verlag, Düsseldorf, 1991.
[14.18]   Mattei, D.: New Version of IGES Supports B-REP Solids. Mechanical Engineering, January 1993, pp. 50–52.
[14.19]   Kiesel, R.; Rheinbay, J.; Leber, M.: Optimierung des CAD-Datenaustausches durch firmenübergreifende Zusammenarbeit. Konstruktion. 45, 1993, pp. 217–220.
[14.20]   Anderl, R.: STEP – Schritte zum Produktmodell. CAD-CAM-Report, No. 8, 1992, pp. 48–57.
[14.21]   Grabowski, H.: Das Produktmodellkonzept von STER, VDI-Z, 131, No. 12, 1989, pp. 84–96.
[14.22]   ISO/WD 10303-214 – Core Data for Automotive Mechanical Design Processes. September 1994.
[14.23]   Grabowski, H.; Anderl, R.; Polly, A.: Integriertes Produktmodell. Beuth Verlag, Berlin, Wien, Zürich, 1993.
[14.24]   Scharf, A.: 3D-CAD beschleunigt die Konstruktion. VDI-N, 1996, No. 20, p. 11.
[14.25]   Anderl, R.: CAD-Schnittstellen. Carl Hanser Verlag, Munich, Vienna, 1993.
[14.26]   Ziebeil, E.: Produkt- und Werkzeugentwicklung – Der Weg zum Kunststoffteil. Braun AG, Kronberg, December 1997.
[14.27]   Dressler, E.: Computer – Graphik – Markt 1996/97 – Ein systematischer Leitfaden durch die Branche. Dressler Verlag, Heidelberg, 1996.
[14.28]   Vajna, S.; Weber, C.: CAD/CAM-Systemwechsel – Chancen, Risiken, Strategien und Erfahrungen. Springer, VDI-Verlag, Düsseldorf, 1997.
[14.29]   Boshoff, E.: Integration von FEM – Berechnungen in den CAD-gestützten Konstruktionsprozeß durch bidirektionalen automatischen Geometrieaustausch. Dissertation, Aachen, 1997.
[14.30]   Lynen, W.: Partner für die Geräteentwicklung. F&M, 105 (1997), 6, pp. 424–427.
[14.31]   Michaeli, W.; Menzenbach, D.; Müller D.: CAD-Systerne in Spritzgießbetrieben. Plastverarbeiter, 46 (1995), No. 11, pp. 48–55.
[14.32]   Michaeli, W.; Menzenbach, D.; Schlesinger, K.: Beurteilung und Erweiterung von CAD-Systemen für den Einsatz in Kunststoffspritzgießbetrieben. Abschlußbericht zu einem AiF-Forschungsvorhaben. IKV, Aachen, 1996.
[14.33]   Bittermann, H.-J.: CAD ist zu einem Wettbewerbsfaktor geworden. PROCESS, 3, 1996, pp. 38–39.

[14.34] Sammet, F.: Einführung von CAD-Systemen. CIM Management, 10 (1994), 2, pp. 21–24.

[14.35] CEFE-Kriterienkatalog für CIM-Bausteine. IKO Softwareservice, Stuttgart, 1987.

[14.36] Reisbeck, C.: CAD/CAM – Einführung, Praxis, Auswahl. Hoppenstedt-Technik Tabellen Verlag, Darmstadt, 1990.

[14.37] Eversheim, W.; Dahl, B.; Spenrath, K.: CAD/CAM-Einführung. RKW-Verlag, TÜV-Rheinland, 1989.

15 Maintenance of Injection Molds

Injection molds represent a major investment for plastics processors. They constitute a large position of the company's assets and are the basis for production, economic success, and technical development. For these reasons, injection molds must be in good working order and ready for use.

In practice, however, situations frequently arise in which defects and improper maintenance of injection molds cause major disruptions to current production, occurring particularly during modifications. This reduces the actual working time of the injection molding machines and continually impedes proper, planned production. Against this background, back in 1992, an industrial survey by the Institute for Plastics Processing (IKV) in Germany [15.1] showed that on average almost 7% of possible production time was lost due to damage to injection molds (Figure15.1). Comparison with the results from 1973 clearly show that this figure has more than doubled in twenty years. As opposed to that, the proportion of machine-related downtimes is much lower. Technological developments in injection-molding machines have lowered the proportion by as much as one third.

In view of this situation, it is difficult to understand why injection molding shops, which would frequently have to maintain as many as 1000 molds, still employ the "fire brigade" approach, by which is meant that a mold is only repaired when it has failed. The industrial survey mentioned above [15.1] showed that only 30% of injection molding shops carry out preventive maintenance at fixed intervals. In these shops, again, only one third of the maintenance data is recorded and evaluated systematically. It follows from this that only 10% of injection molding shops perform preventive maintenance founded on a sound database [15.1].

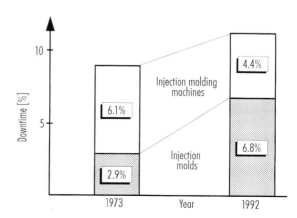

Figure 15.1 Change in production downtimes

The situation just described may be used to illustrate the deficits arising during the maintenance of injection molds (Figure 15.2). The state of the art is such that while damage and cost data are recorded, they often cannot be combined with each other. While such information, which represents invaluable experience for mold-making, is

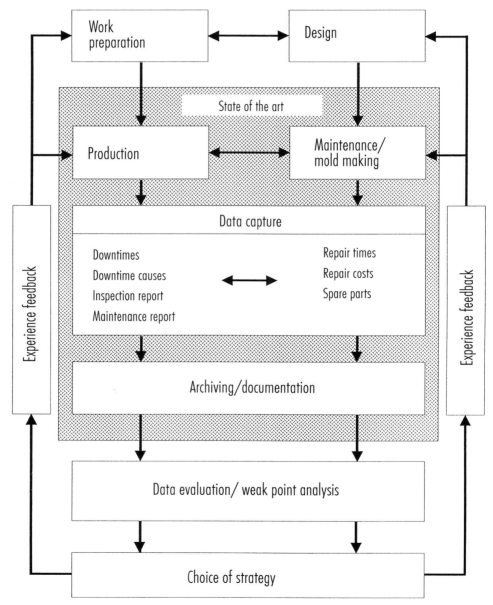

Figure 15.2   State of data acquisition and evaluation

archived, it merely serves documentation purposes. The goal should be, however, to use this invaluable practical experience as a basis for preparatory work and design.

Much as damage to molds is of interest, its causes are even more important. It turns out that, of the most frequent causes of damage, wear comes top of the list. This is followed by set-up errors and operating errors. The similarly relatively high proportion of design errors can, among other things, often be attributed to poor communication between those responsible for mold maintenance and mold design [15.2] (no feedback or archiving).

Since every injection mold is unique, it is not possible to generalize about maintenance. Commonplace are maximum possible standardization, the use of mold standards, easy accessibility and exchangeability of parts on the injection molding machine where possible, and the wear-resistant construction of friction pairings. But there are invaluable hints to be gained for individual molds, particularly from use. It will be shown below how these signs of weak points taken from production can be recorded so as to reduce costs and to optimize processes in production and mold-making.

15.1   Advantages of Maintenance Schedules

Figure 15.3 compares the work processes involved when the "fire brigade" and the preventive strategies are employed, in terms of attainable machine utilization and resultant downtimes.

Examples from shop practice prove the efficacy of performing scheduled preventive maintenance. Constant monitoring of the throughput times of maintenance jobs or actual repair times (Figure 15.4) permit measures to be taken so as to increase efficiency (e.g.,

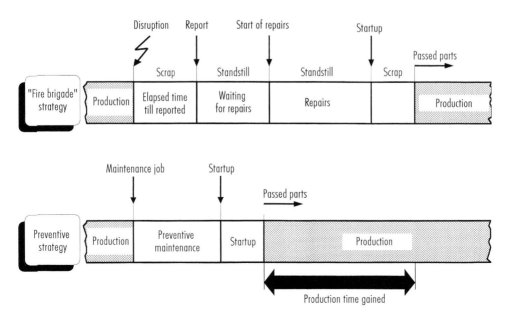

Figure 15.3   Time scheme for application of different maintenance strategies

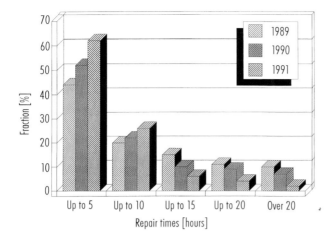

Figure 15.4   Decreasing repairs over the years

investment in new machining equipment) and decrease the throughput times for maintenance and for production through reducing downtimes. This effect is reinforced by purposeful preparation of the measures to be implemented (e.g., provision of equipment and spare parts), much as when set-up preparations are made when molds are changed. As may be seen in Figure 15.4, the proportion of quick repairs has increased in our example while the proportion of longer-lasting ones has receded over the years, as a result. This clearly illustrates the success of the measures implemented.

15.2   Scheduling Mold Maintenance

15.2.1 Data Acquisition

The choice of molds to examine first necessitates the acquisition of detailed data. Since various factory studies revealed that a great deal of acquired data are not used, particular value should be attached to goal-oriented or need-oriented data acquisition. The goal of data acquisition must be the provision of informative maintenance data in the form of feedback to staff in design, work preparation and mold making (Figure 15.2).

Data on mold maintenance is essentially required in two areas. The first is the control and monitoring of the direct and indirect costs that arise. It should be possible to report on all molds, a particular class of molds, an individual mold, or a functional group. The second is selective weak-point analysis, which requires detailed data acquisition. Here, a distinction needs to be made between the damage that occurs and its actual cause [15.4].

The mold data can be stored in a type of lifetime. As shown in Figure 15.5, the data for each individual mold should be recorded in the form of a lifetime data record. Item 1 is the mold identification number. To be able to schedule maintenance measures or intervals, the number of cycles needs to be known as it is a wear-determining factor (item 2). It is also important to establish if the maintenance measure is scheduled or non-scheduled (item 3). For referencing purposes, the functional system where the damage occurred must be noted (item 4). Description of the damage (item 5) and, where possible, the cause (item 6) should be coded for the weak-point analysis. Space is also required for a brief comment. The costs are entered into item 8, separated according to direct and indirect costs, for the purposes of evaluation. The mold lifetime can be kept for all injection molds by a central unit and forms a good basis for informative evaluations.

To illustrate the need for cost-related data acquisition, two evaluations of a mold résumé will now be presented. In the first, the maintenance activities were assigned to the various functional groups. The sum of the activities and the relation to the total instances of damage are shown in Figure 15.6.

In this example, repairs to the demolding system were the most frequent (50%), followed by mold cavities at 14%. The other functional groups sustained much less damage, amounting to less than 10%.

However, reporting mold damage in terms of the number of repairs is not satisfactory. It is important to link each event with the time for repair and the costs incurred. In this example, it made sense to use the available data to weight the damage susceptibility of certain modules according to the number of maintenance hours incurred. This afforded the possibility of making a concrete, value-based evaluation.

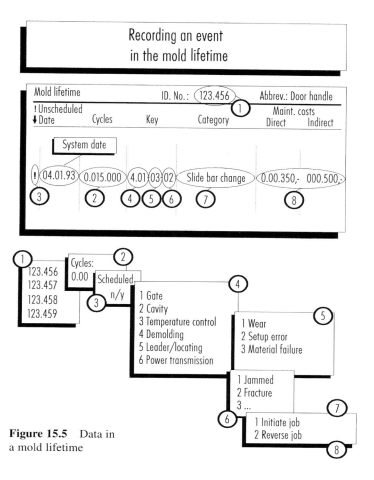

Figure 15.5   Data in a mold lifetime

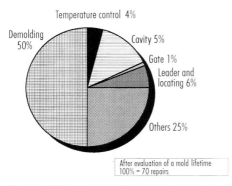

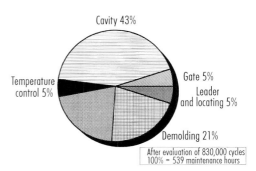

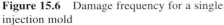

Figure 15.6   Damage frequency for a single injection mold

Figure 15.7   Maintenance hours expended on a single injection mold

Application of this approach to the same injection mold yielded the distribution of maintenance hours that is shown in Figure 15.7. This modified damage distribution is based on a total of 539 maintenance hours for a mold that carried out roughly 830,000 cycles in the production period concerned.

This analysis differs enormously from that based on the number of activities. While mold cavity and demolding still constitute the most damage, their ratio is now reversed:

demolding: previously 50% – now 21%,
cavity: previously 14% – now 43%.

This reversal is logical considering that an ejector can generally be replaced very quickly, but a repair to what often is a polished or chrome-plated mold cavity is relatively time-consuming. From the economics point of view and for the purpose of establishing a work-benefit ratio of maintenance measures, the analysis shown in Figure 15.7 must be considered to be more informative.

Conventional data acquisition using forms still serves a purpose, especially if it is only a temporary measure. A company will resist the unavoidable effort involved until it recognizes the advantages that this approach has to offer [15.5]. Although computer support should be the long-term goal, despite the considerable work involved for evaluation, forms can be used with great effect in a pilot project or for multiple instantaneous records. Generally, however, there is no extra work involved for the company as most already perform data acquisition, even if this does not always satisfy the criteria for an evaluation.

15.2.2 Data Evaluation and Weak-Point Analysis

A major goal of data acquisition and evaluation is to illustrate the failure modes of injection molds. This goal is served by the answers to the various questions, such as:

– What are the most common types of damage?
– Which functional system of an injection mold is most frequently affected by damage?
– Which molds are the most susceptible?
– What are the most common causes of damage?
– Which types of damage cause the most trouble?

The financial effects will not be discussed in detail here. Instead, the focus will be on technical aspects and possible consequences. A determination of the proportion of the most serious types of damage and their causes can reveal, for instance, that only five types account for more than 50% of all failures [15.6]. This provides those responsible in mold making with a direct starting point for eliminating the weak points.

A "Pareto Principle" can be derived from this relationship; it states that a small number of monitored types of damage will incur by far the most costs [15.7]. Also known as the "ABC method", this can be illustrated as shown in Figure 15.8. This tool can considerably reduce the amount of work involved in that it restricts attention to the greatest causes of costs incurred by molds on the one hand (Figure 15.8, top) and investigates only the most important types of damage for these on the other (Figure 15.8, bottom).

To make acquisition and evaluation of the various types of damage ascertained as easy as possible, a numbering system should be employed for the various types, just as was done for the various mold parts. For the sake of clarity, initially no more than 10 types of damage should be identified per functional system. Implemented as a numbering system, this means that ejector fracture would have a two-digit number (e.g., 41 where 4 = demolding system and 1 = ejector fracture). For five functional systems, this would allow fifty different types of damage to be described. The particular advantage of this is unambiguous identification of damage during data acquisition – employees are not using

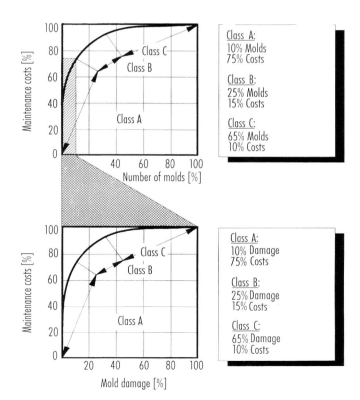

Figure 15.8   ABC analysis of damage

their own descriptions for the same type of damage. Furthermore, the numbering system allows direct classification and access to the corresponding work plan. It also helps to smooth the transition to a computer-based support system.

15.2.3 Computer-Based Support

Crucial to the use of a computer system are the functions provided for data evaluation, which must suit the case at hand. A peculiarity arises from the fact that injection molds are not fixed permanently in one place, but rather have to be tracked as movable inventory. A computer system must therefore be able to accommodate the respective status with more detailed information (e.g., in the mold department for repair until approx. …) This information serves production control for planning subsequent production orders as well as maintenance in the planning of preventive measures.

From the point of view of the prime aim of a maintenance analysis, a computer system must be in a position to provide the following functions:

– acquisition of all relevant data in a mold lifetime,
– support for ABC analyses for all molds, mold groups, functional systems, and mold damage,
– comprehensive support in the evaluation of lifetime data,
– presentation of percentage types of damage for a single mold or mold group,
– possibility of classifying damage within a functional system,
– presentation of the proportion of a certain type of damage in a mold group,
– presentation of the maintenance measures for the service life in cycles,
– comparison of intervals between occurrences of a particular type of damage,
– presentation of the frequency of the damage that has occurred with the goal of weak-point analysis,
– tracking of repair times, comparison of in-house/external share, etc.

The ultimate goal of the evaluations must be to eliminate primarily those weak points that incur the highest costs. An example of such a presentation is shown in Figure 15.9. This allows the costs of the different functional systems of a specific mold to be compared. If a functional system becomes noticeable because of extremely high maintenance, it must be possible to call up more detailed information on the proportions of the various types of damage. The special advantage of this presentation comes to the foreground when an evaluation can be performed separately on the basis of direct and indirect costs. In this case, those weak points whose direct costs were not high enough to cause concern can be uncovered if they lead to indirect costs in the form of lost profit contributions due to equipment downtimes.

15.3   Storage and Care of Injection Molds

Injection molds have a limited service life (Table 15.1). Appropriate measures can greatly extend this, however. Such measures can be classified on the basis of:

– maintenance,
– storage, and
– care.

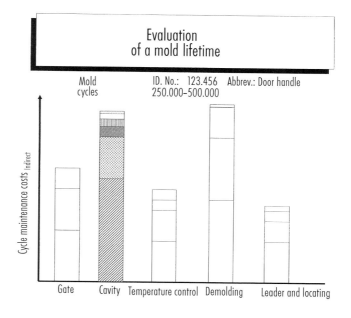

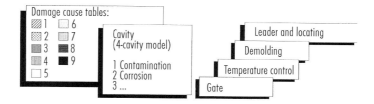

Figure 15.9 Weak point analysis based on mold lifetime

Table 15.1 Numbers of molded parts obtainable with various mold materials [15.8]

Material		Attainable number
Zinc alloys	Casting	100,000
Aluminum	Casting	100,000
Aluminum	Rolled	100,000–200,000
Copper-beryllium	Surface hardened	250,000–500,000
Steel		500,000–1,000,000

To be able to quickly fall back on ready-to-use molds, the following demands on storage and care must be fulfilled:

– every mold must be stored along with one molded part and a mold card in its own, easily accessible space in the mold store,

– only ready-to-use, complete, clean molds may be stored. The purpose of also storing a molded part (usually the last one from previous production) and a mold card bearing the article number and the mold number is to allow the mold to be uniquely identified.

The mold card should also bear all the information needed for setting up the mold and starting up the injection molding machine. Information in this category includes the following:

– mold design (split, sliding split, unscrewable mold etc.),
– dimensions of the mold and the molded part,
– mold mounting equipment,
– injection molding machine suitable for production,
– shot weight (injection volume),
– suitable plastic,
– rules on material pretreatment,
– processing temperatures,
– mold temperature and heat-control medium (water, oil, etc.),
– cycle times,
– injection pressure, follow-up pressure, dynamic pressure,
– injection speed,
– screw speed,
– cylinder equipment (sliding shut-off nozzle, non-return valve),
– maintenance intervals,
– number of pieces produced.

This list could be extended and thus matched to the special needs of a factory. Instead of on a mold card, much of this information, such as the settings for the injection molding machine, could be stored on external data storage media that could be read into the control unit prior to production startup.

Mold changes can only be performed quickly if the molds are ready for use when they leave the stores and can go into production without the need for major assembly or cleaning work. Every mold must therefore be a self-contained unit, i.e., it must not be made of parts that are required for other molds. Parts or groups of parts that are "loaned" or "borrowed" often disappear or are needed elsewhere just when the mold is scheduled for use. The consequences are unnecessary, incalculable, and often time-consuming downtimes.

Cleaning work also delays the start of production. It should therefore be kept to a minimum. This means that special care has to be taken of the molds (discussed later) and imposes specific demands on the store, its cleanness and particularly the ambient conditions. Damp and unheated rooms promote corrosion. Once rust has begun to attack the mold, maintenance becomes very time consuming and very expensive. Often it is impossible. The mold store should therefore be kept at a constant temperature where possible, and dehumidified. Not much equipment is required for this, and it soon pays for itself.

Important to the accessibility of the molds is also the size of the store. It is essentially determined by the vehicles available in the factory (e.g., forklift) and the maneuvering space.

When a job is complete, the mold may only be returned to storage when its suitability for future use has been checked. The last parts produced with it can provide an indication of its condition. They must be examined for dimensional stability and closely scrutinized. This will provide information about the state of the mold surface, the level of seal in the mold parting line (perhaps flash formation on the molded part) and the working order of the ejectors, ejector bushes, etc. If no deficiencies are found, the maintenance work then takes the form of the general care measures described below.

Maintenance of Cooling Lines

Cooling lines must be cleaned thoroughly to eliminate scale, rust, sludge, and algae.

Since these deposits decrease the diameter of the channels, measuring the flow rate is a way of checking the system. A pressure-controlled valve is installed between mold and water line and a defined pressure drop is set, which has to be the same for each examination. If the flow rate was measured with the new mold, a comparison with any new measurement after a production run provides information about the degree of clogging of the cooling channels.

For cleaning, the cooling lines are usually flushed with a detergent because mechanical removal of the deposit is generally not feasible due to the geometry of the system. Detergents and special cleaning equipment are marketed by several producers [15.9, 15.10]. A solution of hydrochloric acid (20° Be) with two parts water and a corrosion inhibitor has been successfully used.

The nipples, bridges, bolts and feed lines (tubes) outside the mold are also checked for damage and replaced where necessary, provided they stay on the mold.

Before the mold is stored, water has to be removed with compressed air and the system dried with hot air.

Care and Maintenance of the Mold Surfaces

After the end of production, the mold must be carefully cleaned of any adhering plastic residue. The work is independent of the type and amount of molding material. It is advisable to use soap and water for removing material remnants and other deposits. The mold then has to be dried carefully.

Rust spots from condensed water or aggressive plastics have also to be removed before storage. Depending on the degree of chemical attack, abrasives for grinding and polishing (car polish) may be suitable.

Removal of residual lubricants from movable mold components is also part of the cleaning operation. Degreasing detergents for this are available on the market.

Care and Maintenance of the Heating and Control System

This work is particularly important for hot-runner molds. After each production run, heater cartridges, heater bands, and thermocouples should be checked with an ohmmeter and the results compared with those on the mold card. Accidental grounding should be investigated, too. The control circuits are easily tested with an ammeter installed in the circuit.

A check should also be made to ensure that lines, connections, insulation, and main lead cleats are in proper working order.

Care and Maintenance of Sliding Guides

The guides on movable mold parts require particularly careful cleaning and must be washed with resin-free and acid-free lubricants. Also check the level of seal in the cylinder in the case of hydraulically actuated slides and cores.

Care and Maintenance of the Gate System

Start checking at the nozzle contact area, which is subjected to very high loads during operation. Check also any special nozzles belonging to the mold. In the case of temperature-controlled gates that are not generally demolded with every shot, it is necessary – to an extent depending on the plastic processed – to flush the gating system until the end of production with a plastic that has wide processing latitude.

Care Prior to Storage

At the end of each maintenance work, the mold has to be carefully dried and lightly greased with noncorrosive grease (petrolatum). This is especially important for movable parts such as ejector assembly, slides and lifters, etc. For extended storage, the mold should be wrapped in oil paper. Greasing and wrapping of the mold in oil paper is crucial when the mold store does not satisfy the demands above and below.

All observations and maintenance work are recorded on the mold card [15.11, 15.12].

15.4   Repairs and Alterations of Injection Molds

Injection molds can be subjected to extreme conditions during operation. This gives rise to wear symptoms that are due to rolling, sliding, thrusting, and flowing movements. A survey of the various kinds of wear, their causes and symptoms is provided in Figure 15.10.

	Type of wear Initial conditions	Characteristic	Manifestation, progression, results
With and without lubrication (metals, plastics, solids)	Sliding friction	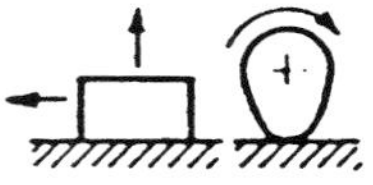	Seizing, cratering, grooving, running, clearance, chatter marks
	Rolling wear, with and without slippage	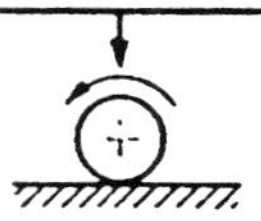	Pitting, peeling, spalling, rippling, seizing, grooving
	Wear by shock	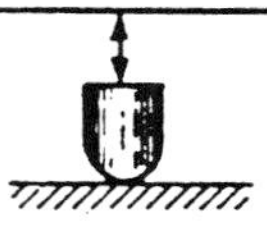	Break out, peeling, pitting
	Vibrational wear	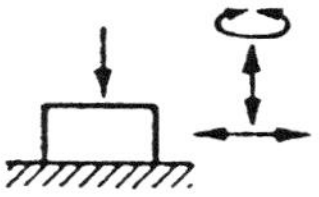	Roughening, seizing, oxide fluttering, fretting
Erosive Abrasive wear	Particle, sliding and rolling friction wear	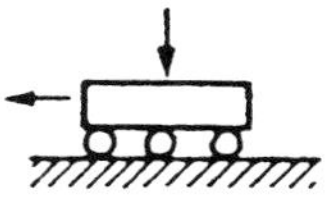	Grooving, break out, rolling tracks
	Sliding friction wear	 Counter-particle furrowing Particle furrowing a) b) 	a) Grooving, break out, embedding, smoothing b) Flat grooves, washout
	Hydroabrasive wear, radiation wear, other erosive wear		Waves, cavities, piercing, washout

Figure 15.10   Overview of types of wear [15.13]

The consequences of wear are dimensional inaccuracy, flawed surfaces and flash on the molded part. Before the damage can be repaired, the cause must be determined. Remedial measures require a detailed knowledge of the cause of damage. The following are possible:

– simple mechanical finishing,
– replacement of parts or modules,
– deposition of material.

Leaky parting planes are typical injection molding damage. When this is not very extensive, it can be eliminated by grinding. However, this is limited by the tolerances imposed on the molded-part dimensions.

Minor damage to the mold surface (pits) that can be attributed to impact can be remedied by reboring, remilling, and then setting pins or wedges. Once the flaw has been treated, the mold is heated and the drill hole or groove closed with a cold insert (slightly overdimensioned). The repaired spot is then rendered flush with the mold surface by grinding or polishing.

It is important to use the same type of material for this repair work, as the repaired area should have the same material properties as the rest of the mold surface.

Damage to functional and mounting parts, such as guide pins and bushings, ejectors, locating flanges, nozzles, etc., should not be repaired. These are normally standard parts (see Chapter 17) available in various dimensions and can thus be replaced cheaply. Doing this means that the molds will function perfectly and avoid any major risks.

The repairs described so far will often be inadequate and material will have to be deposited because, e.g., edges or corners have broken off. Welding is necessary in such cases.

Repair welds to injection molds should always be preceded by heating to keep thermal stress and the formation of internal stress as low as possible.

Preheating avoids compression and shrinkage in the weld zone and, above all, prevents heat from being dissipated so quickly from the weld area that hardening sets in (as when heated parts are quenched in oil or water).

The preheating temperature (at which the workpiece must be kept during welding) depends on the material to be welded, and in particular on its chemical composition. Steel manufacturers provide details of this.

During welding, the workpiece must be kept at the preheating temperature. When welding is complete, it is cooled to between 80 and 100 °C and then reheated again to the normalizing temperature [15.14].

Welding repairs are performed by the TIG method and welding with coated electrode wires. TIG (tungsten inert gas) offers distinct advantages. The following basic rules must be observed for repair welding:

– The electrode wire material should be of the same composition as the mold material, or at least similar. Ensuing heat treatment of the weld results in equal hardness and structure [15.14].
– The amperage has to be kept as low as possible to prevent reduced hardness and coarse structure [15.14].
– The preheating temperature must be above the martensite-forming temperature. It can be taken from the respective temperature-time phase diagram for the steel. It should not be considerably higher, however, since it increases the depth of burn-in [15.14].
– During the entire welding process, the mold must be kept at the preheating temperature. This is particularly the case for several deposits.

– At edges, the molten material needs to be supported. This can be effected with copper pieces or copper guide shoes that can be water-cooled if necessary.

Very recently, lasers have been used for repair welding of molds. Mostly these are pulsed solid-state lasers, e.g., ND-YAG lasers, with laser capacities of 50–200 Watt for hand welding.

The great advantage of laser welding over "conventional welding" is that low amounts of energy are applied with extreme precision to the welding site. Due to the very short welding impulses (1–15 milliseconds max.), the heated zone is very small, in the order of a few hundred millimeters. Thermal stress on the mold is therefore slight. Laser welding is more or less distortion-free [15.15].

Figure 15.11 shows which welding depths and seam widths are possible with lasers. Only relatively minor damage can be repaired in one working operation.

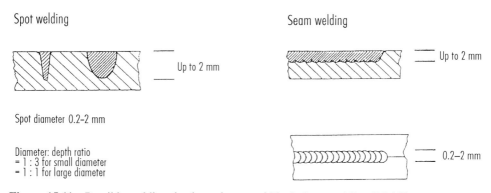

Figure 15.11   Possible welding depths and seam widths in laser welding [15.15]

The electrode wire material is generally < 0.5 mm in diameter, a small portion of which is melted onto the mold with every welding pulse. The wire material is available in different thicknesses and compositions.

The welding process itself is observed through a stereomicroscope fitted with a proper shield.

Due to the expected and actual difficulties inherent in all forge welding techniques, the calls for "cold metal-deposition processes" are understandable. One such process is electrochemical metallizing for depositing all kinds of metals and alloys on almost all metallic materials.

Dimensional corrections up to several tenths of a millimeter are possible with this method on flat surfaces, shafts, and in drill holes [15.16].

The steps required in effecting a repair (Figure 15.12) vary with the type and extent of the damage. Major damage (deeper than 0.5 mm) is first rebored and the hole sealed with pins. Then the damaged area, e.g., minor damage, is ground out in a hollow and sandblasted or electrochemically cleaned with a so-called preparatory electrolyte. An area treated in this way, free of grease and oxide, is optimally prepared for metal deposition.

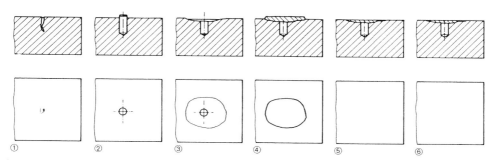

Figure 15.12 Working stages in electrochemical metal deposition [15.16]
1 Flaw; 2 Pins inserted; 3 Cavity ground out; 4 Cavity metallized with rapid-depositing electrolyte; 5 Leveling of metal filling (mechanical); 6 Transition to intact surface, covered with hard finishing layer

The repair area is then sealed off with galvanic sealing tape and the ground-out hollow is filled with a fast-depositing metal such as copper or nickel and mechanically flattened. The damaged area is then ready for sealing flush to the mold surface with an appropriate covering metal [15.16].

Sealing is carried out by soaking a graphite anode surrounded with an absorbent material in the desired high-performance electrolyte and moving it across the area to be coated. Under direct current, the metal is deposited onto the cathode, i.e., the mold surface.

There are also micro-cold or deposit welding devices [15.17, 15.18] on the market that operate on the principle of resistance pressure welding. The most common application of this process is spot welding.

Resistance pressure welding uses the heat generated by the electric current in overcoming the electric resistance at the point of contact with the parts to be welded. At the points of joining, the parts become pasty and are pressed together without the need for additional materials [15.19].

For repair welding of molds, e.g., filling out of hollows, one "part for welding" is replaced, e.g., by a steel tape which covers the hollow. During welding, the electrode is rolled along the steel tape, and pressed at the same time against the area to be repaired. The steel tapes are available in thicknesses of 0.1 to 0.2 mm. For deeper hollows, the process has to be repeated.

For minor repairs, e.g., to edges or corners, the steel tape is replaced by metal powder or metal paste [15.17].

The repaired areas can then be machined afterwards and polished to a high finish. Hardening and coating are also possible.

Metal-depositing processes are risky ways of effecting repairs, require dexterity and good knowledge of material behavior and the actual process employed.

References

[15.1]   Michaeli, W.; Feldhaus, A.; Eckers, C.; Lieber, T.; Pawelzik, P.: Instandhaltung von Spritzgießwerkzeugen – Ergebnisse einer Befragung von Spritzgießbetrieben. Prospectus, IKV, 1992.

[15.2]   Feldhaus, A.: Instandhaltung von Spritzgießwerkzeugen – Analyse des Ausfallverhaltens und Entwicklung angepaßter Maßnahmen zur Steigerung der Anlagenverfügbarkeit. Dissertation, RWTH, Aachen, 1993.

[15.3]   Hackstein, R.; Richter, H.: Optionale Instandhaltung – untersucht am Beispiel von Spritz- und Druckgußmaschinen. FB/IE 24 (1975), 5, pp. 267–273.

[15.4]   Oltmanns, P.: EDV-Unterstützung zur Instandhaltung von Spritzgießwerkzeugen, Unpublished report, IKV, Aachen, 1993.

[15.5]   Mexis, N. D.: Die Verfügbarkeitsanalyse in der Investitionsplanung. Verlag für Fachliteratur, Heidelberg, 1991.

[15.6]   Wilden, H.: Werkzeugkonzeption. In: Der Spritzgießprozeß. VDI-Verlag, Düsseldorf, 1979, pp. 87–109.

[15.7]   Taubert, D.: Wirtschaftliche Bewertungskriterien für die geplante Instandhaltung, VDI-Berichte, No. 380, 1980, S. 13–19.

[15.8]   Rheinfeld, D.: Werkzeug soll in Ordnung sein. VDI-Nachrichten, 30 (1976), 31, p. 8.

[15.9]   Reinigungsgeräte für Kühlkanäle. Kunststoffe, 54 (1974), 3, p. 112.

[15.10]   Spritzgießen-Werkzeug. Technical information, 4.3, BASF, Ludwigshafen/Rh., 1969.

[15.11]   Kundenzeitschrift. Arburg heute, 10 (1979), 16, June 1979.

[15.12]   Oebius, E.: Pflege und Instandhaltung von Spritzgießwerkzeugen. Kunststoffe, 64 (1974), 3, pp. 123–124.

[15.13]   Brandis, H.; Reismann, J.; Salzmann, H.; Spyra, W; Klupsch, H.: Hartschweißlegierungen. Thyssen Edelstahl, Technical report, 10 (1984), M1, pp. 54–75.

[15.14]   Rasche, K.: Das Schweißen von Werkzeugstählen. Thyssen Edelstahl, Technical report, 7 (1981), 2, pp. 212–219.

[15.15]   Schmid, L.: Reparaturschweißen mit dem Laserstrahl. Paper presented at the 8th Tooling conference at Würzburg: "Der Spritzgießformenbau im internationalen Wettbewerb", Würzburg 24. 9. 1997–25. 9. 1997.

[15.16]   Elektrochemischer Metallauftrag. Prospectus, Baltrusch und Mütsch GmbH & Co., KG, Forchtenberg.

[15.17]   Prospectus, Joisten und Kettenbaum GmbH & Co., Joke KG, Bergisch Gladbach.

[15.18]   Fachkunde Metall. Verlag Europa-Lehrmittel, Nourmney, Vollmer GmbH & Co., Haan-Gruiten, 1990.

[15.19]   Prospectus, Schwer & Kopka GmbH, Weingarten.

16 Measuring in Injection Molds

16.1 Sensors in Molds

Crucial to the quality of injection moldings is the state of the melt after plastication and how the mold-filling process went, which is characterized by the change in pressure and temperature.

It is therefore advisable to incorporate appropriate sensors into the molds for monitoring and perhaps control purposes.

On the other hand, installing them incurs additional costs, could even leave marks on the moldings to be made, and could possibly make mold changing difficult. Moreover, the cooling channels and the ejector units take up considerable space, with the result that there is limited choice as to the best place to install them.

For these reasons, it is not standard practice to measure melt temperature in production molds. However, pressure measurements in molds are being used more and more to monitor production (for documentation and scrap monitoring) and are a prerequisite for pressure-dependent switching and for controlling the pressure in the mold [16.1].

16.2 Temperature Measurement

16.2.1 Measuring Melt Temperatures in Molds Using IR Sensors

The thermocouples and resistance thermometers used to measure and control mold wall temperatures are unsuitable for measuring melt temperatures (these are discussed in Chapter 18: Temperature Controllers for Injection and Compression Molds).

Melt temperatures are measured by methods that detect the inherent radiation emitted by the melt in a defined wavelength range.

To this end, a bundle of optical fibers is installed inside the mold, flush with the wall. The inherent radiation emitted by the melt is fed to a radiation-sensitive detector. The electrical output signal from the detector then correlates with the melt temperature.

The radiation emitted to the environment by a body as a result of its temperature depends on its emission power. Ideal blackbody emitters emit all their heat, i.e., they have an emission power of 1. Real bodies have a lower emission power, the amount of radiation emitted being reduced by the amounts lost through reflection and transmission.

Polymer melts inside a closed mold behave almost like ideal blackbodies since the reflected and transmitted radiation emitted from the cavity surfaces impinge on the IR sensor. This is why plastics that have an emission coefficient of 0.9 in air have emission coefficients of 0.99 in molds [16.2–16.4].

Melt temperatures inside molds can therefore be measured with high precision ($\pm 1\%$ of the display value) and speed (response time < 10 milliseconds) and with high reproducibility using optical fibers connected to IR measuring devices. The measurement ranges depend on the wavelength of the optical fiber, lying between 70 and 260 °C, and between 177 and 427 °C [16.2].

16.3 Pressure Measurement

16.3.1 Purpose of Pressure Measurement

Figure 16.1 illustrates the change in pressure within a cavity and the properties of the finished part that are affected by it. The shape of the filling phase has a major influence on the texture of molding, the degree of orientation, the level of crystallinity in the layers close to the surface, and thermal and mechanical stress on the melt.

Complete filling of the contours of a mold cavity, along with the occurrence of flash and possible resultant mold damage depend on how the pressure changes as it rises in the compression phase.

During the holding-pressure phase, the weight and shrinkage of their molding is determined. Just as in the filling phase, the degree of crystallization and orientation of the macromolecules in the inner regions of the molds can be affected here.

16.3.2 Sensors for Measuring Melt Pressures in Molds

The informativeness of the change in pressure inside the mold depends primarily on the position of the installed detector. The pressure sensor measures the pressure during the injection phase, as soon as the flow front passes over the point of installation and throughout the holding-pressure phase until the molding shrinks so much at the point of installation that it loses contact with the cavity. For this reason, pressure sensors should

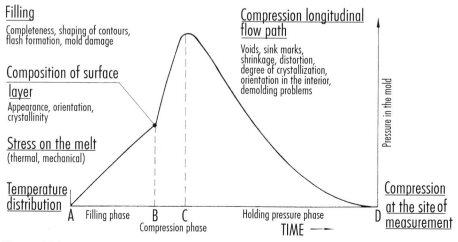

Figure 16.1 How change in pressure affects quality of finished parts

be installed close to the gate as the change in pressure measured there best describes molding formation for the longest possible period. A position close to the gate is also recommended for any intended measures to control the change of pressure inside the mold.

To measure the degree of filling, by contrast, installation away from the gate is best. Therefore, it would be better to position the pressure sensor away from the gate in the case of molds that have very long flow paths or that tend to overpack away from the gate. A range of different designs based on piezoelectric and strain gage detectors are available for direct and indirect measurement of internal mold pressures [16.2, 16.5].

16.3.2.1 Direct Pressure Measurement

In the direct method of measuring internal mold pressure (Figure 16.2), the hole for the detector is bored right into the cavity such that the head of the detector is in contact with the melt.

The head is shaped as a piston or membrane. It transmits the pressure to the measuring element behind it. The head of pressure detectors fitted with a piston can be machined and shaped within certain limits to the contour and texture of the cavity.

Piezoelectric pressure detectors return a charge signal proportional to the mechanical load that is transformed in a downstream amplifier into the corresponding voltage.

Even in the most unfavorable measurement points, system drift by the analyte signal is better than 0.5%/min. Moisture and contamination of the feed lines can exacerbate the drift and it is absolutely vital to keep them clean or clean them as necessary [16.5].

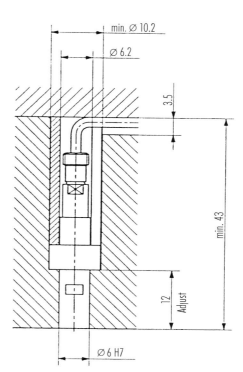

Figure 16.2   Direct measurement of cavity pressure with the aid of a piezoelectric pressure sensor [16.2]

Piezoelectric measuring systems are only suitable for making dynamic and quasi-static measurements and they must therefore be brought into a defined state before every measurement. This can be effected automatically by a signal from the machine control (e.g., "Move forward") prior to the start of injection.

Piezoelectric pressure detectors feature high accuracy and linearity ($< \pm$ 1% of measuring range) and a low temperature drift ($< \pm$ 0.01%/°C); common measuring ranges extend up to 200 MPa at 200 to 350 °C (depending on model) [16.5].

Strain gage detectors (working on the principle of a change in resistance) cannot attain the same high threshold frequencies as piezoelectric detectors but their dynamic behavior is totally adequate for measuring very rapid changes in internal pressure during the mold filling process. On account of their measuring principle and construction, they are suitable for both static and dynamic measurements. Resistance changes brought about by temperature changes in the strain gage elements are compensated because the entire measuring chain, consisting of pressure detector and amplifier, is automatically calibrated before each measurement. Again, this is effected by means of a signal from the machine control prior to the start of injection (see above). The overall error arising from measurements with strain gage detectors is $< \pm$ 1% of the measuring range, and temperature drift varies according to model from $\pm$ 0.02 to 0.04%/°C; common measuring ranges extend up to 150 MPa at 200 °C [16.2].

16.3.2.2 Indirect Pressure Measurement

Additional marks in the molded part can be avoided by using sensors for indirect measurement of the melt pressure placed under ejector pins that are already present anyway. The analyte parameter is the force acting on the ejector; the diameter of the ejector is used for calculating the cavity pressure (Figure 16.3).

During mold construction and installation, a good fit of the ejector pins is necessary to ensure that the measurement is not falsified by frictional forces.

There are two types of design; one is bolt-shaped with piezoelectric or strain gage elements. If the pressure is monitored only during mold proving and not continuously, the sensors can be removed and replaced by "blank sensors" [16.2, 16.5].

The data of the sensor for indirect pressure measurement correspond to the data of the sensor for direct pressure measurement.

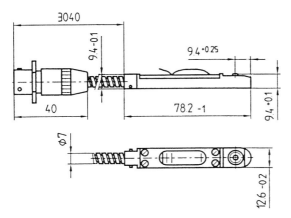

Figure 16.3   Sensor with piezo-electric or DMS element [16.2]

The data quoted for measuring accuracy, temperature dependency and response time, just like those concerning measuring ranges and service temperatures are guide values; the manufacturer's instructions must be observed each time.

16.4 Use of Sensor-Transducer Probes

The sensors described can be used for various processes associated with production. These can be divided into process-optimizing and quality-monitoring mechanisms.

16.5 Process Optimization

Optimum machine setting parameters are crucial to the quality of the molded parts. Optimum process control entails minimizing cycle times and increasing the service lives of machines and molds. It is often not enough to just consider process parameters on the machine side as a means of optimizing the injection molding process. This is because this approach fails to take account of the filling behavior of a mold, for instance, viscosity fluctuations of the raw material, and changes in ambient conditions such as shop temperature and humidity. The more sensors which are fitted to the mold, the more information that can be gained from the production process.

The most useful information about the process of molded-part formation can be obtained from measuring the mold cavity pressure. Optimizing the shape of this process curve can greatly influence molded-part properties such as dimensions, weight, and internal properties, as well as cycle times and mold wear. The measurements also help to speed set-up when molds are used on different machines. Figure 16.4 illustrates various cavity-pressure curves. Of most interest is the linear rise in the mold cavity pressure, synonymous with virtually constant flow front speed [16.11] during injection (e), the switch-over phase from the speed-controlled injection phase to the pressure-controlled holding-pressure phase (n) and the duration and value of the holding pressure. Figure 16.4 shows pressure curves that are obtained when the switch-over point in the mold is altered. Switching too late from the injection to the holding-pressure phase (1) causes the cavity pressure to peak and may lead to overpacking of the cavity, demolding difficulties and damage to the mold, while switching too early may cause the cavity to be filled under holding pressure (2). Incompletely filled molded parts and a high level of orientations are the result. The optimum cavity-pressure curve has neither peaks nor pressure breaks (3). With the aid of the cavity pressure sensor, the optimum point can be determined rapidly. Similarly, the gate-sealing point, the point when the gate is frozen and no more material can be transported into the mold, can also be determined and the holding-pressure time adapted. The residual cooling time is the time elapsed until the molded part detaches from the mold and is equivalent to the time when the mold cavity pressure has reached the zero limit. The whole process can be recorded by placing the mold-cavity pressure sensor close to the gate. The farther the sensor is located away from the optimum position, the less information is obtained about the process.

Aside from manual corrections to the molded-part formation processes performed with the aid of cavity-pressure curves, modern injection-molding control units offer the possibility of using the cavity pressure to switch over from the injection to holding phase. Once a threshold value is exceeded, a corresponding signal is transmitted. This switching process makes for greater reproducibility. Above all, it allows for fluctuations

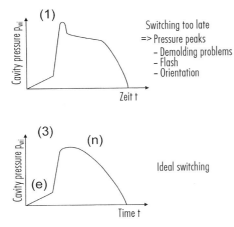

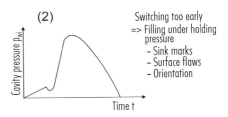

Figure 16.4   Change in pressure at different switching points

in metering. The threshold value must be determined iteratively in trials or with the aid of simulations. Automatic determinations can be made by measuring the change in pressure on-line [16.5, 16.12]. The significant rise in the pressure change serves as the criterion for the determination and the switch-over point from the injection phase to the holding phase is made during the cycle. This process allows the system to respond automatically to fluctuations in materials or the process, without the need for external intervention.

After the mold-cavity pressure, which is the most relevant process parameter during injection molding, temperature measurements may be used to obtain further important information for optimizing and stabilizing injection-molding production. Infrared sensors for measuring the temperature in front of the screw provide information about the actual melt temperatures and uncover possible processing errors due to specification limits that have not been observed. The temperature that has been preset for cooling systems or measured and depicted describes the temperature of the cooling medium as it returns from the mold to the machine. The actual mold-wall temperature, however, can only be measured with the aid of thermocouples mounted just beneath the mold wall in the mold. Recording these temperatures allows not only non-observed specification limits to be identified and compensated but also dynamic process changes, such as improperly set controllers and ambient influences.

16.6 Monitoring Quality

High quality demands imposed on injection-molded parts require intensive testing by the molded-parts manufacturer. Statistical process control (SPC) is standard in companies and uses random samples to make inferences about the process level or quality level and to identify trends in production. The constantly rising demands on injection-molded products and the random controls of SPC frequently make continuous quality control monitoring (CQCTM) unavoidable. Sensors in the mold and machine are used to effect on-line monitoring. The goal is to calculate quality properties direct from the measured process parameters, such as cavity pressure and temperature. This provides the

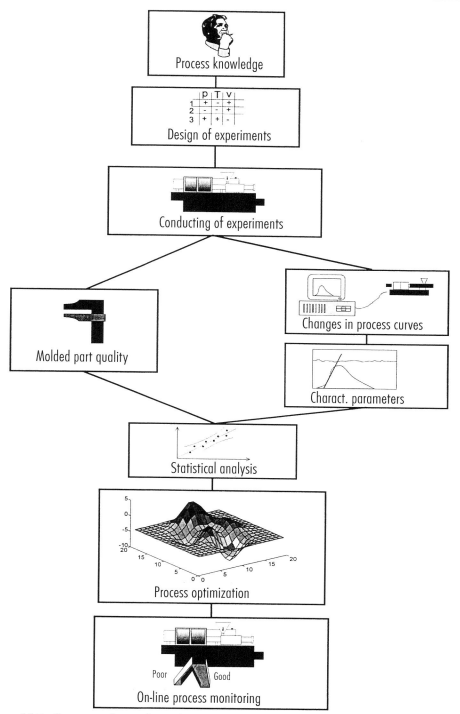

Figure 16.5   Sequence of steps for setting up process models for monitoring quality

possibility of performing a 100% quality control without the need to examine each and every molded part.

The key to the quality calculation is a model that describes the dependencies of molded-part properties and process parameters. The approach for determining such a model is illustrated in Figure 16.5. Starting from existing knowledge of the process, such as that of a machine setter, a statistical experimental plan is drawn up in which alterations to machine settings are set out during trials to be conducted. Statistical experimental design makes it possible to design trials efficiently and to gain a great deal of information from technical processes with a low number of experiments.

In injection molding, process parameters are recorded during the trial phase on the machine and the molded parts produced are removed and labeled so that clear classification of trial parts and measured process parameters is ensured during the modeling phase. When the trials are over, the quality characteristics of the produced molded parts are determined.

Modeling of the interdependencies of the quality characteristics and process parameters is performed off-line. Mostly, statistical methods are used as models. The choice of regression to use depends on the type of quality characteristic to be evaluated. Continuous characteristics, such as weights and dimensions, are evaluated with multiple regression calculations. Logistic regression is used for describing the dependencies of attributive characteristics. This group of quality characteristics includes those which can only be subdivided into classes, such as "Characteristic present" and "Characteristic absent". Burns and sink marks numbers are among these. Increasingly, statistical methods are being replaced by algorithms from research areas of artificial intelligence, such as neural networks. These have the advantage of being easier to apply since the relationships between process parameters and quality characteristics can be learned independently [16.13, 16.14].

With the aid of the models created, the quality of molded parts can be calculated on-line during injection molding from the process parameters measured during the production cycle. Thus, as soon as the cycle is over, information is available on the product just made. This can be used to aim for 100% quality documentation and establish a "reject route" for molded parts that fail to meet specifications.

References

[16.1]   Recker, H.; Spix, L.: Meß-, Steuer- und Regeleinrichtungen. In: Johannaber, F. (ed.): Kunststoffmaschinenführer, 3rd Ed., Carl Hanser Verlag, Munich, Vienna, 1992, pp. 1123–1180.

[16.2]   Company brochure, Dynisco Geräte GmbH, Heilbronn.

[16.3]   Wittemeier, R.: Temperaturmessung im Schneckenvorraum einer Spritzgießplastifiziereinheit mit Infrarotsensoren, Study report, Univ. GH, Paderborn, 1990.

[16.4]   Bluhm, R.: Methoden zur Bestimmung der Massetemperatur - Infrarotsensoren und miniaturisierte Thermoelemente, Paper presented at the 17th Tooling Conference, IKV, Aachen, 1994.

[16.5]   Company brochure, Kistler Instrumente AG, Winterthur, Schweiz.

[16.6]   Sarholz, R.; Beese, U.; Hengesbach, H. A.: Prozeßführung beim Spritzgießen, IKV, 1977.

[16.7]   Stitz, S.: Konstante Formteilqualität beim Spritzgießen durch Regeln des Werkzeug-Druckverlaufs, Kunststoffe, 63 (1973), 11, pp. 777–783.

[16.8]   Rörick, W.: Zur Praxis der Prozeßregelung im Thermoplast-Spritzgußbetrieb, Dissertation, RWTH, Aachen, 1979.

[16.9]   Sarholz, R.: Rechnerische Abschätzung des Spritzgießprozesses als Hilfsmittel zur Maschineneinstellung, Dissertation, RWTH, Aachen, 1980.

[16.10] Hellmeyer, H. O.: Ein Beitrag zur Automatisierung des Spritzgießprozesses, Dissertation, RWTH, Aachen, 1977.

[16.11] Lauterbach, M.: Ein Steuerungskonzept zur Flexibilisierung des Thermoplastspritzgußprozesses, Dissertation, RWTH, Aachen, 1989.

[16.12] Bader, C.: Die Bedeutung der Werkzeuginnendruckmeßtechnik beim Spritzgießen. Reprint at the conference on: Qualitätssicherung und -optimierung beim Spritzgießen, IKV, Aachen, 1997.

[16.13] Gierth, M. M.: Methoden und Hilfsmittel zur prozeßnahen Qualitätssicherung beim Spritzgießen von Thermoplasten, Dissertation, RWTH, Aachen, 1992.

[16.14] Schnerr, O.; Vaculik, R.: Qualitätssicherung mit Neuronalen Netzen. Paper presented at the 18th Tooling Conference, IKV, Aachen, 1996.

17 Mold Standards

Injection molds are always made in accordance with the same rules. Therefore, it should not come as a surprise that their design approach is always similar. This holds particularly true for the basic components. A great number of companies are specialized in manufacturing such basic elements. They produce these elements on a large scale and in a variety that a detailed discussion of these products would be beyond the scope of this chapter. We emphasize, therefore, to contact a supplier of mold standards. Suppliers offer extensive and very informative catalogs. Sometimes the information is also available as a software database. Figures 17.1–17.3 and Table 17.1 show the most common mold standards and their application areas.

Mold standards are parts or modules whose dimensions have been standardized and characterized. In line with the basic design of a mold, they can be classified as standards for the mold structure, the cavity, the gate system, the guides and centering, for heat control, for demolding and for accommodating the mold in the injection molding machine.

The use of mold standards shortens production times and relieves the designer and mold maker of routine work. Advantages are:

– For computer-assisted design, standards can be retrieved from a database, displayed on the screen and employed in the design ("features"). The user has the option to "play" with several variants to select the solution which is the most suitable one in his opinion. This relieves him of routine work.
– The uncertainty in a quotation is smaller because one can calculate with fixed costs for the individual elements.
– Work input into the production of the molds can be reduced by 25–45% through the use of standards. Extensive studies have shown that 55% of the total work is performed by the mold maker himself, 25% by the standards producer, and 20% can additionally be delegated to the standards producer [17.2].
– The mold maker can adjust his machinery to the special requirements of mold making since he has to work primarily on the cavities. He saves on capital expenditure and can operate more effectively and with lower costs.
– Supply of replacement is simplified because standards are interchangeable. There is no need for maintaining an expensive in-house stock.
– Discarded molds can be disassembled and part of their components re-used.

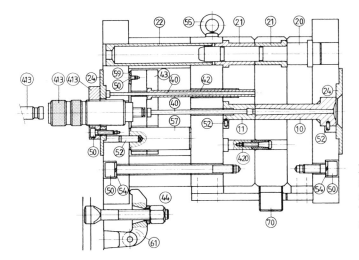

Figure 17.1   Use of mold standards for injection molds [17.1] (See Table 17.1 for explanation of numbers)

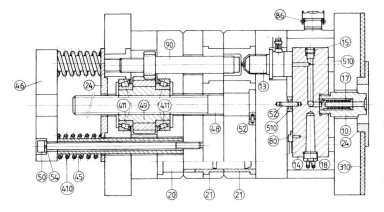

Figure 17.2   Use of mold standards for injection molds [17.1] (See Table 17.1 for explanation of numbers)

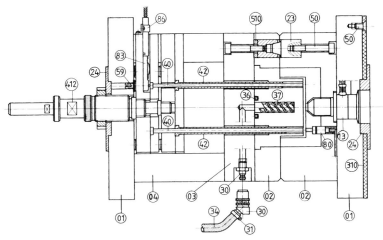

Figure 17.3   Use of mold standards for injection molds [17.1] (See Table 17.1 for explanation of numbers)

Table 17.1   Mold standards and their applications (see Figures 17.1–17.3 for illustrations of parts)

Module	Functional group	Part No.	Name	Application area
0	Mold frame	1	Mounting plates	Circular, rectangular, drilled and undrilled plates, machined on all sides, in various dimensions. Material: steel and different aluminum grades. Used for individual mold constructions into which only the cavities or mold inserts, gating system, cooling system and drill holes for the ejector systems have to be incorporated. The drilled plates contain drilled holes for guide elements and for bolting plates together; these are exchangeable parts.
		2	Mold plate	
		3	Backing plate	
		4	Spacer strip	
1	Gating systems	10	Sprue bushes	Accommodating the drill hole for the sprue, sealing the mold against the injection nozzle
		11	Sprue retaining bush	Demolding the gating system, especially the sprue and the sprue gate
			Sprue bushes with pneumatic sprue removal and corresponding nozzles	Fully automatic sprue removal
		13	Hot runner nozzles, with and without needle valve	Single-point and multipoint gate nozzles lead the melt directly to the cavity. Runners etc are eliminated. Fully automatic production
		14	Heated manifold blocks	Leading the melt in hot runner molds from the injection nozzle to the hot runner nozzles
		15	Closure plug	Sealing a melt channel in hot runner blocks
			Deflector plug	Deflecting the melt in hot runner blocks
		17	Melt filter Filter insert	Retaining contaminants. For injection nozzles and hot runner nozzles
		18	Heating elements, like: Cartridge heaters Band heaters, Flat tubular heating elements, Spiral tube cartridges, Ring heaters with corresponding control devices	Temperature control in hot runner molds, molds with insulated runners and thermoset and elastomer molds

Table 17.1 (continued) Mold standards and their applications (see Figures 17.1–17.3 for illustrations of parts)

Module	Functional group	Part No.	Name	Application area
2	Guide and locating elements	20	Guide bolts	Internal guiding and locating of mold halves and lateral slide bars
		21	Guide bushes	Locating and guiding mold halves and ejector assembly
		22	Locating bush	Locating mold plates and strips relative to each other
		23	Locating unit	Locating the mold plates
			Flat guides	For individual guides with solid lubricant depots
		24	Locating ring	Locating and centring mold halves on the machine mounting plates
3	Temperature control system	30	Couplings and nipples, also as sealing couplings and nipples	Connecting the mold to the temperature control units
		31	Hose connectors Hose clamps Pinch-off sleeves	
		34	Hoses: PVC Hose with fabric insert Vition hose with metal braiding Corrugated metal hose with metal braiding	Depending on temperature of heating medium
			Sealing bolt	Locking heating channel holes, e.g. at deflectors (cross holes)
		36	Seals	O-rings for sealing the temperature control system, Teflon® ribbon, etc.
		37	Spiral cores, single flighted, double flighted	Cooling mold cores
			Cooling finger	Spot cooling at points in barely accessible areas of mold
			Heat pipe	
			Heat sink cartridge	
		310	Thermal protection plates Insulating plates	Avoiding heat dissipation into machine mounting plates in hot runner molds, molds with insulated runners and thermoset and elastomer molds
			Heating elements (see Gating system)	See Gating system

Table 17.1   (continued) Mold standards and their applications (see Figures 17.1–17.3 for illustrations of parts)

Module	Functional group	Part No.	Name	Application area
4	Ejector system	40	Ejector pins	Demolding molded parts and gating systems
			Flat ejectors	
		42	Ejector bushes	
			Stripper plate	
		43	Ejector plate	Accommodating the ejector pins and the retraction forces
		44	Ejector retaining plate	Supporting the ejector pins Application of load into pins
		45	Ejector guide bolts	Guiding, locatings and displacing the ejector assembly Application of load into ejector system on opening of mold
		46	Ejector plate	Tensioning the pull-back springs
			Retaining disc	
			Quick-action coupling	Connection to machines with hydraulic ejector system
		48	Coarse pitch spindles	Demolding molded parts with threads
		49	Coarse pitch nuts	
			Pinions	
		411	Bearings	
			Gear racks	
			Spur gear	
			Idle wheels	
	Ejector retraction/ retraction force	410	Retraction spring	Retracting ejector assembly after demolding. Protection of ejector pins and mold plates
			Ejector plate return pins split/non-split	
		412	Retraction device	
		413	Feedback device	
			Return pin	
	Ejector subassembly		Ratchet rod	Subdivision of ejector movement for molds with several parting lines
		420	Driver	
			Two-stage ejector	
5	Accessories	50	Bolts	For bolting together mold plates, locating ring, thermal protection plates, etc.
			Setscrews	
		52	Straight pins	

Table 17.1   (continued) Mold standards and their applications (see Figures 17.1–17.3 for illustrations of parts)

Module	Functional group	Part No.	Name	Application area
			Ball catch	Stopping movable mold parts, e.g. jaws
		54	Springs/spring elements	Compression and disc springs for e.g. ejector retraction, bolt security, etc.
	Transport aids	55	Ring bolts	Lowering lifting tool and securing molds during transport
	Spacer segments	57	Support rolls Support strips	Partial support for mold plates and for bridging large distances bewteen the mold plates with minimal thermal conduction
		59	Headed dowels and support discs	Supporting the ejector assembly
		510	Spacing rings	Supporting the hot runner blocks Compensating plate thicknesses
6	Clamping systems/rapid clamping systems	60	Mechanical and hydraulic clamping systems	Mounting the molds on the machine mounting plates
		61	Clamps/brackets	
7	Slide bar mechanism	70	Mechanism comprising tie rod and slide rod	Long slide bar distances that otherwise are attainable only with hydraulic or pneumatic cylinders
			Locking cylinders	Locking and pulling lateral slide bars and core pins
8	Measuring and control devices	80	Heat sensors	Temperature control and monitoring in hot runner molds with insulated runners and thermoplastics and elastomer molds
			Electronic temperature controllers	
			Pressure sensors	Monitoring and recording cavity pressure and pressure in the hydraulic system of the injection molding machine
		83	Sensor	
			Recording device	
			Computer	
		86	Plug-in connectors	Connecting sensors to measuring and recording devices
9	Mold inserts	90	Cores	Core pins for swappable cores, e.g. unscrewing cores

Table 17.2 Selected suppliers of mold standards, hot runners and accessories, pressure and temperature sensors as well as mold changing systems in North America. These suppliers offer not only tried and true parts but also systems tailored to specific needs.

Company	City, State
American MSI	Moorpark, CA/USA
Bohler	Arlington Heights, IL/USA
D-M-E	Madison Heights, MI/USA
DMS	Windsor,Ontario/Canada
Dynisco	Gloucester, MA/USA
Enerpac	Milwaukee, WI/USA
Engel	Guelph, Ontario/Canada
Eurotool	Gloucester, MA/USA
Ewicon	East Dundee, IL/USA
East Heat	Elmhurst, IL/USA
Gammaflux	Sterling, VA/USA
Gunther	Buffalo Grove, IL/USA
Hasco Internorm	Chatsworth, CA/USA
Hasco Mold Bases	Arden, NC/USA
Hotset	Battle Creek, MI/USA
Husky	Bolton, Ontario/Canada
Incoe	Troy, MI/USA
Kistler	Amherst, NY/USA
Manner	Tucker, GA/USA
Mold-Masters	Georgetown, Ontario/Canada
Plasthing	Mishawaka, IN/USA
Stäubli	Duncon, SC/USA
Thermodyne	Beverly, MA/USA
Uddeholm	Wood Dale, IL/USA
Watlow	St. Louis/MO/USA

References

[17.1] Catalog of Standards, Hasco, Lüdenscheid, Germany.
[17.2] Heuel, O.: Werkzeug – F M E A für Spritzgießwerkzeuge. Kunststoffberater, 1, 2 and 3, 1994.

18   Temperature Controllers for Injection and Compression Molds

18.1   Function, Method, Classification

Temperature controllers have the function to bring up molds connected to them to processing temperature by circulating a liquid medium and keep the temperature automatically constant by heating and cooling.

The basics of temperature control are shown in Figure 18.1.

The heat-exchange medium in the tank (1) with built-in cooler (3) and heater (2) is delivered to the mold (10) by a pump (4) and returned to the tank.

The sensor (9) measures the temperature of the medium and passes it on to the main-control input (7). The controller adjusts the temperature of the heat-exchange medium, and, thus, indirectly the mold temperature.

If the mold temperature rises above the set value, then the magnetic valve (5) is actuated and cooling initiated. Cooling takes place until the temperature of the medium, and with it that of the mold, have reached the set value again.

If the mold temperature is too low, heating (2) is activated in analogy to cooling.

Temperature controllers can be classified as follows: devices for operating with water and heat-transfer oil. Devices for operating with water generally have an initial temperature of 90 °C or of roughly 160 °C if is pressurized with water. Those for heat-transfer oil without pressurization have an initial temperature of 350 °C.

Figures 18.2 and 18.3 illustrate a temperature controller for water and oil operation up to 90 and 150 °C.

Table 18.1 presents a summary of properties of additional equipment employed for temperature control.

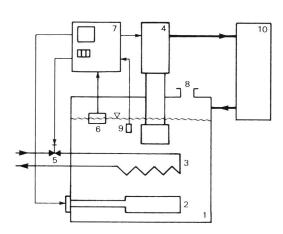

Figure 18.1   Method of temperature control
1 Tank, 2 Heater, 3 Cooler, 4 Pump,
5 Magnetic valve for cooler, 6 Level control, 7 Main control, 8 Inlet,
9 Temperature sensor, 10 Mold

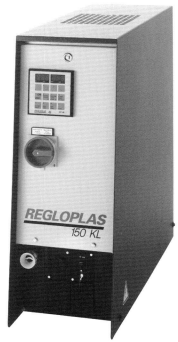

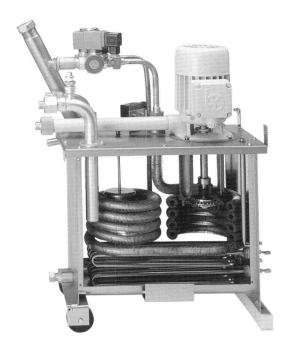

Figure 18.2   Temperature controller for water or oil operation

Figure 18.3   Temperature controller for water or oil operation (housing removed)

Table 18.1   Equipment for temperature control (summary)

System	Characteristics
Cooling tank	For cooling only (water) Control depending on skill of operator Mold temperature depending on temperature and pressure of line water Processing conditions (cycle time, coolant temperature, breaks) are not considered
Cooling-water thermostat	Automatic control considers processing conditions For cooling only (water) Control independent of skill of operator Use is limited to temperatures above temperature of line water
Refrigerator	For cooling only (except special design) Use independent of water supply No water consumption
Temperature control devices	For cooling and heating Control independent of skill of operator because of automatic control Mold temperature independent of water temperature and pressure Processing conditions are considered Heating up to processing temperature possible Use is limited to temperature above temperature of line water

18.2   Control

18.2.1   Control Methods

There are three methods to control the mold temperature:

Control of the heat-exchange medium (Figures 18.4 and 18.7).
The sensor measures the temperature of the medium in the equipment. Figure 18.6 presents the response of this control type in various stages. The characteristics are as follows:

– The temperature set at the controller and required for reproducible production is not measured. Therefore, depending on the disturbance variables, different temperatures can occur in the mold, even though the same value is selected.
– Variations of the temperature in the mold may be relatively great because the disturbance variables (9) affecting the mold are not directly considered and controlled (such as cycle time, injection, melt temperature, etc.).

Direct control of the mold temperature (Figures 18.5 and 18.8).
   The temperature sensor is in the mold. This results, in most cases, in a much better stability of the mold temperature than with the control of the medium temperature.
   Figure 18.8 shows the response of the temperature control of the mold in various stages. The primary characteristics of this control are:

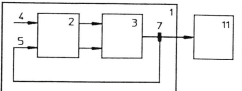

Figure 18.4   Control of heat-exchange medium

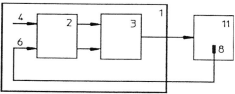

Figure 18.5   Control of mold temperature

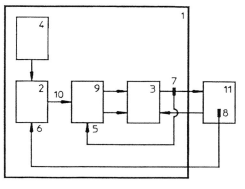

1 Temperature controller, 2 Controller, main controller PI, master controller, 3 Heating/cooling system, 4 Set value, 5 Actual value, 6 Actual mold temperature, 7 Temperature sensor for heat transfer medium, 8 Temperature sensor for mold, 9 Auxiliary controller (P/D), 10 Correcting set value, 11 Mold (consumer)

Figure 18.6   Cascade control

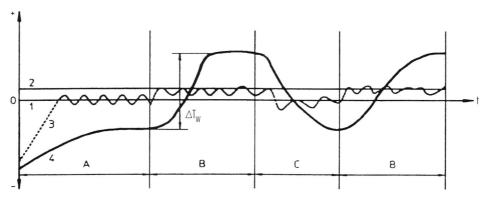

Figure 18.7   Control of coolant temperature with PID controller

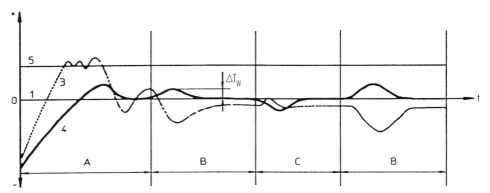

Figure 18.8   Control of mold temperature with PID controller

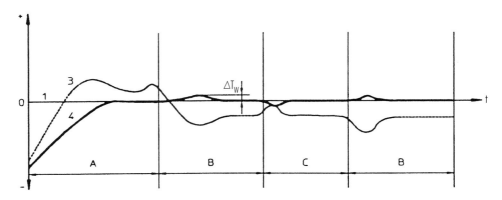

Figure 18.9   Control of mold temperature with PID/PD cascade control
1 Set value for heating, 2 Set value for cooling, 3 Temperature of coolant,
4 Mold temperature, 5 Limit temperature for coolant,
ΔT_W Difference in mold temperature
A Heating, B Injection (cooling), C Shut down (heating)

– The temperature set with the controller corresponds with the mold temperature.
– Disturbance variables (9) affecting the mold are considered and leveled out. This results in very small temperature variations of the mold.
– Processing data are reproducible.

Cascade control
(Figures 18.7 and 18.9)
This is a combination of the two different types of control presented so far. By means of two coupled controllers (2) and (9), both the medium and the mold temperature are controlled. Mold temperature constancy is further improved.

18.2.2 Preconditions for Good Control Results

The following preconditions have to be met to obtain good stability of the mold temperature:

1. Properly dimensioned channels in the consumer (arrangement, heat-exchange surface, diameter, circuits) (see also Chapter 8).
2. Use of temperature-controlled devices with properly dimensioned heating, cooling, and pumping capacity.
3. Correct adaptation of the controller to the controlled system (mold), i.e., actuation by means of control measurement.
4. Correct positioning of the temperature sensor in the mold in the case of regulation of mold temperature and cascade control (see Section 18.2.2.4).
5. The heat carrier should have good heat-transfer properties to carry appropriate quantities of heat in a short time.

These preconditions are discussed in more detail in the following sections.

18.2.2.1 Controllers

The controllers employed in temperature-control equipment are three-point controllers with the positions "heating – neutral – cooling" (quasi-steady controllers). For special applications steady controllers are used. Heating or cooling are not controlled in an in/out mode but in a "more/less" mode depending on the capacity demand.

An interface (e.g., RS 232, 485) enables software dialog with the processor of the injection molding machine. The machine provides the data to be set, calls the parameters for the controller, orders the mold to empty, asks for the status of equipment (e.g., breakdown), etc.

18.2.2.2 Heating, Cooling, and Pump Capacity

Insufficient heating capacity prolongs the heating-up phase and levels out disturbances imperfectly or too slowly. Oversizeing can make the control circuit oscillate. During start-up, a temperature overshooting may occur.

The capacity of the pump determines the temperature gradient of the mold or the heat-exchange circuit. The greater the capacity, the smaller is the temperature difference of the heat carrier in the mold between inlet and outlet. On the other hand, a large discharge causes a large pressure drop in the mold so that pumps with high discharge pressure are

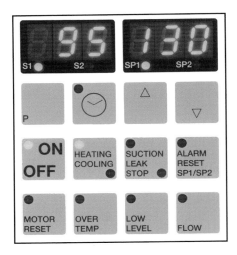

Figure 18.10   Microprocessor controller

needed. The consequence is that the temperature difference should be kept as small as necessary, but not as small as possible (Section 18.2.2.5) (< 5 °C).

18.2.2.3 Temperature Sensors

One distinguishes two kind of temperature sensors:

– Resistance sensors. The resistor consists of a platinum wire, which has a well defined resistance over a wide temperature range. This principle is based on the temperature dependence of the electric resistance of metals. With increasing temperature the resistance increases and vice versa.

Main features: Very good long-term stability, simple connections (no special connectors needed). Measurements are absolute values, independent of ambient influences.

– Thermocouples (e.g., Type J/Fe-CuNi). If two wires of dissimilar metals are brought into intimate contact, a voltage is generated which depends on the temperature of the junction and the kind of metals used. The voltage is proportional to the difference between the temperature of the control point (mold) and the outside, ambient temperature.

Main features: Special connection lines and connectors necessary; punctiform measurement, inexpensive, service life dependent on production quality.

18.2.2.4 Installation of Temperature Sensors in the Mold

For the installation of sensors in the mold one has to pay attention to the following criteria:

– The most suitable position of the sensor in the mold depends, among others, on the configuration and the design of the mold as well as on the location of the cooling channels.

– The sensor should be located in a place where temperature plays a decisive role, e.g., dimensions with close tolerances, regions with a tendency to warpage, or high demands on mechanical properties.
– The sensor should be placed in a defined distance from the cavity wall because the largest temperature variations during a cycle occur there (Figure 18.11), and this could interfere with the response of the controller.

The temperature variations are caused by physical conditions (criteria: mold material, processed material, temperature) and cannot be influenced by the temperature controller. Immediately before injection the cavity wall has the controlled temperature $T_{W\,min}$. When the hot plastic melt comes into contact with the cooler cavity wall, a contact temperature is generated, which is between the temperature of the cavity wall and that of the melt. It decreases continuously during the cycle. The contact temperature $T_{W\,max}$ depends on the heat penetrability of the mold and the plastic. The amplitude of the temperature variation ΔT_W decreases with increasing distance from the cavity-wall surface.

Calculation programs for determining the sensor distance from the mold cavity surface are obtainable from institutes and software companies. As a rule of thumb, a distance of 0.5 to 0.7 d is recommended (d = diameter of sensor).

18.2.2.5 Heat-Exchange System in the Mold (see also Chapter 8)

The surface of channels of the heat-exchange system in the mold has to be dimensioned so that the generated heat can be carried away from the heat carrier of the temperature controller or the necessary heat can be supplied. The larger the channel surface, the smaller is the temperature difference between medium and mold and the faster the temperature variations are leveled out.

Compared with water as a heat carrier, molds operated with a heat-transfer oil needed a 2 to 3-times larger channel surface because of the smaller coefficient of heat transfer. Small channel diameters cause a large pressure drop in the mold. This calls for equipment with expensive pumps (high discharge pressure) or a dividing of the channel system into several parallel circuits to reduce the pressure drop.

Figure 18.11   Temperature of cavity wall versus time
1 Injection
2 Ejection:
$T_{2\,min}$ Minim. temperature of cavity wall,
$T_{2\,max}$ Maxim. temperature of cavity wall,
T_2 Average temperature of cavity wall (important for cooling),
T_4 Temperature of demolding,
t_C Cooling time,
t_c Cycle time.

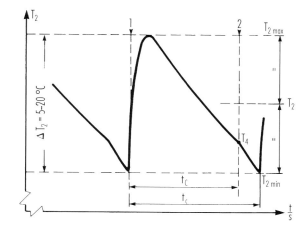

Below, the characteristics of heat-exchange systems in series or parallel are explained (Figure 18.12).

With an arrangement in series the individual circuits are passed through by the heat carrier one after the other. With a parallel arrangement they are passed through after flowing through a distributor. With an arrangement in series the main problem may become an impermissible great temperature difference ΔT between the first and the last circuit, depending on the application.

Because of the great pressure drop, a higher discharge pressure is needed to convey the required amount of heat carrier through the channel system and a relatively high pump capacity may be needed.

The main problem of the parallel arrangement is the difference in flow rate between the individual circuits even from only slightly different flow conditions (e.g., unequal cross sections, lengths, number of bends of the channels). The branch with the smallest resistance to flow experiences the best heat exchange (smallest Ap). A correction of unequal conditions with a hand-valve control cannot be recommended. An improvement may result from a parallel layout in series.

18.2.2.6 Keeping the Temperature as Stable as Possible

It is usually better to maintain a stable temperature by controlling the temperature of the circulating heat carrier, but not by changing the flow rate. The reason for this is the increase in the temperature gradient from a reduction of the flow rate. This can lead to a nonuniform heat dissipation in the mold. Besides this, a reduction of the flow rate results in a poorer coefficient of heat transfer. This means a reduced heat exchange in the mold.

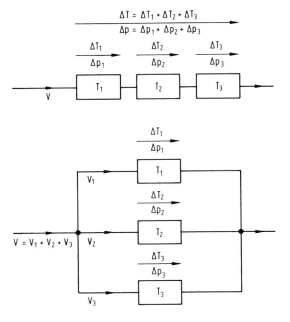

Figure 18.12   Control in series and parallel control

18.3 Selection of Equipment

The selection of temperature-control equipment depends primarily on processed material, the weight of the mold, the desired heat-up time, the amount of material to be processed per unit time (kg/h), the permissible temperature differential in the mold, as well as pressure and flow-rate conditions in the mold (Table 18.2).

The discharge pressure of the pump can only be determined with a diagram "pressure drop as a function of flow rate". If there is no such diagram, the pressure drop can be estimated from experience with similar molds.

As already mentioned in Section 18.2.2.5, water should be preferred as the heat carrier if possible because of its better heat-transfer properties. It should be used for mold temperatures up to 90 °C. Employing equipment for pressurized water allows its use up to temperatures of about 140 °C. Beyond 90 or 140 °C respectively, heat-transfer oil has to be employed as heat carrier.

Table 18.2   Criteria for selection of equipment

Specification	Objective
Processed material	Mold temperature
Mold temperature	Heat exchange medium (water/oil)
Mold weight, heating time	Heating capacity
Amount of processed material per unit time	Cooling capacity
Temperature gradient of mold	Discharge of pump
Pressure drop versus flow rate	Discharge pressure of pump

18.4 Connection of Mold and Equipment – Safety Measures

For reasons of safety and reliability of operation, maintenance, and avoidance of leaks the following items should be observed:

- Only plugs with tapered pipe threads should be employed in heat-exchange circuits.
- Only pressure- and heat-resistant hoses should be used. Twisting, too small bending radii, compression, etc. have to be prevented.
- Cooling should always be connected for safety reasons.
- Heat-insulated lines should be employed for greater distances between equipment and mold.
- Periodic examinations of the circuit (controller, connections, mold) for leaks and proper function.
- For a change-over from water to oil in suitable equipment, one has to proceed cautiously: hazard of an accident during heating from excessive vapor pressure of water remnants in the circuit.
- Periodic change of the heat carrier.
- Use of synthetic heat-transfer oil because of little tendency to form deposits.

– Heat exchanger or expansion tank have to be placed somewhat higher than the mold if the latter is very large and contains a proportional quantity of heat carrier. This prevents a slow return of the content of the mold into the expansion tank and over-flowing during a shutdown. A return is almost inevitable because of tiny leaks at connections in the circuit which permit air to enter. If a higher placement is not feasible, then there are the following options:

  – Size of the expansion tank has to be adequate to accommodate all returning fluid during the shutdown.
  – Mounting shut-off valves at inlet and outlet at the mold, which can be closed during shutdown.
  – Use of magnetic valves, which close when the equipment is turned off.
– Reduction of cross-sections in the circuit, only if needed, and then close to the mold. A measure for the size of the connections are the connectors of the control equipment. This permits the full use of the pump capacity.

18.5   Heat Carrier

In the following, the characteristics of water and heat-transfer oil are compared. Water offers the following advantages over heat-transfer oil:

– Cheaper, cleaner, and ecologically safe. Water from leaks in the circuit can be discharged into the sewer system without further precautions.
– Very good thermal properties, such as heat transfer (high heat transfer coefficient α), heat capacity (high specific heat capacity c), and thermal conductivity (high thermal conductance f).
– Comparatively low channel surface area required for heat supply and dissipation.

Disadvantages are:

– low boiling point.

In the case of tap water:

– Danger of corrosion and scale development in the heat-exchange circuit leading to greater pressure drop and a deterioration of the heat exchange.
– Detrimental substances dissolved in water (nitrites, chlorides, iron, etc.) are increasingly precipitated at elevated temperatures.

Preventive measures depending on shop conditions are:

– use closed-loop cooling circuit,
– removal of solids with strainers,
– periodic flushing of equipment and mold with a scale remover,
– treatment of the circuit with a corrosion inhibitor,
– correct water quality.

To avoid damage to the cooler of the temperature-control device and in the temperature-control circuit (device and attached consumers), the water used must meet the following requirements:

Hardness	10 to 45 ppm equivalent calcium carbonate,
pH value at 20 °C	6.5 to 8.5,

Cl-ions   max. 150 mg/l,
Sum of chlorides and sulfates   max. 250 mg/l.

Potable water often meets these requirements. If the water is too soft (distilled water, rain water), it can cause corrosion. Corrosion is also promoted by the prescuce of chlorides. Water that is too hard promotes scale and sludge development.

Heat-transfer oils do not exhibit the disadvantages of water. Because of their higher boiling point they can be used up to more than 350 °C, depending on the type.

Disadvantages of heat-transfer oils are:

– poorer heat-transfer characteristics. Optimum heat transfer can only be obtained under defined flow conditions. The flow rate (l/min) has to be precisely adjusted to the specific heating capacity (W/cm^2),
– inflammable under certain conditions (there are not yet flame retardant oils),
– aging (oxidation, increase in viscosity),
– costs.

Synthetic heat-transfer oils exhibit better solubility of products from the aging process, which are primarily formed near the temperature limits. This reduces the danger of undesirable deposits considerably. Of course one has to pay a higher price than for a mineral oil.

18.6 Maintenance and Cleaning

Maintenance and cleaning of the heat-exchange system (device, mold, connecting lines) must be performed regularly as otherwise "creeping" deterioration (slow reduction of cross-sections of channels by scaling) may occur.

To avoid production downtimes on expensive production equipment, prevent accidents, and maintain the temperature-control device at a high level of operational safety, the periodic controls and maintenance work prescribed in the manufacturer's manual must be performed.

Deposits of rust and scale adversely affect the heat exchange between mold and circulating water, if the latter is the heat carrier.

Contamination and deposits can constrict the flow-through cross-section and result in a higher pressure drop in the mold so that, after some time, the pump capacity does not suffice for a troublefree operation anymore. Periodic examination and cleaning is therefore indispensable. Adding a corrosion inhibitor is a commendable means of prevention.

Deposits in the mold from heat-transfer oil can also impair heat transfer and result in higher pressure drop. Adding a detergent before changing the oil or preventively during the whole time between changes is likewise suggested. The oil should be examined periodically. This is best done by the supplier with an analysis. Dark color is by itself no quality criterion.

References

[18.1]   Handbuch der Temperierung mittels flüssiger Medien. Regloplas AG, 5th Ed., St. Gallen, Hüthig Verlag, Heidelberg, 1997.

19 Steps for the Correction of Molding Defects During Injection Molding

In the following a number of molding defects, which are mostly caused by faulty mold design, are listed along with steps to remedy them. A first summary is presented with Figure 19.1. It shows a point system, which has been developed to remove visible defects from acrylic moldings [19.1]. However, it can also be applied to other thermoplastic materials.

Melt leaks between nozzle and sprue bushing.

Orifices of nozzle and sprue bushing are misaligned. Check locating, ring, and alignment. Contact pressure between nozzle and sprue bushing is insufficient. Increase pressure. Check uniform pressure with thin paper. Proper contact results in uniform indentation. Radii of nozzle and sprue bushing do not match. Nozzle orifice is larger than orifice of sprue bushing.

Sprue cannot be demolded; breaks off.

Sprue has an undercut because
1. radii of nozzle and sprue bushing do not match,
2. nozzle and sprue bushing are misaligned,
3. nozzle orifice is larger than orifice of sprue bushing,
4. sprue bushing is insufficiently polished, grooves in circumferential direction. Polish mold, round sharp corners.

Sprue is not yet solidified because
1. sprue bushing is too large and sprue too thick,
2. area of sprue bushing is inadequately cooled, check mold temperature.

Part cannot be demolded.

General: cooling time too short; mold overloaded; undercuts too deep; rough cavity surface. Reduce injection speed and holding pressure; eliminate undercuts; repolish cavity.

a) Molding sticks in cavity.

Sprue or runner system has undercuts. Check radii of nozzle and sprue bushing. Check surface of cavity for imperfections. Repolish it and round sharp corners. Cavity temperature is too low. Formation of vacuum. Check taper. Consider venting during demolding

b) Molding is demolished during demolding.

Undercuts too deep. Ejectors are placed in locations which are detrimental to proper transmission of ejection forces resulting in unacceptable pressure peaks. Check cavity surface for imperfections. Formation of vacuum. Check taper.

Distortion of molding.	Poor location of gating. Poor type of gate. Nonuniform mold temperature. Unfavorable cross-sectional transitions. Considerable difference in wall thickness requires more than one cooling circuit. Wrong mold temperature. Nonuniform shrinkage. Adverse switch-over point from injection to holding pressure. Poor location of ejectors.
Burned spots at the molding.	Melt temperature is too high. Material is overheated in narrow gates. Mold is insufficiently vented. Adverse flow of material. Residence time too long. Select smaller plasticating unit.
Darkening at the molding.	Residence time of the material in the barrel is too long. Barrel temperatures too high. Rate of screw revolutions too high. Gates too small.
Dark (black) specks.	Impurities in the material; wear in the plasticating unit. Use of a corrosion- and wear-resistant plasticating unit.
Discoloration near gate.	Mold temperature too high. Enlarge runners and gate. Cold-slug well needed.
Brittle moldings.	Raise mold temperature. Enlarge runners and gate. Cold-slug well needed. Thermally damaged or heterogeneous melt. Humid material.
Molding has mat or streaky surface.	Mold surface inadequately polished. Poor location of gating. Poor type of gate. Increase size of runners and gate and polish. Cold-slug well needed. Mold too cold (condensed humidity). Overheating of material from insufficient venting or unfavorable flow path. Lower rate of screw rotation; pre-dry material.
Color streaks.	Poor mixing or separation of molding material. Raise rate of screw rotation, raise back pressure. Modify temperature profile of barrel. Reduce feed hopper cooling. Use another screw.
Bubbles, humidity streaks.	Material is too humid, dry sufficiently.
Formation of ruts and clouds; Flaking (scaling) of molding surface.	Poor gate position. Raise mold and melt temperature. Raise injection speed. Check plasticating unit for wear. Check gate. Excessive temperature differential between melt and mold. Raise mold temperature. Enlarge runners and gate. Contamination from a different material.
Molding with poor surface gloss. Incomplete mold filling.	Check mold temperature. Enlarge runners and gate. Round sharp corners. Repolish mold. Runners too long or too small or both. Enlarge runner system. Flow in cavity is restricted. Open gates. Core shifting. Insufficient mold venting. Mold and melt temperature too low. Raise injection speed and/or pressure. Insufficient feeding of material, no cushion.

Molding with sink marks and shrink holes.	Runners and gate too small. System solidifies before holding pressure has become effective. Check mold temperature. Raise against shrink holes, lower against sink marks. Lower melt temperature. Check cushion, lengthen holding-pressure time, raise holding pressure. Lower injection speed.
Mold flashes.	Poor fit at parting line. Matching surfaces damaged by e.g., remnants of material or excessive pressure. Rework surfaces. Clamping force is insufficient because projected area of part is too large. Core shifting has caused a considerable differential in wall thickness causing flash on one side and insufficient filling on the other. Mold temperature too high. Injection speed and/or pressure too high. Advance switch-over point. Mold is not sufficiently "rigid". Platens of the clamping unit bend. Reinforce platens, change machine. Mold cavity is not dead center in mold.
Molding with visible knit lines.	Poor position of gating. Objectionable type of gate. Detrimental gate and runner cross sections. Enlarge gate. Poor mold venting. Nonuniform mold cooling. Raise mold temperature. Raise injection speed. Wall thickness of molding too small. Choose cascade gating.
Cold slug	Nozzle temperature too low, retract nozzle from sprue bushing sooner. Reduce cooling of sprue bushing, enlarge nozzle bore.
Jetting	Relocate gate that material is injected against a wall. Enlarge cross section of gate.
Off-sized molding. *a) with uniform shrinkage*	Too high a mold and melt temperature increase shrinkage. Longer cooling time; higher injection pressure and increased holding-pressure time decrease shrinkage.
b) with nonuniform shrinkage	Poor heat exchange, gate freezes too soon. heterogeneous melt.
Stress cracking.	Sharp corners and edges in the mold. Processing with inserts: preheat inserts, avoid sharp corners and edges.
Rippled molding surface.	Reduce mold temperature. [19.1 to 19.9]

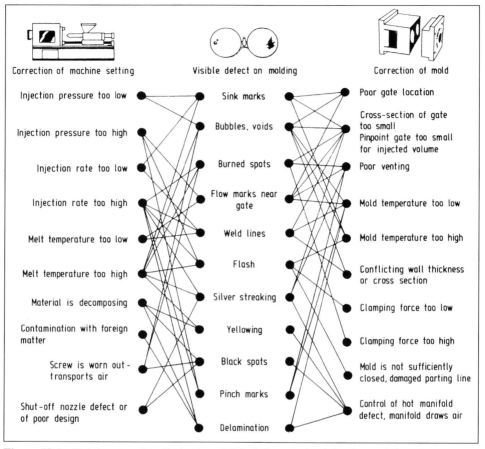

Figure 19.1   Point system for eliminating visible defects during injection molding of acrylics [19.1]

References

[19.1]   Punktesystem zur Beseitigung von sichtbaren Spritzgießfehlern bei der Verarbeitung von Resarit Acrylformmassen (PMMA). Table by Resart-Ihm AG, Mainz, 1987.

[19.2]   Barich, G.: Häufig auftretende Fehler bei der Spritzgießverarbeitung von Thermoplasten und ihre möglichen Ursachen. Plastverarbeiter, 33 (1982), 11, pp. 1361–1365.

[19.3]   Spritzguß-Hostalen PP. Handbook, Farbwerke Hoechst AG, Frankfurt, 1965.

[19.4]   Strack-Normalien für Formwerkzeuge. Handbook of Standards, Strack-Norma GmbH, Wuppertal.

[19.5]   Mink, W.: Grundzüge der Spritzgießtechnik. Kunststoffbücherei. Vol. 2. Zechner + Hüthig, Speyer, Wien, Zürich, 1966.

[19.6]   Schwittay, D.: Thermoplaste – Verarbeitungsdaten für den Spritzgießer. Publication, Bayer AG, Leverkusen, 1979.

[19.7]   Spritzen – kurz und bündig. Publication. 4th Ed., Demag Kunststofftechnik, Nürnberg, 1982.

[19.8]   Verarbeitungsdaten für den Spritzgießer. Publicaton, Bayer AG, Leverkusen, 1986.

[19.9]   Poppe, E. A.; Leidig, K.; Schirmer, K.: Die TopTen der Spritzgießprobleme. DuPont de Nemours GmbH, Bad Homburg.

20 Special Processes – Special Molds

20.1 Injection Molding of Microstructures

The injection molding process may be used to produce on the one hand extremely small parts with molded-part weights of less than 1000th of a gram and, on the other hand, parts with structured areas each measuring just a few square micrometers. Both applications enable mass production of molded parts intended specifically for microsystems technology.

The following description pertains to macroscopic parts with microstructured areas. This mold technology has already been used to produce structures with minimal width of 2.5 µm and a height of 20 µm [20.1]. The so-called aspect ratio (ratio of the maximum height to the minimum lateral dimensions of a structure) has a value of 8 in this case. This illustrates the difference from CD production, which also involves the production of microstructures from plastic. The smallest lateral dimension occurring in the production of compact discs is approx. 0.6 µm and the maximum structure height is approx. 0.1 µm, this giving an aspect ratio of far less than 1.

Figure 20.1 shows an injection-molded demonstration structure. The roughly 1100 pillars of POM each have a width across of 80 µm for a height of 200 µm and are spaced 8 µm apart.

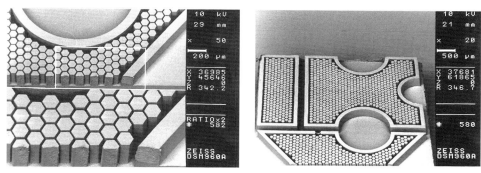

Figure 20.1   Injection molded demonstration structure

20.1.1 Molding Technology and Process Control

The quality of injection molded parts is critically dependent on the molding technology. Already during the design of a microinjection mold, particular attention must be paid to the filigree structure of the cavities themselves. Additionally, the injection molding of microstructured parts requires special process control (Figure 20.2).

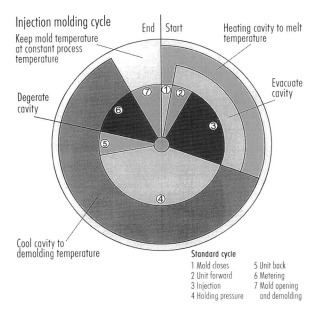

Injection molding cycle

Keep mold temperature
at constant process
temperature

End | Start

Heating cavity to melt
temperature

Evacuate
cavity

Degerate
cavity

Cool cavity to
demolding temperature

Standard cycle
1 Mold closes   5 Unit back
2 Unit forward   6 Metering
3 Injection   7 Mold opening
4 Holding pressure   and demolding

Figure 20.2   Cycle for
the injection molding of
microstructured parts

The molds used nowadays for this special kind of injection molding have the following
features [20.2–20.7]:

– They are fitted out with a variotherm heating system, i.e., the mold is heated before the
  actual injection process to approximately the melt temperature and then cooled to the
  demolding temperature. It is often reported in the literature that the maximum mold
  temperature should be roughly 40 K below the melt temperature of the plastic
  employed [20.7] (but this does not appear generally acceptable).
– They are fitted out with a vacuum unit that evacuates the mold in the vicinity of the
  mold cavity. This is necessary for avoiding the "Diesel effect" in the blind holes of the
  cavities during injection and to support the mold-filling process. In some cases, a
  pressure of less than 1 mbar is required [20.5, 20.7].
– High requirements continue to be imposed on the tolerances of the closing joint, on
  required ejector pins, and on the actual mold inserts. During production, tolerances of
  ≤ 1 μm must be observed [20.1].

The sample mold shown in Figure 20.3 was made primarily from standard mold parts
[20.8]. The modular structure permits rapid, cost-effective adaptation to different
molded-part geometries.

   The outer, water-cooled heating circuit keeps the entire mold at a constant temperature (for instance: T = 60 °C). The variotherm unit comprises a spiral inner heating insert,
through which oil flows from the inside to the outside, and the electric heater. The
counterpiece on the fixed mold half is fitted with cooling channels running perpendicularly to each other for production reasons. Air gaps ensure thermal isolation from the
rest of the mold structure. The internal heating unit is kept constantly at the demolding
temperature of the plastic to be processed (for POM, T = 120 °C), with the initial
temperature of the oil being much higher. The temperature may be monitored with the
aid of a separate thermocouple.

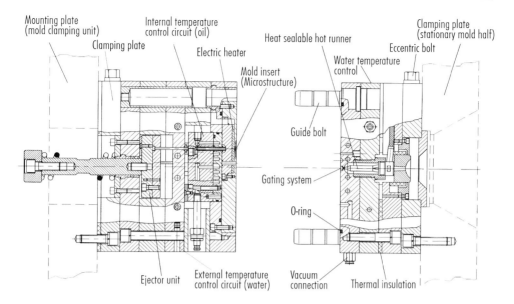

Figure 20.3   Mold for injection molding microstructured parts

The electric heater directly behind the mold cavity provides temporary heating of this area to the desired higher temperature. After injection, this heater is switched off, and the internal temperature-control circuit removes heat until the injection molded part can be demolded (Figure 20.4). The electric heating system consists of a 9-mm-thick, circular brass plate with a diameter of 70 mm and three recessed heating cartridges, each providing 200 W (surface output: 6.2 W/cm^2). Figure 20.5 shows the basic structure of the mold cavity and heating element area.

The necessary vacuum equipment is built according to the circuit shown in Figure 20.6. Due to the low volume for evacuation, instead of the vacuum pump usually employed, a vacuum ejector with two operating steps is used that can generate vacuum close to that of a medium vacuum. An independently operating vacuum controller can be used for infinitely regulating the vacuum [20.16, 20.17].

During the closing movement, the mold is initially in the vicinity of the recessed O-ring. Via a drill hole on the fixed side, the vacuum is drawn on the one hand and the mold is vented prior to the mold-opening movement on the other.

Since, with a conventional hot runner structure, melt would be drawn in prior to injection on account of the vacuum, this mold is fitted with a shut-off hot-runner nozzle [20.9].

Demolding is effected via ejector pins that are each vacuum tight.

20.1.2 Production Processes for Microcavities

A large number of standard techniques have been scaled down for the production of micromechanical and optical parts. But new processes have also been developed. The

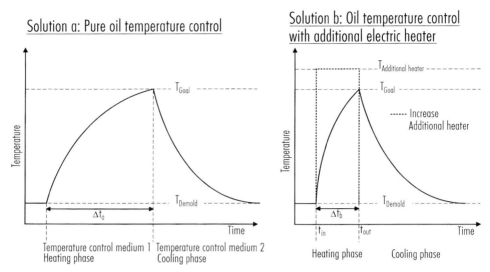

Figure 20.4   Change in temperature with time in variotherm molds

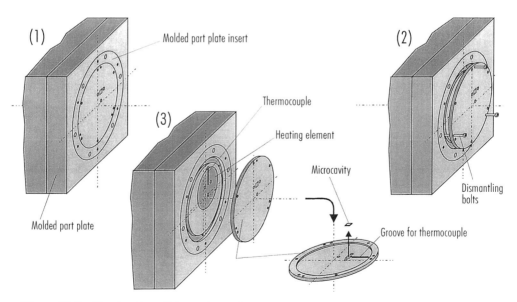

Figure 20.5   Construction of the cavity region

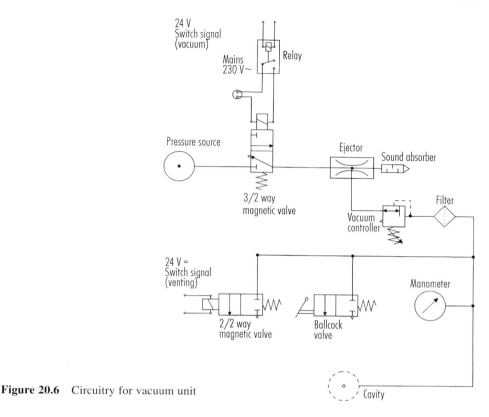

Figure 20.6   Circuitry for vacuum unit

first of these groups includes micromachining, microelectric discharge machining and laser removal. Silicon technology is used in electrical engineering and was the first technology to be adapted for the manufacture of micromechanical parts. A new specialty process that was developed in the 1980s was the LiGA technique. All processes are capable of generating parts or part structures in the micrometer range. The resultant structured parts may be used directly or serve as mold inserts for mass production.

The processes mentioned above will now be analyzed separately for their strengths and limitations as tools for the production of microcavities. Generally, it is necessary to employ special tooling machines that, for example, are highly stabilized and that through the extremely high speeds involved in ultraprecision machining are capable of producing the specified precision. No details will be provided of machine technology or process control for the tooling machines. The reader is referred to the pertinent literature [20.10–20.15].

20.1.2.1 Silicon Technology

Silicon technology was developed to a very high degree in the 1970s for the field of electrical engineering.

Silicon is generally structured by wet or dry etching [20.13, 20.18]. Etching can be further subdivided into isotropic and anisotropic types (Figure 20.7). Isotropic etching occurs in all directions. Wet-chemical, anisotropic etching, in contrast, exploits the fact that the activation energy for etching a monocrystal by certain etchants depends on the orientation of the various crystal planes. The (111) plane has the highest activation energy and is therefore attacked the most slowly by the etchant. It is even possible to virtually halt etching completely by generating etch-stop layers. This is effected by doping areas of the monocrystal with high boron concentrations.

Anisotropy (Figure 20.7) and etching rate depend on the composition and temperature of the etchant [20.18]. The disadvantage of this process lies in the limited design freedom, which is caused by the lattice structure of the crystal. Round pillars cannot be generated by this process, for example.

As requirements on the fineness of the structures become more and more stringent, reactive dry etching is establishing itself as a structuring technique [20.19]. This technique generates three-dimensional structures by bombarding a substrate with atoms, ions, or free radicals in a plasma discharge. The resultant fragments can be converted into the gas phase and thus removed by admixing a reactive component. The anisotropy as well as the selectivity of the process depend on the type and quantity of the reactive component employed.

To make the process more powerful, dry etching is often combined with wet chemical processes in a method known as surface micromachining [20.20]. The disadvantage of this process is that it is very limited in terms of structure depth, and practically only vertical edges can be produced. A comprehensive explanation of the physical and chemical processes is provided in [20.14, 20.21].

As is the case for the LiGA technique described next, subsequent electrodeposition and molding of the resultant cavity (SiGA technique) can yield mold inserts that can be used for the mass production of parts made of plastic. Direct use of the Si structures is not possible as they are not mechanically strong enough.

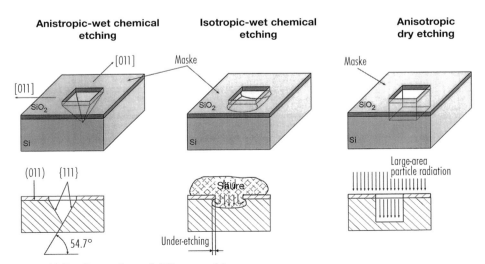

Figure 20.7   Comparison of different etching processes

20.1.2.2 The LiGA Technique

The LiGA technique (from the German for Lithography, Electrolytic Forming and Mold-Making) is a relatively new process for producing microstructures (Figure 20.8). The first step consists of applying a 1-mm-thick resist of polymer (usually PMMA) onto a metallic substrate. By means of a specially developed mask [20.13] and parallel X-rays from a synchrotron source with a wavelength of roughly 0.2 to 0.5 nm, the structure pattern is transferred to the resist [20.22, 20.23]. The energy-intensive radiation causes the molecule chains to shorten in the exposed areas of the plastic. These areas are removed with a solvent. The remaining PMMA structure serves as a basis for the subsequent electroplating step. In an electroplating bath, a metal layer "grows" on the substrate, replacing the polymer that has been dissolved away. The residual polymer is dissolved to yield a metallic secondary structure that represents the negative of the primary structure and may be used as a mold insert for mold making.

With the aid of this process [20.24], it is possible to produce $2^1/_2$-dimensional structures with minimal dimensions of less than 2 μm and a surface roughness of R_a of ≤ 10 nm. The height of the structure depends primarily on the duration of exposure in the

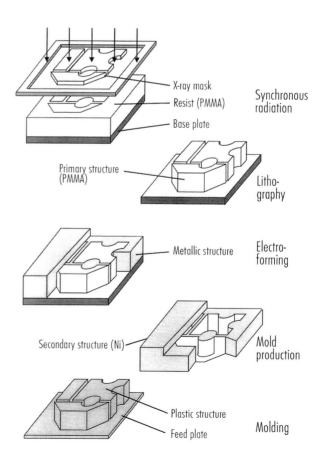

Figure 20.8   The LiGA process

lithographic step. This step can yield lateral accuracies of 0.2 µm at a structure height of 0.5 mm.

The extensively researched steps of lithography and electroplating are not acceptable in terms of costs for large series. For example, the electroplating time for a 1-mm-thick layer is roughly one week. Only the last step of the LiGA technique, namely molding by reaction casting, vacuum deep-drawing, or polymer injection molding makes mass production economical [20.26, 20.27].

20.1.2.3 Laser LiGA

Laser LiGA is inexpensive relative to LiGA. It replaces the "lithography by synchrotron radiation" step with the use of an excimer laser. The two steps of "exposure and development" of the PMMA resist constitute one process stage since the primary structure can be generated directly with the aid of the laser. However, due to the larger wavelengths of the laser, this technique can only generate smaller aspect ratios (< 10) and larger structures (> 1 µm) than those of the LiGA. The surface quality cannot be compared directly with that of the LiGA technique [20.28, 20.29]. Furthermore, absolutely parallel side walls cannot be produced, but this, in fact, aids demolding after injection molding.

20.1.2.4 Laser Removal

Laser removal in the macrorange of metals is used particularly in the production of forging dies, plastic injection molds, and die cast molds as an alternative to machining processes and electric discharge machining.

A knowledge of the possibilities afforded by microstructuring of metals is crucial for mold-making for microstructural injection molding. There are three ways in which these materials can be removed by means of a laser:

– Materials such as steel can be evaporated by means of highly energetic laser pulses (Figure 20.9, left). The point of removal is flushed with inert gas to prevent the purged material from re-condensing on the surface and to retain the original surface quality [20.30–20.32].
– It is also possible to melt the material with a laser and then to blow it away, as basically happens in the case of laser cutting [20.31, 20.32] (see also Section 2.9).
– The laser is used to initiate selective oxidation at the workpiece surface and thus to trigger surface erosion. Oxygen is admitted to effect local combustion, which releases additional energy (Figure 20.9, right). If the process parameters are properly set, thermal stresses are induced in the cooling zone, with the result that the oxidized material detaches itself from the unchanged base material. This process is highly suitable for mold-making because the attainable surface quality is very high, just as in the case of electric discharge finishing [20.32, 20.33].

Lasers can be used to process all materials. Radiation sources for laser removal include CO_2 lasers, Nd:YAG lasers and excimer lasers. Table 20.1 summarizes important information about the scope and limits of the various instruments for the microstructuring of metals.

Detailed information of the physical processes that occur during laser removal is contained in Herziger [20.38].

Laser beam melt removal

Inclined
processing head

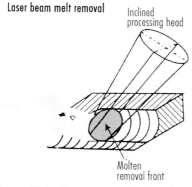

Molten
removal front

Laser beam oxide machining

Vertical processing
head

Steam
jet

Detaching
metal

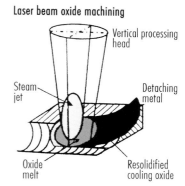

Oxide
melt

Resolidified
cooling oxide

Figure 20.9   Laser processing (carving)

Table 20.1   Lasers suitable for mictrostructuring [20.30–20.32, 20.34–20.37]

	CO$_2$-Laser	Nd: YAG-Laser	Excimer-Laser
Characteristic wavelength	10.6	1.06	0.157(F$_2$)*** – 0.351 (XeF)***
Removal rate [mm^3/min] for moderate laser power	35(200–2,500)* 300(160–800)*	10–80(200)** 300	N.I. 3(F2)*** – 200(XeCl)***
Attainable surface quality	R$_a$ = 20–70 μm	R$_a$ = 10–40μm (9μm)**	R$_a$ = 1.3 μm
Attainable processing stage	20 μm to several mm	20 μm to 4–6 mm	N.I.
Mold accuracy	N.I.	Several μm at approx. 100 μm structure size	N.I.

* = Oxygen as process gas; ** = Compressed air as process gas; *** = Laser gas; N.I. = No information

20.1.2.5 Electric-Discharge Removal

Already in the 1970s, reproduction of surface detail to an accuracy of 1 μm and less at roughness heights of 0.1 μm was being reported for microerosion [20.39].

A distinction is drawn between the basic processes of cutting and machining by electric discharge (spark erosion) [20.36] and, more recent, EDM milling [20.40–20.42] and grinding [20.43] (Figure 20.10). For spark erosion, the material must have a minimum conductivity of 0.01 to 0.1 S/cm [20.31].

Details of the process will not be discussed here. For more information, the reader is referred to the pertinent literature. The essential characteristics are summarized in Table 20.2 to provide an overview of the current possibilities and limits of spark erosion with respect to microremoval.

Table 20.2   Possibilities of spark-erosive methods [20.40, 20.42, 20.43, 20.44–20.47]

Spark erosive	Cutting	Sinking	Milling	Grinding
Smallest electrode diameter	20 µm	2.5 µm	N.I.	N.I.
Attainable tolerances	±2–3 µm	Roundness flaws: 100 nm Straightness flaws: 500 nm	Generally: ±10 µm Flatness: ±5 µm bei 150 x 100 mm² plate	N.I.
Producible geometries	Land widths < 40 µm Internal radii: 20 µm Aspect ratio: 1,000	Smallest drill diameter: 50 µm Depth: 100 µm Aspect ratio: > 100	N.I.	Minimal width: 40 µm Aspect ratio: > 15
Surface roughness	R_a = 150 mm	R_a = 200 nm R_{max} = 100 nm	R_a = 1µm	N.I.
Typical application	Apertures	Cavities	Cavities, narrow and deep gaps	Straight paths

N.I. = No information

Table 20.3   Possibilities of micromachining [20.48, 20.53–20.56]

	Microturning	Microgrinding	Micromilling
Smallest mold diameter	Tolerance on the turning lathe: < 20 nm	Grinding pin I Grain on pin Disc width Disc grain	N.I.
Attainable tolerances	Mold deviation: < 30 nm	N.I.	N.I.
Producible geometries	Cylindrical parts with Diameter: 8 µm Length: 111 µm Width at the groove tip: 1 µm	Grooves with Width: 15 µm Depth: 500 µm	Lands with Width: 1.5 µm Depth: 200 µm
Surface roughness	R_{max} = 5 nm R_a = 2 nm	R_a < 1 µm	R_a ≤ 10 nm
Typical application	Cut glass mirror, Shaft	Microheat exchanger	Microheat exchanger

N.I. = No information

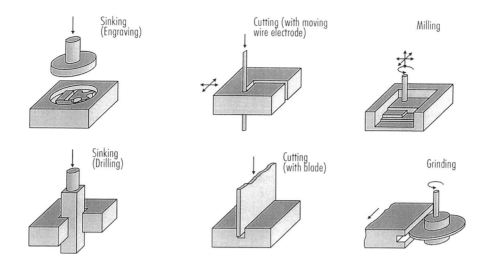

Figure 20.10   Variants of spark erosion [20.36, 20.40, 20.43]

20.1.2.6 Micromachining

Like spark erosion, the classical turning, milling, and grinding techniques have been adapted for microprocessing. The tools for turning and milling are generally monocrystalline diamonds, which can be ground very sharply to produce cutting edges with roundness errors of less than 20 nm [20.48]. The diamond tools are generally rectangular, trapezoidal or semicircular [20.49] and are soldered to steel holders or mounted in sintered metal for clamping into the machine [20.50].

The state of the art in turning, grinding, and milling is summarized in Table 20.3. The data in the table pertain usually to brass. Much higher dimensions result for steel because it can only be worked to a limited extent, if at all, by diamond [20.51, 20.52]. It should be added that the smallest drill diameter at the moment is 50 μm [20.53].

20.2   In-Mold Decoration

In-mold decoration is a processing technique with which decorated molded parts can be produced by injection molding in a few working steps. The molded parts consist of a thermoplastic support and a decorative material. The latter is usually a film or textile.

Figure 20.11 illustrates the process of in-mold decoration. The decorative material is inserted between the two halves of an injection mold (moving and stationary mold halves). The mold is then closed, this causing the decorative material to be immobilized in the parting line. During injection, the plastic disperses itself in the cavity and combines with the decorative material. After freezing, the decorative molded part is demolded.

Application fields for this technique involving textile decorative material are primarily to be found in the automotive industry. These include molded parts, such as A, B and C pillar trim in cars, side-door trim, trunk covers, parcel shelves, dashboards, cable duct covers, and front panel trim [20.58].

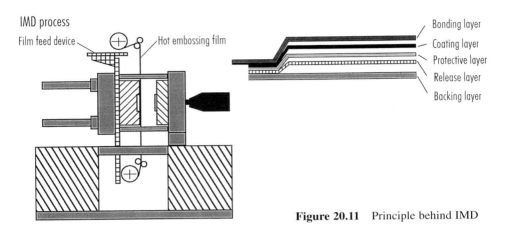

Figure 20.11   Principle behind IMD

Molded parts decorated with film are also primarily used in the automotive sector. Typical applications are back-lit parts in the dashboard, as well as molded parts that must have a high quality surface or special color effects [20.63, 20.64]. Applications in ornamental hub caps and fenders are currently being researched (Table 20.4).

The insertion of decorative material and the need to inject plastic melt behind it call for modified mold designs. There are basically two ways in which the decorative material can be arranged: on the stationary half or the moving half of the injection mold. If it is on the stationary half, the melt must flow through a suitably designed gate across the parting line right through to the moving half. This can be accomplished by a curved tunnel gate or a hot runner manifold [20.58].

However, it is more usual for the decorative material to be on the moving half. In the case of a highly curved pillar trim, this means that the core of the injection mold is not, as is usual, on the moving half, but instead on the stationary half. During freezing, the injection molded part shrinks onto the core, clinging to the stationary half when the mold opens. This results in the major difference over conventional molds: ejectors for in-mold decoration must be integrated on the stationary half. Machine manufacturers, therefore, offer specially designed injection molding machines in which the hydraulics and actuation for the ejector package are located on the stationary side. A positive side-effect is that the ejector pins do not press against the decorative material, which might otherwise cause damage [20.60].

Many in-mold-decorated molded parts are elongated parts. Due to the flow-path/wall-thickness ratio, these parts need to be injected through several gates. If it is not possible to install the gate system in the parting line, gating can be achieved with a three-plate mold in which a floating plate is located between the stationary and moving mold plates [20.58].

Table 20.4   Evaluation of decorative material insertion systems [20.59]

	Decorative material		Tenter technique
	Specially cut	From the roll	
Advantages	Low decorative material wastage Parts removal and insertion of decorative material possible in one pass; high reproducibility	Simple system, short handling time Punching in mold possible	Defined position in the case of decorative material patterning, partial pretensioning or slippage of the decoration possible
Disadvantages	Accommodation of thin decorative materials through handling difficulties; prestressing of decorative material complicated	Partial prestressing difficult; in some cases, high wastage Complicated molded part geometries barely possible	External equipment for tenter feed necessary, elaborate system overall
Applications	Geometrically simple molded parts Decorations showing patterns or revealing threads	Simple, flat molded parts	Complicated geo-metries/high thermo-forming ratio; decoration materials with pronounced patterning
Comment	Most frequent type of application	Used, e.g., for A-pillar production	Rarely used

The problem of multiple gating is much easier to solve through recourse to a hot runner. For this reason, hot-runner technology is linked directly with in-mold decoration, e.g. [20.60, 20.61] by the cascade injection technique (Figure 20.12).

Further distinctive features arise from the need for handling the decorative material. The mold plates must offer enough space to accommodate the material. Transfer and locating aids, as well as clamping elements, have to be integrated into the mold. Due to the additional sliders for holding, clamping, and locating the decorative material, the susceptibility of the mold to wear must be borne in mind [20.62].

A key aspect of in-mold decoration is making a fold in the injection mold. If this fold is created during injection molding and not in a separate stage afterward, major economies can be made.

One way to generate a 180° fold is to employ a sliding split mold. As the mold closes, slide bars traverse inwards from the side; their surface is curved enough to produce the necessary groove. The decorative material is inserted into the groove, forming the fold. For demolding, the slide bars have to be retracted first.

Another way to design the fold is to insert into the injection mold a piece of decorative material that folds over the edges of the molded part. The decorative material is made of face fabric and foam layer. Folding of the material produces a double layer of foam at the edge, with the foam backs lying against each other. During injection, melt squeezes in between the two layers. This serves to ensure that the melt is completely enveloped in decorative material at the edge of the molded part, i.e., a fold is formed. Here, again, pre-made-up decorative material must be used.

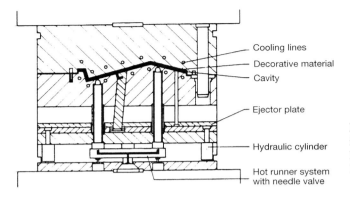

Figure 20.12 Mold for in-mold decoration
Photo: Georg Kaufmann, Busslingen/ Switzerland

A 90° fold can be produced with the aid of a mold that essentially consists of three plates [20.65, 20.66]. This ensures that the undercut formed is demolded. Furthermore, the textile is stretched with the aid of a tenter frame.

Figure 20.13 shows one mold design that is notable primarily for the fact that it generates the fold during formation of the molded part (180°) and offers the possibility of an integrated textile trim [20.58, 20.68].

The mold consists of a moving half, stationary half, and a third plate between them. A semicircular flute in the plate accommodates the textile. The decorative material is inserted between the stationary half and the third plate. The mold is then closed until the third plate is pressed against the textile on the stationary half by the pressure springs. The decorative material is prestressed by spring-loaded ejector pins in the stationary side (not shown in Figure 20.13), so that the textile is immobilized and can slip into the cavity in a controlled manner.

Then, the core integrated in the stationary half traverses and prestretches the textile. This causes enough decorative material to enter into the cavity and ensures that melt is

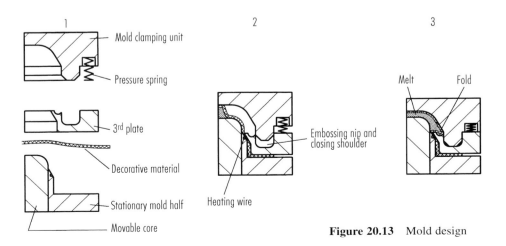

Figure 20.13 Mold design

not injected through it. When the core retracts, the mold closes further. The provision of vertical flash faces allows both conventional injection molding and injection compression molding to be employed. Injection compression molding also allows sensitive decorative material to be processed.

20.3   Processing of Liquid Silicone

Liquid silicone rubbers (LSRs) differ considerably from conventionally crosslinking elastomers in terms of processing and material properties. The raw material is a low-viscosity, two-pack system that is supplied in 20-liter and 200-liter containers by the manufacturer [20.69] The two components have to be mixed before the system becomes reactive, and addition-crosslink under the influence of heat without releasing by-products. The crosslinking rates of 3–7 mm/s wall thickness are much higher than those of conventional elastomers [20.70–20.73]. The material is processed in injection molding machines fitted with extra equipment for conveying the raw components. Because the material, which is kept at 20–40 °C in the plastifying section, swells extensively in heated molds at up to 240 °C, it is necessary to underfill the mold so as to avoid flash formation.

The low viscosity during the filling process and the extremely high crosslinking rates impose stringent demands on the mold design. Thus, fully automatic, machining-free production of LSR molded parts requires greater precision in mold making. If, during thermoplastic processing, gap widths of 0.01 to 0.02 mm are sufficient for avoiding flash formation, the tolerable gap widths of LSR molds range from less than 0.005 to 0.01 mm [20.74–20.75]. To maintain the narrow tolerances, after every operation, the mold plates have to be stress-free annealed and, after hardening, freed of residual stresses by repeated normalizing [20.76]. Due to the high mold temperatures, high-temperature, hardenable steels are used for mold plates and inserts. [20.77–20.80].

To avoid flash formation, LSR molds are made to be very rigid despite the relatively low cavity pressures of 20 MPa max. The aim is to keep the design as simple as possible and to more or less eliminate slide bars and split molds due to the problems with sealing them. The critical mold areas for mechanical design are in the vicinity of the locating ring aperture on the stationary half and in the vicinity of the ejector system on the moving half. Dimensioning of the mold-plate thickness is based on that of thermoplastics molds. The high cavity pressures occurring during thermoplastics processing are compensated by the lower gap tolerances in LSR molds [20.73].

20.3.1 Evacuation

Because LSR crosslinks so quickly, high injection speeds are necessary, especially for molds with high flow-path / wall-thickness ratio, in order that flow errors due to material starting to crosslink during mold filling may be avoided. However, high injection speeds readily cause air pockets in the flow front. For this reason, and to prevent residual air being trapped, if the mold does not part close to the end of the flow path, the cavities have to be evacuated to 200–800 mbar. Evacuation takes place on the machine via a so-called intermediate stop program (also known as embossing or venting program) [20.77, 20.81]. It is carried out during the closing process by closing the mold halves initially to within a few tenths of a millimeter. Special seals recessed in the mold parting line bridge the gap and seal off the resultant mold cavity against the environment. Then the cavities and gates are evacuated via a drill hole in the parting plane. Only after evacuation is complete does the clamping mechanism of the injection molding machine build up the maximum locking force and the injection process is started.

20.3.2 Gate

The cavities are usually gated with a pin-point gate (0.1–0.8 mm diameter) or film gate (0.08–0.12 mm film thickness) unless sprueless injection is performed (see Cold Runner Technique). To improve the economics of injection molding, the predominantly small LSR molded parts (part weight frequently less than 10 g), multicavity molds containing from 4 to 16 cavities (up to 128 for very small parts) are used. Natural, symmetrical balancing is preferred for multiple gating, so as to avoid different degrees of fill during partial filling. Even minor differences in degree of fill can lead to underfilled parts or flash formation [20.73].

20.3.3 Demolding

Fully automatic demolding is hampered by the adhesive, rubber-elastic properties of LSR. For this reason, usually several demolding systems within the mold are required, which demold the part stepwise. In some cases, external devices are used, such as brushes and grippers [20.77, 20.82, 20.83]. Demolding systems inside the mold may be of the passive or active types. The passive type include selective setting of the roughness of the cavity surface to exploit the adhesion of the LSR material for positioning the molded part on the desired mold half. Greater roughness, as generated by etching, sandblasting or eroding, lowers the adhesion while lower roughness increases it [20.76, 20.77, 20.84]. It must be remembered that the molded parts tend to cling to the cavity and not to the core, due to the thermal expansion.

Common active demolding devices are mold wipers and mushroom ejectors with compressed air support. The valve-like mushroom ejectors are raised by compressed air (see Figure 12.47). Air fed between the molded part and cavity surface detaches the part from the wall and forces it out of the cavity. The current of air, however, causes highly convective heat release and, in certain circumstances, may lead to unpermissible interference with the temperature homogeneity of the mold. Conventional, cylindrical ejector pins are only used for composite parts if the force can be introduced via a rigid insert [20.73].

20.3.4 Temperature Control

A heating capacity of roughly 50 W/kg mold weight is needed for achieving a uniform temperature field at high mold-operating temperatures. Strip heaters and conical cartridge heaters are preferred as they, unlike cylindrical cartridges, have a heat transfer independent of the drill hole tolerance [20.77]. This has a positive effect on the homogeneity of the heat introduction and the cartridge life time. To reduce heat release to the mold platens and the surrounding air, the cavity platens need to be insulated with epoxy resin platens.

20.3.5 Cold-Runner Technique

Due to high raw material costs and elaborate gate separation, the use of cold runners in LSR processing is state of the art and highly advanced [20.85]. Aside from open cold-runner systems with reduced sprue waste, mainly cold runners with needle valves for direct gating are used, with the result that production can be sprueless and hence totally free of sprue waste. Cold-runner systems are either mold-dependent or machine-dependent. Machine-dependent systems with up to 8 individual nozzles serve as standardized heads, instead of the shut-off nozzle; they are mounted on the plasticating unit and protrude into the mold [20.86]. The advantages lie in the possibility of multiple use for different molds and the attendant lower costs and in easy changes in the event of disruptions. Figure 20.14 shows a cold-runner adapter specially developed for LSR molds that is mounted on the machine nozzle [20.87]. It protrudes with its sprues down as far as the cavities to inject into these direct.

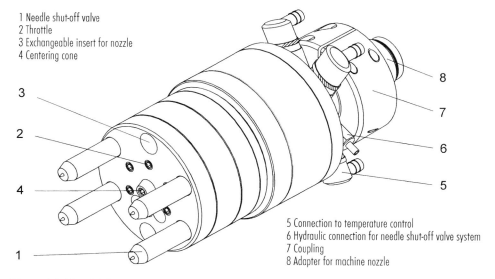

1 Needle shut-off valve
2 Throttle
3 Exchangeable insert for nozzle
4 Centering cone

5 Connection to temperature control
6 Hydraulic connection for needle shut-off valve system
7 Coupling
8 Adapter for machine nozzle

Figure 20.14   4-fold cold runner head for permanent mounting on the plasticating unit [20.85]

20.4   Injection-Compression Molding

The injection-compression molding process can be roughly divided into two steps, namely injection and compression (Figure 20.15). During the injection phase, a precisely metered amount of plastic melt is injected into the mold, which has been opened by a compression nip. Since at this point the volume of the cavity is greater than the volume of the molded part later, the injected melt forms a cake. In the compression phase, this cake is forced into the cavity by the closing moving mold half and is shaped. This presupposes, however, that the sprue is closed at the start of the compression phase in

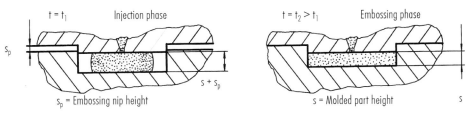

Figure 20.15   The compression-molding technique

order that the melt may be prevented from flowing back out of the cavity. It is not possible to apply a holding pressure [20.88].

In industrial practice, the injection-compression molding process can be varied to suit the particular application. Thus, in simultaneous compression, the compression movement starts during the injection process. Moreover, the compression movement may in part be effected by movable mold inserts that act on part of the surfaces or on the entire cross-sectional area (Figure 20.16). In certain cases, the mold cavity is filled completely so as to compensate the shrinkage in volume (Figure 20.16, variant c). For thin-walled parts, however, the compression process described at the outset has become the established process [20.90, 20.93].

Due to the special processing sequence and the positive properties of the molded parts, injection-compression molding has already attained major technical importance in

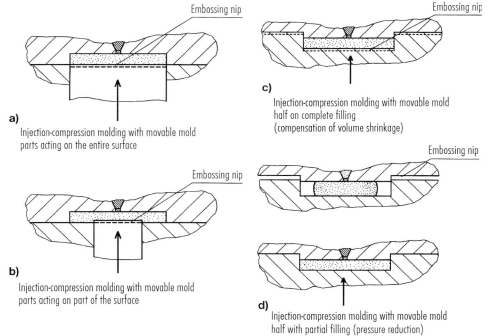

Figure 20.16   Injection-compression molding

thermoset and elastomer processing [20.89, 20.90]. In the processing of thermoplastics, the technology has so far mostly been applied to the manufacture of thick-walled molded parts that must satisfy high demands on dimensional stability (e.g., for optical lenses). However, even technical molded parts with large diameters and low wall thicknesses (e.g., membranes for microphones and loudspeakers) are increasingly being made by injection-compression molding [20.91, 20.92]. Studies performed on these kinds of membranes and others have revealed that molded parts with wall thicknesses less than 0.25 mm and faithful surfaces can be produced straightforwardly in short cycle times by varying certain process parameters [20.92, 20.93].

Furthermore, it is possible to reduce the mold cavity pressure with the injection-compression molding process. This may, in certain circumstances, mean that an injection molding machine with lower clamping force can be used for a similar molded part when this process is used. This lowering of pressure is of more interest, however, for the in-mold decoration of parts with textile decorative materials (see Section 20.2) [20.94]. Given optimized process parameters, the use of injection-compression molding can greatly increase the quality of the molded part [20.95, 20.97].

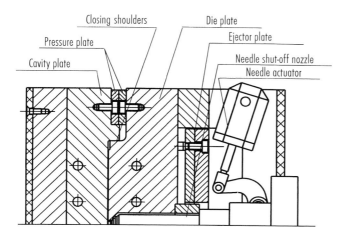

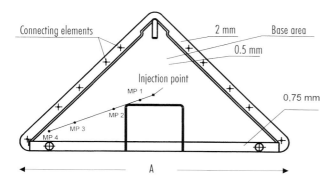

Figure 20.17   Mold design and part

These advantages of injection-compression molding are offset by the greater technical outlay relative to injection molding. The considerable associated costs of mold and machine or accessories would appear to make injection-compression molding economical only for a large numbers of parts. Also, injection-compression molding cannot be used for molded parts of arbitrary geometry since, due to the vertical flash faces, there are geometric restrictions [20.89, 20.93] that primarily greatly limit the size.

Figure 20.17 shows an injection-compression mold that meets the essential requirements of this process [20.96]. The mold is designed as a positive mold and has a hot runner shut-off nozzle intended for mechanically sealing the sprue. With this design it is not necessary to shorten the cycle time. The movable mold plate should be guided precisely so as to avoid wear on the vertical flash faces [20.92].

The vertical flash face clearance in the injection-compression molding of thermoplastics is not as critical as in the case of thermosets and elastomers. Nevertheless, narrow limits must be observed, so that flash can be ruled out. See Chapter 7 for guideline values concerning gap widths for venting of different thermoplastics.

Particular attention should also be paid to mold-temperature control. The narrow vertical flash face clearance requires that the pre-determined temperatures for both mold halves be scrupulously observed since exceeding a certain mold temperature difference between the moving and stationary mold half can lead to flash, and staying below it may lead to collisions.

References

[20.1]   Rogalla, A.: Analyse des Spritzgießens mikrostrukturierter Bauteile aus Thermoplasten. Dissertation, RWTH, Aachen, 1998.

[20.2]   Michaeli, W., Rogalla, A.: Kunststoffe für die Mikrosystemtechnik. Ingenieur-Werkstoffe, 6 (1997), No. 1, April, pp. 50–53.

[20.3]   UETP-Mems Training in Microsystems. Materials for Microstructures, IMM, Mainz, 1995.

[20.4]   Nöker, E.; Keydel, L.: Herstellen von Mikrostrukturkörpern aus Kunststoffen. Kunststoffe, 82 (1992), 9, pp. 798–801.

[20.5]   Nöker, E.: Perspektiven und Möglichkeiten des Spritzgießens in der Mikrotechnik. In: Paper at the conference on: "Hochleistungsspritzgießen". SKZ, Würzburg, 1995.

[20.6]   Ruprecht, R.; Hanemann, T.; Piotter, V.; Hausselt, J.: Fertigung von Kunststoff-Mikroteilen für optische und fluidische Anwendungen. Swiss Plastics, 19 (1997), 4, pp. 5–8.

[20.7]   Weber, L.: Formenbau in der Mikrotechnik. Darmstädter Erfahrungsforum Werkzeug- und Formenbau, Darmstadt, December 1995.

[20.8]   Normalien für Spritzgießwerkzeug. Publication. Hasco Normalien, Hasenclever GmbH & Co., Lüdenscheid, 1994.

[20.9]   Spear System. Product Catalog. Seiki Corporation, Japan, 1990.

[20.10] Bacher, W.; Feit, K.; Harmening, M.; Michel, A.; Mohr, J.; Stark, W.; Stölting, J.: LIGA-Abformtechnik zur Fertigung von Mikrostrukturen; KfK-Nachrichten, 23 (1991), 2/3, pp. 84–92.

[20.11] Beitz, W.; Küttner, K.-H.: Dubbel – Taschenbuch für den Maschinenbau. 15th Ed., Springer-Verlag, Berlin, Heidelberg, New York, Tokyo, 1986.

[20.12] König, W.: Fertigungsverfahren – Abtragen. Vol. 3. 2nd Ed., VDI-Verlag, 1990.

[20.13] Menz, W.; Bley, P.: Mikrosystemtechnik für Ingenieure. VCH-Verlag, Weinheim, 1992.

[20.14] Seidel, H.: Der Mechanismus des Siliziumätzens in alkalischen Lösungen. Dissertation, Univ. Berlin, 1986.

[20.15] Seif, O.: Konstruktion und Erprobung von Formeinsätzen für das Spritzgießen von Mikrostrukturen. Unpublished report. IKV, RWTH Aachen, 1997.

[20.16] Automatisieren mit Pneumatik. Catalog, FESTO Pneumatik, Esslingen, 1996.

[20.17] Vakuum-Technik - Komponenten für die Automatisierung. Catalog, J. Schmalz GmbH, Glatten, 1996.

[20.18] Büttgenbach, S.: Mikromechanik – Einführung in Technologie und Anwendungen. B.G. Teubner-Verlag, Stuttgart, 1991.

[20.19] Liebel, G.: Ionenstrahlätzen. Feinwerktechnik & Meßtechnik, 95 (1987), pp. 436–441.

[20.20] Kowanz, B.; Ehrfeld, W.: Systemtechnische Analyse und Entwicklung eines Ventils in LIGA-Tech. Report No. 4886, KfK, Karlsruhe, 1991.

[20.21] Seidel, H.: Naßchemisches Tiefertätzen. In: Mikromechanik. Heuberger (ed.), A. Springer-Verlag, Berlin, 1989.

[20.22] Becker, E.; Ehrfeld, W.: Das LIGA-Verfahren. Phys. Bl., 44 (1988), 6, pp. 166–170.

[20.23] Menz, W.: Die Mikrosysterntechnik und ihre Anwendungen. In: Spektrum der Wissenschaften, Dossier, 4: Mikrosystemtechnik, 1996.

[20.24] Becker, E. W.; Ehrfeld, W.; Münchmeyer, D.; Betz, H.; Heuberger, A.; Pongartz, S.; Glashaziser, W.; Michel, H. J. V.; Siemens, R.: Production of Separation – Nozzle Systems for Uranium. Enrichment by a Combination of X-Ray Lithography and Galvanoplastics. Naturwissenschaften, 69 (1982), pp. 520–523.

[20.25] Ehrfeld, W.; Lehr, H.: LIGA-Method – Deep. X-Ray Lithography for the Production of Three-Dimensional Microstructures from Metals, Polymers and Ceramies Radiation Physics and Chemistry Pergamon Press, 45 (1995), 3, pp. 349–365.

[20.26] Götz, W.: "Mikrostrukturtechnik". In: Industrieanzeiger, 1992.

[20.27] Rogner, A.; Eicher, J.; Peters, R.-P.; Mohr, J.: The LiGA-Technique – what are the new opportunities. Reprint: MME '92, Micromechanics Europe 1992, 3rd Workshop an Micromachining, Micromechanics and Microsystems, Leuven, Belgien, 01./02.06.1992.

[20.28] Arnold, J.; Dasbach, U.; Ehrfeld, W.; Hesch, K.; Löwe, H.: Combination of Excimer Laser Micromachining and Replication Processes Suited for Large Scale Production ("LaserLIGA"). Lecture. EMRD Spring Meeting, Straßbourg, France, May 23–27, 1994.

[20.29] Arnold, J., Ehrfeld, W.; Hesch, K.; Möbius, H.: Kostengünstige Serienfertigung von Mikrobauteilen durch "Laser-LIGA". F&M, 103 (1995), 1–2, pp. 48–49.

[20.30] Brinkmann, U.; Basting, D.; Endert, H.; Pippert, K.: Licht in der Mikrofertigung Warum an Excimer-Lasern kein Weg vorbeiführt. F&M, 103 (1995), 6, pp. 304–308.

[20.31] Westkämper, E.; Freytag, J.: Slab im Vorteil – Laserstrahlabtragen: Leistung gesteigert. Industrie Anzeiger, 35/1992, pp. 64–66.

[20.32] Witjes, G.; Hildebrand, P.; Kuhl, M.; Wrba, P.: Der direkte Weg – Laserabtragen im Werkzeug- und Formenbau rechnet sich. Form + Werkzeug, March 1996, pp. 32-35.

[20.33] Eberl, G.; Hildebrand, P.; Kuhl, M.; Sutor, U.; Wrba, P.: Neue Entwicklungen beim Laserabtragen Laser und Optoelektronik, (1992) 4, pp. 44–49.

[20.34] Eversheim, W.; Klocke, F.; Pfeifer, T.; Weck, M.: Herstellen von Mikrobauteilen. In: Aachener Werkzeugmaschinen-Kolloquium '96; Wettbewerbsfaktor Produktionstechnik – Aachener. Perspektiven, VDI-Verlag, Düsseldorf, 1996, pp. 4–194.

[20.35] Jiang, C. E.; Lau, W. S.; Yue, T. M.; Chiang, L.: On the Maximum Depth and Profile of Cut in Pulsed Nd:YAG-Laser Machining, Analysis of the CIRP. Vol. 42, 1993, 1, pp. 223–226.

[20.36] König, W.: Fertigungsverfahren. Vol. 3: Abtragen. 2nd Ed., VDI-Verlag, 1990.

[20.37] Pfeufer, V.: Präzise Lichtblitze – Mikrobearbeitung mit Excimerlaser. F&M, 104 (1996), 7–8, pp. 532–536.

[20.38] Herziger G.; Wester R.; Petring, D.: Abtragen mit Laserstrahlung Laser und Optoelektronik, No. 4, 1991, pp. 64–69.

[20.39] Genath, J.: Die Mikrofunkenerosion erzielt hohe Genauigkeiten. VDI-Nachrichten, No. 6, 7.2.1973.

[20.40] Bahle, W.: In den richtigen Bahnen – Elektroerosives Fräsen erweitert die Erodiertechnik. Form und Werkzeug, May 1995, pp. 37–38.

[20.41] Balleys, F.: Steigende Bedeutung – EDM-Fräsen. Schweizer Maschinenmarkt. SMM, No. 19, 1995, pp. 36–37.

[20.42] Maag, J.: Formenwerkzeuge mit Funken fräsen. Industrieanzeiger, 41/1995, pp. 44–46.

[20.43] Wolf, A.; Ehrfeld, W.; Lehr, H.; Michel, E.; Richter, T.; Gruber, H.; Wörz, O.: Mikroreaktorfertigung mittels Funkenerosion – Funkenerosives, Strukturierungsverfahren für mikrofluidische Systeme. F&M, 105 (1997), 6, pp. 436–439.

[20.44] Li, H.; Masaki, T.: Micro-EDM; Society of Manufacturing Engineers SME, Technical Paper No. MS91-485, Grand Rapids, Michigan, USA, 1991.

[20.45] Heinzelmann, E.: Funkenerosion: Neue Qualität dank Mikrotechnik; Technische Rundschau TR, No. 48, 1995, pp. 24–27.

[20.46] Masuzawa, T.; Fujino, M.: A Process for Manufacturing Very Fine Pin Tools. Society of Manufacturing Engineers. SME, Technical Paper No. MS90-307, Chicago, USA, 1990.

[20.47] Masuzawa, T.; Kuo, C.-L.; Fujino, M.: A Combined Electrical Machining Process for Micronozzle Fabrication. Annals of CIRP. Vol. 43 (1994), 1, pp. 189–192.

[20.48] Obata, K.; Oka, K.; Chujo, T.: Development of Ultra Cutting Tool. Society of Manufacturing Engineers. SME, Conference: Superabrasives '91, Chicago, Illinois, USA, June 11–13, 1991, Paper-No. MR 91–194, pp. 12–38.

[20.49] Schubert, K.; Bier, W.; Linder, G.; Seidel, D.: Profilierte Mikrodiamanten für die Herstellung von Mikrostrukturen. Industrie Diamanten Rundschau. IDR, 23 (1989), No. 4.

[20.50] Klocke, F.; Koch, K.-F.; Zamel, S.: Technologie des Hoch- und Ultrapräzisionsdrehens. Zeitschrift für wirtschaftlichen Fabrikbetrieb, 90 (1995), 5, pp. 217–221.

[20.51] Moriwaki, T.; Shamoto, E.: Ultraprecision Diamont Turning of Stainless Steel by Applying Vibration Annals of CIRP. Vol. 49 (1991), 1, pp. 559–562.

[20.52] Weck, M.; Bispink, T.: Drehmaschinen zur Herstellung hochpräziser Bauteile: Übertragung der Erkenntnisse aus der Ultrapräzisionstechnik auf die Stahlfeinstbearbeitung; Symposium report, TH, Darmstadt 1991.

[20.53] Westkämper, E.; Hoffmeister, H.-W.; Gäbler, J.: Spanende Mikrofertigung – Flexibilität durch Schleifen, Bohren und Fräsen F&M, 104 (1996), 7–8, pp. 525–530.

[20.54] Tönshoff, H.; v. Schmieden, W.; Insaki, L.; König, W.; Spur, G.: Abrasive Machining of Silicon Manufacturing Technology – CIRP, Annals. Vol. 39 (1990), 2, pp. 621–635.

[20.55] Weck, M.; Wieners, A.: Maschinen für die Hoch- und Ultrapräzision. TECHNICA, 1995, 13/14, pp. 14–19.

[20.56] Weck, M.; Vos, M.: Gedrehte und gefräste Mikrostrukturen VDI-Z (1995), No. 7/8, pp. 33–35.

[20.57] Witjes, G.; Hildebrand, E.; Kuhl, M.; Wrba, P.: Der direkte Weg – Laserabtragen im Werkzeug- und Formenbau rechnet sich. Form + Werkzeug, March 1996, pp. 32–35.

[20.58] Galuschka, S.: Hinterspritztechnik – Herstellung von textil-kaschierten Spritzgußteilen. Dissertation, RWTH, Aachen, 1994.

[20.59] Bärkle, E.; Rehm, G.; Eyerer, P.: Hinterspritzen und Hinterpressen. Kunststoffe, 86 (1996), 3, pp. 298–304.

[20.60] Jaeger, A.; Fischbach, G.: Maschinentechnik und Prozeßführung zum Dekorhinterspritzen. Kunststoffe, 81 (1991), 10, pp. 869–875.

[20.61] Bürkle, E.: Hinterspritzen und Hinterpressen von Textilien beider Innenausstattung von Automobilen. In: Textilien im Automobil. Reprint of Conference. VDI-Gesellschaft Textil und Bekleidung, Düsseldorf, 1992, pp. 37–57.

[20.62] Bürkle, E.: Hinterpressen und Hinterprägen – eine neue Oberflächentechnik. In: Innovative Spritzgießtechnologien – ein Beitrag zur Konjunkturbelebung, Würzburg, December 1–2, 1993, pp. 7–28.

[20.63] Manz, B.: Spritzgießteile mit farbigem Oberflächendekor. Kunststoffe, 85 (1995), pp. 1346–1350.

[20.64] Wank, J.: Spritzgießteile mit Oberflächen-Dekor: Vergleich der Verfahren. Plastverarbeiter, 44 (1993), 7, pp. 70–76.

[20.65] Zwei getrennte Fertigungsschritte in einem: Textilmelttechnologie. Plastverarbeiter, 44 (1993), 7, pp. 80–82.

[20.66] Spritzgießform. Offenlegungsschrift. DE 4033297.

[20.67] Verfahren zum Herstellen von kaschierten Kunststoffformteilen. Publication, DE 4114289 C2.

[20.68] Becker, H.-G.: Konstruktion eines Versuchwerkzeuges für die Hinterspritztechnik. Unpublished report, IKV, Aachen, 1992.

[20.69] Bayer Silicone Silopren LSR. Daten – Fakten – Ratschläge zur Verarbeitung. Technical Information, Bayer AG, Leverkusen, 1995.

[20.70] Batch, G. L.; Macosko, C. W.: Reaction kinetics and injection molding of liquid silicone rubber. In: Rubber Chemistry and Technology, 64 (1991), 2, pp. 218–233.

[20.71] Pradl, F.: Flüssigsilikonkautschuk – Charakteristik, Verarbeitung und Anwendung. In: Kautschuk + Gummi. Kunststoffe, 38 (1985), 8. pp. 408–409.

[20.72]   Steinberger, H.: Flüssiger Siliconkautschuk – eine Alternative zu festem Silikon-kautschuk. In: Paper given at the Conference: Silikonelastomere, SKZ; Würzburg, 1993.

[20.73]   Walde, H.: Beitrag zum vollautomatischen Spritzgießen von Flüssigsilikonkautschuk. Dissertation, RWTH, Aachen, 1996.

[20.74]   In der Dichtung liegt Wahrheit – Präzisionsspritzgießmaschinen verarbeiten Flüssig-silikon. In: KPZ (1991), 12, pp. 26–27.

[20.75]   Elastosil, LR: Eigenschaften, Verarbeitung, Typen, Hilfsmittel und spezielle Informationen. Technical Information. Wacker Chemie GmbH, Munich, 1990.

[20.76]   Spritzgußverarbeitung von Flüssigsilikonkautschuk. Technical Information. Wacker Chemie GmbH, Munich, 1991.

[20.77]   Steinbichler, G.: Spritzgießen von Flüssigsilikonkautschuk. In: Kunststoffe, 77 (1987), 10, pp. 931–935.

[20.78]   Huber, A.: Flüssigsilikonverarbeitung – Verfahrenstechnik oder Maschinentechnik – wo liegt die Problematik? In: Paper given at the Conference on: "Verarbeitung von flüssigen Silikonkautschuken" im Spritzgießverfahren, SKZ, Würzburg, 1984.

[20.79]   Engel-LIM-Verfahren. Technical Information. Engel GmbH, Schwertberg, 1990.

[20.80]   Steinbiehler, G.: Rationelleres Spritzgießen von Flüssigsilikon. In: Maschinenmarkt, 91 (1985), 77, pp. 1508–1511.

[20.81]   Spritzgießmaschinen der Baureihe CI/C2. Technical Information. Krauss-Maffei Kunststofftechnik GmbH, Munich.

[20.82]   Kaltkanalwerkzeuge für Babysauger aus LSR. Technical Information. Emde Industrie-technik GmbH, Nassau, 1993.

[20.83]   Bürstvorrichtungen für Spritzgießmaschinen. In: KPZ, (1991), 3, p. 15.

[20.84]   Schray, J. G.: Vergleiche bei der Verarbeitung fester und flüssiger Silikonkautschuke im Spritzgießverfahren. In: Paper given at the Conference on: "Fester und flüssiger Silikonkautschuk", SKZ, Würzburg, 1991.

[20.85]   Masberg, U.: Die Kaltkanaltechnik in der Eiastomerverarbeitung. In: Kautschuk + Gummi. Kunststoffe, 43 (1990), 9, pp. 810–816.

[20.86]   Kaltkanalsysteme, standardisiert. Technical Information. EOC Normalien GmbH & Co. KG, Lüdenscheid, 1996.

[20.87]   Emmerichs, H.; Giesler, D.: Angußsysteme für die LSR-Verarbeitung. In: Kunststoffe, 86 (1996), 11, pp. 1678–1680.

[20.88]   Jürgens, W.: Untersuchungen zur Verbesserung der Formteilqualität beim Spritzgießen teilkristalliner und amorpher Kunststoffe. Dissertation, RWTH, Aachen, 1968.

[20.89]   Keller, W.: Vorteile in Marktnischen, Spritzgießen-Übersicht über die gängigen Spezialverfahren. Kunststoffe, Synthetics, 9/95, pp. 32–42.

[20.90]   Knappe, W.; Lampl, A.: Zum optimalen Zyklusverlauf beim Spritzprägen von Thermo-plasten. Kunststoffe, 74 (1984), 2, pp. 79–83.

[20.91]   Friesenbichler, W.; Ebster, M.; Langecker, G.: Spritzprägewerkzeuge für dünnwandige Formteile richtig auslegen. Kunststoffe, 83 (1993), 6, pp. 445–448.

[20.92]   Gissing, K.: Spritzprägen von dünnwandigen technischen Formteilen aus Thermoplasten. In: Neue Werkstoffe und Verfahren beim Spritzgießen. Paper given at the Conference: VDI Gesellschaft Kunststofftechnik, Düsseldorf, 1990, pp. 119–131.

[20.93]   Friedrichs, B.; Friesenbichler, W.; Gissing, K.: Spritzprägen dünnwandiger thermoplastischer Formteile. Kunststoffe, 80 (1990), 5, pp. 583–587.

[20.94]   Galuschka, S.: Hinterspritztechnik – Herstellung von textil-kaschierten Spritzgußteilen. Dissertation. RWTH, Aachen, 1994.

[20.95]   Michaeli, W.; Brockmann, C.: Injection Compression Moulding – A Low Pressure Process for Manufacturing Textile-Covered Mouldings. Society of Plastics Engineers, Proceedings of the Annual Technical Conference (ANTEC), Toronto/Can, 1997.

[20.96]   Conze, M.: Konstruktion eines Versuchswerkzeuges zur Herstellung dünnwandiger Formteile mit Hilfe des Spritzprägeverfahrens. Unpublished Report; RWTH, Aachen, 1996.

[20.97]   Kuckertz, M.: Vergleich zwischen Spritzgieß- und Spritzprägeverfahren hinsichtlich der Eignung zum integrierten Kaschieren von Dekormaterial. Unpublished Report, IKV; RWTH, Aachen, 1996.

Index